Sino-Russian Oil and Gas Cooperation

Sino-Russian Oil and Gas Cooperation

The reality and implications

KEUN-WOOK PAIK

Published by the Oxford University Press
for the Oxford Institute for Energy Studies
2012

OXFORD

UNIVERSITY PRESS

Great Clarendon Street, Oxford OX2 6DP

Oxford University Press is a department of the University of Oxford.
It furthers the University's objective of excellence in research, scholarship
and education by publishing worldwide in

Oxford New York

Auckland Cape Town Dar es Salaam Hong Kong Karachi
Kuala Lumpur Madrid Melbourne Mexico City Nairobi
New Delhi Shanghai Taipei Toronto

with offices in

Argentina Austria Brazil Chile Czech Republic France Greece
Guatemala Hungary Italy Japan Poland Portugal Singapore
South Korea Switzerland Thailand Turkey Ukraine Vietnam

Oxford is a registered trade mark of Oxford University Press
in the UK and in certain other countries

Published in the United States
by Oxford University Press Inc., New York

British Library Cataloguing in Publication Data

Data available

Library of Congress Cataloguing in Publication Data

Data available

Cover designed by Cox Design Limited
Typeset by Philip Armstrong, Sheffield
Printed by Berforts Information Press

ISBN 978-0-19-965635-6

1 3 5 7 9 10 8 6 4 2

To the memory of my late father Jee-Kwon Paik

CONTENTS

LIST OF TABLES

LIST OF MAPS AND FIGURES

Maps

Figures

ACRONYMS AND UNITS OF MEASURE

API	American Petroleum Institute
APR	Asia–Pacific Region
ASOC	Anglo-Siberian Oil Company
bbl	standard barrel of crude oil (159 litres)
bcf	billion cubic feet
bcm	billion Cubic Metres
bn	billion
boe	Barrel of Oil Equivalent
bt	billion tonnes
btoe	billion tonnes of oil equivalent
btsce	billion Tons of standard coal equivalent
CAGR	Compound annual growth rate
CBM	Coalbed Methane
CDB	China Development Bank
CEC	China Electricity Council
CEP	Chubu Electric Power
CGA	China Gas Association
China Gas	China Gas Holdings Ltd
CHPP	Combined Heat and Power Plant
CIC	China Investment Corporation
CIECC	China International Engineering Consulting Corporation
CIF	Cost, Insurance and Freight
CITIC	China International Trust and Investment Corporation
CNG	Compressed Natural Gas
CNPC	China National Petroleum Corporation
CNPC E&D	CNPC Exploration and Development Company Limited
CNPCI	CNPC International Limited
CNODC	China National Oil and Gas Exploration and Development Corporation
CNOOC	China National Offshore Oil Corporation
CNUOC	China National United Oil Corporation
COSL	China Oilfield Services Limited
CPCIA	China Petroleum and Chemical Industry Association

CPIC	China Petroleum International (Exploration & Development) Company
CPPE	PetroChina Petroleum Pipeline Company
CPPEI	China Petroleum Planning and Engineering Institute
CPPLB	Petroleum Pipeline Bureau, CNPC
CR Gas	China Resources Gas Group Limited
CREPC	Chinese-Russian Eastern Petrochemical Company
CUCBM	China United Coalbed Methane Corporation
DME	Dimethyl Ether
DOI	Declaration Of Intentions
DRC	Development Research Centre, the State Council
EAGC	East Asia Gas Company
EBRD	European Bank for Reconstruction and Development
EGP	Eastern Gas Programme
ELG	Energy Leading Group
ENL	Exxon Neftegaz Limited
EPC	Engineering, Procurement and Construction
ERI	Energy Research Institute, NDRC
EOR	Enhanced Oil Recovery
ESOGC	East Siberian Oil and Gas Company
ESGC or VSGK	East Siberian Gas Company
ESPO oil pipeline	East Siberia Pacific–Ocean oil pipeline
FES	Commission on the Fuel and Energy Sector
FOB	Free on Board
FS	Feasibility study
FSO	Floating Storage and Offloading
FSU	Former Soviet Union
FTS	Federal Tariff Services
FYP	Five Year Plan
GAIL	Gas Authority of India Limited
GDP	Gross Domestic Product
GIIC	Guanghui Industrial Investment Corporation
GIV	Gazprom Invest Vostok
GW	Gigawatt
HOA	Heads of Agreements
IEA	International Energy Agency
IGNG	Institute of Oil and Gas Geology, Siberian Division, Russian Academy of Sciences
IMF	International Monetary Fund

IOC	Irkutsk Oil Company
IOCs	International Oil Companies
IRR	Internal Rate of Return
Japex	Japan Petroleum Exploration
kWh	kilowatt-hour
JBIC	Japan Bank for International Cooperation
JOGMEC	Japan Oil, Gas and Metals National Corporation
JPDO	Japan Pipeline Development Organization
JSPC	Japan Sakhalin Pipeline Company
LNG	Liquefied Natural Gas
LSAW	Longitudinal Submerged Arc-Welded
LIBOR	London Interbank Offered Rate
m^3	cubic metre
mb/d	Million Barrels Per Day
Mcf	Thousand cubic feet
mcm	million cubic metres
MEDT	Ministry of Economic Development and Trade
METI	Ministry of Economy, Industry and Energy
MFE	Ministry of Fuel and Energy
MIE or Minpromenergo	Ministry of Industry and Energy
MIIT	Ministry of Industry and Information Technology
MLNR	Ministry of Land & Natural Resources, China
mmbtu	Million metric British Thermal Units
MMCF	Million cubic feet
mn	million
MNR	Ministry of Natural Resources, Russia
MOF	Ministry of Finance
MPa	MegaPascal
mt	Million Tonnes
mtce	million tonnes of coal equivalent
mtsce	Million Tonnes Standard Coal Equivalent
mtoe	Million Tonnes of Oil Equivalent
mtpa	million tonnes per annum
NAGPF	Northeast Asian Gas & Pipeline Forum
NDRC	National Development and Reform Commission
NEA	National Energy Administration
NEB	National Energy Bureau
NEC	National Energy Commission
NIGEC	National Iranian Gas Exports Company
NIOC	National Iranian Oil Company
NPRC	National Petroleum Reserve Centre

NPT	Nadym–Pur–Taz
OIES	Oxford Institute of Energy Studies
ONGC	India's Oil & Natural Gas Corporation
OPEC	Organization of Petroleum Exporting Countries
PetroChina	PetroChina Company Limited
PJ	petajoule
PKI	PetroKazakhstan Inc.
PM	Prime Minister
PPP	Purchasing Power Parity
PRC	People's Republic of China
PSA	Production Sharing Agreement
RCA	Russian and Central Asia
RES	Russia's East Siberia
ROA UES	Russian Unified Energy System Correct
Rosnedra	Federal Agency for Subsoil Use
Rosprirodnadzor (RPN)	Federal Natural Resources Management Service
Rostekhnadzor	Federal Service for Environmental, Technological and Nuclear Supervision
RZD	Russian Railways Co
SETC	State Economic and Trade Commission
SAFE	State Administration of Foreign Exchange
SALM	Single Anchor Leg Mooring
SASAC	State Assets Supervision and Administration Commission
SDPC	State Development Planning Commission
SEIC	Sakhalin Energy Investment Company
SEO	State Energy Office
SEP	Sichuan-to-East pipeline
SKV gas pipeline	Sakhalin–Khabarovsk–Vladivostok gas pipeline
SIDANCO	Siberian Far East Petroleum Company Limited
SINOPEC	China Petroleum and Chemical Corporation
SLOC	Sea Lanes of Communication
SMNG	Sakhalinmorneftegaz
SNIIGGIMS	Siberian Research Institute of Geology, Geophysics and Minerals
SNG	Synthetic Natural Gas
SOC	Sakhalin Oil Company
SODECO	Sakhalin Oil and Gas Development Co Limited
SOMO	State Oil Marketing Organization, Iraq
SPR	Strategic Petroleum Reserves

SSAW	Spiral Submerged Arc-Welded
STN	Sakhatransneftegaz
SUEK	Siberian Coal and Energy Company
tcf	Trillion cubic feet
tcm	Trillion Cubic Metres
Towngas	Hong Kong and China Gas Co Limited
TNK	Tyumen Oil Company
toe	Tonnes of Oil Equivalent
Trans-Asia Gas	Trans-Asia Gas Pipeline Company Limited
UES	Unified Electricity System
UGSS	Unified Gas Supply System
UNIPEC	China International United Petroleum & Chemicals Co Limited
UNDP GTI	United Nations Development Programme Greater Tumen Initiative
UOG	United Oil Group Limited
VAT	Value Added Tax
VCNG	Verkhnechonskneftegaz
VE	Vostok Energy
VIOCs	Vertically Integrated Oil Companies
VLCC	Very Large Crude Carriers
VNIGRI	All-Russian Petroleum Research Exploration Institute
Xinao Gas	Xinao Gas Holdings Limited
WEP	West–East Gas Pipeline
YNAD/YNAR	Yamal-Nenets Autonomous District or Region
YPCL	Shaanxi Yanchang Petroleum (Group) Co Limited
YTZ	Yurubcheno-Tokhomskaya Zone
ZPGC	Zhuhai Pipeline Gas Company

ENDNOTES

The endnotes following each chapter contain additional details and references. References with named authors are abbreviated to Author's name, date of publication, and, where appropriate, page number(s). In references to books, chapters of books, and journals the date given is the year; for other periodicals it is the day, month, and year. The full versions of these references are shown in the Bibliography. References to articles without a named author are also found in the endnotes but do not appear in the Bibliography. These authorless references come mostly from the publications listed below. References which are purely web addresses, without more details, also appear in the endnotes. Multiple references in a single endnote are separated by semi-colons.

References from newspapers and bulletins without named authors come largely from the following publications:

AFP	*Agence France Presse*
AP	*The Associated Press*
AT	*Asia Times*
CD	*China Daily*
CER	*China Energy Report, Dow Jones*
China ERW	*China Energy Report Weekly, Interfax (until 2 June 2010)*
China EW	*China Energy Weekly, Interfax (from 3 June 2010)*
China OGP	*China Oil, Gas & Petrochemicals*
CSJ	*China Securities Journal, Xinhua News Agency*
EDM	*Eurasia Daily Monitor*
FT	*Financial Times*
IGR	*International Gas Report, Platts*
MT	*Moscow Times*
NYT	*The New York Times*
OGJ	*Oil & Gas Journal*
RPI	*Russian Petroleum Investor*
Russia & CIS OGW	*Russia & CIS Oil and Gas Weekly, Interfax*
WSJ	*The Wall Street Journal*

ACKNOWLEDGEMENTS

When in March 1983 I started my own long march of research into Sino-Russian oil and gas, I did not dream that it would last three decades. I started to work on this, my second, book in January 2007 when I joined the Oxford Institute for Energy Studies (OIES); it is a complete update of my first book, *Gas and Oil in Northeast Asia: Policies, Projects and Prospects*, which was published by Chatham House in 1995. The five years which it has taken to produce the present book may be the culmination of my thirty-year research career. I doubt if I will dare to embark on another major book project in the near future. This one has been like a second PhD project. But it was ultimately a pleasure and I feel privileged and honoured to have been able to write it.

The starting point of this book was a meeting in autumn 2006 when Chris Allsopp, Director of OIES, and Jonathan Stern, Chairman of its Gas Programme, invited me to work on the project. The inspiration, guidance, and encouragement they have given me has been an essential stimulant and I am at loss to find the right words to express my heart-felt thanks to both of them. In particular, Professor Stern has put huge efforts into improve the book during its lengthy editing period. My special thanks should also be given to the selected reviewers of the draft of the book. They include Chris Allsopp, Jonathan Stern, Simon Pirani, Bobo Lo, Tim Gould, Xia Yishan, Xu Yihe, Masumi Motomura, Ivan Sandrea, Kent Calder, and Mikkal Herberg. Their comments were important in pointing out my faults.

Many scholars, professionals, and officials with knowledge of the subject have shared with me their invaluable knowledge and expertise, and the project is hugely indebted to them. I have also benefited from the abundant information and analysis obtained from conferences, workshops, and seminars organized by OIES, Chatham House, the China Institute for International Studies, the National Bureau of Asian Research, the United Nations Development Programme's Greater Tumen Initiative, the Japan Bank for International Cooperation, the Northeast Asian Gas and Pipeline Forum, the Deutsche Bank, the Energy Intelligence Group, the CWC group, and Vostok Capital. The author would like to express deep thanks to the librarians of the Energy Institute (Chris Baker), Chatham House (Malcolm Madden, Sue Franks, Linda Bedford), and the Institute of Energy Economics, Japan. Their help was invaluable.

I am pleased to acknowledge financial support from TNK–BP and to convey my special personal thanks to Alastair Ferguson, then VP responsible for gas business development for TNK–BP. I am also grateful for the generous financial and material support which I received from the Japan Oil, Gas and Metals National Corporation, the Japan External Trade Organization in London, the Korea Energy Economics Institute, and the Korea Development Institute.

At OIES, I am hugely indebted to Kate Teasdale for enduring all the difficulties of coordinating this book's publication process, and to Bob Sutcliffe and Catherine Gaunt for their devotion to maximizing the editing quality of this book. I also am especially grateful to David Sansom whose work has improved the quality of the maps. The result, of course, is solely my responsibility.

Finally, I would like to express my special thanks to two most important women, whom I love beyond all words: my mother Kee-Hee Lee who taught me the wisdom of patience and overcame two strokes to see the completion of this project, and my wife Soo-Hyun Kim who put up with many troubles (including what she called 'old man's grumpiness') during the last five and a half years.

PREFACE

A generation has passed since the government of the world's most populous nation made its momentous decision to pursue industrialization on a scale never seen before. China already has, on some calculations, the world's largest economy, and certainly consumes more primary energy than any other country; its need for imports of energy is vast and growing.

Twenty years ago, the rulers of the USSR, the other former centre of world communism, politically collapsed, leaving a rusting industrial economy in a state of crisis; but the legacy of the new regime which emerged from this collapse was massive reserves of oil and gas. Dr Keun-Wook Paik's book chronicles how, from these two starting points, China and Russia have approached the building of a new relationship, a relationship which has major consequences for the economy and politics of both Asia and the world. This book concentrates on aspects of the new Sino-Russian relationship connected with the countries' respective oil and gas sectors.

Dr Paik's account is based on unrivalled knowledge gained from 25 years specializing in this subject. In this book, after describing the separate development of the energy sectors of both countries during these years, he concentrates on understanding the many oil- and gas-related issues which enter into the new relationship between China and Russia – one which has had both conventional and unconventional aspects and which has produced both collaboration and conflict.

Although China has made considerable efforts to make the most effective use of its modest energy resources, the question of large-scale imports of oil and then gas have been at the centre of the Chinese leaders' concerns. They have employed important policy initiatives, aimed at extending the sources from which they import oil and gas, thereby demonstrating that China has more alternatives sources of supply than was first supposed. The interconnection of two powers (China and Russia) has been converted into a more complex one which includes the Central Asian republics, which have become important energy suppliers to China, as well as countries where the export of Chinese capital has begun to play an important role in promoting energy exports to China. These include not only Russia and the Central Asian republics but also countries much further afield.

An unresolved mystery runs through Dr Paik's story: why, despite ample preparations, has no contract been signed between China and Russia for large-scale exports of the massive gas reserves of Russia's Far East to the cities and factories of the world's biggest energy user? Both countries say they want this to happen, and on numerous occasions agreement seems to have been imminent. Dr Paik's detailed narrative suggests that the reason the two powers have failed to shake hands on a price and volume is that they both continue to believe that they can increase their own, or reduce the other's, negotiating power, by changing some aspect of their energy policies.

A settlement of the great question of whether, when, and on what terms high volumes of gas will begin to flow from Russia's Far East to China's most industrialized provinces may still not be imminent. But, if and when it comes, it will profoundly affect not only Sino-Russian relations, but also international relations and the world's hydrocarbon market – either because of the consequences of the exports if they do occur, or because of the alternatives if they do not. However this is decided, the readers of this book will be among the best equipped to understand what happens and why.

Christopher Allsopp,
Oxford Institute for Energy Studies

INTRODUCTION

This book contains a study of Sino-Russian energy relations, in particular those which concern oil and gas. This is a topic which has so far received very little in-depth study. Plenty of high quality accessible research on Sino-Russian relations has been published but hardly any of it specifically focuses on oil and gas cooperation. This study aims to begin to fill this relative gap regarding a subject which has huge implications not only for the future of Sino-Russian relations but also for world oil and gas markets and for global climate change. Emphasis will be placed first on how energy cooperation has advanced during the last two decades, and second on expectations for the coming decades. Light will be shed on how and why gas and oil have evolved differently within overall Sino-Russian energy relations. We shall ask why, in the last two decades, Sino-Russian cooperation has had such disappointing results for the Northeast Asia region, whether or not this seems likely to improve in the 2010s, and what will be the consequences of success or failure in energy cooperation.

To answer these questions the book is divided into eight chapters. The first sets out an overview of Sino-Russian relations, first in general and then with special emphasis on the energy factor. Chapters 2 to 5 review oil and gas developments in eastern Siberia and the Sakhalin islands in the Russian Federation, and the rapid expansion of China's oil and natural gas industry. Since Sino-Russian oil and gas cooperation, together with the Sino-Central Asian Republics' oil and gas cooperation, have been the centrepiece of China's transnational pipeline development strategy and of its energy supply diversification strategy, Chapters 6 and 7 give special attention to China's approach to foreign oil and gas developments, focusing on the Central Asian republics, describing and explaining the many ups and downs of Sino-Russian oil and gas cooperation during the last two decades. Chapter 8 summarizes the author's verdict on Sino-Russian oil and gas cooperation up to now, and foresees its future.

After a brief overview of Sino-Russian relations during the last two decades, Chapter 1 focuses on the energy-related aspects of relations between the two counties. A summary of the main lines of energy cooperation in the 1990s, highlighting the vast oil and gas potential of East Siberia and the RFE (Russian Far East) and China's bold initiative towards East Siberia, prefaces a first look at actual energy cooperation in recent decades. From the 1970s until 2000 the story is one of unfulfilled initiatives for pipeline infrastructure development. There has been a gulf between the huge potential of the region and the disappointing amount of actual development.

Chapter 2 centres on Russia. After a survey of the role of oil and gas in the economy, the chapter examines the main oil initiatives in East Siberia and the Far East frontier region. These include the East Siberian exploration programmes and the role of the Federal Agency for Subsoil Use (Rosnedra) in understanding the true scale of proven oil and gas reserves in the region. The chapter goes on to describe the main ESPO production centres: the major fields in the Irkutsk region, Sakha Republic and Krasnoyarsk Territory (in particular the Vankor field), and finally the vast Sakhalin offshore developments (Sakhalin-1, 2, 3, 4, and 5). This survey reveals the huge potential scale of oil supply from the Russian Far East to China and the rest of Asia.

Chapter 3 concentrates on the gas industry in Russia and its dominant enterprise, Gazprom. It reviews the development of Gazprom's Asia export policy and the impact on Gazprom of the 2008 global crisis. We describe and assess the Eastern Dimension of the Unified Gas Supply System (UGSS), the East Programme, and the projections for the Gas Industry Development Strategy up to 2030. Special attention is given to the gasification programme in the Asian regions of Russia and to the development of the SKV (Sakhalin–Khabarovsk–Vladivostok) gas pipeline. All this amounts to a comprehensive and detailed review of the three main Gas Supply Sources for the Eastern UGSS – Kovyktinskoye (or Kovykta) Gas, Sakhalin offshore gas, and Sakha Republic gas – from which we can obtain a clearer view of their relative importance in the future development of Russia's gas supply to China.

Chapter 4 moves to the question of energy policy in China and how it has been changing. A general survey is followed by a close look at the restructuring of the government's energy bureaucracy and the country's oil industry restructuring. A summary of the Third National Oil and Gas Reserves survey is a good starting point for looking at both onshore (in particular the Tarim basin) and offshore production capacity and for explaining the rapid expansion of China's oil import needs. The attention then shifts to China's domestic oil pipeline network and its close links with import via pipelines from Russia and Kazakhstan. After an overview of the oil refining industry the chapter highlights the importance of strategic reserves development, a central feature of China's future energy security.

Chapter 5 presents a comprehensive review of the rapid expansion of China's gas industry. Beginning with domestic gas, it describes the main centres of production – the Tarim, Ordos, and Sichuan basins – followed by the offshore gas fields, and then outlines the potential of coalbed methane (CBM) and shale gas. After examining the country's domestic gas demand, the chapter analyses in detail the sensitive issue of gas pricing, explaining why a rapid reform of gas pricing has not been possible. Since the city gas sector will be the largest user of the expanded national production of gas in the coming decades, this chapter examines some of the main firms in this industry – the Hong Kong and China Gas Co. Ltd (Towngas), China Gas Holdings Ltd (China Gas), Xinao Gas Holdings Ltd (Xinao Gas), and PetroChina Kunlun Towngas. The imminent rapid growth of the city gas sector is closely connected with China's nationwide pipeline network development (in particular WEP I, II, III, IV, and V). The WEP network is in turn closely related to the trans-national gas pipelines designed to link up with gas supplies from Russia, the Central Asian Republics, and Myanmar. The chapter then moves on to the expansion of LNG. Both the first phase of LNG expansion, based on Guangdong and Fujian LNG, and the second, based on Shanghai, Zhejiang, Jiangsu, Dalian, and Shandong, illustrate the difficulties which China faces with regard to LNG supply. Delays in the development of Sino-Russian gas cooperation may lead to increased LNG use in China. Thus the expansion of the LNG industry can only be understood in relation to pipeline development.

Chapter 6 examines China's policies towards international expansion in the energy sectors, paying particular attention to the major M&A deals during the 2008–10 period. The central issue discussed in this chapter is the approach towards oil and gas supply from the Central Asian Republics, in particular natural gas from Turkmenistan and crude oil from Kazakhstan. The characteristics of the Central Asian Republics' supply options to China explain why the Beijing authorities have made a clear choice to maximize imports from Central Asia while making a different choice in relation to Sino-Russian oil and gas cooperation.

Chapter 7 tells the story of Sino-Russian cooperation during the period 1993–2010. In order to interpret this we need to look at the main forward and backward steps concerning cooperation in both the oil and gas sectors, at a number of merger and acquisition deals, and at other steps taken towards strategic partnership. The nature of cooperation in the oil sector explains the collapse of the dream of building the Angarsk–Daqing pipeline (see Chapter 7), as well as the origin of the ESPO saga. The completion of the first stage of the ESPO pipeline project now strongly suggests that the second stage will also go ahead, but the ultimate scale of the pipeline is unknown, since the pertinent exploration work is not yet done. Turning attention to the gas sector it can be seen how the Kovykta gas project became a casualty of the Asia policy of Gazprom, which has complete control of Russian gas exports to Asia. Disputes over asset ownership, the helium issue, and export prices all prevented meaningful negotiations between Kovyktagas and Chayandagas and potential customers in north-east Asia, in particular China and South Korea. But the study concludes that the ultimate obstacle was and is the paralysed negotiation between Russia and China over the export border price. Finally, a string of merger and acquisition deals and strategic partnership agreements reveals what went wrong and how in the end a more successful deal was arrived at. Based on its analysis of the events of the last two decades, this chapter ends with a prognosis about how Sino-Russian oil and gas will evolve over the decades to come.

Chapter 8 aims to answer the key interrelated questions which run through the preceding chapters. These are:

• What are the prospects for strengthened Sino-Russian oil and gas cooperation in the foreseeable future?
• More specifically, since Sino-Russian cooperation did not show particularly positive results during the 2000s, why should the 2010s prove more successful?
• What benefits would greatly strengthened cooperation in the oil and gas sector bring both for Russia and for China?
• Finally, however Sino-Russian oil and gas cooperation evolves, what effects will it have on the energy situation in the rest of the world?

CHAPTER 1

THE ENERGY FACTOR IN SINO-RUSSIAN RELATIONS

The turbulent break-up of the Soviet Union after 1990 transformed the range of potential answers to numerous questions in international relations, not least of these being the future of energy relations between Russia, with its vast multiform reserves of energy, and China whose unprecedented economic growth was already generating a gargantuan appetite for energy. It was widely expected that the symbiosis between Russian energy supply and Chinese energy demand would result in massive exports from the former to the latter. In some ways that indeed is part of the story which has evolved; but it has not done so without many surprises and complications.

A Brief Review of Sino-Russian Relations

For China the collapse of the former Soviet Union (FSU) in 1991 came as a major shock. No sooner had Boris Yeltsin been elected to the newly created post of president of the Russian Soviet Federative Socialist Republic than he seized more power, in the aftermath of a failed coup that had attempted to topple the reform-minded Gorbachev. The reaction of the Chinese, however, created no tangible obstacle to the transformation of Sino-Soviet relations into Sino-Russian relations. During the first half of the 1990s, however, the relationship between these two superpowers did not regain the previous level of strategic partnership. Yet, during the second half of the 1990s the relationship began to re-gather strength, even though it was not until the 2000s that a new Sino-Russian relationship really emerged.

Understanding Sino-Soviet relations is still critically important for understanding Sino-Russian relations. The Sino-Soviet relationship, characterized by economic dependence and military alliance, began in 1950 after the Chinese Revolution and broke down, in James Bel-lacqua's words, 'over ideological disputes between Mao Zedong and Nikita Khrushchev, culminating in a messy divorce of the two Communists in 1960'.[1] It had lasted only a decade. Less than two decades later, in the short time between Mao's death in 1976, and 1978, China was transformed. The ascendancy of Deng Xiaoping opened the way to a reconciliation between the leaders in Moscow and Beijing, but this opportunity was grasped only slowly. The normalization talks which began in 1982 did not achieve any momentum. Division in China, and then incompetence in the Soviet leadership were partly responsible, but on both sides ideologues trumped experts. Between 1986 and 1988 Mikhail Gorbachev's new thinking laid the foundation for normalization, but this was treated with suspicion even before the backlash emanating from Beijing's 1989 internal crackdown.[2] The fragile Sino-Soviet relationship, however, evolved into the Sino-Russian strategic partnership established by the agreement signed by the Russian president, Boris Yeltsin, and his Chinese counterpart, Jiang Zemin, in April 1996. This was upgraded with the signing of the Treaty of Good-Neighborliness and Friendly Cooperation between the Chinese President, Jiang Zemin, and his Russian counterpart, Vladimir Putin, on 16 July 2001.[3] This twenty-year strategic treaty laid a solid foundation for stronger relations between the two super powers. From 2003 there was a notable quickening of the pace of improvement in relations and a broadening of the scope of bilateral ties, which reached a new peak in the years 2006–8. Gilbert Rozman finds the successive agreements during the years 1976 to 2008 to be like snapshots, each one showing some improvement over its predecessor, even though momentum was at times lost in the intervening years.[4]

This pattern of interaction between two major powers is perhaps not surprising, given how much China and Russia have in common. Russia's economy flagged seriously in the 1990s, but during the 2000s, prior to the onset of the 2008/09 global financial crisis, both countries were experiencing robust economic growth fuelled by their exports – Russia's of oil and gas, and China's of just about everything else. At the same time, both nations characterize themselves as 'developing countries', although China's main task is the struggle to lift hundreds of millions people out of poverty, while Russia's is the attempt to diversify its economy away from a heavy dependence on raw material exports. The governments of both nations are stable, authoritarian, and administratively centralized (even though China has devolved considerable decision-making powers to the local provinces), yet both governments have problems controlling their distant regions. China and Russia are both proud, sensitive countries, strongly conscious of their global position, status, and degree of influence. They are equally conscious of their impact on global affairs and often vote the same way in the UN Security Council, of which both are permanent members with a veto.[5]

The nature of the new Sino-Russian relationship is increasingly influenced by the recent growth of China's importance in the global arena. During the two decades since the collapse of the FSU, the most important change in the world balance of power has been the rise of China to the point where the USA has had to relinquish its briefly held position as the world's sole superpower, and to share joint superpower status with China.[6] China's rapid economic growth shows no sign of any slow-down yet. Its status as the hub of world manufacturing[7] and as the world's biggest foreign currency holder[8] has given the country an unprecedented position in modern history (See Table 1.1). China's vast spending on imported oil, gas, and mineral resource assets was an unavoidable part of its growth, which depends on a reliable supply of energy, in particular of oil and gas. However, the biggest beneficiaries of China's capital spending have been African countries, followed by the Central Asian Republics, and Latin American countries. Purchasing assets in Russia has not been China's priority, since Russia has rigidly stood out against Chinese acquisition of Russian upstream oil and gas assets. The Russian Federation had to struggle to survive the dark post-communist decade of the 1990s, but at length it managed to restore its oil and gas industry, and to become reborn as a major energy power during the 2000s.

Table 1.1: Comparison of China and Russia, 2010

	People's Republic of China	*Russian Federation*
Area	9,640,011 km² (3,717,813 sq. mi)	17,075,400 km² (6,592,800 sq. mi)
Population	1,341,150,000	143,905,200
Population Density	140/km² (363/sq. mi)	8.3/km² (21.5/sq. mi)
Capital	Beijing	Moscow
Largest City	Shanghai	Moscow
Government	Unitary socialist republic	Federal semi-presidential republic
Official language	Chinese	Russian
GDP (nominal)	$5.878trillion	$1.479 trillion
GDP (PPP)	$10.084 trillion	$2.812 trillion
GDP (nominal) per capita	$4,393	$10,440
GDP (PPP) per capita	$7,536	$19,840
Human Development Index	0.663	0.719
Foreign exchange reserves	US$2.622 million	US$501 billion
Military expenditures	$78 billion	$39.6 billion

Sources: World Bank, *World Development Indicators 2011*; United Nations Development Programme, *Human Development Report 2011*.

Hiski Haukkala and Linda Jakobson have argued that:

[t]he simultaneous ascent of Russia and China is increasingly greeted in
the West with suspicion and even growing trepidation. China's rise has
been seen as the emergence of a new robust rival; Russia's as the possible
prelude to a renewed Cold War. The basic assumption is that the rise of
new powers can only be to the detriment of the established ones – notably
the United States, but also the countries of the European Union.

...

More often it is China and Russia together that are tagged as the main
culprits in the imminent unravelling of the current liberal world order...
The suspicions of the United States and the EU extend to both of the
Eastern powers but fears of Washington are centred on China, while the
EU's main concern is Russia.[9]

Haukkala and Jakobson conclude that:

... these current fears of China and Russia are overstated because the
Sino-Russian challenge is an uneven and asymmetrical one, meaning that it
is highly unlikely to be an existential threat to the West. The asymmetries
between China and Russia [they conclude] hardly make them ideal partners
in an alliance against the West.

These authors add that:

... both China and Russia need the current world order, even though
Russia's recent moves have been much more resentful towards the West,
than those of China.[10]

During the last decade Sino-Russian cooperation has appeared in
various guises: the 2001 bilateral Treaty of Good-Neighborliness and
Friendly Cooperation, the six member Shanghai Cooperation Organization, an increasing number of combined military exercises aimed at
combating terrorism and secessionism, a great deal of cooperationist
rhetoric, and a few real acts of cooperation. Yet, as Haukkala and
Jakobson conclude:

Sino-Russian friendship and strategic partnership are a façade concealing
tensions that will undermine the future relationship of the two powers, at
least when it comes to mounting a united and sustained assault against the
current world order.[11]

Chinese elites do not see Russia as a major factor in China's becoming
a world power but are more concerned about their country's status
in relation to the USA, which in turn, after the successful Chinese
economic reforms, now sees China as having greater importance than
Russia. In addition, Chinese and Russian elites both retain memories

of centuries of armed conflict between the two countries, some of it extremely brutal.

What stops a deeper rapprochement between the two countries is that Russia's China policy is contradictory. On the one hand, as Haukkala and Jakobson argue:

> Russia is enticed by the economic prospects of China while it is virtually locked into an enduring (inter-) dependence with Europe. Given the location of its major hydrocarbon reserves, as well as the existing pipeline and other transport infrastructure geared towards the West, it would take decades and astronomical costs to reverse their direction.

On the other hand, they say:

> Russia has been China's biggest arms supplier for years, while the main threat scenario that the Russian General Staff is preparing for is an armed attack from China, with a growing consensus being that in the wake of such an event Russia would have no other means to defend itself but to go nuclear immediately. Adding to Russian frustration is the fact that it does not have a similar role in Chinese strategic thinking, where the main adversary is the United States.[12]

This discussion clearly shows the decline in Russia's status in the global scene, but that is the hard reality. The relative status of China and Russia in the global scene has shifted markedly and the gap between them is widening. Russia sees energy as a tool to restore its status. Bobo Lo explains that:

> Sino-Russian relations are driven by interests and by not ideology and this is both a strength and a weakness. On the one hand, it enables Moscow and Beijing to escape some of the baggage of the past, focusing instead on what unites rather than what divides them. On the other hand the problems of international relations may also be a source of new tensions and divisions between the two powers.[13]
>
> …
>
> The Sino-Russian relationship is defined by tangible interests and realities of power. And herein lies its greatest source of vulnerability. China's rise as the next global superpower threatens Russia, not with the military or demographic invasion which many fear, but with progressive displacement to the periphery of international decision-making. Although it is fashionable to bracket Russia and China together as emerging powers, along with India and Brazil, the trajectory of their development portends different fates. The aggregate bilateral balance of power – economic, political, technological, strategic – has already shifted in Beijing's favour, and its disparity with Russia will only become more marked with time. More than any other single factor, it is this growing inequality in an uncertain world that will inhibit the development of a genuinely close partnership.[14]

Russia's specialists, fully aware of this imbalance of power between Russia and China, see the necessity of new strategy, not just for China, but for Asia as a whole. Sergei Karaganov, chairman of the Presidium of the Council on Foreign and Defense Policy, has pointed out that:

> [t]he main force holding us back from pursuing a reasonable and purposeful Asian policy is ignorance, misunderstanding of the opportunities, and myths about the real state of affairs in that region… But if the current economic trends persist, it is very likely that Russia east of the Urals and later the whole country will turn into an appendage of China – first as a warehouse of resources, and then economically and politically. This will happen without any 'aggressive' or unfriendly efforts by China; it will happen by default. The geopolitical implications of such developments are obvious. There will be no chances for Russia to play the 'Chinese card'. Beijing will rely on Moscow, whose real sovereignty over the eastern territories will be de facto wearing thin … A modern strategy – so-called Project Siberia – should be internationally oriented from the outset. It should combine Russian political sovereignty with foreign capital and technologies. And not just from China, but also from the U.S., Japan, the EU states, South Korea and the ASEAN countries, all of which are keenly interested to ensure there should be no exclusive dominance of China east of the Urals … The project should be aimed at making Russian eastern regions one of the resource and food bases of rising Asia. A provider of goods with a relatively high degree of added value.[15]

His points indirectly tell us where Russia's priority lies in its Asian Strategy development.

During President Medvedev's visit to China in late September 2010, Dmitri Trenin, director of the Carnegie Moscow Center, writing in the *China Daily*, raised an interesting question:

'Is there', he asked, 'a chance of the Sino-Russian relationship suddenly turning sour?' and answered 'Not under the present circumstances and the current leaderships in the two countries'. But, he added:

> a word of caution is needed. Both countries are seeing a rise in nationalism – defensive in post-Soviet Russia and more assertive in China, which is feeling its new strength and has not forgotten the humiliation it was subjected to in the past. The two phenomena are understandable from a historical perspective, but the governments in Moscow and Beijing need to ensure that the national feelings of their citizens are converted into constructive patriotism, rather than destructive xenophobia. Despite his obvious importance, Dmitry Medvedev will be only one of about 1 million Russians who come to China each year. It is ordinary people, tourists and business travellers, academics and artists, who form and shape modern Sino-Russian ties. What started as party-to-party and then became state-to-state is now

being transformed into a people-to-people relationship. This is, by far, the strongest bond between the neighboring countries.[16]

Trenin's point was that, despite growing inequality between the two countries, the Sino-Russian relationship is now becoming genuinely bilateral. The growth in crude oil trading, coal trading, and nuclear power plant development means that both parties are deriving strong concrete benefits from their relationship. The biggest contribution to the Sino-Russian relationship will come from cooperation in the gas sector, since its implications go well beyond energy trading between the two countries. But whether there is an upgrading of Sino-Russian gas cooperation during the present decade is something which will have to be decided by the leaderships of the two states.

The Energy Factor in Sino-Russian Relations

The importance of the energy factor in a strengthened Sino-Russian relationship cannot be over-emphasized. However, the pace of progress in Sino-Russian energy cooperation has been slow and encumbered by mutual mistrust and differing policy agendas, in spite of recent advances in cooperative energy engagement and of public overtures by the leadership of the two countries.[17] Both countries recognize the vital importance of deepening engagement where both energy and power are concerned.[18] (This book, however, will focus on oil and gas cooperation and power sector cooperation based on nuclear power plant development – the initiative to supply surplus electricity from Russia to the north-eastern provinces of China should not distract from that focus.)

Normally it is safe to assume that an increased trade volume reflects an improved relationship between the two parties, even though some exceptions exist. Sino-Russian trade has increased dramatically since 2000. Trade in terms of value reached a peak of $58.8 billion in 2008, making China Russia's single largest trading partner in 2009, if its trade with the EU is disaggregated on a state-by-state basis. More troubling for Russia, however, is the fact that beginning with the September 2008 collapse in global oil prices, Sino-Russian trade, measured in dollar terms, actually fell, reaching $38.8 billion in 2009. This is less indicative of a dramatic fall in trade volumes than it is attributable to a dramatic decrease in the per-barrel price of oil since 2008, given that oil dominates Russia's trade with China. The fact is that Russia needs Chinese trade, particularly in terms of energy resources, more than China needs Russia. In 2009 Russia was placed only fourteenth on the list of China's trading partners, accounting for only 1.7 per cent of its total trade with foreign partners. The combined fall in global oil (and energy) prices, coupled with the disproportionate relationship in trade, suggests that Russia is not exporting less oil to China, but simply that the value of its hydrocarbon exports has declined over time due to an upward movement of the Chinese yuan against the Russian ruble, and to a secondary devaluation of oil prices overall, as noted above.[19] According to both the Federal State Statistics Service (Roskomstat) and Chinese customs authority data, foreign trade turnover of Russia with China reached $55.44 billion in 2010, and consequently China overtook Germany to become Russia's largest trading partner in 2010.[20]

On 22 November 2010, the launch of yuan–ruble trading became a reality, and the ruble became the sixth currency to be officially traded in China. Russia has been pushing for a greater role for the ruble on global financial markets, and Beijing is also seeking a greater international role for the still tightly controlled yuan. Russian analysts and market players agreed that, while significant in the long run, in the short term volumes and market impact would probably be limited. There is a big potential, but the current volume of trade is not yet big enough to deliver a significant impact.[21] This is a symbolic initiative, possibly prior to a huge impact at a later stage.

As for the role of energy in Sino-Russian relations, Bobo Lo put it that:

> … energy has become a central plank of the bilateral relationship and of the two countries' foreign policies more generally. Moscow regards control of oil and gas resources as its most effective means of power projection in the post-bipolar age. For Beijing the quest for energy has become an all-encompassing priority, the engine of China's modernisation. Sino-Russian energy cooperation is emblematic of the potential, but also of the shortcomings of their partnership. It offers a vision of the future as the most plausible avenue for taking relations to the next level, with political, strategic, as well as economic, benefits. However, progress has been slow … Such difficulties reflect Moscow's reluctance to become too China-dependent in terms of market, as well as concern that Russia is turning into a resource-cow for Chinese modernisation. For its part Beijing has reacted to the Kremlin's erratic behaviour by widening the search for new suppliers of energy.[22]

Russia's President Medvedev has had to tread a fine line between carving himself a separate identity as a politician, and not alienating the kingmaker who stands between him and a second presidential term. His criticism of Prime Minister Putin has been carefully structured and oblique – notably calling the economy he inherited 'primitive' for its reliance on raw material exports. Prime Minister Putin has supported President Medvedev's efforts at modernizing the economy but his backing has appeared tempered by a desire to keep hold of all levers of control.[23]

Bobo Lo points out that at first sight the Sino-Russian energy relationship appears to be based on an almost ideal complementary model: on one side is the world's biggest exporter of oil and gas, and on the other, the world's second largest consumer of energy after the USA. Add to this flourishing bilateral relationship a public commitment to developing energy ties, and there seems no impediment to the closest possible partnership based on convergent interests. Energy has become emblematic of the evolution of the countries' relationship, from the largely political partnership of the 1990s to today's more pragmatic and business-like interaction. Despite this favourable context, however, the energy relationship has been dogged by problems. The most fundamental aspect is that Moscow and Beijing have very different understandings of energy security. To the Russian planners, it means security of demand. Oil and gas represent over 60 per cent of Russia's exports in value terms and over half of federal budget revenues. Consequently a loss of overseas markets would be catastrophic for economic prosperity and political stability. By contrast, China's perception of energy security is centred in the more conventional understanding of security of supply. These polarized understandings of energy security translate into an imperfect complementarity. To the Chinese planners, crude oil imports have been the priority, and to the Russian planners, natural gas the ultimate prize for cooperation with China.[24]

In short, the approaches made towards energy cooperation by Russia and China were very different, and the two sides' expectations were wide enough apart to require tedious negotiation and continuous compromises. The role of the energy factor in strengthening the Sino-Russian relationship has been quite restricted, even though energy has played a pivotal role in the rise of Russia and China respectively. It is fair to say that the energy factor in the Sino-Russian relationship is still dependent on identifying the common interest between these two energy powers.

A Brief Review of Sino-Russian Energy Cooperation in the 1990s

Strictly speaking, Sino-Russian oil and gas cooperation in the 1990s achieved nothing, except for continuous negotiations and the confirmation of the huge potential of East Siberia and the Far East of the Russian Federation. The summary of Sino-Russian energy cooperation

Table 1.2: Sino-Soviet/Russian oil and gas cooperation, 1989–99

June 1989	Soviet and Chinese specialists working together to develop improved technologies for drilling wells down to 6,000 metres.
July 1989	A Soviet oil delegation went to Harbin, expressed a tentative desire for Heilongjiang to provide technology, labour service, and equipment for the Transbaikal oil transfer depot.
March 1990	A Soviet drilling platform underwent a thorough overhaul in Shanghai.
November 1990	Russia's Ministry of Oil and Gas Construction agreed to supply equipment and skilled technicians for the development of 2 or 3 oil fields in north-west China while the Chinese sent consumer goods and semi-skilled labour to fit out two new oil fields in Western Siberia.
December 1990	'Nanyang' oil field in Henan Province was discovered with the application of 'Dianchang Chafen Fa' invented by a Soviet scientist.
June 1991	Representatives of the Sakha Republic (Yakut) conducted preliminary negotiations with Harbin Gas Company (HGC). The HGC expressed its interest in purchasing 5 bcm/y of gas for Harbin District. As an option arrangement, the HGC would buy 8 bcm/y for the Open Littor Zones.
May 1992	The Sixth Geophysical Prospecting Team under China's Ministry of Geology and Mineral Resources was reported to be participating in oil exploration in Russia's Tyumen region.
July 1992	Zhang Yongyi, vice president of the CNPC, proposed the development of oil in East Siberia to Russia and Japan.
February 1993	Primorskii Krai and Suifenhe signed an agreement to set up a joint venture to provide LPG.
September 1993	CNPC's subsidiary, Daqing oil field, in partnership with Canada's MacDonald Petroleum, entered into negotiations to explore the Mapkob and Yarokotan oil and gas fields in the Irkutsk region. Reportedly, Daqing oil field had gained permission to establish a branch office in Irkutsk, and had rigs drilling exploratory wells in two virgin fields.
November 1994	CNPC and MINTOPENERGA signed an MOU, aiming to construct a trans-boundary long-distance natural gas pipeline running though Inner Mongolia and Hebei Province, terminating in Shandong Province.
June 1997	During PM Viktor Chernomyrdin's visit to Beijing, a governmental framework agreement was signed between Russia and China to export natural gas and electricity from East Siberia to China. The gas export scheme envisaged 25 bcm/y of gas export to China over 30 years.
February 1999	Three important agreements were signed between PM Zhu Rongji and PM Yevgeny Primakov after the fourth regular meeting: i) a preliminary FS on crude exports from Angarsk to Daqing through a 20-30 mt/y capacity pipeline; ii) a FS on the Irkutsk region's natural gas export to north-eastern provinces in China through a long-distance pipeline; iii) a preliminary FS on western Siberia's gas export to Shanghai by a transnational pipeline passing through the Xinjiang Autonomous Region.

Sources: Paik (1996); various issues of *CD* *(China Daily)* and *China OGP.*

during 1989–9 in Table 1.2 shows that there was a flurry of reports on cooperation. The majority were plans and intentions; little was actually achieved in terms of energy production.

China's Bold Initiative towards East Siberia: 1993 and 1997

It was in 1993 that China became an oil importer for the first time. According to *China OGP*, China has imported more than 20 types of oil with API gravities ranging between 30 and 45 since the early 1990s. About 65 per cent of China's crude imports were sourced in Indonesia, Oman, and Malaysia. The rest came from Vietnam, Iran, Dubai, and Russia.[25] It is not surprising then to find China also interested in East Siberian oil and gas development through Sino-Russian cooperation, if not yet on a significant scale. As mentioned in Table 1.2, CNPC took the initiative in judging the possibility of East Siberian oil and gas exploration and development. Considering that China's state oil firms had had virtually no experience in foreign oil and gas exploration and development (including in Russia), the 1993 initiative was a bold step (CNPC's first foreign merger took place in 1992). The important point was, as mentioned above, that 1993 was the year in which China became a net oil importer for the first time, and the impact of this on the Chinese energy planners was marked. Since many Chinese energy specialists had years of experience in Russia, it was quite natural for CNPC to show serious interest in examining the exploration opportunities in remote East Siberia.[26] Around the same time, Russia was also actively promoting plans to develop Irkutsk province's gas fields and export electricity to China. The province's Kovyktinskoye gas-condensate field, discovered in 1987, had estimated total reserves of 600–800 bcm (21.2–28.2 tcf) of gas. In addition, the Verkhnechonskoye field has estimated in-place reserves of 600–650 mt of oil.[27]

In 1993, CNPC established a company named Central Asia Corporation, directed by the CNPC Vice President Zhang Yongyi, to promote the development of a long distance gas pipeline from Turkmenistan to Japan via China. The original plan for this, introduced in 1993, aimed to build an onshore pipeline from Turkmenistan to Tanggu, a port close to Tianjin, where the gas would be liquefied and then shipped to Japan. CNPC was interested in this project as it would play a significant role in piping out Xinjiang's gas to consumers in east and south China.[28] It confirmed that CNPC was simultaneously pursuing the options of gas supply to China from both Russia and the Central Asian Republics.

Between 1993 and February 1999, when three agreements for preliminary feasibility study works on oil and gas pipeline development were signed, Sino-Russian oil and gas cooperation was expanded via a number of key agreements. The first was the MOU, signed by CNPC and MINTOPENERGA in early November 1994, which envisaged the construction of a trans-boundary long-distance natural gas pipeline running though Inner Mongolia and Hebei Province and terminating in Shandong Province. At that time, former petroleum minister Dr Wang Tao was president of CNPC. The trans-boundary pipeline, proposed by Sidanco, was to be designed to transport 20 bcm of gas per annum from the Irkutsk region in East Siberia to the coastal cities of East China.[29] This 1994 MOU could be regarded as the first governmental level agreement relating to Sino-Russian oil and gas cooperation.

Just before Prime Minister Li Peng's visit to Moscow at the end of December 1995, China's main focus was on participating in trans-national pipeline development. CNPC's feasibility study, however, concluded that the gas reserves in East Siberia, intended for export, were not as substantial as claimed by the Russians. The CNPC geologist estimated that the gas in place in Irkutsk amounted to 211 bcm, not enough to guarantee accumulated gas production of at least 500 bcm. Adding to reserves therefore became a top priority for both Russians and Chinese to support the pipeline plan. This was the reason why CNPC began to explore the option of combining the gas reserves of the Irkutsk region with those of the Sakha Republic.[30]

During the state visit by China's Prime Minister Li Peng to Russia at the end of December 1996, the *Xinhua News Agency* report highlighted a 4,070 km pipeline connecting Irkutsk in East Siberia to Rizhao in Shandong province which was high on the diplomatic agenda. The anticipated discussion between Prime Minister Li and President Yeltsin on this project was expected to open the way to a final contract after three years of bilateral negotiations. A detailed discussion was planned between CNPC's President Wang Tao, who was in Prime Minister Li's entourage, and his Russian counterpart from the Ministry of Fuel and Energy.[31]

In fact, the first Russia–China inter-governmental agreement for collaboration in the energy sector was signed in 1996. The two countries set up a joint commission to arrange regular meetings between their prime ministers, in order to oversee directly the development of bilateral cooperation. There were eight sub-committees under the joint commission, concerned with economics and trade, energy, technology, nuclear energy, aerospace, transportation, banking, and information technology. In the case of oil and gas projects, the governments had the final say over the negotiations. The SDPC (later changed to the NDRC) of China and the Russian Ministry of Energy are designated as the government representatives on each side. But the negotiations were between Chinese and Russian oil and gas companies. On the Chinese side, CNPC is the only company involved in negotiations on both crude and natural gas pipelines.[32]

Another important agreement was signed in 1997, when petroleum security became a priority issue. It was decided to diversify away from the Middle East rather than assume that the Gulf region would remain stable. Three decisions – on price reform, overseas E&P, and reduced dependence on Middle East imports –converged in 1997 as a necessary prelude to making large foreign investments possible. In the space of three weeks in June 1997, CNPC struck a series of deals with Kazakhstan, Venezuela, and Iraq, involving total investment of $5.6 billion. In the same month, a governmental framework agreement between Russia and China was signed during Prime Minister Chernomyrdin's visit to Beijing to promote the export of natural gas and electricity from East Siberia to China. The two countries agreed that a 3,000 km pipeline would be built, and that Russia would be committed to supplying an annual volume of 25 bcm for 25 to 30 years.[33] At the same time, China was also contemplating an 8,000 km gas pipeline running from Turkmenistan via Uzbekistan and Kazakhstan to Lianyungang in China's Jiangsu province and terminating at Japan's Kita Kyushu.[34]

In late September 1997, Li Peng visited Kazakhstan to sign a $9.5 billion deal to develop Uzen and Aktyubinsk oilfields and to build two oil pipelines, one running for 3,000 km to China, and the other, 250 km in length, going through Turkmenistan to the Iranian border.[35] CNPC's bold initiative to buy out oil assets in Kazakhstan confirmed China's intention to supply equity oil for the planned Kazakhstan–China oil pipeline. This initiative was later repeated during the 2000s when China decided to import Turkmenistan gas to China (See Chapter 6).

Two months later, Russia's Deputy Prime Minister Boris Nemtsov and his Chinese counterpart Li Lanqing, meeting in Beijing, tried to make progress on other energy projects and signed a technical memorandum regarding the construction of a Kovyktinskoye gas pipeline.[36] At the end of December 1997, a multilateral memorandum between five countries (Russia, China, Mongolia, South Korea, and Japan) was signed at a meeting in Moscow. It called for a coordinated approach to improving reserve estimates, further developing fields, defining the market for developed gas, and carrying out a feasibility study for the transnational pipeline project.[37] Separately from the Five Party initiative on the Kovykta gas project, Gazprom announced a $15 billion plan to construct a pipeline to transport gas from deposits in Krasnoyarsk Krai and the Yamal-Nenets Autonomous Region to Shanghai over a

30 year period beginning in 2004 (this was the starting point of the Altai project).[38]

The five country feasibility study, initiated at the end of December 1997, spectacularly collapsed during the Irkutsk meeting on 24 December 1998. At that time, Japan took the initiative on the question of feasibility by offering major funding for the study, but both Russia and China questioned Japan's ability to play such a coordinating role. Japan's proposed contribution to the feasibility study was not acceptable to either Russia or China. Japan was left without leverage as it could not offer a sizable gas market for the long-distance gas pipeline. After the collapse of the five country feasibility study, gas pipeline negotiations have taken place only between the two parties Russia and China.

Russia showed no interest in China's initiative in using the equity oil and equity gas option in Russia's East Siberia oil and gas development in the 1990s, a stance which did not change at all during the 2000s. The most that Sino-Russian oil and gas cooperation could achieve during the 1990s was the three feasibility studies for oil and gas pipeline development, signed in February 1999 by the Chinese and Russian prime ministers. During the 1990s, China's highest priority was to know whether the proven reserves were sufficient to justify the building of long-distance pipelines from faraway East Siberia. Russia, knowing that China's own proven oil and gas reserves were too small to satisfy the soaring level of demand, was over-confident about oil and gas pipelines, assuming that China would have no choice but to pursue the pipeline development urgently. A brief look at the oil and gas potential in East Siberia, Russia's Far East, and in China itself in the 1990s reveals why China was so anxious to secure oil and gas from Russia's frontier areas.

Oil and Gas Potential in East Siberia and RFE, the early 1990s

North-east Asia is a huge region – its nearly 20 million square kilometres of land area, dominated by China and Russian Asia, is equivalent to that of the USA and Canada combined. Of the seven countries in the region, the largest is China, extending 5,200 km from east to west and 5,500 km from north to south, with 9.6 million square kilometres. The second largest is Russian Asia, composed of the RES (1.6 million square kilometres) and RFE (6.22 million square kilometres). Despite the region's sheer size, the population of East Siberia and RFE in 1989 was merely 5.24 and 7.95 million respectively. The total size of other north-east Asian countries is a mere 2.2 million square kilometres, including Mongolia (1.6 million square kilometres), the Korean peninsula (0.2 million square kilometres), Taiwan (0.04 million square kilometres), and Japan (0.4 million square kilometres).[39] In the wake of the collapse of the cold war, cooperation among the north-east Asian countries was explored for the first time. Within the region there are substantial fossil fuel resources, but their uneven distribution indicates that cooperation in oil and gas development within the region could produce common benefits, and the target area for cooperation in fossil fuel development was East Siberia and the Russian Far East.

Despite these great potential oil and gas reserves, the RFE region had contributed almost nothing to cumulative oil output, and a negligible amount to gas output. The reality of the RFE's oil and gas industries reflected the long delay in the region's oil and gas development. What was urgently needed to redress this delay was timely and massive foreign investment since Russia itself lacked the capacity to invest in Far East oil and gas development.

Oil and Gas Potential in China in the early 1990s

In the early 1990s estimates showed that China had a total of 246 sedimentary basins covering 5.5 million square kilometres, of which 4.2 million square kilometres were onshore and 1.3 million square kilometres offshore. Total oil reserves were estimated at 78.75 billion tonnes (bt), of which 62.6 bt were onshore and 16.7 bt offshore, while gas reserves were estimated to be 33.3 tcm (1176.6 tcf). The Chinese authorities were not reluctant to release figures on Xinjiang Province, especially the Tarim basin oil and gas reserves. In an interview with *World Oil* in late 1989, Wang Tao, then president of CNPC, suggested that a third of all potential oil resources and half of the total gas resources in China were located in the far west. In the early 1990s Chinese geologists estimated that the geological reserves of the Tarim basin, discovered on 22 September 1984, were as high as 74 billion barrels of oil and 283 tcf of natural gas, equal to about one-sixth and one-quarter of the national totals. *China OGP* reported that the Tarim basin, similar in size to France, contained 19.76 bt of crude and 8.39 tcm of natural gas. CNPC's focus on China's western frontier was necessary, as China had to take this step to cope with the production decline from China's super-giant oil field, Daqing in Heilongjiang province.

At that time CNPC projected that by the year 2000, Xinjiang's recoverable oil reserves would total 14 billion barrels (including heavy oil), compared to China's total reserves which in 1994 had been estimated at 22–25 billion barrels. More inflated non-Chinese estimates placed the basin's identified oil reserves in the range 18–18.45 bt. Offshore reserves were put by CNOOC at 1.2 bt of oil and 180 bcm of gas. Earlier in the 1990s CNOOC had suggested figures of 870 million tonnes and 133 bcm, but these figures seemed to refer to recoverable reserves.

In China, natural gas had been largely neglected, and production was only 2–3 per cent of total energy consumption, compared with Asia's average of 8–9 per cent. While crude oil production rose by 30 per cent during the 1980s, gas production fell sharply in the early part of the decade, and by 1990 had only regained its 1980 level. In 1994 production had reached 16.6 bcm, but China was aiming to produce 20–22 bcm in 1997 and 25–30 bcm by 2000. In 1993 two groups of gas fields were producing more than 2 bcm, and more than 80 per cent of China's total production came from the five major gas fields. The largest Sichuan gas fields cluster alone accounted for 43 per cent of production (6.5 bcm). Sichuan's importance was heightened by the fact that it has the highest ratio (90 per cent) of commercial sales to production, and accounts for more than 60 per cent of national sales.

Gas production costs, however, were high because of complex geology and high sulphur contents (averaging 0.27 per cent by weight). In fact, in Sichuan province, exploration cost per metre rose from 2,285 yuan in 1991 to 4,200 yuan in 1993 (far above the planned cost of 2,941 yuan), while over the same two year period the number of exploration wells decreased from 40 to 21 (compared with the planned number of 29). In the Weiyuan gas field in central Sichuan province, more than half the wells had been closed or had suffered production cuts due to water incursion. Since gas provides fuel or feedstock to one third of industrial operations and many residential customers in Sichuan province, the field's large production was scarcely able to satisfy the heavy demand in the region itself.[40]

Unfulfilled Initiatives for Pipeline Infrastructure Development in the 1970s, 1980s, and 1990s

In order to realize the already mentioned oil and gas potential in East Siberia and the Russian Far East, the Soviet and then Russian authorities prepared a series of ambitious export schemes. The first proposal, aimed at securing Japanese participation in the construction of a gas pipeline from the Yakutsk gas fields to the Soviet Far Eastern coast, was made by N. K. Baibakov, Gosplan chairman, in November 1968. The proposal partly overlapped another, also from Baibakov in 1968, for Sakhalin offshore gas exports to Japan. On 12 February 1970 the Soviet Union proposed constructing a one metre diameter gas pipeline linking Yakutian gas reserves into the Sakhalin–Hokkaido system. The line's route would be Yakutsk–Khabarovsk–Sakhalin–Hokkaido. Two other proposals regarding Yakutian gas exports were made in unexpectedly quick succession in 1970 and 1972: the first, from the Soviet Union's Prime Minister Aleksei Kosygin, was for the construction of a pipeline from Yakutsk to Magadan along with a gas liquefaction plant at Magadan; and the second, from the US companies El Paso Natural Gas and Bechtel, and Japan's Sumitomo Shoji, was for the construction of a 3,600 km, 142 cubic metre gas pipeline from Yakutsk to Nakhodka. The Yakutian project, however, lapsed because each of the parties had a different preference for the pipeline route. A meeting of all parties in Paris in November 1974 to sign a General Agreement on the exploration of Yakutian natural gas failed to reach an accord. To make matters worse, the Soviet side proposed Ol'ga as the liquefaction plant site instead of Nakhodka. Although in 1979 the Soviet side finally agreed to the Yakutsk–Ol'ga pipeline route, the Yakutsk project was suspended in the wake of the Soviet invasion of Afghanistan in December 1979.

In January 1989 the forgotten project was revived with an ambitious plan by Chung Ju-Yung, founder of the Hyundai Group, to lay a gas pipeline grid from Yakutsk to South Korea through North Korea. Around the same time, US lawyer John Sears and Uebayashi Takeshi of Tokyo Boeki (a subsidiary of the Mitsubishi group) visited Moscow to present a similar idea. These proposals in effect constituted a revival of the Vostok Plan since they included plans for gas pipelines from the Sakhalin and Yakutian gas fields through Primorskii Krai and across the Korean Peninsula and the Tsushima Straits to southern Japan.

At the end of 1991, a consortium composed of Tokyo Boeki and the US company Far East Energy reached an accord with the Yakut-Sakha Republic to conduct a feasibility study for the development of Yakutian oil and gas resources within the Vostok project. The agreement was followed by a related contract, signed with the Russian government on 18 March 1992, for the appraisal of a 4,000 km gas pipeline from southern Yakutia to Wakkanai (Hokkaido) via Sakhalin, and a 5,000 km trunk line to northern Kyushu via Vladivostok, North and South Korea. But the study, which was due to be completed by May 1993, was suspended.

At the beginning of 1991, close to the end of the Soviet era, another very ambitious plan for accelerating Soviet Far East gas development and export was prepared. A joint report by the USSR Ministry of Geology, the Russian Republic's Committee on Geology and Utilization of Energy and Mineral Resources, the USSR Ministry of Oil and Gas Industry, the State Gazprom Concern, the USSR Academy of Sciences, and the Russian Technological Academy together outlined a proposal about the 'Concept of Developing Yakutian and Sakhalin Gas and Mineral Resources of East Siberia and the USSR Far East'. This became commonly known as the 'Vostok (East) Plan'. According to this, by 2005 the region would produce about 15.7 mt/year of gas for the Russian Far East and 13.3 mt/y for export. Of this, South Korea and Japan would both receive 6.0 mt/y and North Korea 1.3 mt/y. Key elements of the plan would be the construction by 1995 of a 3,230 km gas pipeline from Sakhalin across Russian territory and North Korea to South Korea, and by 2000 a 3,050 km line from Yakutsk to Khabarovsk. Interestingly, however, the Vostok Plan did not envisage any gas export to China at that time.

The Vostok plan would have remained as a mere dream without the changes that occurred in north-east Asia after the early 1990s. After the demise of the cold war, new links across north-east Asia were established between countries previously in economic isolation from each other. The new developments during the first half of the 1990s included the Sino–Russian rapprochement, the establishment of diplomatic relations between the FSU and South Korea, and between China and South Korea, and the admission to the United Nations of both Koreas. These changes transformed international relationships in north-east Asia. Geo-political re-alignments have taken place, and geo-economic patterns have begun to assume much greater importance than in the cold war period.

The new geo-political alignments have resulted from a greater recognition of the region's multipolarity and a clear abandonment of the previously bipolar East–West policy framework by which policy approaches were largely circumscribed. The changes have been partly brought about by the diminished role and influence of the superpowers. A re-alignment of the balance of power has occurred among the four major powers, namely the USA, Russia, Japan, and China. With the end of the USA–USSR cold war China and Japan became the major regional powers in the area, although at first China's rise to G2 status was not foreseen even though its rapid economic advance was generally expected.

During the second half of the 1990s, an attempt was made to construct a long-distance gas pipeline connecting Kovykta gas field in the Irkutsk region with Beijing. Although this initiative, the decade's last, was made by a consortium including the five countries – Russia, China, Mongolia, South Korea, and Japan – it failed to find sufficient common interest among the participating countries. Since the initiative was Japanese (coming from JNOC and Sumitomo) the Chinese and Korean companies questioned whether the country which was to be the final market for most of the gas should be the dominant player. In the end this pipeline was also abandoned, as the prospect for Sino-Russian gas cooperation became the pivotal question in long-distance gas pipeline development in the region. This new failure was thus a clear indication that, given the geographical and economic realities, the obvious choice was to strengthen Sino-Russian oil and gas cooperation. This was the reason why the three important feasibility studies for both a crude oil pipeline and a natural gas pipeline from Russia to China were undertaken in the late 1990s. The consequences of Sino-Russian oil and gas cooperation during the 2000s would affect the direction of the cooperation in the coming 2010s and 2020s. The impact will not be confined to the two countries and will eventually be extended to the north-east Asian regional market, and beyond to the global market. The following chapters aim to review how much of that imperative has been achieved in the development of oil and gas in Russia's East Siberia and Far East regions during the 2000s, and to explore how far Sino-Russian oil and gas cooperation can go in the coming decades.

CHAPTER 2

THE DEVELOPMENT OF THE OIL INDUSTRY IN EASTERN SIBERIA AND THE RUSSIAN FAR EAST

The deep industrial slump which followed the fall of communism changed perceptions of the Russian economy. While Soviet planning had always emphasized heavy industry for home consumption, in the new Russian economy a more dynamic sector of the economy was the export of raw materials, especially oil and gas. Concern grew, however, about how to avoid the resource curse and to make sure that the country's riches were turned into sustainable economic growth.

After a brief review of the role of oil and gas in the Russian economy, this chapter analyses the 'resource curse' argument and Vladimir Putin's 'resource nationalism'. The main focus of the chapter is on Russia's initiatives towards East Siberia and the Russian Far East's frontier oil development during the 2000s. This chapter also aims to understand the level of exploration and development in the main bases of the East Siberia–Pacific Ocean (ESPO) pipeline, including the Irkutsk region, the Sakha Republic, and Krasnoyarsk Territory, together with the extent of Sakhalin offshore oil exploration and development during the 2000s. Understanding the capacity of the production bases will provide a clue to understanding the limits to the expansion of oil production in East Siberia and the Far East.

The Role of Oil and Gas in the Russian Economy

The collapse of the oil price in the wake of the 2008 global economic crisis exposed the vulnerability of Russia's excessive dependence on oil and gas very clearly. Since that time, the Russian leadership has become very interested in diversifying the economy from natural resources into high-technology industries and the creation of new jobs. But this will be a difficult task. Following the price increases of the early part of the decade, the ratio of Russia's oil and gas exports to GDP rose above 19 per cent in the two years 2005 and 2006. In 2008 Russia exported US$161 billion of crude, US$80 billion of oil products, and US$69 billion of natural gas. The combined figure of US$310 billion amounted to 66 per cent of total exports and 18.5 per cent of GDP. By 2009, however, economic crisis and rapidly declining oil prices had changed the economic and budgetary situation drastically. Combined oil and gas exports fell to US$191 bn (15.5 per cent of GDP).[1] As shown in Table 2.1, the total of oil and gas exports in 2010 was $254 bn. Although the Russian economy is far stronger than when President Putin took power in 2000, it is also even more reliant on hydrocarbon production.

As shown in Table 2.2, budget revenue from the oil and gas sector as a percentage of GDP in 2004 was 7.5 per cent (5.9 per cent from oil and 1.6 per cent from gas). By 2008 revenues from the two sectors had reached 11.1 per cent of GDP (9.7 per cent from oil and 1.4 per cent from gas). In 2009, however, the figure was 7.9 per cent, close to the 2004 level. It is noteworthy that the gas sector's contribution was no more than 14–15 per cent of that of the oil sector between 2006 and 2008. A report by Boris Nemtsov and Vladimir Milov pointed out that in 2007 Gazprom paid only slightly more than $7 in tax on each barrel of oil/gas while oil companies paid $40 per barrel.[2]

Table 2.1: Russia's Oil and Gas Exports

	Oil and Gas Exports (US$bn)	Share of Total Exports (%)	Percentage of GDP (%)
1998	27.9	32.2	10.4
1999	31.0	36.6	15.8
2000	52.8	46.1	20.3
2001	52.1	46.1	17.0
2002	56.3	46.4	16.3
2003	74.0	49.2	17.1
2004	100.2	55.2	16.9
2005	148.9	61.6	19.5
2006	190.8	63.3	19.3
2007	218.6	62.1	16.9
2008	310.1	66.2	18.5
2009	190.7	63.2	15.5
2010	254.0	63.5	17.3

Sources: 1998–2003 period: E. T. Gurvich, Macroekonomicheskaya ostenka roli rossiiskogo neftegazovogo sektora, *Voprosy Ekonomiki* (a journal of the Institute of Economics of the Academy of Sciences) 2004, no. 10, quoted in Ellman (2006); 2004–2008 period: GDP (Russian rubles) from Russia's Federal Statistics Service, average exchange rate of US dollars against rubles from Bank of Russia, and total export revenues from Russia's Federal Customs Service.

Note: The author is grateful to Dr Liu Xu of the Slavic Research Centre, University of Hokkaido for translating and compiling the contents of both this and the next table from the original sources mentioned.

Table 2.2: Consolidated Budget Revenue from the Oil and Gas Sector[3] (billion rubles)

	2004	2005	2006	2007	2008	2009	2010
GDP	17,048.0	21,665.0	26,621.3	32,988.6	41,540.4	39,016.1	44,491.4
Budget revenue from oil and gas sector	**1,285.6**	**2,321.7**	**3,114.9**	**3,081.1**	**4,606.8**	**3,062.8**	**3,991.7**
% of GDP	*7.5*	*10.7*	*11.6*	*9.3*	*11.1*	*7.9*	*9.0*
Oil Sector	1,005.9	1,990.3	2,681.3	2,690.2	4,026.1	2,638.5	3,713.3
% of GDP	*5.9*	*9.2*	*10.0*	*8.1*	*9.7*	*6.8*	*8.4*
Gas Sector	279.7	331.4	433.6	390.9	580.7	424.3	278.4
% of GDP	*1.6*	*1.5*	*1.6*	*1.2*	*1.4*	*1.1*	*0.6*
(1) Excise on goods	**176.4**	**119.3**	**119.1**	**129.7**	**138.3**	**143.4**	**160.9**
Oil products	138.4	114.7	119.1	129.7	138.3	143.4	160.9
Gasoline	71.0	79.6	85.0	95.0	103.9	109.8	124.9
Gas oil	27.3	33.0	32.0	32.9	32.9	32.3	34.4
Others	2.1	2.1	2.1	1.8	1.5	1.3	1.6
Natural gas	38.0	4.6	-	-	-	-	-
(2) Mineral extraction tax	**491.1**	**885.9**	**1,135.8**	**1,166.8**	**1,670.9**	**1,016.2**	**1,361.3**
Crude oil	428.6	801.4	1,038.4	1,070.9	1,571.6	934.3	1,266.8
Natural gas	58.9	79.2	89.9	88.3	90.5	75.0	85.1
Gas condensate	3.6	5.3	7.5	7.6	8.8	6.9	9.4
(3) Export duty	**618.1**	**1,316.5**	**1,860.0**	**1,784.6**	**2,797.6**	**1,903.2**	**2,469.5**
Crude oil	353.6	871.4	1,201.9	1,151.5	1,784.8	1,134.1	1,672.4
Oil products	81.7	197.5	314.4	330.5	522.6	419.8	603.8
Natural gas	182.8	247.6	343.7	302.6	490.2	349.3	193.3

Source: GDP, excise on goods and mineral extraction tax from various issues of *SEP* and *Rosstat monthly*. Export duty from Ministry of Finance of RF. See: www.countdownnet.info/archivio/analisi/russian_federation/199.pdf (last accessed 30 September 2011);

www.gks.ru/wps/wcm/connect/rosstat/rosstatsite.eng/.

The large size of oil and gas exports tends to put upwards pressure on the ruble, hurting the competitiveness of other sectors and threatening Russia with a case of 'Dutch disease'. Oil and gas revenues flow through the economy and drive up prices and also wages. Since 2000 the authorities have tried hard to reduce inflation, but it still stood at 9 per cent at the end of 2006. The Kremlin seemed determined to repeat the 1998 financial crisis. The combined total of the oil stabilization fund (over $100 bn) and foreign exchange reserves (close to $350 bn) reached the equivalent of two years' imports in spring 2007; and government debt was a mere 10 per cent of GDP.[4]

Before the 2008 global financial crisis, cheap lines of credit from foreign banks were a major source of funding for long-term investment.

Elena Sharipova, senior economist at Renaissance Capital, remarked: 'Our entire financial system was based on foreign money, not on domestic savings. When the inflow was reversed, this made us tremendously vulnerable'.[5] Russia's status as a leading exporter of oil and gas shielded it less than before. When in 2009 crude prices fell to around $65 a barrel, the Russian budget was barely able to break even. Another source of capital which dried up at the same time was western equity markets. In 2008 Russian companies expected to raise $50 billion from initial public offerings, but only $2.5 billion was raised.[6] Chris Weafer from the investment bank Uralsib Financial Corporation remarked that 'the growth we experienced during the past eight years was propelled by $1.3 trillion in oil and gas revenues but this was not accompanied by any changes in the institutional infrastructure. This is the reason why it has all collapsed so quickly. We didn't have any foundations'.[7]

The ruble fell roughly a third against the dollar between the end of summer 2008 and 2 February 2009, when it was 36.35 to the dollar. Between these dates the Central Bank spent more than US$200 bn defending the ruble at a gradually depreciating rate. As resources became scarcer, however, a consensus emerged that the government should use its reserves to support the banking system rather than to bail out individual oligarchs like Oleg Deripaska (the tycoon to whom the government lent $4.5 bn to save his 25 per cent stake in Norilsk Nickel from falling into the hands of foreign banks). Russian companies were due to repay $140 bn of foreign debt in 2009 – a large proportion of the country's reserves, which stood at $388 bn.[8]

As energy markets shrank, the tactics that the Kremlin had used to build Gazprom's awesome economic and political power, thereby restoring Russian influence in the world, began to backfire, slashing both profits and influence. The strategic goal of Vladimir Putin in the eight years of his first period as president was to dominate natural gas supplies to Europe, and to control the pipelines that delivered them.

But, wrote Andrew Kramer in the *New York Times*:

> ... in his zeal to monopolize gas supplies, Mr Putin, now prime minister, committed Gazprom to long-term contracts with Central Asian countries for gas at a cost far in excess of current world prices ... In a cruel twist, the company has also found itself forced to close its own wells in Russia, which produce gas for a fraction of the cost of that from Central Asia, in order to balance its supply with declining world demand.[9]

The 2008 global economic crisis exposed how the Russian strategy to revive its superpower status had led it into a dangerously excessive degree of dependency on the oil and gas sectors. Nonetheless, Russia appears to have weathered the 2008 global crisis relatively well. During 2009 real GDP increased by 8.1 per cent – the highest rate since the fall of the Soviet Union. In addition the ruble remained stable, and inflation continued to be moderate (11.9, 13.3, and 8.8 per cent in the three years 2007, 2008, and 2009 and an estimated 8–8.5 per cent in 2010),[10] while investment began to increase again. Oil and gas, however, continue to dominate exports, so the country remains highly dependent upon the price of energy.

The Resource Curse and Putin's Resource Nationalism

During the first eight years of the Putin presidency, Russia's economy grew at 7 per cent a year in real terms. Total GDP doubled, climbing in dollar terms from the twenty second to the eleventh largest in the world. Compared on a purchasing power parity basis it is the sixth largest. In 2007, Russia's GDP surpassed that recorded for 1990, meaning it had overcome the devastating recession of the 1990s. Per capita GDP was $19,840 in 2010, the thirtieth in the world in PPP terms.

Russia's economy is highly resource-based. Fuel and metals together accounted for an estimated 65 per cent of value added in industry in 2000. In 2003, hydrocarbons, metals, and other raw materials accounted for 76 per cent of total exports, equivalent to 31.5 per cent of GDP. This undoubtedly qualifies the country as having a resource-based economy but not a 'typical' one. The typical resource-dependent economy is less developed, with a large agrarian sector, low levels of urbanization, and low overall levels of education. Russia's situation is closer to that of a highly industrialized economy following the discovery of major new resource wealth, such as the Netherlands or the United Kingdom in the 1970s and 1980s. The 'resource shock' in Russia resulted not from the discovery of new resources but from the adjustment of relative prices at the start of the post-Soviet transition. The relative prices of primary raw materials, having been held at artificially low levels under central planning, soared after prices were freed and foreign trade was liberalized. This triggered a radical reallocation of the resource rents derived from Russia's primary sector, a reallocation that is still being contested.[11]

Marshall Goldman has pointed out that:

> ... the increase in the value of the ruble relative to other currencies – precipitated both by the rise in the price of oil and by the sharp increase in production and export of Russian petroleum after 1999 – also gave rise to what can be called the 'Russian disease'. Not only does a booming export market for energy resources have an adverse impact on domestic

manufacturers but the appearance of a large and expanding petroleum sector inevitably triggers a ferocious struggle to win control of the oil-producing fields, at least in countries where the state allows the private ownership of energy-producing companies. Partly as a result of this struggle for control, whenever petroleum and gas industries begin to dominate a country's economy, democratic institutions often seem to weaken if not collapse. Venezuela is one of the more recent examples of this connection.[12]

A contributor to a Chatham House round table meeting in 2009 pointed out:

> the resource curse in Russia is unique in the developed world … Lack of competition leads to a growth in costs. In 2000, one litre of petrol cost $0.95, and in 2009 the price was $1.15. In 2000, the average cost of one barrel of oil equivalent was $3.80. This figure was about the same for Yukos and Rosneft. The cost of producing one barrel in 2009 stood at $10.80.[13]

To understand the core of Russian policy in the natural resource sector, it would be very helpful to review a fairly comprehensive policy statement by Putin. In 1997, before he was appointed Prime Minister and then elected President, Mr Putin defended a Candidate of Sciences (kandidat) thesis at the St. Petersburg Mining Institute and subsequently (in 1999) published an article in the Institute's journal outlining his views on a natural resource policy for Russia. It was the lead article in an issue of the journal devoted to the fuel and energy complex. His basic message was that Russia's mineral resources, and particularly its hydrocarbons, would be the key to the nation's economic development for the foreseeable future. To guarantee the most effective exploitation of the country's enormous mineral wealth, the state must regulate and develop the resource sector. This could best be done, according to Putin, by fostering large firms capable of competing on equal terms with Western transnational corporations. While the policies should be influenced by market mechanisms, the interests of the Russian state and people (and of Russian corporations) must be protected.[14]

The resource sector, Putin argues, is too important to be left entirely to market forces:

> … regardless of whose property the natural resources and in particular the mineral resources might be, the state has the right to regulate the process of their development and use.

In doing this the state acts in the interests of society as a whole, and also helps property owners to resolve their conflicts through compromise:

> … unfortunately, when market reforms began the state lost control of the resource sector. However, now the market euphoria of the first years of economic reform is gradually giving way to a more measured approach,

allowing the possibility and recognizing the need for regulatory activity by the state in economic processes in general and in natural resource use in particular ... a contemporary strategy for the rational use of resources cannot be based exclusively on the possibilities offered by the market.

In sum, Putin emphasizes the importance of the resource sector for Russia's economic and geostrategic revival, notes the need for mixed forms of property without specifying the optimal mix, asserts the primacy of state interests, and advocates fostering large, vertically-integrated firms that will further those interests.[15]

A spokesman for Putin, Dmitri Peskov, has said, according to Andrew Kramer in the *New York Times*, that the energy market 'was, is and will remain a strategic sphere for Russia' and that government leaders in Moscow should be versed in the topic. On Mr Putin's deep personal knowledge of the oil business, he said that the prime minister showed a similar attention to detail in other matters. On policy, 'Peskov denied that the Kremlin used exports for political purposes but made clear that Russia's overarching goal is to prevent the West from breaking its monopoly on natural gas pipelines from Asia to Europe'.[16]

John Grace has argued that the rise of a strong Russian oil industry is the most influential development in the world oil market since the empowerment of OPEC in the 1970s. He also pointed out that oil policy both mirrors and influences larger developments in Russian politics. The rise of a more monocratic state under the Putin presidency was closely related to oil-driven prosperity. The crackdown on the nation's leading oil producer, Yukos, and its erstwhile chief executive, Mikhail Khodorkovsky, furthered two basic goals of the Putin administration: crushing the opposition to centralization of authority, and reversing the privatization of national political power by a small, extraordinarily rich group of men known as the oligarchs.[17] The attack on Yukos and Khodorkovsky[18] highlighted Putin's determined effort to reign over these upstart oligarchs and at the same time to renationalize their firms and refashion them into state companies and national champions.[19] For Richard Sakwa,

> The rise and fall of Yukos is one of the great dramas of our age. Built on the efforts of the preceding Soviet generation, the company symbolized both the robber baron phase of Russian capitalist development and then its move into the age of corporate capitalism ... and the affair is at the heart of our understanding of the complexities of contemporary Russia, which themselves reflect the contradictions of the modern world.[20]

Andrei Illarionov, sometime economic advisor to Putin, criticized the Yukos attack because

> The sale of the main oil-producing asset of the best Russian oil company
> ... and its purchase by Rosneft company, 100 per cent owned by the state,
> has undoubtedly become the scam of the year ... When the Yukos case
> began, everybody was asking what will be the rules of the game ... Now
> it is clear that there are no rules of the game.[21]

The collapse of Yukos also meant the collapse of the Angarsk–Daqing
crude pipeline proposal which had been very aggressively promoted by
Khodorkovsky from the late 1990s. This pipeline became a casualty of
the ESPO pipeline project development, and it also affected TNK–BP's
ambitious plan to export Kovykta gas to China (Gazprom agreed to take
over TNK–BP's equity in Rusia Petroleum but the transaction never
materialized). The Russian authorities' stance towards the Sakhalin-2
project confirmed that Moscow would not tolerate the massive increase
in the cost of Sakhalin-2, from \$10 bn to \$20–22 bn, since this would
prevent the Russian participants from seeing any profit until the Shell
consortium recouped the costs. Shell's consortium had underestimated
the development cost, and as time went on, costs ran out of control
due to the harsh development environment, but the consortium did
not inform the Moscow authorities of this. As Goldman pointed out,
the Russians ceased to be supplicants once they were able to revitalize
their energy sector. The change in status led Putin and those around
him to find ways to regain control over all the mineral assets, energy,
and metals that had slipped from state control during the Yeltsin era.[22]

In a *Financial Times* interview, Andrei Illarionov, the Kremlin's economic adviser, said that:

> ... the school of thought that natural resources are a very special com-
> modity is spreading very fast. In the 1900s it was widely believed that
> natural resources could be privately owned and therefore private companies
> accumulated reserves. But over the past several years, a near-consensus has
> emerged that natural resources should belong to the state – not to private
> citizens and private companies but the state. This is now very clearly
> understood and that is why any decision by foreign investors to acquire
> natural resources is expected to be discussed with the state.[23]

This interview indirectly confirmed the level of state intervention in
decision making relating to natural resources business in Russia.

Simon Pirani has remarked to the author that an important factor
that determined Putin's actions was that when he came to office, Russia
was the only large oil producer, apart from the USA, without a strong
national oil company; the privatizations of the Yeltsin era had been
unprecedentedly chaotic and damaging to the state; and the state had
ceased to carry out basic functions, most significantly the collection
of sufficient taxes to cover the budget. In other words, Putin was

motivated not only by resource nationalism but by a desire to sort out the unsustainable and chaotic legacy of the 1990s slump.[24] The author shares this view.

Russia's Initiative towards East Siberia and the Far East's Frontier Oil Development

In 1987 Russia, at that time the world's largest oil producer, reached its peak production of 11.484 mb/d (18.9 per cent of the world total). Twenty years later, with production of 9.978 mb/d, it ranks as the second largest producer, only 4.2 per cent behind Saudi Arabia's 10.413 mb/d. During the second half of the 1990s, however, the Russian oil sector had experienced a protracted crisis caused by the disruption of former Soviet commercial and technological links and an economic slump at home.

The process by which Russia's vertically integrated oil companies were created began in 1992 when a major sell-off was launched by President Yeltsin's decree mandating the conversion of state-owned oil and oil refining enterprises into stock companies. As a result, the first three vertically integrated oil companies (VIOCs) – Rosneft, Lukoil, and Surgutneftegaz – were created. A fourth, Tatneft, was established by a separate decree from the President of Tatarstan. Two years later, several other new companies were created as spin-offs from Rosneft – Sidanco, Slavneft, Opaco, TNK, Sibneft, the Eastern Oil Company (Vostochnaya Neftenaya Kompaniya), Komi TEK, and Bashneft.[25]

The majority of Rosneft spin-offs were frail and helpless from the beginning, and were acquired by much larger VIOCs like Yukos,[26] Lukoil, and Surgutneftegaz. Yukos took over the Eastern Oil Company, the East Siberian Oil and Gas Company, and the Lithuanian Mazeikie Nafta. Lukoil bought Archangelskgeologiya, KomiTEK and a number of oil refineries and petrol stations. The only spin-offs from Rosneft able to survive and flourish independently were TNK and Sibneft. By 2004 there were eight VIOCs, producing 95 per cent of Russia's crude oil and more than 70 per cent of its refined products.[27] Russia's crude output growth slowed markedly in the middle of 2004 and from early 2008 began to fall.[28] To restore oil production, a major restructuring of the oil industry was urgently needed.

The dominance of the top five VIOCs – Rosneft, Lukoil, TNK–BP, Surgutneftegaz, and Gazprom Neft – continued up to the second half of the 2000s, as shown in the Appendix to this chapter (Table A.2.2). With the exception of Lukoil, these big companies became the driving forces of oil development in East Siberia and the Far Eastern frontier. As shown in Table 2.3, the role of West Siberia in Russia's oil production is critical. In 2008, oil production from West Siberia was 68 per cent of the total and gas production 93 per cent, while East Siberia accounted for only 0.3 per cent of oil and 0.8 per cent of gas. If the figure for the Far East is combined with that of East Siberia, the combined percentages rise to 2.9 for oil and 2.2 for gas. In the wake of the ESPO development, however, a very large increase is expected in oil production from East Siberia and the Far East.

Table 2.3: Russia's Oil Production by Region, 2008

Regions	Oil mt	% of total
European Regions	**141.9**	**29.0**
West Siberia	**332.3**	**68.0**
Khanty–Mansi Autonomous Area	277.6	56.8
Yamal–Nenets Autonomous Area	39.2	8.0
Tomsk Region	10.5	2.1
Novosibirsk Region	2.1	0.4
Omsk Region	1.5	0.3
south of Tyumen Region	1.4	0.3
East Siberia	**1.4**	**0.3**
Krasnoyarsk Region	0.1	0.0
Irkutsk Region	0.5	0.1
Sakha Republic	0.8	0.2
Far East	**12.9**	**2.6**
Sakhalin Islands, Offshore of Sea of Okhotsk Chukchi Region	12.9	2.6
Siberia and Far East, Total	**346.6**	**71.0**
Russia, Total	**488.5**	**100.0**

Source: Kontorovich and Eder (2009).

That picture will change greatly in coming decades, as Table 2.4 suggests. Academics from the Russian Academy of Sciences have projected that annual oil production from East Siberia and the Far East will reach 69 mt in 2020 and 100 mt in 2030. The production figures of Table 2.3 and Table A.2.4 are not the latest ones, but can be used for making a revealing comparison with the projected figures in Table 2.4. Total oil production figures of Russia's official energy strategy 2030 are at least 100 mt lower than the figures of 561 mt in 2020 and 636 mt in 2030 in Table A.2.4, yet the scale of oil production from East Siberia and Far East shows virtually no difference.[29] The 2008 figures in Tables 2.3 and 2.4 were the benchmarks for the projections in Table 2.4. The absence of change in the East Siberia figures confirms the reliability of those benchmarks.

Table 2.4: Oil Production in Siberia and the Far East up to 2030 (mt/y)

	2005	2008	Phase I*	Phase II**	Phase III
North, north-west	24.5	29.1	32–35	35–36	42–43
Volga Region	52.7	54.1	49–50	44–45	34–36
Urals	49.2	52.6	45–47	36–41	25–29
Caucasus, Caspian Sea Region	4.9	4.8	7–11	19–20	21–22
Tyumen Region	320.2	319.0	282–297	275–300	291–292
Tomsk Region	14.1	13.7	12–13	11–12	10–11
E. Siberia	0.2	0.5	21–33	41–52	69–75
Far East	4.4	13.8	23–25	30–31	32–33
Total	**470.2**	**487.6**	**486–495**	**505–525**	**530–535**

*   The first implementation phase of the Strategy ends in 2013–15; ** the second ends in 2020–2.

Source: Ministry of Energy, Russian Federation.

Based on the massive expected increase in production, Moscow aims to export to north-east Asian countries and others in the Pacific market almost 110 mt of oil by 2020 and 124 mt by 2030. Over 60 mt of the 2020 figure and over 90 mt of the 2030 figure will be from East Siberia and the Far East. (See Table 2.5)

The most urgent task in the development of the East Siberia and Far Eastern frontier oil resources is the construction of a long-distance crude oil pipeline network. This is why the Russian government has assigned the highest priority to the completion of the first stage development of the ESPO oil pipeline which will play a pivotal role in Russia's large scale oil supply to the north-east Asian region in the coming decades. It remains to be decided, however, whether the East Siberia and Far East reserves are large enough to justify the planned additional 1.6 mb/d capacity of the second stage of the development of ESPO (See Chapter 7).

East Siberia's Exploration Programmes

In parallel with the development of the ESPO oil pipeline, the Russian authorities started to identify sources of supply by means of comprehensive exploration programmes. According to the Ministry of Natural Resources (MNR), by 2004 around 46 oil, gas, oil and gas, and oil and

Table 2.5: Oil Exports to the Pacific Market: 2010–2030 (mt/y)

	2010	*2015*	*2020*	*2025*	*2030*
From W. Siberia	25.0	35.0	46.0	38.0	29.0
WS-K-C pipeline	2.0	5.0	10.0	10.0	10.0
Rail Z-M	12.0	10.0	5.0	5.0	5.0
ESPO OP	11.0	20.0	31.0	23.0	14.0
to China	5.0	10.0	15.0	10.0	10.0
to Japan	2.0	3.0	5.0	3.0	1.0
to Korea	3.0	5.0	7.0	7.0	2.0
to OCPM	1.0	2.0	4.0	3.0	1.0
From E. Siberia	7.0	27.0	41.0	53.0	66.2
Rail Z-M	1.0	2.0	2.0	2.0	2.0
ESPO OP	6.0	25.0	39.0	51.0	64.2
to China	3.0	12.0	20.0	25.0	30.0
to Japan	1.0	5.0	5.0	7.0	9.5
to Korea	2.0	7.0	10.0	14.0	19.2
to OCPM	0.0	1.0	4.0	5.0	5.5
From Far East*	16.3	17.6	22.1	26.1	29.1
from De-Kastri	8.8	9.5	12.3	13.0	13.8
to China	2.6	2.9	3.7	3.9	4.1
to Japan	0.4	0.5	1.7	2.0	2.1
to Korea	2.2	2.4	3.1	3.3	3.5
to OCPM	3.5	3.8	3.8	3.9	4.1
from SST**	7.5	8.1	9.8	13.1	15.3
to China	1.5	1.6	2.6	3.9	4.4
to Japan	4.5	4.6	4.5	4.5	4.4
to Korea	1.5	1.7	2.1	3.5	5.3
to OCPM	0.0	0.1	0.6	1.2	1.3
From RPMT***	48.3	79.6	109.1	117.1	124.3
to China	27.1	43.5	58.3	59.8	65.5
to Japan	7.9	13.1	16.2	16.5	16.9
to Korea	8.7	16.1	22.2	27.8	30.0
to OCPM	4.5	6.9	12.4	13.1	11.9

Note:
WS–K–C pipeline = oil pipeline West Siberia–Kazakhstan–China (Omsk–Atasu–Alashakou);
Rail Z–M = railway Zabaikalsk–Manchuria with three options (Zabaikalsk–Manzhouli, Naushki–Sukhe-Bator; Grodekovo–Suifenghe to China); ESPO OP = ESPO Oil Pipeline (including pipe-bend Skovorodino–Daqing), ports of the Far East in particular Nakhodka;
OCPM = other countries of the Pacific Market;
Far East* = Far East (Sakhalin, Kamchatka);
SST** = Southern Sakhalin (Prigorodnoe) terminal:
RPMT*** = Russia to the Pacific Market, Total.

Source: Kontorovich and Eder (2009).

gas condensate fields had been identified in East Siberia. Total recoverable C1 and C2 oil reserves in East Siberia and the Sakha Republic are 1.12 billion tonnes. The extractable resources are 4.7 billion tonnes of C1 oil reserves, 5.7 tcm of C1 and C2 recoverable gas reserves, and 14.6 tcm of recoverable resources.[30]

On 11 November 2004 the Russian government approved a programme (to last until the end of 2020) entitled the 'Long-term State Programme to Study Subsoil and Replenish Russia's Mineral Resources'. It was prepared by the MNR in accordance with the 'Basic Principles of State Policy regarding the Use of Mineral Resources and Subsoil', approved by the government in April 2003.[31]

At the beginning of 2005, only 85 per cent of the oil and gas condensate produced in Russia was offset by an increase in reserves. Pessimists said that, at the current rates of production and exploration, in the absence of profound changes, commercial oil reserves in Russia would be depleted by 2015 and gas condensate reserves by 2025. The cost of the exploration programme over a period of 15 years is expected to be 1,784 billion rubles in the form of expenses for prospecting and prospecting–appraisal operations, evaluation and exploration, and research and scientific–procedural support. Only 10 per cent of this amount is to be funded from the federal budget; it is assumed that the rest will come from mining and oil companies and others engaged in subsoil exploitation.[32]

The government has increased pressure on operating subsoil users to accelerate investigation and development of oil reserves within the ESPO project, while also measuring offers of new licenses within the limits set out in the 'Programme of Geological Study and Granting Use of Hydrocarbon Raw Materials in East Siberia and the Republic of Sakha (Yakutia)'. The MNR approved this Programme in July 2005. Production in 2005–6, however, amounted to 21.6 mt, only one eighth of the 187 mt of oil envisaged in the Programme. Officials began to fear that the project of the century would not be provided with sufficient raw material.[33]

According to the government's licensing programme for East Siberia, the first blocks auctioned were to be the most promising and best prepared. Beginning in the middle of 2005, MNR planned to auction 39 promising blocks, of which 14 were in the Irkutsk Region, 13 in the Sakha Republic, 10 in Evenkia, and two in the Krasnoyarsk Territory. These blocks contain explored reserves of 128 mt of oil and 1.7 tcm of gas, according to the Ministry of Natural Resources. The total estimated and proven reserves of 1.2 bt of oil and over 2.7 tcm of gas should make these blocks attractive to potential investors. The most lucrative blocks are Chayandinsky, Taas–Yuryakhsky, Verhne–Vilyuchansky, Tympuchikansky, Mogdinsky, Vostochno–Sugdinsky, Baikitsky, Tukolano–Svetlaninsky, and Chambinsky. In 2006, the government aimed to transfer 40 more blocks to subsoil users, with reserves estimated by the MNR of 24 mt of oil and 141 bcm of gas, and potential reserves estimated at over one billion tonnes of oil and 3.3 tcm of gas.[34]

The licensing programme developed by the MNR for East Siberia covers an area of over 600,000 square kilometres, comprising 213 prospective blocks of between 1,000 and 5,000 square kilometres, each to be distributed by an auction, although it is likely that certain blocks in East Siberia will be transferred without auction procedures. Some blocks may be listed in the registry of the Federal Fund of Reserve Fields.[35]

Rosnedra and geological exploration
Rosnedra, according to its head, Anatoly Ledovskikh, planned to boost the annual level of Federal budget spending on geological exploration of the subsoil to 16.5 billion rubles ($0.59 billion) in 2006, compared with 5.2 billion and 10.7 billion rubles in 2004 and 2005. He added that in the subsequent three years demand for geological exploration operations would be as high as they had been in Soviet times. Subsoil users in 2006 were expected to put significantly more funds into geological exploration than in previous years. The ratio of geological exploration efforts by the state and by private subsoil users is 1 to 8. Overall then, a sum of up to 120 billion rubles (some US$4.5 billion) was due to be sunk in geological exploration in Russia in 2006,[36] but confirmation of this spending is not available.

In 2006, MNR put up 42 subsoil blocks for auction in East Siberia, Irkutsk region, Sakha Republic (Yakutia), and Krasnoyarsk Territory (Evenki Autonomous District). These blocks included some that had not been planned for auction before 2005. The East Siberian auctions revealed two notable features. First, the high level of interest in the auctions near Ust-Kut in late 2005 and early 2006 (since it was the nearest ESPO point to the Yakut and Krasnoyarsk fields) cooled off as soon as information on the pipeline route moving north was available. Second, Rosneft demonstrated that it would stop at nothing in fighting for blocks which it expected to be lucrative.[37]

In 2006–7 the number of subsoil use auctions carried out by Rosnedra and its regional managements, and the number of companies contesting for these rights, both increased. In 2007, there were 11 subsoil users in Evenki Autonomous District, five in Krasnoyarsk Territory, 22 in Irkutsk region, and nine in Sakha Republic. The total size of the license blocks held by these 47 subsoil users was 37,682 square kilometres. The Jewish Autonomous region in Russia's Far East was the only territory to conduct auctions in the summer of 2008.[38] During the first five months of 2008, officials held 51 auctions, and Rosnedra had a prospective list of 254 subsoil blocks. But by the end of August, Rosnedra had announced auctions for only 20 blocks. Legislative changes had substantially complicated the organizational procedures for auctions, and it became impossible to conduct an auction until all affected land users had provided written approval. According to the Programme of Geological Study and Granting Use of Hydrocarbon Materials in East Siberia and the Republic of Sakha (Yakutia), as shown in Table 2.6, the total amount of federal budget money allocated for exploration in East Siberia during the 2007–20 period is estimated at 135.8 billion rubles.

Table 2.6: Structure of Charges to the Federal Budget for Exploration in East Siberia, 2007–2020 (billion rubles)

	Total	*Regional Geophysical Work*	*Parametrical Drilling Programmes*	*Search and Estimation of New Objects*	*Research and Development*
2007	4.200	2.787	1.243		0.170
2008	5.100	3.300	1.300		0.500
2009	7.200	4.450	1.300	0.800	0.650
2010	9.500	3.600	1.500	3.500	0.900
2011	10.500	4.550	1.500	3.500	0.950
2012	11.500	5.000	2.000	3.500	1.000
2013	11.500	4.400	2.500	3.500	1.100
2014	11.500	3.900	3.000	3.500	1.100
2015	11.500	3.400	3.500	3.500	1.100
2016	11.500	3.300	3.850	3.200	1.150
2017	11.500	3.300	4.050	3.000	1.150
2018	11.500	3.300	4.050	3.000	1.150
2019	11.500	3.300	4.050	3.000	1.150
2020	11.500	3.300	4.050	3.000	1.150
2008–20	135.800	49.100	38.650	37.000	13.050

Source: Programme of Geological Study and Granting Use of Hydrocarbon Materials in Eastern Siberia and the Republic of Sakha (Yakutia), quoted in Chernyshov (2007), 17.

In 2008 it was expected that geological survey investment in East Siberian deposits would grow by 30–40 per cent. By 2009, the volume of government financing was expected to reach 6.5 billion rubles ($265 million), a 65 per cent increase over 2007. According to an addendum to the existing Programme, investments for the Russian subsoil

geological survey would increase 1.5 times. Once the government rati-
fied the document, the total volume of state financing would amount to
15–16 billion rubles annually.[39] Considering that Russian oil companies
reduced the number of new wells completed in 2009 to 4,835 (down
3.7 per cent) after five consecutive years of growth,[40] however, it is
very likely that the global financial crisis in 2008–09 had significantly
reduced the investment spending on the geological survey in 2009.

Eastern exploration programme approved
On 27 March 2008 the Programme for Reproduction of the Mineral
Raw Material Base up to 2020, prepared by the MNR, was approved by
the Russian cabinet. According to this programme, the Russian government will invest $22.6 billion in geological prospecting (a substantial
increase from the $9 billion estimate in the old programme of 2005) and
estimates that subsoil user companies will invest seven to eight times as
much (approximately $169 billion). Over the following 12 years, subsoil
users should invest approximately $150 billion.[41] (See Table 2.7.)

A serious obstacle to subsoil development in East Siberia was the
decline in exploration volume compared to the Soviet era. At that time,
geological exploration amounted to between 350,000 and 420,000 metres of deep drilling a year. From 1985 to 1990 surveyors discovered vast
oil and gas deposits in East Siberia. During 2006 and 2007, only 57,000
metres of drilling was carried out, revealing only four gas condensate
deposits. For the preceding 15 years, there had been no discoveries at
all. Arkady Yefimov, general director of the Siberian Scientific Research
Institute of Geology, Geophysics and Mineral Resources, has pointed
out that it takes about a month to drill a well in West Siberia while it
takes about a year in East Siberia.[42] This explains why extra funding
is needed for East Siberia's comprehensive exploration programmes.

Table 2.7: Investment in Exploration, 2005–20

Region	Total Investment ($billion)	State investment (%)	Private investment (%)
Volga–Urals OGP	10.088	94.3	5.7
Timano–Pechora OGP	6.958	96.0	4.0
West Siberian OGP	83.975	98.5	1.5
East Siberian OGP	26.138	79.5	20.5
Far Eastern OGP	1.250	93.0	7.0
Continental Shelf	28.200	93.3	6.7

Note: OGP = oil and gas province

Source: Ministry of Natural Resources, quoted in RPI May 2008, 13.

In 2004 the state approved a raw material base Programme to begin in 2005. According to the Minister of Natural Resources, Yury Trutnev, during the years from 1990 to 2003, the state was not actually involved in subsoil investigation and minerals research. During the first three years of the Programme, some distortions occurred, both in territories and by type of raw materials, which made it necessary to update the programme, chiefly by increasing the amount of finance. Programme activities during 2005–7 resulted in the opening of 194 oil and gas deposits. During the period, license auctions for subsoil use earned the state $5.527 billion ($1.606 billion in 2007). According to government information reports, between 2008 and 2020 the Programme could result in gains of subsoil value of $8.208 trillion. Each ruble ($0.04) invested via the budget in geological exploration is planned to result in 70–100 rubles of increased revenue.[43] According to Trutnev's report, the main expenditures on geological exploration were to begin in 2010. The state would sharply increase investments in developing East Siberia, from $2.75 billion to $5.375 billion. Some 20 per cent of the funds expended would come from the budget (normally, in other areas, the total is only 1–5 per cent). Given the absence of any infrastructure, geological exploration in East Siberia is five to six times more expensive than in other regions and so a greater share of state investment is justified.[44] The highest priority was to be given to long distance crude oil pipeline development in the Russian Far East (to be discussed in detail in Chapter 7).

ESPO Pipeline's Main Production Bases

Arkady Yefimov, general director of the Siberian Scientific Research Institute of Geology, Geophysics and Mineral Resources, has given a strong warning that an output level of 80 mt a year in East Siberia cannot be reached until 2025, yet the Ministry of Industry and Energy (MIE) wants to commission the second stage of the ESPO pipeline in 2015–17. By this time, East Siberian deposits should provide nearly 56 mt to ESPO. Yefimov pointed out that increasing oil extraction to a level of 56 mt by 2020 and keeping it at that level for 30 years adds up to a cumulative total of 1.5 bt of oil, yet by 2008 the volume of proven reserves had only reached 0.52 bt of oil. He considered that, in order to attain 1.5 bt of oil reserves, the state should transfer almost 200 blocks to subsoil users for development. But by the beginning of 2008 only 70 deposits had been distributed. To start the second stage of ESPO, 30 mt/y is necessary.[45]

By 2008, the total crude oil production from the companies listed in

Table 2.8 reached only 1.36 mt, though the figure will change significantly in the coming years. Rosneft applied to supply 25 mt annually to the pipeline from Vankor beginning in 2009, but this required the construction of an oil pipeline from Purpe to a connection point with ESPO. Verkhnechonskneftegaz plans to provide oil to the pipeline from the Verkhnechonsky deposit, and Surgutneftegaz expects its Talakanskoye deposit in the Sakha republic to supply 2 mt/y of crude during the first stage. Urals Energy (Cyprus) will contribute 0.7–0.8 mt from the Dulisminskoye deposit. The total volume from these fields in 2011 will reach nearly 12 mt.[46]

Table 2.8: East Siberia's Oil and Gas Production by Company, 2008

Company	Oil mt	%
Lenaneftegaz (supervised by Surgutneftegaz)	0.5976	44.1
Ust–Kut Neftegaz (supervised by Irkutsk Oil Co)	0.2775	20.5
Verkhnechonskneftegaz (supervised by TNK–BP and Rosneft)	0.1593	11.7
Yakutgazprom	0.0798	5.9
Irelyakhneft	0.0669	4.9
Dulisma (supervised by Urals Energy)	0.0557	4.1
Taimyrgas (supervised by Norilsk Nickel)	0.0492	3.6
Vostsibneftegaz (supervised by Rosneft)	0.0280	2.1
Danilova (supervised by Irkutsk Oil Co)	0.0159	1.2
Taas-Yuryakh Neftegazodobycha	0.0101	0.7
Vankorneft (supervised by Rosneft)	0.0084	0.6
Alrosa-Gaz (supervised by Alrosa)	0.0044	0.3
Norilskgazprom (supervised by Norilsk Nickel)	0.0032	0.2
Sakhatransneftegaz	0.0002	0.0
Suzun (supervised by Gazprom and TNK–BP)	0.0002	0.0
Branches vertically integrated oil and gas mining and smelting companies	0.8503	62.7
East Siberia, Total	**1.3564**	**100.0**
Russia, Total	**488.486**	
Percentage share of East Siberia in the Russian Federation	0.3	

Source: Kontorovich and Eder (2009).

As already mentioned, producing 56 mt by 2020 and maintaining that level for 30 years requires cumulative production of 1.5 bt of oil. By 1 January 2007 it was estimated that the total extractable oil reserves of category C1 and C2 (probable reserves) in the southern territories of East Siberia and the Republic of Sakha amount to 1.2551 bt, distributed as follows: i) 689.7 mt in Krasnoyarsk region, including the Evenki Autonomous Region; ii) 235.6 mt in the Irkutsk region; iii)

329.8 mt in the Republic of Sakha. The estimates put total extractable C1 oil reserves at 554.2 mt and C2 at 701 mt. Extractable resource volumes for categories C3 and D1 in the southern regions of the Siberian platform are 4.6444 billion tonnes, comprising 2.1762 billion tonnes in the Krasnoyarsk region, 2.025 billion tonnes in the Irkutsk region, and 0.4432 billion tonnes in Sakha. The potential undistributed fund of oil reserves is 0.1324 billion tonnes for C1+C2 reserves and 2.8524 billion tonnes for C3+D1 resources. Thus the largest part of the oil reserves and resources is concentrated in the Irkutsk region and the Republic of Sakha.

There could, the Russian government believes, be four oil production centres: the Yurubcheno–Tokhomskoye and Kuyumbinskoye deposits would form the Yurubcheno–Kuyumbinsky centre; the Talakanskoye and Verkhnechonskoye deposits would form the Talakano–Verkhnechonsky centre; the Sobinskoye and Paiginskoye would form the Sobinsko–Teterinsky Centre, and the Sredne-Botuobinskoye, Taas–Yuryakhskoye, Irelyakhskoye, Machobinskoye, Stanakhskoye, Mirninskoye, Iktekhskoye, and Verkhnevilyuchanskoye deposits would form the Botuobinsky centre.

On 1 January 2007, the estimated crude oil resource bases of the proposed centres for C1+C2 reserves and C3+D1 resources respectively stood at:

• Yurubcheno–Kuyumbinsky centre – 675.9 mt and 1.451 bt;
• Talakano–Verkhnechonsky centre – 13.8 mt and 361.3 mt;
• Sobinsko–Teterinsky centre – 430.4 mt and 2.5437 bt;
• Botuobinsky centre: 135 mt and 166.9 mt.

By 2015 the development of these base deposits will allow an annual oil extraction level of 30 mt annually. If the satellite sites are included, the figure could reach 36–37 mt by 2020.[47]

At first, the MNR thought that in order to provide sufficient volume for ESPO, it was necessary to source quantities jointly from East and West Siberian deposits. As planning proceeded, however, it has seemed preferable to base the plans on oil extracted within West Siberia, with the addition of the Bolshekhetskaya zone, until oil production levels in East Siberia are sufficient to fill the pipeline. The estimated extractable oil reserves of the Bolshekhetskaya zone amount to 256.1 mt. In 2007 only the Suzunskoye deposit was ready for development, and other deposits are still at the geological prospecting stage. Experts put the hypothetical total volume of oil extraction at all Bolshekhetskaya deposits at 16 mt/y. To achieve this, however, a new pipeline from the Vankorskoye deposits to Purpe, the location of a main oil line, was needed. The construction of this 550 km long pipeline was completed as planned. It is calculated that to supply 24 mt/y of oil to ESPO it is necessary to provide between 8 and 18 mt/y of Bolshekhetskaya sourced crude oil.[48]

One of the main problems faced by East Siberian developers is that of providing oil field services. Except for Surgutneftegaz, all the large oil companies working in the region report this problem; there are too few drilling contractors, their services are of poor quality, and there is a lack of seismic contractors and geophysical works. Other problems reported include remoteness from the traditional centres of oil and gas extraction, insufficient development of infrastructure, seasonal restrictions of work, and the lack of qualified personnel.[49]

The Irkutsk region
Verkhnechonskoye (developed by TNK–BP & Rosneft). The Verkhnechonskoye oil and gas condensate field was discovered in 1978 and Rusia Petroleum obtained a license to develop the field in 1992. In 2008, estimates of reserves stood at more than 159.5 mt of C1 oil, 42 mt of C2 oil, 0.4 mt of C1 condensate, 2.9 mt of C2 condensate, 56 bcm of C1 gas, and 105 bcm of C2 gas.[50]

In late 2005, Rosneft joined the Verkhnechonskoye project, whose majority partner is TNK–BP. The purchase of an equity share in VCNG, the project operator and license holder of the Verkhnechonskoye oil and gas condensate deposit, was a partial fulfilment of Rosneft's strategic goal of increasing its presence in East Siberia. On 13 January 2006 the shareholders of VCNG, on the proposal of Rosneft which in the previous October had acquired 25.94 per cent of the VCNG shares for US$230 million from Interros, met to elect its board of directors, on which Rosneft gained two seats. The other shareholders are TNK–BP with 62.71 per cent and the ESGC with 11.29 per cent.

In December 2005, Rosneft acquired the Vostochno–Sugdinskaya acreage, adjoining the Verkhnechonskoye field, at auction, setting a record for the amount paid for a single oil-and-gas license (US$260 million). In February 2006, Rosneft won the bidding for three promising blocks located near the Vankorskoye field, paying 5.4 billion rubles for them. Rosneft planned further expansion in the Irkutsk region, and acquired ESGC's stake in VCNG. Under an accord with the Irkutsk Regional administration, ESGC was authorized to sell 11.29 per cent of its Verkhnechonskneftegaz shares. The amount thereby raised (with deductions of the 170 million rubles paid by the administration for the shares of an additional issue by the ESGC and of taxes) will be funnelled into the Regional budget.[51]

According to a TNK–BP report, the firm has purchased an additional

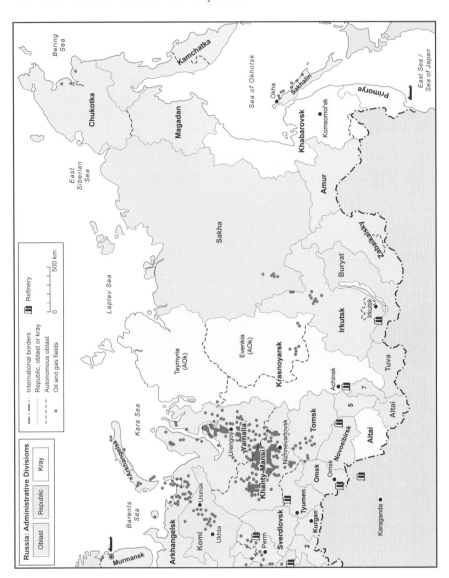

Map 2.1: East Siberia and Russia's Far East.

Source: *Petroleum Economist* (with minor revisions).

5.6 per cent of the capital of VCNG from ESGC, increasing its share to 68.4 per cent. Under an agreement reached between TNK–BP and EGSC, Rosneft should also purchase 5.6 per cent of VCNG from EGSC.[52] On 15 October 2008 the first vice president of Rosneft, Sergei Kudryashov, in comments relating to a dispute over an 11.29 per cent stake in VCNG, said the company has not decided against increasing its stake in VCNG.[53] TNK–BP bought its stake in July 2007, while Rosneft had yet to officially announce the purchase of its stake (which was confirmed later). On 18 September the Irkutsk region's arbitration court held preliminary hearings in a suit filed by the region's property management agency against ESGC, which was set up on an equal basis by TNK–BP and the Irkutsk region, seeking the invalidation of the board of directors' decision to sell the VCNG stake.[54]

The Development Scheme. In August 2005, the TNK–BP board of directors decided to earmark US$270 million to fund the pilot commercial phase of the Verkhnechonskoye oil and gas condensate field development. A pilot scheme for commercial development (PCD) envisaged the construction of surface field facilities, a central oil gathering station, and other infrastructure, all to be completed by the end of 2006. It was planned to drill 20 wells, of which 13 would be development and seven injection holes. The early oil project provided deliveries of the first oil to ESPO by the end of 2008, about 1.0–1.5 mt/y. It also involved transport infrastructure, including the construction of a 120 km pipeline and a year-round road from Verkhnechonskoye to the Talakanskoye deposit. The early oil project would allow the acceleration of preparatory work on the full-scale development of the deposit. Investment at this stage could approach US$600 million.

Part of the allocated funds would be invested in the construction of an almost 600 km field pipeline to link the Verkhnechonskoye field with the Ust-Kut station on the Baikal–Amur Main Railway. A rail terminal would be built there to receive and offload Verkhnechonskoye oil. TNK–BP foresaw developing the field within the PCD regime for the following five to six years. Full-scale development was planned for the period between 2012 and 2027, with annual production at around 7–10 mt per year. In 2006, KCA Deutag Drilling GmbH was scheduled to drill four new development wells at the field while OOO Neftegaz Engineering was to build the 200 mm, 571 km oil pipeline by October 2006.[55]

At the end of May 2007, VCNG completed operational well number 1002 at Verkhnechonskoye. Drilled by the German company KCA Deutag, the well reached a depth of 1,793 metres and utilized the controlled-directional method, with a vertical offset of 638 metres. Drilling of a second well began in early 2007. VCNG said it would reach peak extraction of 9 mt by 2015–17, three to five years later than previously estimated. Extraction of 1.7 mt of early oil by 2009 was planned. Rosneft was not happy with the delay in reaching peak production and wanted to accelerate full-scale development.[56]

In October 2007, TomskNIPIneft (a subsidiary of Rosneft) issued a DOI (Declaration of Intentions) for full-scale development of the Verkhnechonskoye oil and gas condensate deposit, and in the following month this was approved by public bodies in the Karagansky district of the Irkutsk region. According to the DOI, during its calculated lifetime (29 years), the project would extract 150.77 mt of oil, reaching its maximum annual extraction level (9.49 mt of oil and 0.868 bcm of gas) in 2017. The capital expenses of the development of the deposit were expected to increase significantly from $4 billion (based on the 2005 estimate) to $8–13 billion.[57]

The economic viability of the Verkhnechonskoye project is very sensitive to the price received and the level of oil extraction. According to the DOI, a 10 per cent reduction in price or in the extraction level would make the project inefficient. The efficiency calculations were based on two assumptions: that 70 per cent of the oil would be sold in the domestic market at 8,799 rubles (US$355) per tonne including VAT (the remaining 30 per cent being for export at US$80 per tonne), and that the netback price is 7,555 rubles per tonne including VAT. Another factor, in which a small change could render the Verkhnechonskoye project unprofitable, is capital expenses. According to the DOI an increase of 12.3 per cent in capital expenses would make development of the deposit unprofitable.

Calculations done by TomskNIPIneft have shown that the development has a low level of economic efficiency; the IRR is 16.5 per cent if internally financed and 9.9 per cent if financed by credit; netback pricing reduces the return to 7 per cent. For internally financed investments, the pay off period is 13.3–17 years (at discount rates of 10–15 per cent), while with the use of credit and netback pricing the project does not pay off at all.

There are grounds for believing that drilling costs will grow in the future. For full-scale deposit development, the attraction of new contractors and significant purchases of new drilling equipment is required, most likely leading to a rise in drilling costs. Douglas-Westwood (UK) estimated that the average cost of drilling in East Siberia will grow to $3,000 per metre by 2011. TomskNIPIneft also warned in the DOI about a possible rise in costs.[58]

In 2008, VCNG planned to invest 13 billion rubles ($521 million) in development, a 62.5 per cent increase compared with 2005. Since December 2005, when trial operations began, three reactivated and 18 drilled wells have emerged on the deposit, 13 of them drilled in 2007. In 2008, the company planned to drill 29 more wells. The average production rate of these wells is about 120 tonnes per day. In order to increase this, VCNG revised its drilling programme in the third quarter of 2007. The company has defined the list of wells subject to drilling based on modern horizontal end technology. Horizontal sections of up to 500 metres in length inside an oil bearing will allow a reduction in the total number of wells and consequently moderate the environmental impact.

The general well fund at the deposit is planned to consist of 1,306 wells (including 42 reserve wells), of which 938 are operational and 368 injection wells. The use of directional drilling will reduce the surface area of the field construction, thereby also reducing capital expenses, and at the same time increasing drilling speed. The plan called for 159 multiple well platforms, including 12 platforms by 2009 within the framework of the early oil project. Drilling is planned to occur up to and including 2019. Connected to each multiple well platform will be an infrastructure comprising a highway, an oil pipeline system to gather well production, and a high pressure water pipeline.

According to the October 2007 DOI, well production will be prepared by the construction on the deposit of four booster pump stations (BPS) with preliminary water gathering units (PWGU), two oil treatment facilities (OTF), as well as gas-compressor stations for the preparation of gas for transport via pipeline to Ust-Kut and an installation to refine oil into gasoline. The plans assume the construction of six water injection stations (WIS 1–6) to maintain reservoir pressure. These would be located close to the BPS, PWGU, and OTF units. The high-pressure water pipeline network will extend about 325 km, while the system of oil and gas gathering pipelines will total some 335 km, and in-field oil pipelines will measure 98 km. The Giprovostokneft Institute in 2007 prepared a plan for an 85 km feeder pipeline to ESPO.[59]

The revised DOI expected that the extraction capacities at the Verkhnechonskoye deposit would begin by the end of 2008 – at an initially planned ESPO start-up volume of 3 mt/y (0.5 mt higher than that foreseen a year earlier). The total revenues lost for a year of idle time will be 22–31 billion rubles. The risk of additional project expenses also exists in relation to equipment on allocation, processing, clearing, storage, and transport of the helium contained in the gas from the deposit. In the revised DOI, TomskNIPIneft has removed reference to the possibility of helium extraction. However, in the programme to create a uniform East Siberian and Far Eastern system of extraction, gas transport, and supply, with a view to possible exports to the Chinese and other Asia–Pacific regional markets, Gazprom relies on helium extraction, including in the Verkhnechonskoye deposit.

The DOI anticipates that exports will not exceed 30 per cent of extraction. However, it is unlikely that regional domestic consumers will be able to absorb the remaining 70 per cent of ESPO capacity. The cumulative capacity of Khabarovsk, Komsomol, and that projected for the Perevoznaya Bay refineries adds up to 22 mt a year, which is 73 per cent of the first stage pipeline capacity (of 30 mt/y) and 27.5 per cent of the overall annual capacity (80 mt) once the second stage of ESPO is operational. Specific plans for the necessary increase in regional oil refining (up to 35–36 mt/y) are still unfinished.[60]

On 15 October 2008 TNK–BP sent oil from the Verkhnechonsk oil and gas condensate deposit to the ESPO pipeline. The n-23 oil pump station was commissioned by TNK–BP's Chief Operating Officer, Tim Summers, and Rosneft's First Vice President Sergei Kudryashov. TNK–BP and Rosneft have invested $1.0 billion so far, and the total development costs are expected to reach $4–5 billon. The development plan is to drill 650 wells, 450 production and 200 injection wells. In 2008, around 30 wells were expected to be drilled. Starting in 2013, the Verkhnechonsk field is to produce around 7 million tonnes annually. In total the field is expected to remain in exploitation for 20–30 years.[61] 2009 was the first full calendar year of production at the field and oil output in that year was 8.6 million barrels (1.2 mt). Thirty-eight new production wells were brought into operation during 2009, and by the end of the first year the number of wells in production had risen to 59.[62]

Nikolay Savostyanov field. Rosneft was to make further discoveries: in January 2010 it announced the discovery of:

> …a large new oilfield within the Mogdinsky and East Sugdinsky license areas in Katangsky District of the Irkutsk Region. Rosneft had obtained licenses for both plots through auction in 2006. The field was named after Nikolay Savostyanov who was the head of the Principal Division for Oilfield and Site Geophysics (Glavneftegeophysica) of the USSR Ministry of Oil Industry between 1976 and 1990 and then of Rosneft's Department for Geophysical Operations during the years 1993–1997. The field's primary recoverable reserves under C1+C2 categories exceed 160 million tons, enough to consider the field to be a strategic one. The field is located approximately 80 km from Verkhnechonskoye Oil & Condensate Field, in which Rosneft is an active participant, and 150 km from the ESPO pipeline.

Exploration in the area started in the 1980s but the first commercial oil was recovered only in 2009 as a result of exploration drilling. In 2010 Rosneft intends to continue exploration in the license blocks of the Irkutsk Region, particularly to run a 2D seismic survey of 2,930 km and a resistivity survey of 3,700 km. Drilling of four new wells is also scheduled.[63]

This comprehensive exploration work may pave the way to sizable additional production.

Dulisminskoye (developed by Urals Energy). Urals Energy owns the transport company LTK (with a 99 per cent holding), along with the oil production companies Dulisma (99 per cent) and Taas–Yuryakh NGD (35.329 per cent). Dulisma is developing the Dulisminskoye deposit in the Kirensky area of the Irkutsk region (with C1+2 recoverable reserves of 180 million barrels of oil and 62 bcm of gas). An independent valuation by DeGolyer and MacNaughton has estimated proven and probable reserves of oil and condensate at 109 million barrels, along with possible reserves of 8.7 million barrels of oil and condensate and 4.9 bcm of gas. The assets of Taas–Yuryakh include an oil refining complex and the license to develop the Sredne-Botuobinskoye oil and gas deposit with approved reserves totalling 53 mt of oil and 130 bcm of gas.[64]

Urals Energy also owns Petrosakh (with a 97.16 per cent holding). Established in 1991 as a Russia–USA joint venture, Petrosakh developed the Okruzhnoye field on the eastern coast of the Sakhalin Islands. In 1993 and again in 1997 Petrosakh received a 20 year oil production license for the field. In 1995, Petrosakh commissioned a terminal to load tankers, and a system for injecting associated gas into the production layer followed in 1999. In September 2000 the Alpha-Eko Industrial Group purchased Petrosakh, which in 2001 received a geological study license for the Sakhalin-6 shelf project (the Pogranichny block, directly offshore and parallel to Okruzhnoye). Two years later Petrosakh acquired a 480 square kilometres 3D seismic survey programme covering the central portion of the Pogranichny Block. In 2004 it obtained an additional 65 square kilometres of 3D seismic survey in two programmes covering the northern and southern areas of the near coastline transition zone. Estimated Sakhalin-6 reserves are about 1 bt of conditional fuel, with potential reserves in Petrosakh license blocks totalling 240 mt.

Also in 2004, Urals Energy took over Alpha's 97.16 per cent interest in Petrosakh. The main risks for Petrosakh are its assets on the shelf, since all shelf reserves are considered strategic and so, under newly enacted Russian law, their development will be in the hands of Gazprom and Rosneft.[65] On 18 April 2006, Urals Energy, registered in Cyprus, announced the acquisition of both the Russian oil company Dulisma (which holds a production license for the Dulisminskoye oil and gas condensate field, strategically located north-west of Lake Baikal, along the planned route of the ESPO oil pipeline in the Irkutsk region) and of the Lenskaya Transportanaya Kompaniya, an oil transportation company. The cost of the deal was US$148 million. At the end of March 2006 preliminary estimates by the US auditing company DeGolyer and MacNaughton indicated that the proved and probable resources of the Dulisminskoye field comprise 14.57 mt of oil, potential resources of 11.63 mt of oil and condensate, and 54 bcm of natural gas.[66] At the end of April 2008, Urals Energy decided to sell its extracting subsidiaries in the Republic of Komi to begin the accelerated development of the Dulisminskoye deposit in the Irkutsk region and Sredne-Botuobinskoye in the Republic of Sakha (Yakutia).[67]

North Mogdinsky, Bolshetirsky and Zapadno–Yaraktinsky (developed by JOGMEC & IOC). After many years' preparation, JOGMEC decided to initiate an oil and gas exploration project in Irkutsk with IOC (Irkutsk Oil Company),[68] a company founded in 2000 by Marina Sedykh (then general director) and Nikolay Buinov (currently board chair). Through affiliated structures the firm owns licenses for oil and gas recovery at the Yaraktinskoye, Markovskoye, Danilovskoye, and Ayanskoye deposits, and is also working on six hydrocarbon blocks in the northern part of the region. In 2006, the firm extracted 166,000 tonnes of oil (rising to 200,000 tonnes in 2007) and its 2006 revenues amounted to 1.4 billion rubles.[69]

In 2007, JOGMEC and IOC set up a joint venture named IOC-North (51 per cent held by IOC and 49 per cent by JOGMEC) for development of the North Mogdinsky oil and gas block in northern Irkutsk. At the end of April 2008, IOC announced the beginning of exploration at the block, which covers 3.747 square kilometres. The JV's subsoil use rights were renewed by a 25 year license. The North Mogdinsky block is one of IOC's most remote assets. The block is located 1,000 km from Irkutsk and 150 km from ESPO. JOGMEC has spent $95.8 million on seismic surveys and the drilling of exploration wells.[70]

JOGMEC's vice president, Hironori Wasada, has said that the joint venture with IOC resulted from a political decision. IOC-North has a similar capital structure to that of its parent, the En+ Group. The joint venture is engaged in geological exploration on the North Mogdinsky block which has reserves of 15 mt of oil and 50 bcm of gas. Wasada said that JOGMEC had taken the least studied block and consequently needed to conduct a large amount of exploration.[71] IOC has a number of subsidiaries, such as Ustkutneftegaz which owns licenses for the Yaraktinskoye and Markovskoye oil and gas condensate deposit, the Oil Company Danilovo which is developing the Danilovskoye oil and gas condensate deposit, and IOC–Neftegazgeologiya which owns the license for the Ayanskoye gas condensate deposit, the Ayansky block, and the Potapovsky area.[72]

As the largest independent oil producer in the Irkutsk region, IOC planned to spend as much as 4.5 billion rubles between 2008 and 2012 to develop the following sites: the Ayanskoye deposit and site and the Potapovsky, Zapadno–Yaraktinsky, Bolshetirsky, Angaro–Ilimsky, and Naryaginsky sites. IOC planned to invest nearly 12 billion rubles up to and including 2012 for E&P at the Danilovskoye, Yaraktinskoye, and Markovskoye deposits. Most of this work was planned to go to its subsidiary IOC–Service and about 20–30 per cent of it was to be assigned to outside contractors. The main infrastructure on IOC's production sites was already constructed and IOC aimed at commissioning more substantial oil preparation facilities at Yaraktinskoye in 2008. Construction of a permanent pipeline connection to ESPO was promoted in 2009. By 2011 additional booster pump stations, electrical transmission lines, internal oil and gas pipelines, roads, and power stations were to have been constructed on the Yaraktinskoye and Markovskoye deposits and an installation for the demercaptanization of oil to have been placed on Markovskoye.[73]

In March 2008, IOC negotiated a purchase, worth $85 million, of IOC share capital by the EBRD and in May IOC confirmed EBRD plans to use the funds to purchase an 8.15 per cent interest in the holding company that controls IOC. The investment will go towards restructuring the credits and performance of the environmental programme, including a project for gas injection into the production layer at the Yaraktinskoye deposit.[74] During the Baikal Economic Forum, the CEO of IOC, Marina Sedykh, announced that the firm may increase oil supplies to the ESPO pipeline to 2.5 mt/y in 2012, twice as much as previously agreed. The increase will be facilitated by the boost in the firm's oil reserves. In 2008 it added 31 mt of oil of C1+C2 categories to its existing reserves, 6.5 mt at West Ayanskoye and 24.5 mt at Yaraktinskoye.[75]

In September 2008, JOGMEC's Vice President Wasada, revealed that a joint venture between IOC and JOGMEC would commence drilling in the North Mogdin section of the Irkutsk region in 2009. After studying seismic findings in 2008, the drilling of the first exploration well took place in 2009. This section has an area of 3,747 square kilometres and is located 1,000 km north of Irkutsk and 150 km from the ESPO pipeline.[76] In 2009, JOGMEC decided to set up a joint venture named INK-Zapad that aims to develop the Bolshetirsky and Zapadno–Yaraktinsky blocks in the Irkutsk region. JOGMEC will own 49 per cent and IOC 51 per cent of the venture, which planned to spend 15 billion yen on exploration in the period up to 2013.[77]

According to Japan's Shinsuke Kitagawa, general director of the natural resources and fuel department at the NREA under METI:

> ... uncertainty in the Middle East is growing. There is rapid economic growth in the Asian countries. Therefore, Japan should increase its sources of energy and pay more attention to the Far East and East Siberia, which could become long-term reliable suppliers to Japan. As a result of the ESPO pipeline it is very probable that Japan will decrease its oil imports from the Middle East ... Japan is now actively cooperating with Russia in the Sakhalin projects.[78]

These remarks indirectly confirm that Japan has a serious interest in importing East Siberian crude oil, even though it has no interest in investing in the second stage ESPO pipeline. JOGMEC has also expressed its interest in working jointly with Rosneft to develop deposits with extractable reserves of not less than 100 million barrels (13.6 mt), preferably near the ESPO network. In March 2008, Rosneft signed a framework agreement on cooperation with the NREA; JOGMEC had already expressed its interest in working in East Siberia and on the Sakhalin shelf. However, JOGMEC's vice president Wasada has said that

> ... the reality is that up to now Japanese investors approach entry to the Russian market with caution caused by the absence of transparent rules for foreign investors. The events surrounding the Sakhalin-2 project only lowered confidence among Japanese companies. Today, therefore, no Japanese company wishes to begin operations in Russia in the short run.[79]

In October 2010, the JV between JOGMEC and IOC announced that it had found oil and gas in three East Siberian blocks. The discoveries were made in the Severo–Mogdinsky, Zapadno–Yaraktinsky and Bolshetirsky blocks. The blocks together cover almost 11,900 square kilometres. At the Severo–Mogdinsky block, crude oil flows were confirmed at the first well spudded in June 2009 and the second in April 2010. Two wells have been drilled in Severo–Mogdinsky, where recoverable reserves to date are put at 14.8 mt (108 million barrels). Total recoverable reserves in the block, where light, sweet oil has been found, are expected to amount to 50 mt (370 million barrels). As for the Zapadno–Yaraktinsky licensed block, gas production was confirmed during testing at the exploration well, with a gas-flow rate of 117,000 cubic metres per day and gas condensate rate up to 27.4 tons (243 bbl) per day. The estimated amount of investment for exploring these blocks is some $300 million through 2014. Once commercial production is started, the crude produced is expected to be transported to the Japanese market and other Asian markets through ESPO.[80] It is not clear whether private Japanese companies will step in when the JOGMEC venture moves to the commercial production level, but the venture's oil production is in a good position from which to take advantage of ESPO development.

The Sakha (Yakutia) Republic

In late February 2004, the government of the Sakha Republic proposed a new route to Nakhodka, via Nizhnyaya Poima, Yurubcheno, Tokhomskoye field, Verkhnechonskoye field, Talakanskoye field, Chayandinskoye field, Lensk, Olekminsk, Aldan, Neryungri, Tynda, Skovorodino, Nakhodka, for crude oil supply. At a conference on the development of transportation infrastructure in the Far East, held in Khabarovsk on 26 February 2004, President Putin charged Vyacheslav Shtyrov, president of Sakha Republic, with the task of submitting development studies on a new pipeline.

In May 2004, MNR decided on the list of oil and gas blocks of the Sakha Republic that would be put up for tender in the 2005–6 period. For a long time Yakutian authorities had been trying in vain to sell their oil and gas resources in order to finance the region's development. But tender and auction deadlines were repeatedly postponed, and the licensing delays led to a drastic change in the economics of East Siberia projects and the list of parties wishing to participate in tenders.

Yakutia's government was about to expand the resource base for the Eastern oil pipeline in three ways. The initial plan was to sell 16 already explored but unassigned fields, which would permit switching the category of their reserves from C2 to C1. It had then planned to conduct auctions in 2004–5 to issue exploration-cum-production licenses for 12 geological blocks which had a strong possibility of containing oil and gas fields. However, nine fields and 10 blocks offered by Yakutia were excluded from the final list prepared by MNR. Part of Yakutia's proposal clashed with the Programme of Comprehensive Exploration and Development of Oil & Gas Reserves and Resources of Russia's East Siberia and the Far East for 2004–20, which was being prepared by the Siberian Research Institute of Geology, Geophysics and Minerals (SNIIGGIMS), and these differences needed to be reconciled.[81]

Talakanskoye (developed by Surgutneftegaz). Talakanskoye is the biggest oil field discovered in the Republic of Sakha. Proven recoverable resources of the Tsentralny block (part of the field) alone amount to 113.7

mt of crude oil and 31.8 bcm of gas. The Tsentralny block in the Talakanskoye field had already been put up for sale on two occasions, in April 2001 and December 2002. In 2001 Sakhaneftegaz, together with Yukos, took part in the auction and agreed to pay US$501 million bonus, but Sakhaneftegaz did not have the financial resources to proceed. At that time, the list of those vying for the Talakanskoye field became very long, as the race was joined by companies like Yukos, Sibneft, Surgutneftegaz, Rosneft, Gazprom (through its joint venture, Sevmorneftegaz), TNK, and Total (France). But in December 2002, the Talakanskoye tender was called off. The MNR had proposed selling not only the Tsentralny block but also two other blocks of the Talakanskoye field for $900 million, but the MEDT would consent to sell only the Tsentralny block for a starting price of $150–200 million. In the autumn of 2003 Surgutneftegaz was awarded a front-end engineering license to the Tsentralny block and this gave it the opportunity to obtain the Talakanskoye field, the license to which had previously been held by Lenaneftegaz (a subsidiary of Sakhaneftegaz, in which Yukos held 50.6 per cent of the shares).[82]

In the spring of 2004, Yukos, Surgutneftegaz, and the Republic of Sakha signed an agreement for the sale of assets at the Talakanskoye field and shares in Lenaneftegaz to Surgutneftegaz. Even though this transaction never took place, Lenaneftegaz operated with Surgutneftegaz after June 2004. During 2004 and the first half of 2005, Surgutneftegaz conducted large-scale seismic research, drilled four wells with horizontal sections, and completed 6,600 metres of exploitation drilling. At the end of 2004, there were 33 units of operating well stock at Talakan. At the June 2005 annual meeting of Lenaneftegaz, the shareholders decided to sell the assets they held in Talakanskoye field to Surgutneftegaz. These consisted of 38 wells, a number of warehouses, three residential buildings, and social and cultural facilities, all of which were sold to Surgutneftegaz for 1.35 billion rubles (approximately US$48 million). After winning a court battle with the state for the Talakan license and after acquiring three nearby blocks, Surgutneftegaz sought to acquire the property of Lenaneftegaz in order to make more effective use of its existing holdings.[83]

Surgutneftegaz holds 12 licenses in the Republic of Sakha (Yakutia) and the Irkutsk and Krasnoyarsk regions. These consist of the following: three E&P licenses on deposits in the central block of the Talakanskoye, Verkhnechonskoye, and Alinskoye deposits in the Republic of Sakha; five licenses allowing investigation as well as E&P, one in the Republic of Sakha, three in the Irkutsk region, and one in the Krasnoyarsk region; and four search licenses in the Republic of Sakha.

Up to the beginning of 2007 the firm had spent 2.3013 billion rubles on capital expenses for exploratory work, along with 818.3 million rubles for seismic study. In 2007–10, the company planned to spend a further 16.1885 billion rubles on exploratory drilling and 1.6402 billion rubles on seismic surveys. Total spending on geological exploration in East Siberia from 2004 to 2010 would total 21.826 billion rubles. Between 2004 and 2006, Surgutneftegaz drilled 34,000 metres of exploratory wells in East Siberia, planned to drill 33,170 metres in 2007, and a further 162,370 metres up to 2010. In other words, by the end of 2010 the company was intending to have drilled nearly 200,000 metres of wells. By 2009, it was planning to shoot 7,430 line kilometres of 2D seismic survey and 80 square kilometres of 3D seismic survey.[84]

In late July 2008, Surgutneftegaz announced that the Talakan oil field was completely ready for commercial oil extraction and pumping to the Transneft pipeline system. The company president, Vladimir Bogdanov, said that in the previous two-and-a-half years his company had invested 101.3 billion rubles to bring Talakan on-stream, and that another 47 billion rubles was earmarked for equipment purchases. He added that the outlook in East Siberia was good since only about 7 per cent of the fields had been studied. These included one gas field with over 100 bcm of reserves and two oil fields. Surgutneftegaz aims at raising annual production at the Talakan oil field to around 6 million tonnes in 2016. In 2008, a total of 52 production wells were in operation at the field, with a daily production of 4,000 tonnes. In 2009, annual oil production at Talakan was targeted to be around 2 mt. During the period 2004–8, annual production at Talakan came to 1.4 mt of oil. In 2009, total combined cumulative production at Talakan was estimated to be at 21.2 mt.[85]

In late January 2009, *OGJ* reported that:

> ... Russia's subsoil use agency, Rosnedra, was reported to have set a starting bid price of 1.66 billion rubles ($50.3 million) at an auction for the rights to the East Talakan field in the Sakha Republic. The business daily *Vedomosti* reported that Surgutneftegaz is likely to get the rights to East Talakan, although Gazprom is also a leading candidate. The field holds proven and probable reserves of 9.9 million barrels of oil, 22.9 bcm of gas, and 0.2 mt of gas condensate.[86]

Surgutneftegaz plans to invest 276 billion rubles in the development of oil and gas fields in East Siberia during the period 2009–14. Between 2004 and 2009 it had already spent 102 billion rubles on exploration and production drilling, on the construction of a road to the Vitim–Talakanskoye field, and on power lines, pipelines, infrastructure, and social facilities. In particular, Surgutneftegaz has built a gas-powered electrical generator with a 12 megawatt capacity, and a gas turbine power station with a capacity of 96 megawatts. The company is now building the third phase of this gas turbine power station, with a capacity of 48 megawatts and a compressor station.

Surgutneftegaz carried out 92,800 running metres of exploration drilling and built 47 exploration wells between 2004 and 2008, and opened three fields – Verkhne–Peleduiskoye, Severo–Talakanskoye and Vostochno–Alinskoye. A sum of 10.5 billion rubles was spent on geological exploration in this period. The Alinskoye field was due to be put into operation in 2009. By the end of 2008, planned construction work included a 30 km road from the Talakan field to the Alinskoye field, 35 km of power line, a pumping station, and several pipelines. Surgutneftegaz continues to expand its presence in Sakha, where operations are currently being carried out in 12 licensed areas. In addition to the 12 licenses in the region, four new areas were acquired in 2008, and in 2007 two licenses were received for the opening of oil fields (Vostochno–Alinskoye and Severo–Talakanskoye).[87]

Krasnoyarsk Territory
According to a detailed exploration carried out in 2000, the Krasnoyarsk region holds about 10 per cent of all Russian hydrocarbons, coming second (after Tyumen) in order of importance for oil production. Hypothetical resources are 8.2 bt of oil and gas condensate, 23.6 bcm of freestanding gas, and 638 bcm of gas dissolved in oil. In total, this is half of the hydrocarbon resources in Eastern Russia. In 2000, subsoil site allocations were assigned, making the start of oil and gas production possible. Primarily this was for the Bolshekhetsky and Yurubcheno–Tomsk oil and gas bearing areas, where there are a number of gas locations.[88]

At the Third Annual International Conference on Mergers, Acquisitions and Licensing in the Russian Oil and Gas Industry, held in Moscow at the end of May 2005, Alexei Kontorovich (director of the IGNG, Novosibirsk) gave a detailed report on prospects for the Predyeniseiskaya Sub province basin which lies along the left bank of Siberia's Yenisei River, at the dividing point between West and East Siberia. The basin extends from the Tomsk Region in the south to the Yamal-Nenets Autonomous District (YNAD) in the north, and includes areas in Krasnoyarsk Territory, the Tomsk Region, and in Khanty-Mansiysk and Yamal-Nenets Autonomous Districts. In 2004, IGNG, jointly with other geological institutes and regional committees of the MNR, developed the Vostok Programme, later approved nationally by the MNR, to study the entire territory west of the Yenisei River.[89]

At this time there was no consensus as to which state company should be the driving force of oil and gas exploration and development in Krasnoyarsk Territory. During 2005 the Governor of the Territory, Alexander Khloponin, stated that the Evenkia Autonomous District would benefit most if Gazprom headed the development in the district. In September 2005, however, German Gref, minister of MEDT, argued that the operator in the Yurubcheno–Tokhomskaya Zone (YTZ) should be Rosneft.[90]

In April 2006, the region set out a strategy for the oil and gas sector designed to accelerate the overall socio-economic development of Krasnoyarsk territory, along with the Taimyr (Dolgano-Nenets) and Evenki Autonomous districts. Their entry into the unified Krasnoyarsk region followed in January 2007 and in July of that year the regional administration adopted the first departmental target programme ('The Development of Crude Oil and Natural Gas in the Territory of the Krasnoyarsk Region for the Period 2007–2010'). This document put forward the policies to be followed by the territorial administration (with federal support) to govern the activities of subsoil users in the exploration, extraction, and transportation of oil.[91]

According to Alexander Yekhanin, head of Krasnoyarsknedra, ESPO construction has resulted in an increasing readiness to finance oil and gas exploration in the region. In 2007, the Krasnoyarsk region scheduled 11 blocks on the approved list for subsoil auctions, and all of them found buyers. The 2007 auctions resulted in a contribution of over 3 billion rubles (about $125 million) to the federal budget. Plans in 2008 called for conducting auctions on a further 17 oil and gas blocks, and the schedule called for auctioning off another 52 blocks by the end of 2010. Slavneft planned to invest $3 billion in exploration and the preparation of deposits for commercial extraction in the Krasnoyarsk region. In this way Slavneft would become one of the biggest players in the Krasnoyarsk oil fields. In the year 2010–11, Slavneft planned to produce 1.2 mt of oil from the region.[92]

The importance of the territory was underlined by a visit from Prime Minister Putin in 2009, described by *OGJ*:

> On his visit to the town of Igarka in Krasnoyarsk Territory, Putin said that when it reaches its estimated capacity the region would provide more than 115 mt/year of additional oil and condensate. On national television Putin said Vankor is highly significant as the first step in the implementation of the large scale, strategic project for the integrated development of hydrocarbon deposits in the north of Krasnoyarsk Territory and the Yamal-Nenets Autonomous Area ... He added that many oil deposits are already subsidized as a result of a zero rate of subsoil resources extraction

tax and that this should be extended as the norm in the whole of the Yamal-Nenets Autonomous Area. He emphasized the importance of this plan by confirming that the Yamal-Nenets Autonomous Area and the north of Krasnoyarsk Territory contains 67 per cent of Russia's natural gas, 15 per cent of its oil, and 60 per cent of its gas condensate.[93]

The enterprise, however, is not without problems. Around 10 oil production companies have already requested the territorial subsoil management for KMAR–Yugra to suspend their license because, given the existing prices for oil, extraction is simply unprofitable. According to estimates by the Alpha Bank, since January 2009, due to the devaluation of the ruble and a decrease in export duties, each barrel of oil has yielded as much as $9 in profits. Unlike the big companies, small ones sell production in the domestic market; when the price falls below $45 per barrel, it is unprofitable for them to produce.[94] Krasnoyarsk's potential is huge but it will be some time before sizable oil and gas production is seen in this frontier area. For the time being, Vankor field, located in northern Krasnoyarsk, will be a key supplier of crude oil to the ESPO pipeline.

Vankor field (developed by Rosneft). The project to produce oil on a large scale in the Vankor field is certain to advance rapidly. It is important to Rosneft as a source of output growth for the company;[95] it will be a major user of the oil transport capacity of the ESPO pipeline, and it has great socio-economic importance to Krasnoyarsk Territory and its development.

The Vankorskoye oil field was discovered in 1991, while drilling in the southern part of the field. It is made up of two license blocks located in two territorial sub-divisions of the Russian Federation – in the Turukhansky area of Krasnoyarsk Territory (Vankorskoye acreage, license holder ZAO Vankorneft), and in the Dudinsky area of the Taimyr (Dolgan-Nenets) Autonomous District. In late April 2003, Rosneft purchased 97.46 per cent of the shares of the ASOC, direct and indirect owner of development licences for Vankorskoye field and the Severo–Vankorsky prospect. It purchased the shares at one pound sterling each (a 100% premium over the market price).[96] Based on this purchase, Rosneft began an in-depth study on the Vankorskoye field and in 2004 it proved that the originally delineated Vankorskoye field and the northern Severo–Vankorskoye block were linked in a single, closed, double-domed structure, with the dimensions 15 by 37 km. In 2004 operators drilled three wildcat wells within the Vankorskoye and Severo–Vankorskoye license blocks, two at Vankorskoye and one at Severo–Vankorskoye.[97] Recoverable oil reserves in the C1 category at the Vankorskoye field rose to 41.5 mt, and in the C2 category to 185.4 mt.

Gas reserves in the C1 and C2 category total 28.2 bcm and 61.6 bcm respectively. Oil resources in the C3 category are estimated at 98.4 mt.[98] Further development of the field is in the hands of Rosneft's subsidiary Vankorneft, which plans to drill primarily horizontal wells at Vankor, 75 per cent of which will have more efficient 'smart' completion techniques.

Up to around the mid-2000s, all the potential investors interested in the Vankor project favoured the northern route to Dickson, since it would permit exporting up to 100 per cent of Vankor oil without blending it with the lower-quality crude of the Transneft system. Both Shell, which until 1999 had held an option to buy a part of Vankor project, and Total, which had been trying to enter into the project since 2001, supported the northern route. The Krasnoyarsk Territory authorities have always supported the northern route because it would allow them to develop the poorer northern areas of the Territory and exploit a larger number of Krasnoyarsk's fields.

Political decision on the route. Since 2003, when Rosneft took control of the Vankor project, Rosneft also favoured the route to Dickson. However, Russia sacrificed the economic efficiency of the Vankor project to political objectives. When difficulties arose in obtaining uncommitted oil volumes, a decision was made to alter the oil export route from Vankor. Instead of the more efficient northern route, oil would be diverted into West Siberian oil pipelines and then into ESPO. This southern route will reduce the revenues generated by Rosneft from the sale of oil by $15 billion and the project's IRR will drop from 18 per cent to 13 per cent (See Table 2.9). But oil export to Asia, as an alternative to the European market, had become Moscow's priority.

Table 2.9: Economic Indicators of the Vankor Project using the Northern and Southern Routes

	Northern Option – to Dickson and onward along the Northern Sea Route to Europe	*Southern Option – to Transneft's oil pipelines in the Purpe area*
Oil Production, mt	234.9	234.9
Price for Urals, $/bbl	19.0	19.0
Price for Vankor Oil, $/bbl	20.5	20.5
Total Revenues, $mn	47,270	31,847
Capex, $mn	5,447	4,779
Opex, $mn	4,746	3,906
IRR, %	17.98	12.99
Payback period, years	14.7	33.0

Source: Glazkov (2006b), 21.

The oil trader Crown Resources was the first to draw public attention to the problem of discrimination against Russian oil in Europe in 2001. Data supplied to it by the auditing firm Andersen led to the conclusion that from 1996 to 1999 Russian oil flowing via the Druzhba oil pipeline was selling at $1.9 per barrel less than the same oil delivered by tanker. In 2004 Lukoil protested against an increase in discounts for Druzhba (up to between $3 and $7 per barrel), as well as high discount for the Russian Urals compared to North Sea Brent ($7 per barrel), for which there was no reason whatsoever. In April 2005 the Transneft president, S. Vainshtock, claimed that the key reason for this discrimination was the 'overfeeding of Europe with Russian oil' and in August Putin instructed the government to draw up measures to overcome this situation. In 2006, the MEDT estimated that Russia's dependency on exports to the European market, accounting for nearly 96 per cent of its oil exports and one hundred per cent of gas exports, would lead to the loss of between $5 billion and $8 billion per year as a result of price differences. The state decided to use tariff policy to stimulate the use of ESPO as an alternative to exporting to Europe.[99]

In 2003 Rosneft calculated that for the Vankor project to be economically viable the total reserves should be 250–300 mt of oil; the next year they raised that figure to 350–400 mt. In 2004, Rosneft obtained, on privileged non-competitive terms, prospecting licenses at 10 blocks in Krasnoyarsk Territory and in the Taimyr Autonomous District (recently merged into the Krasnoyarsk Territory). These blocks had combined hypothetical resources of 700 mt of oil. In late March 2005, Rosneft signed a one year, $13 million contract with the Canadian engineering company SNC-Lavalin for design work at Vankor. The contract included the oil pipeline route and a project to construct a marine terminal in the Dickson area. The terminal was to have an annual capacity of about 18 mt and a tank farm with holding capacity of 0.4 mcm.[100] In June 2006 construction of the Vankor–Purpe pipeline began. It has been designed to pass through Krasnoyarsk and the YNAD. In Purpe (in the Purovsk area of Yamal), the pipeline links Vankor with the trunk pipeline network of Transneft.[101]

In 2007, Rosneft planned the largest of its capital expenditures in East Siberia, an investment of 490 billion rubles. This sum did not include expenses at newly purchased sites in the Yurubcheno–Tokhomsky zone (southern Krasnoyarsk region). Most of the anticipated work is in the north of the Krasnoyarsk region, where expenses for development of 16 sites in the Vankor project (reserves of 490 mt and resources of 575 mt) add up to 480 billion rubles, including 13.5 billion rubles spent on geological exploration.[102]

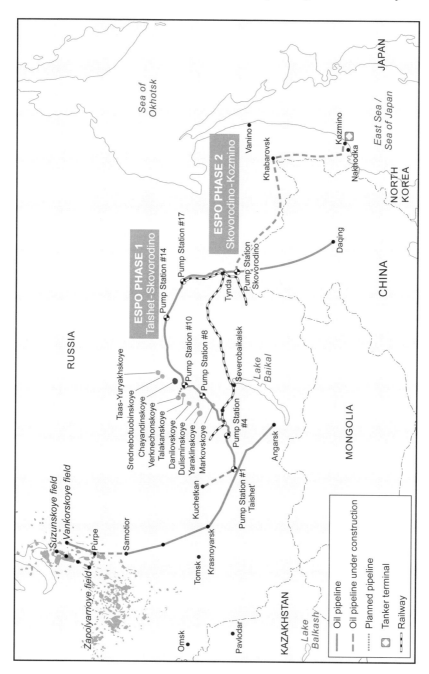

Map 2.2: Vankor field and ESPO

Source: Author, based on maps from *RPI* and *Platts*.

Vankor Development Plan.[103] As a result of Rosnedra's key part in filling the ESPO pipeline, its Central Commission on Deposit Development took the unusual step of approving a full-scale development project without any trial operation. Instead of traditional vertical wells, Rosneft planned to drill horizontally. Plans call for 137 operational wells on the deposit, along with 90 water-injection and eight gas-injection wells. Tenders were to have been announced for the drilling of seven exploratory wells in 2007, 135 wells in 2008, and 242 wells in 2008–10. (In fact, by the end of 2009 a total of 142 wells had been drilled at Vankor, including 119 production and injection wells. Total capital expenditure for development of the field in 2005–9 were $6.5 bn.)[104]

Table 2.10: Vankorneft's Operating Highlights

	2004	*2005*	*2006*	*2007*	*2008*	*2009*
Exploration drilling, 1,000 m	8.6	17.0	10.9	8.9	4.1	2.9
2D Seismic, Linear km	0	400	350	0	0	0
3D seismic, sq. km	170	200	0	0	0	0
Production drilling, 1,000 m				18	78	142	277

Source: Rosneft 2011 (www.rosneft.com/Upstream/ProductionAndDevelopment/eastern_siberia/vankorneft/).

The second location which Rosneft intends to develop is a group of license sites in the Irkutsk region (Vostochno–Sugdinsky, Mogdinsky, Sanarsky, Danilovsky, Preobrazhensky, as well as Zapadno–Chonsky and Verkhneichersky through Vostok Energy its joint venture with CNPC), and in the Kulindinsky field in Krasnoyarsk region which has hypothetical resources of 472 mt of oil and gas condensate. According to Rosneft, between 2008 and 2012, they will shoot 10,100 line kilometres of 2D seismic exploration and conduct 73,000 metres of exploratory drilling at a total cost of 7.0 billion rubles.[105]

In 2008, Rosneft held licenses to 14 blocks adjacent to the Vankor field – Sovetsky, Lebyazhy, West Lodochny, East Lodochny, Nizhnebaikhsky, Polyarny, Samoyedsky, Baikalovsky, Peschany, Protochny, Vadinsky, Tukolandsky, Pendomayakhsky, and North Charsky. According to a 2007 estimate of total prospective resources by DeGolyer and MacNaughton, the blocks contained 3.9 bn barrels (532 mt) of crude oil and more than 180 bcm of gas. Rosneft's own official figure on these prospective resources, however, is much more conservative: 2.5 bn barrels (341 mt) of oil and 126 bcm of gas. Of primary importance to Rosneft are the blocks closest to the main Vankor field where infrastructure development is being completed. The 556 km Vankor–Purpe pipeline will connect the Vankor field with the oil fields of Purneftegaz.[106]

According to Rosneft's information services, in May 2009, a new oil and gas condensate field was discovered on Baikalovsky License Block in the Taimyr Autonomous Area in Krasnoyarsk Territory. The block, one of 14 adjacent to Vankor, is located 80 km north of Dudinka. The testing of well no. 1 on this block resulted in natural flow production of hydrocarbons from the 2,000–2,700 metres range with daily gas production over 60,000 cubic metres and daily oil and condensate production over 25 cubic metres. The preliminary estimate suggests the field's minimum reserves are 55 mt of oil and 99 bcm of gas.

On 21 August 2009 an official ceremony was held, with Prime Minister Putin in attendance, to mark the beginning of commercial production of crude oil at Vankor. The launch of Vankor marked the start of the practical implementation of sub-soil development in East Siberia, whose resources will provide the foundation for long-term production growth in Russia, compensating for the exhaustion of reserves at fields in traditional oil producing regions, in particular western Siberia, where about 70 per cent of Russia's oil is currently produced. In short, Vankor has become a hub of development for East Siberia and the Far East.[107]

The price tag of Vankor development was not small. By August 2009, total capital investment by Rosneft at Vankor had exceeded 200 billion rubles. But at present levels of international prices for crude oil, project implementation will provide total tax revenues of 4.5 trillion rubles to all levels of the Russian state budget, representing nearly one half of Russia's entire annual budget. The Vankor project is one of the largest in modern Russia, involving more than 150 equipment suppliers and including 65 manufacturing plants. A total of 450 contractors and subcontractors have worked on the project, and 12,000 construction workers have been employed using 2,000 vehicles at the peak of construction work. Vankorneft employs about 2,000 full-time personnel. A total of 425 development wells are scheduled to be drilled at Vankor, of which 307 will be horizontal.[108]

Vankor will serve as an anchor supply source for the ESPO pipeline. The initial projection of Vankor's production (made in 2008) was that the volume would be 2.5–3 mt in 2008, 8–9 mt in 2009, 14–15 mt by 2011, and possibly reaching 30 mt in 2014.[109] According to Rosneft's 2010 annual report, Vankor began its commercial production in July 2009 and the production recorded was 92.9 million barrels (roughly 12.7 mt) in 2010, 3.5 times greater than the figure for 2009.[110] The company has also launched a workshop for the production of diesel fuel, a gas treatment unit, and three oil pumping stations. Average daily crude oil output at the field in July 2010 exceeded 36,000 tons (over 263,000 barrels), and there were 102 producing wells at 13 well pads. The development plan for Vankor now foresees a production plateau in 2014 when output should reach 25 mt[111] much lower than the initial projection of 30 mt peak production in 2014.

Yurubcheno–Tokhomskoye by Rosneft and Kuyumbinskoye by Slavneft. In addition to Vankor field, there is another major oil field in the southern part of Krasnoyarsk Territory named the Yurubcheno–Tokhomskoye field. Yurubcheno–Tokhomskoye field in the Evenki Autonomous District is set to be another of the main sources for the ESPO oil pipeline. The exploration and production licenses to the Yurubcheno–Tokhomskoye field, as well as an exploration license to the Agaleevskoye gas condensate field in East Siberia, belong to OJSC ESOGC which was established in April 1994 to succeed the state companies Eniseineftegazgeologia and Eniseigeofizika. ESOGC, formerly owned by Yukos, was acquired by Rosneft at an auction held in May 2007. At the end of June 2009, ESOGC received the approval of the State Expert Examination Committee for the design of integrated field facilities at the pioneer section of the field. Rosneft plans to develop the Yurubcheno–Tokhomskoye field in several stages:

> ... in the preparatory stage production drilling will be started, and necessary field infrastructure and a 600-km pipeline to Taishet (the starting point of the ESPO pipeline) will be built. Due to the complex geological structure of the area and the lack of transport infrastructure, Rosneft argues that full-scale field development requires substantial investment and should therefore receive tax benefits.[112]

According to the Krasnoyarsk Territory's main economic development and planning department, the field has recoverable C1 reserves of 64.5 mt of oil, C2 reserves of 172.9 mt and C1+C2 reserves of 387.2 bcm of gas. (According to another projection figure by the Administration of Krasnoyarsk Territory, Yurubcheno–Tokhomskoye's geological oil reserves are 897.8 mt and the extractable reserves are 359.2 mt).[113] During the initial stages of production in 2009–15, 3.5 mt of oil are to be produced at the field annually. Once the field reaches commercial maturity, 20 mt of oil will be produced annually. Vostsibneftegaz, a subsidiary of Rosneft which is technically the license holder, will be the contractor for the development work. Rosneft called for tenders for the field's development. The winner, Tomsk NIPI, will be in charge of developing systems for external transportation to the field and was to fulfil this work between 1 January 2009 and 25 December 2010. The projected start of commercial oil production was 2012 but Rosneft has postponed it to 2013.[114]

Slavneft is developing the Kuyumbinskoye deposit, with (mainly) confirmed commercial oil reserves of 78 mt (and estimated recoverable oil reserves of 150 mt). It plans to conduct trial operations between 2007 and 2011, and it appears that some preparation work has indeed been done. By mid-2005 Slavneft had drilled two new exploratory wells and performed workovers at five old wells at Kuyumbinskoye field. Meanwhile, annual production at the field has been roughly 15,000 tonnes per year.[115] Slavneft claims to be the second largest investor in Eastern Russian oil and gas reserves. Its affiliated company Slavneft–Krasnoyarskneftegaz[116] owns licenses for investigation and extraction of hydrocarbons at eight license blocks in the south of the Krasnoyarsk region – Kuyumbinsky, Tersko–Kamovsky, Kordinsky, Baikitskuy, Tukolano–Svetlaninsky, Abrakupchinsky, Chambinsky, and Podporozhny. The company estimates capital investments on development of the sites by 2025 at 185.668 billion rubles, including 14.325 billion on exploration works during the 2007–2011 period.

During the 2009–11 period, Slavneft planned 52 wells at the deposit. Trial operations included 44 extracting wells, including 34 horizontal ones, two double-barrelled, and eight controlled-direction injection wells (including two for gas return and two exploratory). Full field construction at Kuyumbinskoye requires the construction of 250 km of modular pipelines, 92 km of main pipelines, 80 cluster well platforms, 641 wells, and 8 DNS processing and separating facilities. Oil production at the Tersko–Kamovsky block should have begun in 2011; at Kordinsky, Baikitsky and Tukolano–Svetlaninsky in 2013; at Abrakupchinsky in 2014; and at Chambinsky and Podporozhny in 2015. Plans call for the laying of a pipeline from Karabula to ESPO by 2013.[117]

Rosneft and Slavneft are planning a joint effort to construct, in two stages, a pipeline which will connect the Yurubcheno–Kuyumbinsk deposits to the ESPO. The first stage would connect to the Karabula station (in the Boguchansk district) and the second to the settlement of Nizhnyaya Poima with a connection to the main Transneft system.[118]

The main producing fields for the ESPO oil pipeline can be summarized in Table 2.11. By 2015–16, the four main fields could supply at least 43.5 mt. When Yurubcheno–Tokhomskoye's production reaches 20 mt/y, the total will reach 60 mt/y. It is a significant volume, but is not big enough to fill the 1.6 mb/d (80 mt/y) of capacity.

Table 2.11: Main Producing Fields for ESPO[119]

		Peak Production timing (year: mt/y)
Vankor, Rosneft	Oil: C1 + C2, 41.5 + 185.4 mt, C3, 98.4 mt; Gas: C1 + C2, 28.2 + 61.6 bcm	2014: 25
Verkhnechonskoye, TNK–BP	Oil: C1 + C2, 159.5 + 42 mt; Gas: C1 + C2, 56 + 105 bcm	2015–17: 9
Talakanskoye, Surgutneftegaz	Oil: C1+ C2, 113.7 mt; Gas: C1 + C2, 31.8 gas	2016: 6
Yurubcheno-Tokhomskoye, Rosneft	Oil: C1 + C2, 64.5 + 172.9 mt; Gas: 387.2 bcm	2009–15: 3.5 After commercial production: + 20

Source: Various sources quoted in the section on ESPO's main production bases in this chapter.

Sakhalin Offshore Oil Development[120]

Research performed by **VNIGRI**[121] has identified a number of offshore liquid hydrocarbon fields with significant potential. Of the hundred potential hydrocarbon accumulation zones identified in five regions (the Pechora shelf, the North Sakhalin shelf, the South Kara shelf, the Russian sector of the Caspian Sea, and the Baltic oil and gas bearing area) over 30 are unquestionably top-priority areas for exploration for liquid hydrocarbons. Fifteen zones, with a concentration of nearly 8 bt of in-place liquid hydrocarbon resources, have already been confirmed. In Sakhalin shelf, the most significant is the Odoptinskaya zone.[122]

Sakhalin-1

The Sakhalin-1 project, as described by its website:

> … includes three offshore fields, Chayvo, Odoptu and Arkutun Dagi. Exxon Neftegaz Ltd (ENL) is the operator for the multinational Sakhalin-1 consortium, with a 30 per cent equity stake. The partners include affiliates of Rosneft (RN-Astra with 8.5 per cent and Sakhalinmorneftegaz-Shelf with 11.5 per cent), the Japanese consortium SODECO with 30 per cent, and the Indian state-owned company ONGC Videsh Ltd with 20 per cent. The Sakhalin-1 project's potential recoverable resources are 2.3 bn barrels of oil and 17.1 tcf of gas (or 307 million tonnes of oil and 485 bcm of gas).[123]

In late January 2004, at its meeting in Yuzhno-Sakhalinsk, the Authorized

State Body (ASB) for the Sakhalin-1 project endorsed a budget for 2004 of almost $1.37 billion. The overall volume of investment in the first phase of Sakhalin-1 was to total roughly $5 billion. Capital investment over the entire life of the project may amount to $12 billion. On 9 February the Russian MNR endorsed a positive conclusion reached by the State Ecological Examination Agency regarding the feasibility study for construction of facilities as part of the first phase of the Sakhalin-1 project. On 12 April ENL announced that the Russian Federation approved the Technical and Economic Substantiation of Construction (TESC) for Phase 1 of the project. This was a milestone which allowed the Sakhalin-1 consortium to begin construction of the Sakhalin-1, phase 1 facilities, enabling ENL to keep to its plan to produce the first oil from Chayvo in 2005.

The construction of a new oil terminal in De-Kastri port on the Tatar Strait coast began early in 2004. The Russia–Turkey joint venture Enka-Technostroiexport and the Russian firm Trest Koksokhimmontazh carried out the construction. The latter built two tanks, with a total holding capacity of 100,000 cubic metres, for storing crude oil from the Sakhalin-1 project at the terminal before being loaded into tankers for transport.[124] In 2005, Sakhalin-1 project participants began extraction of early oil for domestic market delivery, and commenced delivery of associated gas via pipeline to Khabarovsk.[125]

The De-Kastri oil shipment terminal opened on 4 October 2006. The Khabarovsk Territory Governor, Viktor Ishayev, announced the completion of the construction of a 24 inch pipeline from onshore processing facilities at the Chayvo field which was expected to offload approximately 12.5 mt of oil annually. The De-Kastri terminal oil storage tanks and offloading facilities could accommodate tankers of 110,000 deadweight tonnes. Year-round oil transportation by conventional tankers would be possible due to the availability of escorting ice-breakers. Trials involving the Primorye double-hull tanker of 105,000 deadweight tonnes escorted by two ice-breakers were conducted in the waters of the Tatar Strait, Aniva Bay, and La Perouse Strait in the winter of 2002. Oil production under the Sakhalin-1 project began in the third quarter of 2005.[126] In February 2007, the project entered into full-scale production, with a production capacity of 250,000 b/d.[127] The first phase of Sakhalin-1 consists of an onshore drilling rig with numerous extended-reach wells and an offshore drilling and production platform. Standing 70 metres tall, the Chayvo land-based drilling rig, Yastreb, is the largest and most powerful land rig in the industry, designed to withstand earthquakes and severe Arctic temperatures.[128]

The Sokol crude from the Sakhalin-1 project has an API gravity of 37.9 and sulphur content of 0.23 per cent, making it light sweet crude with a high yield of middle distillates and gasoline.[129] It is very similar to Sakhalin-2's Vityaz crude, but slightly lighter, according to the data. Vityaz has regularly traded at premiums of between $4 and $7 a barrel compared with Oman prices, and is usually above the price of Murban. It was expected that the brunt of the impact of supplies from Sakhalin-1 would be borne by Abu Dhabi's Murban crude, the grade most similar to Sokol, which would probably have to seek buyers as Japanese refiners with limited growth prospects cut their existing imports to make room for two-thirds of the output of the new Russian grade. Japan had been the biggest buyer of Murban, buying about 40 per cent of the UAE's total output. Murban, coveted for its high yield of distillates such as jet-kerosene used for heating in winter, was valued at about $3.60 higher than benchmark Dubai, but traders warned that this premium would fall with the availability of Sokol. The first Sokol cargo went to ExxonMobil's joint venture Tonen General refinery in Japan. All of ExxonMobil's 30 per cent share, of about 75,000 barrels per day, was expected to follow the same route. SODECO's 30 per cent share of output was also expected to go to Japan, which already imported over 90 per cent of Vityaz crude oil. Even if 150,000 barrels per day of Sokol headed to Japan, however, this would reduce its dependence on the Middle East by less than 4 per cent. For the remaining 100,000 barrels per day, shipping restrictions vied with long-haul sales, as the De-Kastri terminal in Khabarovsk was only able to accommodate Aframax vessels of up to 110,000 tonnes. At one time, it was expected that India's ONGC would get marketing rights to Rosneft's 20 per cent under a loan deal, and might face political pressure to bring equity oil back home.[130]

Crude production from the Chayvo oilfield was expected to fall to around 150,000 b/d in 2010, down from about 165,000 b/d in 2009, as production steadily declines from its peak. The Sakhalin-1 project includes the phased development of three fields. The third field in the development, Arkutun-Dagi, is expected to start production by 2014 and is believed to have peak capacity output of about 80,000 b/d, but nothing has been finalized.[131]

Sakhalin-2
Sakhalin-2 is the world's biggest integrated oil and gas project.[132] It was built from scratch in the harsh subarctic environment of Sakhalin Island in the Russian Far East. Phase 1 involved first oil production from an offshore platform, Molikpaq, installed at the Piltun–Astokhskoye field in 1999. Phase 2 included the installation of two further platforms, 300

Map 2.3: Sakhalin Offshore Blocks

Source: Stern (2008), 231.

km of offshore pipelines connecting all three platforms to shore, more than 800 km of onshore oil and gas pipelines, an onshore processing facility, an oil export terminal, and the construction of Russia's first liquefied natural gas plant.[133]

The project's website provides more details:

The Molikpaq offshore platform was installed during Phase 1 in 1998. It was a converted drilling rig that was first used in Arctic offshore waters of Canada. In 1998, the Molikpaq was towed from the Beaufort Sea in the Canadian Arctic across the Pacific Ocean to Korea where it was upgraded for the Sakhalin II Project. It was then towed from Korea to Russia where a steel 'spacer', manufactured by Amur Shipyard was fitted to the bottom of the Molikpaq so that it could be used in the deeper water offshore Sakhalin

Island. The structure was specifically built to operate in severe ice conditions … Till December 2008 the PA-A platform was the heart of the Vityaz Production Complex, which also included the doublehull Okha FSO vessel, a Single Anchor Leg Mooring (SALM) buoy and a sub-sea pipeline. The Vityaz operations included drilling, production and offloading of oil and the associated support and exploration activities. Oil production occurred only during a six-month period when ice cover was not prohibitive to navigation and SALM operation. During the ice period, from approximately December to May, the FSO was disconnected from the SALM and left the Okhotsk Sea. As the PA-A platform has no oil storage facilities, oil production was suspended till the end of the ice period.[134]

In 1999, Vityaz crude oil sales to Asia–Pacific countries were 1.06 million barrels. The annual sales in the years from 2000 to 2006 were 12.4 mb, 15.0 mb, 10.8 mb, 10.3 mb, 11.7 mb, 12.1 mb, and 11.6 mb. Initially the majority of Vityaz crude was supplied to South Korea, but from 2003 a substantial volume went to Japan. In 2005 the volume allocated to Japan was over 10 mb.[135] On 22 July 2009 Sakhalin Energy exported its two hundredth oil cargo. The *Sakhalin Island*, a PRISCO-owned tanker on long-term charter to the company, delivered some 100,000 tonnes of oil from the Company's offshore facilities in the Sea of Okhotsk to a refinery in South Korea.[136]

According to the Sakhalin Energy website:

… by the end of 2009, only a year after production was launched on Piltun–Astokhskoye-B (PA-B), the platform produced 10 million barrels (1.4 million tonnes) of oil, while also achieving Top Quartile drilling perform-ance. By putting its sixth well into production the platform's production capacity reached nearly 60,000 b/d of oil per day, making it the second best Piltun well drilled to date. By the end of 2009, Sakhalin Energy had delivered 59 oil cargoes and 81 LNG cargoes, as compared to the initial targets set of 53 and 55 cargoes respectively.[137]

and *Reuters* reports that:

… according to Sakhalin-2, condensate associated with gas production that feeds the Sakhalin-2 LNG will account for almost a third of the Vityaz blend by year-end, after the second LNG train is brought on stream. Sa-khalin Energy's spokesperson said that 'by the end of 2009, it is expected that the condensate ratio in the mix will rise to around 30 per cent'. The API gravity for Vityaz has already risen from around 34.3 degrees when it came on stream in 1999 to 38–39 degrees, yielding lighter products such as gasoline and naphtha. The quality of Vityaz could improve further when the second LNG train comes online later in 2009.[138]

Sakhalin-3, 4, 5, and 6

Sakhalin-3 block became a prime target for Rosneft and Gazprom, as IOCs are blocked from entry to this promising block. Rosneft and SINOPEC are involved in the exploration of the Veninsky licensed block of the Sakhalin-3 project. The Sakhalin website reports, in 2006:

> ... the first prospecting well was drilled, tested and liquidated at the South Ayashskaya structure of the Veninsky block. As a result, prospective oil and gas-bearing strata were revealed and tests confirmed the presence of hydrocarbons. Drilling was carried out with the help of the Kantan-3 floating semi-submersible drilling rig owned by the Shanghai Offshore Drilling Company.[139]

According to Rosneft, drilling of North Veninskaya well No.2 at the Veninsky area in 2009 enabled a more accurate estimate of reserves at the North Veninskoye field, which was discovered in 2008 thanks to drilling of a first well. Russian C1 and C2 reserves at the North Veninskoye gas condensate field are estimated at 34 bcm of gas and 2.8 mt of condensate. Drilling of Veninskaya well No.3 discovered the modestly-sized Novoveninskoye oil and gas condensate field.[140] In summer 2009, Putin invited Shell to help develop the giant Sakhalin-3 and Sakhalin-4 projects.[141]

In the case of Sakhalin-4 and Sakhalin-5, Rosneft chose BP as its strategic partner in 2003. Project operational management rests with the joint venture Elvary Neftegaz formed on Sakhalin in 2003 for work on the Kaigansko-Vasyukansky site, contained within Sakhalin-5. On 22 November 2006, Rosneft and BP signed a joint stock operational agreement on the development of Vostochno-Shmitovsky (Sakhalin-5) and Zapadno–Shmitovsky (Sakhalin-4) license blocks. BP was to finance the exploration works on both sites, including drilling of six wells down to commercial depths, subsequently recovering expenses from Rosneft's extraction share. The subsoil users of Vostochno–Shmitovsky and Zapadno–Shmitovsky are 100 per cent Rosneft affiliated companies (Vostok-Schmidt Neftegaz and Zapad-Schmidt Neftegaz). In addition, on 29 November 2006 Rosneft and BP signed a memorandum providing for joint research on the Russian Arctic regional basins and the defining of mutual hydrocarbon exploration interests.[142]

In February 2008, however, BP announced the closure of its Sakhalin Island office after it decided to scrap further drilling at Sakhalin-4 and Sakhalin-5. BP said its joint venture with Rosneft, Elvary Neftegaz, would only carry out geological exploration at Sakhalin-4 and Sakhalin-5, where drilling had not been seen as successful.[143] Disappointed with the exploration results of the Sakhalin-4 contract area, Rosneft and BP decided to abandon the project. The five-year exploration license for the acreage expired in November 2008, and the partners decided to return it to the government rather than apply for an extension, after the two wells drilled on the block were dry. The Rosneft–BP venture also explored the neighbouring Sakhalin-5 area, where hydrocarbons were discovered in quantities insufficient to make their development economic.[144]

Within the Sakhalin 5 project, Rosneft and BP jointly carried out exploration activities at the East Schmidt and Kaigansko–Vasyukansky licensed blocks on Sakhalin shelf. During the period 2004–7, a significant number of 2D and 3D seismic, engineering, and environmental studies were carried out at the East Schmidt block. As a result, 12 prospective structures have been revealed. The license was extended until 2013.[145]

There have been reports that some European companies were interested in the Sakhalin-3 project. Antonio Brufau, CEO of Repsol YPF made it clear that the company was interested in participation in the Sakhalin-3 project. Repsol applied for 25 per cent of the deposits in the Veninsky license block. If the negotiations succeed, Repsol will cooperate with Rosneft. It is estimated that Repsol could acquire the target equity by payment of $250–400 million.[146] Another report suggested that the Norwegian state oil and gas company Statoil was also interested in the Sakhalin-3 project. Gazprom Neft, the oil subsidiary of Gazprom, started activities under a Memorandum of Understanding signed on 2 April 2007 with Statoil. On 17 July 2007 Gazprom Neft asked Statoil to take part in the geological prospecting and further development of the Lopukhovsky block on the Sakhalin shelf (on 5 June the two companies had created a working group to study joint opportunities). The invitation became possible after the commission of Rosnedra under the MNR decided to extend the license terms for geological exploration at the site until 2010. The previous term for exploration of the block had expired on 31 May 2007. In September 2006 the Federal Natural Resources Management Service (Rosprirodnadzor) of the MNR inspected Gazprom Neft's performance under the license agreement and found a number of infringements. Gazprom Neft held the opinion that Lopukhovsky was not a priority but changed its stance after Gazprom's decision to take up the controlling stake in the Sakhalin-2 project.[147]

On 31 May 2007 Gazprom's deputy CEO, Alexander Medvedev, announced that the firm wanted to become the largest subsoil user on the Sakhalin shelf, remaining interested in the Sakhalin-3 and Sakhalin-4 projects, as well as the development of the Lopukhovsky block, which covers 3,500 square metres, with hypothetical reserves of 130 mt of oil and 500 bcm of natural gas. The site belongs to TNK–Sakhalin,

75 per cent of which belongs to TNK–BP. In 2005, Gazprom Neft purchased TNK–Sakhalin by payment of $70 million (according to industry estimates). The remaining 25 per cent is held by the Sakhalin Oil Company (SOC), which is controlled by the Sakhalin regional administration. According to Alexei Romanov, SOC's general director, in 2006 Gazprom Neft had revised their interpretation of the data from the Lopukhovsky block's geological exploration. During the 2003 field season, 3D seismic prospecting shot an area of 2,335 square kilometres, about 75 per cent more than called for by the license obligations. The data analysis conducted in 2004–5 by TNK–BP, with involvement of experts from BP, concluded that geological conditions at the site were essentially less favourable than earlier thought. However, the review of the results by Gazprom Neft was more optimistic. Gazprom Neft presented the MNR with a new programme of geological exploration for 2007–10 and a justification for an extension of the license.[148]

Unlike Sakhalin-3, 4, and 5, Sakhalin-6 failed to attract big companies. The quest for major discoveries has been left to smaller, more adventurous firms. In 1997, Petrosakh was licensed to produce oil from the Okruzhnoye field for a period of 20 years. In 2001, Petrosakh was licensed for the geological study of the Pogranichnoye Block, which lies directly offshore parallel to the existing Okruzhnoye field. In the summer of 2002, Petrosakh also acquired a 480 square kilometres 3D seismic programme covering the central portion of the Pogranichnoye Block.[149] In January 2004, Rosneft withdrew from the Sakhalin-6 block, disappointed by its exploration results on a block initially believed to contain reserves of up to 2.2 billion barrels of oil.[150] In November 2004, Urals Energy acquired full control of Alfa Group's 97.16 per cent ownership interest in Petrosakh. In September 2011 Urals Energy announced that it was plugging and temporarily abandoning its Well #51 at Petrosakh on Sakhalin Island because of difficult drilling conditions.[151] Since three blocks in Sakhalin-6 were not yet awarded,[152] it is safe to say that Sakhalin-6 block remains to a great extent an unexplored frontier.

Conclusion

The key question for East Siberia and the Russian Far East's oil development is: what is the ultimate capacity of oil exports from the region to north-east Asian economies? The answer to that question requires an understanding of the real production capacity of the Irkutsk region, Sakha Republic, and the Krasnoyarsk region, the three main production bases for ESPO. This chapter has attempted to supply that understanding. The answer has a direct link with the energy strategy of Russia. In 2009, the Russian government approved a new Energy Strategy for the period up to 2030, and abandoned the strategy which it had approved in 2003. According to the revised strategy, by 2030 oil production will reach 530–535 mt and crude and petroleum product exports will grow to 329 mt. Crude production from East Siberia will account for 14 per cent of the total or 69–75 mt. Of this, 21–33 mt applies to the first phase, roughly until 2015, and the extra 41–52 mt to the second phase.[153] It is a real question whether this ambitious projection is realistic or not. In the judgement of the author, the target figure of 69–75 mt by 2030 is realistic, but only as long as the intensification of exploration work in East Siberia is continued.[154]

According to an Energy Ministry source, Russia can achieve annual production of 550 mt by 2020 only by putting new fields in the Yamal-Nenets Autonomous District and the north Krasnoyarsk territory on stream, for East Siberia does not contain enough resources on its own. The Yamal–Krasnoyarsk programme includes the Vankor field in the Krasnoyarsk territory, the Tagulskoye, Suzunskoye, and Vostochno–Messoyakhinskoye–Komsomolskoye fields in Yamal-Nenets district and some others. On 5 April 2010, the Russian government's fuel and energy commission approved a programme for the comprehensive development of the Yamal-Nenets Autonomous District and the north of Krasnoyarsk territory.[155]

Grigory Vygon, the director of the Department of Economy and Finance of the Russian Ministry of Natural Resources and Ecology, has said that East Siberia is a promising new region for future oil production, depending upon a number of important decisions, including a change in the tax system. Presently the resource base in the area of the ESPO export pipeline is such that loading for the first and second

Table 2.12: Needed Resource Base in the ESPO Area

	Capacity (mt)	Implementation Deadline	Needed reserves (mt)
ESPO 1 (Taishet–Skovorodino)	30	2010	600
ESPO 2 (Skovorodino–Kozmino)	30	2014	600
ESPO 2 (increasing capacity)	50	2016*	1,000
ESPO 3 (increasing capacity)	80	2025*	1,600

* before postponement of date

Source: Russian Ministry of Natural Resources and Ecology, quoted in Vygon (2009), 10.

stages is available. There are, however, a number of problems. Many deposits located in East Siberia and in the Republic of Sakha, especially the deposits of the Vankorskoye and Yurubcheno–Tokhomskoye centres, are far from transport infrastructure. Hence production from these deposits will not move into the pipeline immediately. In addition, there is uncertainty about transport tariffs. The tax system is also a disincentive to investments in the region. The auction system in East Siberia is practically inoperative and geological exploration is scarce. To improve the situation, the abolition of export duties for East Siberia is under study. In Vygon's opinion, to exclude the possibility of there being firms which have hydrocarbon deposits but no way of exploiting them economically, it is necessary to create a complex programme for the development of East Siberia, in which all activities should be coordinated – programmes of geological study, the development of transport infrastructure, and plans for oil production.[156] This idea confirms that there is a problem in coordinating a long term state programme which includes the development of transport infrastructure and oil production.

The main fields mentioned and discussed earlier as ESPO production bases indicate that the combined production of the main fields in East Siberia, the Sakha Republic, and Krasnoyarsk Krai will not be big enough to fill the 1.6 mb/d pipeline, even though the figure of 1.0 mb/d can be easily achieved. The demand for ESPO crude from north-east Asia's three major oil importers (China, Japan, and South Korea) is, however, very strong and the Moscow authorities should not hesitate to go ahead with the second stage of the ESPO project, as long as the proven reserves are big enough to reach the 1.6 mb/d capacity. The combined peak production capacity of Vankor, Verkhnechonskoye, Talakan, and Yurubchonskoye fields will be close to 45 mt by the mid-2010s, but the figure of 50 mt can easily be met by adding the production from the oil fields close to the first phase section of the ESPO. Even if Yurubcheno–Tokhomskoye reaches its peak production in the early 2020s, the combined volume will be well below 80 mt.[157] Considering that only a limited number of frontier areas in East Siberia and Krasnoyarsk Krai have been explored, there is a good chance of finding more sizable oil fields in the coming decades. If, however, there is no breakthrough in finding oil fields with more than 50 mt proven reserves, Moscow's top decision makers will face a serious dilemma and the diversion of a sizable volume of crude supply from West Siberia will have to be continued.

The implications of Russian oil supply to the Asia–Pacific region through the ESPO are large. For the first time, Russia has managed to establish a long-distance pipeline network connecting its frontier oil fields in East Siberia to the Russian Far East port of Kozmino, near Nakhodka. It took four decades to transform this dream into reality. This oil export network in the Asian part of Russia will help Russia's entry to the gigantic Asia–Pacific economy. Secondly, ESPO pipeline development was designed to satisfy the lower (not the maximum) level of crude supply to both China and north-east Asian countries, but the volume of crude supply should be increased to maximize the impact of ESPO crude specifications in the Asia–Pacific oil market. The demand for high-quality ESPO crude will grow, and the expansion of the supply will be dependent on the results of an expanded exploration programme and a tax holiday for East Siberian oil exploration and development. It remains to be seen whether ESPO's upsurge will contribute to reducing the Asian price premium in the coming years.

Appendix to Chapter 2

Table A.2.1: Russian Oil Industry Indicators

	OO (mt)	DC* (mt)	NCOE** (mt)	ED (km)	DD (1,000 km)	IWC (1,000)
1995	311	146	100	1,079	9.9	39.9
1996	303	130	105	1,026	6.8	36.6
1997	307	129	110	1,007	7.0	36.7
1998	304	124	118	798	4.3	35.0
1999	305	126	111	793	4.9	33.1
2000	323	124	133	1,014	8.3	31.9
2001	348	122	140	1,144	9.0	31.5
2002	380	124	155	721	7.7	36.1
2003	421	123	175	681	8.6	36.3
2004	459	124	207	583	8.4	36.8
2005	470	123	204	627	9.2	30.0
2006	481	129	217	723	11.6	27.2
2007	491	126	214	867	13.8	25.8
2008	488	139	205	852	14.6	25.4
2009	494	135	211	464	14.1	25.9

Note: i) OO = Oil Output; ii) DC = Domestic Consumption; iii) NCOE = Non-CIS oil exports; iv) ED = Exploration Drilling; v) DD = Development Drilling; vi) DIC = Idle Well Count.

*   Domestic consumption includes refinery fuel and losses.

** Reliable Non-CIS oil export data only includes volumes bypassing the Transneft system from 1998 onwards.

Source: RC (2010), 14.

Table A.2.2: Russia's Crude Oil and Condensate Production by Company (mt)

	2003	*2005*	*2007*	*2008*	*2009*
Russia	421.347	469.986	491.306	488.487	494.228
Oil Companies	382.980	432.604	441.118	436.961	438.136
Rosneft	19.568	74.418	101.681	108.345	111.057
Lukoil	78.870	87.814	91.432	90.245	92.179
TNK–BP	61.579	75.348	69.438	68.794	70.236
Surgutneftegaz	54.025	63.859	64.495	61.684	59.634
Gazprom Neft	31.394	33.040	32.666	36.278	35.109
Tatneft	24.669	25.332	25.741	26.060	26.107
Slavneft	18.097	24.163	20.910	19.571	18.894
Russneft	1.985	12.181	14.169	14.247	12.688
Bashneft	12.046	11.934	11.606	11.738	12.234
Yukos	80.747	24.516	8.981	-	-
Gazprom	11.022	12.788	13.154	12.723	12.042
NIP	27.345	24.593	37.035	38.803	44.050

Note: NIP = Non-integrated producers and JVs.

Source: RC (2010), 25.

Table A.2.3: Russia's Oil and Gas Sector Capital Spending ($billion)

	Oil					*Gas*	*Total*
	UP	*RF*	*TN*	*TF*	*S-total*		
1998	2.854	0.431	0.543	0.021	3.849	3.895	7.744
1999	2.352	0.225	0.246	0.015	2.838	2.835	5.673
2000	4.802	0.725	0.429	0.027	5.983	4.538	10.521
2001	6.783	0.998	1.158	0.022	8.961	5.681	14.642
2002	5.530	0.854	0.760	0.063	7.208	5.915	13.123
2003	7.674	0.894	1.628	0.114	10.310	7.450	17.760
2004	9.022	1.411	1.804	0.190	12.332	9.500	21.927
2005	8.664	1.477	2.350	0.220	12.722	10.664	23.376
2006	15.546	2.377	5.338	0.563	24.073	16.242	40.066
2007	21.997	3.432	6.371	0.624	31.768	19.418	51.841
2008	30.810	5.254	6.035	0.228	41.517	24.275	66.602
2009	22.753	7.145	7.383	0.088	37.369	18.803	56.172

Note: UP = upstream; RF= Refining; TN = Transneft; TF = Transnefteprodukt

Source: 1998–2003: RC (2008), 81; 2004–9: RC (2010), 87.

Table A.2.4: Oil Production in Siberia and the Far East up to 2030 (mt/y)

	2010	2015	2020	2025	2030
WS	327	350	352	359	366
ES	8	27	45	55	74
FE	17	19	24	25	26
S1	8	8	7	6	5
S2	8	8	7	7	6
S3–9	0	1	8	11	14
ST	2	2	2	1	1
WKS					
SFE	352	396	421	438	466
RT	488	536	561	600	636

Note: WS = West Siberia; ES = East Siberia; FE = Far East; S1 = Sakhalin 1; S2 = Sakhalin 2; S3–9 = Sakhalin 3–9; ST = Sakhalin Terrestrial; WKS = West Kamchatka Shelf; SFE = Siberia and Far East; RT = Russia Total.

Source: Kontorovich and Eder (2009).

Table A.2.5: Subsoil Users in Eastern Siberia and the Republic of Sakha

	Number of License Blocks	Size of License Blocks, Sq. km	Year in which subsoil user began in region
Evenki Autonomous District			
Vostsibneftegaz	1	5,569	1996
Taimura	1	3	1998
Yukos	4	21,668	2000
Krasnoyarsk Gazprom	1	1,885	2001
Slavneft-Krasnoyarskneftegaz	8	29,327	2002
Khanyaga	2	3,820	2004
ETEK	2	3,113	2005
Krasnoyarskgazdobycha	2	14,616	2005
Kholmogorneftegaz	1	1,600	2005
Gazprom	2	7,116	2005
Rosneft	1	3,596	2006
Sub-Total	**25**	**92,333**	
Krasnoyarsk Territory			
Vostsibneftegaz	1	3,525	1995
Krasnoyarsk Gazprom	2	2,242	2000
Mezhregionalnaya Toplivnaya Kompaniya	4	6,950	2002
Khanyaga	3	3,886	2004
Gazprom	3	14,361	2006
Sub-Total	**13**	**30,964**	

Table A.2.5: continued

	Number of License Blocks	Size of License Blocks, Sq. km	Year in which subsoil user began in region
Irkutsk Region			
Ust-Kut Neftegaz	2	1,366	1996
Rusia Petroleum	1	7,296	1997
Danilovo	1	164	1997
Petromir	3	13,328	1999
Petrosib	2	5,044	1999
Bratskekogaz	1	59	2000
Dulisma	1	1,147	2000
Kovykta Neftegaz	1	3,050	2001
Verkhnechonskneftegaz	1	1,481	2002
Atov-MAG Plus	1	39	2002
SNGK	2	5,100	2003
Gazprom	2	2,676	2003
SibRealGaz	1	3,190	2003
Sayankhimplast	1	457	2003
Baikalgaz	1	2,800	2004
Rosneft	4	11,913	2005
INK-NefteGazGeologiya	5	10,312	2005
Surgutneftegaz	3	6,880	2006
Avangard	2	6,825	2006
Novosibirskneftegaz	2	7,498	2006
Kada-neftegaz	1	5,603	2006
Neftekhimresurs	1	3,594	2006
Sub-Total	**39**	**99,822**	
Sakha (Yakutia) Republic			
Irelyakhneft	2	2,097	1997
Yakutgazprom	2	284	1998
ALROSA-GAZ	1	222	2002
Taas-Yuryakh-Neftegazdobycha	2	1,381	2002
Surgutneftegaz	6	28,597	2003
Kholmogorneftegaz	2	1,558	2005
Sakhatransneftegaz	2	644	2005
Irelyakhneftegaz	1	1,824	2006
Suntarneftegaz	1	1,074	2006
Sub-Total	**19**	**37,682**	

Source: Russian Ministry of Natural Resources, quoted in Baidashin (2007c), 15.

CHAPTER 3

THE DEVELOPMENT OF THE GAS INDUSTRY IN EASTERN SIBERIA AND THE RUSSIAN FAR EAST

Russia's huge resource endowment is a double one. The country not only contains more than 5 per cent of the world's proven oil reserves (outside the Gulf countries only Venezuela has more), but it also has almost one quarter of proven world gas reserves, far more than any other country. Expectations that Russia was ideally placed to fill the shortfall in China's energy supply were even stronger in the case of gas than in that of oil.

This chapter starts with a review of the differences between the gas industry in the European and the Asian parts of the Russian Federation. Based on Gazprom's Asian policy, which focuses on large scale gas exports to Asia, this chapter then explains how the East Gas programme – the blueprint for Russia's Asia export policy – was prepared and developed during the 2000s. The focus then shifts to understanding the capacity of the three main gas suppliers for the eastern unified gas supply system (UGSS), that is, Kovykta gas, Sakhalin offshore gas, and Sakha Republic gas.

The Russian Gas Industry: Divided between West and East

During recent years Russia's oil industry has undergone total reorganization into a number of vertically integrated oil companies. The gas industry has never had a similar experience due to the fact that is it dominated by a single mega giant state-owned corporation, Gazprom. Gazprom State Gas Concern was established in 1989 by the USSR Gas Industry Ministry. In 1993 the Concern laid the foundation of the Gazprom Russian Joint Stock Company, renamed in 1998 the Gazprom Open Joint Stock Company. Gazprom reports that:

> … at the end of 2009, the company's A+B+C1 reserves were estimated at 33.6 tcm and oil and condensate reserves at 3.1 bt. According to the international PRMS standards the company's proven and probable hydrocarbon reserves are estimated at 27.3 bt of fuel equivalent with a value of $230.1 billion. With 17 per cent of global gas production, Gazprom is one of the biggest energy companies in the world and in 2009 its production was 461.5 bcm.[1]

The share of gas production from the Asian part of the Russian Federation is negligible. As shown in Table 3.1, the total production of natural gas from the Irkutsk region and the Sakha Republic is a mere 1.9 bcm, while that of the Sakhalin Islands is 9.2 bcm. But projected gas production from East Siberia and Sakhalin offshore during the 2020s is over 100 bcm/y, an ambitious target, even though enough proven reserves are already confirmed. At present, however, there is no infrastructure to facilitate large-scale gas export. Recent studies have still not included meaningful pipeline investment, although massive investment is needed for the projected gas exports to be realized. As far as natural gas is concerned, Russia has always been two countries: Western Siberia on the one hand and East Siberia and the Russian Far East on the other. There is very little connection between them.

Table 3.1: Russia's Gas Production by Regions, 2008

Regions	*Gas* bcm	*%*
European Regions	**33.7**	**5.1**
West Siberia	**616.7**	**92.8**
Khanty-Mansi Autonomous Area	35.8	5.4
Yamal-Nenets Autonomous Area	575.9	86.6
Tomsk Region	4.4	0.7
East Siberia	**5.3**	**0.8**
Krasnoyarsk Region	3.4	0.5
Irkutsk Region	0.1	0.0
Sakha Republic	1.8	0.3
Far East	**9.2**	**1.4**
Sakhalin Islands, Offshore of Sea of Okhotsk	9.2	1.4
Chukchi Region	0.03	0.0
Russia, Total	**664.9**	**100.0**

Source: Kontorovich and Eder (2009).

As shown in Table 3.2, the final version of Russia's Energy Strategy to 2030, released in late November 2009, foresaw total gas production of 885–940 bcm by 2030, of which East Siberia would account for 45–65 bcm and the Far East 85–87 bcm. That would mean that East Siberia and the Far East together would account for 15 per cent of total production. During Phase I (2013–15) total production would be 685–745 bcm and in the second stage 803–837 bcm. Russia expects its exports of natural gas to be 349–368 bcm by 2030 with exports to Asia–Pacific countries being 11–12 per cent of exports in the first

Table 3.2: Natural Gas Production in Siberia and the Far East up to 2030 (bcm/y)

	2005	*2008*	*Phase I**	*Phase II***	*Phase III*
TGP	**641**	**664**	**685–745**	**803–837**	**885–940**
Tyumen R	**585**	**600**	**580–592**	**584–586**	**608–637**
N–P	582	592	531–539	462–468	317–323
O–TB			0–7	20–21	67–68
B V	3	8	9–10	24–25	30–32
Yamal			12–44	72–76	185–220
Tomsk R	3	4	6–7	5–6	4–5
European R	**46**	**46**	**54–91**	**116–119**	**131–137**
C S R			8–20	20–22	21–22
Stockman D			0–23	50–51	69–71
East Siberia	**4**	**4**	**9–13**	**26–55**	**45–65**
Far East	**3**	**9**	**34–40**	**65–67**	**85–87**
Sakhalin Island	2	7	31–36	36–37	50–51

Note: TGP = total gas production; N–P = Nadym–Purtazovsky; O–TB = Ob–
    Taz Bay; B V = Bolshekhetskaya Valley; C S R = Caspian Sea Region; R =
    region; D = Deposit.
*   By the end of the first implementation phase of the Strategy, 2013–15;
** By the end of the second implementation phase of the strategy, 2020–2.

Source: Ministry of Energy, Russian Federation 2010.

stage, 16–17 per cent in the second stage, and 19–20 per cent in the third stage. LNG would account for 4–5 per cent of exports in the first stage, 10–11 per cent in the second stage, and 14–15 per cent by 2030.

The projection for large-scale exports made by Kontorovich and Eder in the autumn of 2009 indicated that exports would reach 131 bcm by 2020, of which 62 bcm would be to China, and 183 bcm by 2030, of which 93 bcm would be to China (Table 3.3). Under this projection, China and South Korea together would take 70 per cent of Russia's Asian exports in 2020. However, the official projection from the 2030 strategy has significantly reduced the scale of exports from East Siberia due to a reduction of production from that region. In other words, the 2030 strategy has revised the projection shown in Table 3.3 by raising a question about whether such dominance of China as the core market for Russian gas will be acceptable.

The concept of exporting Russian gas to Asia is seen by many as a new one, originating during the 2000s. In fact, however, plans to

Table 3.3: Natural Gas Export to the Pacific Market: 2010–30 (bcm/y)

	2010	*2015*	*2020*	*2025*	*2030*
From W. Siberia	0.0	0.0	20.0	30.0	30.0
to China via Altai	0.0	0.0	20.0	30.0	30.0
From E. Siberia *	0.0	5.0	72.0	90.0	95.0
to China	0.0	0.0	35.0	50.0	50.0
to Japan	0.0	0.0	7.0	10.0	15.0
to Korea	0.0	5.0	20.0	20.0	20.0
to other countries of Pacific Market	0.0	0.0	10.0	10.0	10.0
From Far East (A) **	13.7	20.5	38.6	52.0	57.9
to China	1.4	1.8	3.0	3.8	4.0
to Japan	8.2	10.8	14.0	14.6	15.0
to Korea	2.7	3.6	5.2	5.9	6.0
to other countries of Pacific Market	1.4	1.8	3.2	4.1	4.3
Sub-total	13.7	18.1	25.5	28.4	29.3
From Far East (B) ***					
to China	0.0	0.0	3.9	6.2	8.6
to Japan	0.0	0.2	2.1	3.7	4.3
to Korea	0.0	0.8	3.3	6.2	7.1
to other countries of Pacific Market	0.0	1.4	3.8	7.4	8.6
Sub-Total	0.0	2.4	13.1	23.5	28.6
From Russia	13.7	25.5	130.6	172.0	182.9
to China	1.4	1.8	62.0	90.0	92.5
to Japan	8.2	11.0	23.1	28.3	34.3
to Korea	2.7	9.5	28.5	32.1	33.2
to other countries of Pacific Market	1.4	3.2	17.0	21.5	22.9

*   Gas main East Siberia – the Far East (including pipeline to China, Korea, LNG plants and terminals);

**   includes Sakhalin, Kamchatka, LNG terminal in Prigorodnoe, and new Sakhalin LNG terminals;

*** includes Gas main 'Sakhalin–Vladivostok–Nakhodka' with taps to China, Korea; and new LNG terminals in the Far East (Primorskii Krai and Kamchatka)

Source: Kontorovich and Eder (2009).

export Soviet gas to Asia both by pipeline and converted into LNG, as well as plans to export LNG to North America, date back to the 1960s. During the Soviet era, gas export projects to Asian markets failed to make progress due to a mixture of political, commercial, and institutional obstacles. Up to 1990 Japan was seen as the only realistic Asian market for Soviet gas, but the absence of a peace treaty between Japan and the Soviet Union after World War II was a serious political obstacle to closer trade relations. An equally serious obstacle to any improvement in bilateral relations has been the territorial dispute over four small islands in the Kurile chain, which were occupied by the Soviet Union after World War II. Japan's tendency to follow the political lead of the USA reinforced the lack of trust between Japan and the USSR during the cold war era, and even in the first decade of the twenty-first century the improvement in relations has been very slow. Political problems translated into a lack of enthusiasm on the part of Japanese gas utilities to import Russian supplies. The attitude of these buyers to Soviet and then to Russian gas contrasted sharply with their much greater keenness to develop LNG trade with south-east Asian and Middle Eastern countries.[2]

Gazprom began to see the potential of the Asian gas market in 1997, and at the 2003 Tokyo Gas Conference the Gazprom CEO, Alexey Miller, announced his vision of the construction of a Unified Gas Supply System (UGSS) in the Asian part of Russia, and its connection with the UGSS in West Siberia; he predicted the completion of the Asian UGSS by the autumn of 2007. However, continuously rising oil prices had made Gazprom overconfident that the massive proven gas reserves would serve as sufficiently reliable collateral against the necessary large-scale borrowing from the banks.

The year of 2008 then witnessed an unprecedented global financial crisis with a huge impact. According to Rosstat, Russia's total gas production fell by 20 per cent, while Gazprom produced 25 per cent less gas during the first half of 2009 compared to the first half of 2008. Implicit Russian gas consumption decreased in the first half of 2009 by 10 per cent. The fall in consumption, however, is modest compared to the expected decrease in GDP of around 10 per cent and in industrial production of around 17 per cent, reflecting a low elasticity of gas consumption to output. Russian export volumes to Europe and the CIS went down by 40 and 50 per cent respectively.[3] The collapse of gas demand in the wake of the global financial crisis gave the Gazprom planners space to review the demand factor with a more long-term perspective, and in this context the importance of Asian market development became clearer. This is the reason why Gazprom started the construction of the Sakhalin–Khabarovsk–Vladivostok pipeline project and gave priority to the development of the Sakha Republic's massive gas reserves, despite the 2008/09 global crisis. This is the core of Gazprom's Asia policy, and it is a totally separate initiative from its previous main activity of exporting towards the European region.

Russia's Gas Production: Main vs New Producing Fields

While the Asian part of Russia is still at a very early stage of natural gas-related infrastructure development, the European part of Russia is witnessing a rapid decline of production from the main producing fields in West Siberia. According to Jonathan Stern, Gazprom's production from currently producing fields and those [in 2005] expected to be brought into production would peak in the late 2000s and decline gradually to less than 530 bcm by 2010. In the 2010s, it was expected that this decline would accelerate, due to depletion of the three main fields, and production would fall to around 340 bcm by 2020.[4] In order to maintain production at around 530 bcm, Gazprom would need 70 bcm of new production capacity by 2015 and 186 bcm by 2020.[5] Stern pointed out that Russia's gas production had been sustained for more than 20 years by three supergiant fields in the NPT region of West Siberia. Due to the expected decline of the production from the three fields, however, the Moscow authorities had to prepare the road map for gas production in the coming two decades. Stern explains:

> Of the undeveloped fields for which Gazprom holds development licences … it has three main development options; the Ob and Taz Bay fields, the Yamal Peninsula fields and Shtokman field in the Barents Sea. The Ob and Taz Bay fields are estimated to have a production potential of up to 82 bcm/y, but production from the first of these has yet to be developed and the Kamennomysskoye will not start until 2015.[6]

So the key question for Gazprom is the timing of investments to open up its next generation of supergiant fields on the Yamal Peninsula. Bovanenkovo, the first of these Yamal fields, is expected to start production in 2012. The output from Bovanenkovo will reach as much as 140 bcm by 2014.[7] As for Shtokman, Gazprom has said that Phase I production will start in 2013 and Phase II in 2014, but Stern considered both dates as over-optimistic by at least three to four years.[8]

The strategic direction of development for Yamal-Nenets Autonomous Area (YNAR) oil and gas extraction is the development of deposits on the Yamal Peninsula. As of 2008, 26 deposits were already prepared, having gas reserves amounting to more than 10 tcm. Output from deposits on the Gydansk peninsula is included in long-term prospects. Annual extraction rates on the peninsula could total 140 bcm of gas.[9] The extraction of gas in the YNAR will define the condition and direction of energy development in Russia. In 2008, 222 hydrocarbon deposits were open in the region, of which 59 had moved into commercial operations, 19 were preparing for development, and 144 were in exploration. One in every ten of these deposits has very high reserves.

The YNAR provides 54 per cent of energy resource production in Russia. The region extracts 90 per cent of all Russian natural gas. During the 35 years of commercial hydrocarbon development, deposits in the region have provided 13.4 tcm of natural gas and 830 mt of crude oil and gas condensate. However, at the federal programme level, potential YNAR reserves are underestimated. In the state programme to replenish its hydrocarbon raw material base, West Siberia is placed only third in priority, after East Siberia and the Russian Shelf. According to the Energy Strategy of Russia, YNAR has a goal to extract 650 bcm of gas by 2020. Exploration conducted in the region over the past four decades has resulted in the discovery of more than 75 per cent of the gas in Russian gas reserves. As discussed earlier, the important strategic value of Russian gas extraction today rests on the promise of the NPT region of the YNAR, the location of the largest gas deposits in the world. The current explored reserves of the NPT area alone amount to more than 24 tcm.[10]

The magnitude of Gazprom's current production task was highlighted in a 2008 press release which set out targets of 610–615 bcm/y by 2015 and 650–670 bcm/y by 2020, and concluded that 'by 2020 new fields will account for around 50 per cent of Gazprom Group's natural gas production'. This would mean that during the period 2008–20, Gazprom would have to bring on stream more than 300 bcm/y of new production capacity, a staggering task even for a company with established reserves as large as Gazprom's.[11]

An in-depth study of non-Gazprom gas producers in Russia highlighted the fact that forecasts of growing third party gas production in the revised form of Russia's Energy Strategy 2030 can be taken as an expression of the Russian government's long term strategic intent, with the market share of non-Gazprom production (NGP) rising to 27 per cent by 2030. James Henderson pointed out that:

> [t]he impact of NGPs on western regions is likely to be even greater, as Gazprom is likely to dominate in the East, with the result that more than 80 per cent of Russia's gas supply growth for western markets could come not from Gazprom but from third party output.[12]

In this context, it should be borne in mind that Gazprom's long-term production plan envisages a big contribution by production from East Siberia. If the scale of production there reaches over 100 bcm in the coming decades, it will have a large impact on the Asian gas market.

To understand how Gazprom's Asia policy has been systematically developed, it is essential to review a number of major energy schemes ambitiously pursued by the Moscow authorities during the 1990s and

2000s. Until the end of the 1990s the Russian Federation was not in a position to implement a comprehensive energy strategy to accelerate the oil and gas development of the Asian part of the country. However, the first comprehensive programme for the Russian Far East was prepared in August 1987 under the full name of 'Long Term State Programme for the Complex Development of the Productive Forces of the Far Eastern Economic Region, the Buryat ASSR, and Chita Oblast to the year 2000'. The second was announced in the Spring of 1991, with the title 'The Concept of Developing Yakutian and Sakhalin Gas and Mineral Resources of East Siberia and the USSR Far East', and was commonly known as the 'Vostok (East)' Plan.[13] In the subsequent two years further plans were rumoured or announced, but none of them implemented. Both the Moscow and the regional authorities did not have the financial, technical, or labour resources to undertake the massive developments envisaged in grandiose plans.

Review of Gazprom's Exports to Asia Policy[14]

In February 1997 Gazprom's then CEO, Rem Vyakhirev, first announced the firm's intention to formulate its Asian export policy. In June 1997 he stated that:

> … a new stage in constructing the Eurasian system of trans-continental gas pipelines, unprecedented in terms of its length and throughput, will be started by laying gas trunk pipelines in the south of East Siberia from the polar fields of the Tyumen Region, thus connecting the gas pipeline systems of the East and West of Russia.[15]

These remarks confirmed that Gazprom had decided to change its policy towards Sino-Russian gas pipeline development, but they gave no hint as to when and how such a plan would be put into practice. They strongly hinted, however, that Russia was promoting both East and West Siberian gas exports to China.

In the words of Jonathan Stern:

> … in late 1998 Vyakhirev announced that Gazprom was pursuing both the Altai project, which proposed west Siberia gas exports to China's east coast through Xinjiang province, and the implementation of the Baikal project, which envisages East Siberia gas exports to the north-eastern provinces of China. The institutional basis for gas development in East Siberia and the Far East changed fundamentally in June 2002 when the Russian government issued Decree No. 975-r, instructing the Ministry of Energy and Gazprom to draw up a programme for a unified system of production, transmission and distribution of gas in East Siberia and the Far East.[16]

In early June 2003, at the twenty-second World Gas Conference, the keynote speech by Miller, Gazprom's CEO, revealed a blueprint for Gazprom's Asia policy. Authorized by the Government to coordinate the establishment of a united system for gas production and transportation, Gazprom began to develop the main lines of its gas exports to Asia. It was chiefly interested in monopolizing the system of gas exports to Asia, thus repeating the history of Russia's gas exports to Europe.

Gazprom began by adopting a systematic approach to penetrating the Asian gas market, in particular that of China, but in late July 2004 PetroChina's decision to terminate negotiations with foreign companies relating to participation in the West–East Pipeline I, was a temporary setback to Gazprom's plans. Immediately after PetroChina's decision, Gazprom, significantly, held a meeting with a Kogas delegation in Moscow at which Gazprom made it very clear that Kovykta gas could not be exported to both China and South Korea. Instead, Gazprom would construct a pipeline leading to Vladivostok, extending it to South Korea via the offshore route. But this announcement was not taken seriously and Gazprom's proposal was regarded as unrealistic.

In late 2004, those responsible for Gazprom's strategic development prepared a discussion package (the East Siberia and Far East Natural Gas Production and Transportation Options Economic Feasibility Study) to help the Interagency Working Group (composed of institutions such as Gazprom, the Ministry of Energy, and the Ministry of Natural Resources) develop a programme for creating a Unified Gas Production, Transportation and Supply System in East Siberia and the Far East, with potential exports to markets in China and other countries in Asia and the Pacific. According to this discussion package, Gazprom's strategic priority was to be the Sakhalin offshore gas development.

By giving priority to Sakhalin gas, Gazprom could override the main argument used by TNK–BP when lobbying the government. TNK–BP had argued that if Russian gas deliveries to China and South Korea had to start by 2008–10 (since this market window would close by 2014 because Asian countries would satisfy their growing demand for gas by importing LNG) by promoting gas supply to China and South Korea from Sakhalin-1, the field most prepared for exploitation, Gazprom could significantly reduce this problem.[17]

At the Tomsk Forum on pipeline problems on 2 March 2005, an officer from the fuel and energy department of the MIE confirmed that the Russian government would, by June, review four options: the West, Central, East, and TNK–BP routes for trunk pipelines in East Siberia and the Far East. Already in April 2005, however, it was reported that MIE was strongly backing the East option. If the East option, supported by both Gazprom and MIE, was chosen, a significant delay in starting pipeline gas export from Russia's Far East to north-east Asia would be inevitable since more investment would be needed.

During his visit to Beijing in March 2006, Putin signed a number of important agreements covering the oil, power, and natural gas sectors. The agreement (Protocol on the Supply of Natural Gas from Russia to the People's Republic of China) to construct two gas pipelines to China, with capacity to move 60–80 bcm of gas annually, formed the centrepiece of Putin's state visit. It was agreed that gas would flow to China in five years, beginning in 2011. This gave the Moscow authorities no choice but to accelerate the finalization and approval of the East Programme. However, the main obstacle of Russia's gas export to China lay in the continuing delay in the negotiation of a price for gas.

At the end of 2009, Gazprom announced that an agreement on the main terms and conditions for natural gas supplies from Russia to China had been agreed by Gazprom Export and PetroChina International, a subsidiary of CNPC.

> The Agreement defines the basic commercial and technical parameters of Russian gas supplies to Chinese consumers. The parties also agreed on furthering active cooperation in the first quarter of 2010 towards an understanding on other terms and conditions that would subsequently lay the foundation for entering into contracts on gas supply from Russia to the People's Republic of China via both routes.[18]

However, the price negotiation made no progress during 2010 and the summer of 2011 passed without the proposed final signing of the contract.

The Eastern Dimension of the UGSS[19]

Gazprom has played a pivotal role in developing the Eastern Dimension of the UGSS. Since July 2002, by which time Gazprom had acquired the status of coordinator of Russia's eastern natural gas development projects, the company has elaborated a special programme, in cooperation with other interested parties. The first draft of the programme, to create the eastern branch of the UGSS, was prepared jointly by Gazprom and the MIE and was approved in March 2003. Two months later, the 'Energy Strategy for the period up to 2020' was approved by the Cabinet. At the heart of this grandiose plan, according to Energy Minister Yusufov, are energy safety, the maintenance of Russia's geopolitical position in world markets, and a 50 per cent reduction of the energy intensity of the Russian economy.[20] In early June 2003 the Gazprom CEO, Miller, introduced, for the first time, this eastern dimension of the UGSS into his keynote speech at the World Gas Conference in Tokyo.

The Eastern (Gas) Programme (EGP)

Preliminary work

On 15 June 2007 the government's Commission on the Fuel and Energy Sector (FES) instructed MIE to approve the East Programme. In the middle of July, Gazprom and Transneft announced the creation of a working group to consider the construction of a natural gas pipeline which would run parallel to the ESPO crude oil pipeline, reflecting the structure of the East Siberian hydrocarbon deposits which are rich in associated gas and gas condensate. During the FES session, one Gazprom deputy CEO, Ananenkov, said the gas extracted by the Sakhalin-1 project should be used only to meet Russian domestic needs, and later another deputy CEO, Alexander Medvedev, said that the Sakhalin-1 operator's plan to build a gas pipeline to China with an annual capacity of 8 bcm would be unprofitable. Medvedev further announced that it was Gazprom's intention to purchase all Sakhalin-1 gas for delivery to Sakhalin-2 for the production of LNG.

On 10 July 2007, at a Beijing meeting of the energy cooperation sub-committee of the Russia–China commission which regularly arranges meetings of government leaders, the deputy Minister of MIE, Andrey Dementyev, argued that the East Programme would cost 2.4 trillion rubles ($93.2 billion) and that more than 290 billion rubles would be required for exploration work to raise East Siberia and Far East gas reserves by 7 tcm.

Ananenkov of Gazprom said that gas deliveries to China should come primarily from the Sakha gas extraction centre, but added that an effective alternative would be to mix gas from the Chayandinskoye deposit with gas from Sakhalin. He claimed that it was not necessary to provide for a surplus of gas from eastern to western destinations. The plan calls for 35 bcm annually in reserves to be used for this purpose. Ananenkov also announced that Gazprom was prepared to begin work on the Chayandinskoye deposit in 2007–8, adding that 'if we start working there by the first quarter of 2008, we can commission the deposit in 2016'.[21]

Approval

On 3 September 2007, the Russian Minister of Industry and Energy,

Viktor Khristenko, signed approval order number 340, creating the EGP – a uniform system of gas extraction, transport, and supply for East Siberia and the Russian Far East, which also provides for possible exports to the Chinese and other APR markets. Total programme extraction and processing investments to 2020 was planned to total 1.3 trillion rubles (US$51 billion), rising to 2.4 trillion rubles (US$94 billion) if transportation and storage charges were included.[22]

Gazprom developed the idea of the EGP under the aegis of the MIE over a five year period. In 2002, the Russian government had approved a preliminary development of the plan. Initially there were 15 variants of the plans for East Siberia and Far East development, some with and others without use of the UGSS. Gazprom supported the Vostok-50 variant that included the construction of a gas pipeline parallel to the ESPO oil export pipeline. To Gazprom and MIE planners, as shown in Table 3.4, Vostok-50 was the most effective variant, which provided for the export of 50 bcm of gas to APR countries by 2030. This variant also assumes that gas extraction will begin on the Chayandinskoye deposit in 2017.

According to the EGP, full geological exploration of East Siberia and the Far East will require the expenditure of more than 290 billion rubles, and the extraction and processing of gas will cost about 1.3 trillion rubles. In addition, over 800 billion rubles is required for new gas pipelines, the main one going to China and South Korea, paralleling the ESPO project. Gazprom will be the main investor. Currently the firm's capital expenses amount to over 300 billion rubles a year, so the 2.4 trillion rubles over the next 27 years is a realistic figure, according to Alpha Bank analyst, Konstantin Batunin.

During the preparation of the EGP it was decided that, due to their uniqueness, the Kovyktinskoye gas condensate deposit and the Chayandinskoye oil and gas condensate deposit could not undergo development until suitable raw material processing and petrochemical production methods were developed. Anatoly Yanovsky, director of the MIE department of state energy policy, has said that the rates of development at the huge Kovyktinskoye deposit will depend on solving the helium-recycling problem, and this relates to the need for creating petrochemical complexes.

Gas resource development in East Siberia will take place through the creation of four production centres:

- the Sakhalin Centre: based on Sakhalin Island shelf deposits, which will be responsible for gas supply to the Sakhalin, Khabarovsk, and Primorsk areas and the Jewish Autonomous Region, as well as exports

Table 3.4: The Vostok-50 Scenario

	Without UGSS	*With UGSS**
Annual gas production by 2030 (bcm)	120.8	162.3
Sales in Russia	70.8	112.3
Export, including	50.0	50.0
to China	38.0	n.a
to Korea	12.0	n.a
General capital expenditure till 2030 (US$bn)**	60.08	84.8
including exploration	8.32	10.1
gas production	19.04	45.3
gas processing	10.44	45.3
transport capacities	21.37	27.9
storage of gas and helium	0.91	1.4
General capital expenditure till 2030 (US$bn)		
in gas production	29.6	n.a
in gas transportation	21.1	n.a
General income of projects till 2030 (US$bn)		
Accumulated discounted cash flow		
in gas production	1.341	2.557
in gas processing	2.05	2.696
transportation projects	0.827	1.390
Discounted proceeds to budget (US$bn)	20.8	n.a

*   UGSS = Unified Gas Supply System of Russia, currently without connection to Eastern Siberia.
** = prices as of 2006 (exchange rate: $1 = 28.76 rubles)

Source: Eastern Gas programme, Russian Ministry of Industry and Energy, quoted in Baidashin (2007e), 12.

of piped gas and deliveries of pipeline gas and LNG to the APR region;

- the Yakutsk Centre: based on the Chayandinskoye deposit, which will gasify southern areas of the Republic of Sakha (Yakutia) and the Amur area, and provide for pipeline exports to the APR;
- the Irkutsk Centre: located on deposits in the Irkutsk area, which will supply gas to industrial consumers in the Irkutsk and Chita areas, also to the Republic of Buryatia, and, if necessary, will deliver gas to the Russian UGSS;
- the Krasnoyarsk Centre: which will be located on deposits of the Krasnoyarsk region and will supply gas consumers of this area, and, if necessary, gas deliveries to the UGSS.

According to Yanovsky, annual gas delivery estimates for the Eastern

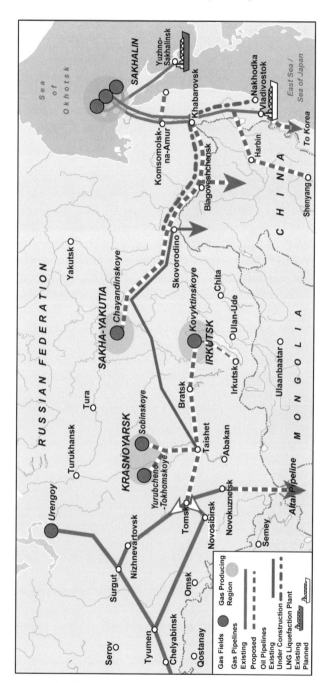

Map 3.1: Eastern Gas Programme

Source: Stern (2008), 252.

Russian domestic market will exceed 27 bcm by 2020 and 32 bcm by 2030 (if the development of gas processing facilities is taken into account, the figure would be 41–6 bcm by 2030). Therefore, the gas portion of the regional fuel balance will increase more than six fold during the period, reaching 38.5 per cent by 2030. To provide the required reserve increases by 2030, a sum of more than 290 billion rubles is required for necessary exploration work. Gazprom has stated that the expected volumes of extraction and gas processing would require investment of nearly $50.92 billion.

Forecasts of demand for Russian gas in APR countries from 2020 call for pipeline deliveries to China and South Korea of between 25 and 50 bcm, while the volume of Russian LNG to the same countries in 2020 would be 21 bcm, rising to 28 bcm by 2030. China, however, is not prepared to buy gas at the European price: hence the years of negotiations which have taken place with CNPC. China has recently agreed a contract for the purchase of a significant volume of Turkmenistan gas at a price of $90 per 1,000 cubic metres.[23]

According to EGP, the initial stage of Sakhalin-1 should deliver gas to consumers on Sakhalin and in the Khabarovsk Territory, utilizing its own deposits and the existing gas transport system. For the gasification of Primorsk and export deliveries of network gas to China and South Korea, the programme planned to construct a Sakhalin–Vladivostok pipeline. The plan also called for the construction of a gas processing plant in the Khabarovsk territory. EGP-stipulated actions for the production of hydrocarbon raw materials and helium in East Siberian and Far Eastern deposits will allow commercial petrochemical production at a volume of not less than 4.5 mt, and gaseous chemical production of not less than 9.1 mt, by 2030.

In February 2010, Gazprom's board of directors approved the work being done by Gazprom to execute the development programme for an integrated gas production, transportation, and supply system in East Siberia and the Far East. The board meeting highlighted a long list of priority projects[24] that should be executed namely:

- the construction of a Sakhalin–Khabarovsk–Vladivostok (SKV) gas transmission system;
- gas supply to Kamchatka Oblast;
- acquisition of Kamchatgazprom;
- the synchronization of commissioning of the Kirinskoye field with the project for SKV construction;
- the continuation of geological exploration at Kirinsky, Vostochno–Odoptinsky, and Ayashsky blocks offshore Sakhalin, the

Zapadno-Kamchatsky prospect, other developments in the Krasno-yarsk Krai and Irkutsk Oblast;

- further development of Gazprom's mineral resource base in the Republic of Sakha (Yakutia), with the aim of bringing on stream Chayanda oil rim in 2014, and the gas deposit in 2016.
- the Yakutia–Khabarovsk–Vladivostok gas transmission system (launched in 2012);
- the necessary preliminary and project work to establish the gas chemical and gas processing facilities in East Siberia and the Far East (already carried out).

This project list forms the cornerstone of the Eastern Gas Programme, and approval by the 2010 board meeting was the first step towards the implementation of the programme.

The Gas Industry Development Strategy to 2030[25]

On 7 October 2008, the MIE's press office (citing a draft Gas Industry Development Strategy to 2030) revealed that the country could boost gas production by between 34 and 50 per cent in 2030 compared with 2007, in other words to between 876 and 981 bcm. (See Tables 3.5 and 3.6.) Gas exports to all destinations could rise by between 69 and 80 per cent to between 415 and 440 bcm, while gas consumption in Russia itself would rise by between 18 and 31 per cent to between 550 and 613 bcm. Capital expenditure by the gas industry between 2008 and 2030 was estimated to be between 13.9 and 16.6 trillion rubles in 2008 prices, not including gasification costs. For these targets to be reached, proven reserves growth of approximately 26 tcm were required, more than half of which would be achieved on the Arctic Shelf (14 tcm) and a quarter in West Siberia (7 tcm onshore).

During the period 2008–30 the expansion plans include: between 4,400 and 5,100 wells; 54–64 gas processing facilities with a capacity of 659–770 bcm annually; 379–422 booster compressor stations with a total capacity of 5,200–6,000 MW; 8–12 platforms; 21,300–27,300 km of linear gas pipeline sections; and 129–159 compressor stations with a combined capacity of 12,500–15,500 MW and capacity to process 195–234 bcm of gas and 26–32 mt of liquid hydrocarbons annually.

This programme is based on the Economic Development Ministry's economic forecasts of an annual average GDP growth of 6.2 per cent in the period to 2030, with the energy intensity of GDP declining by an annual average of 5.0–5.2 per cent. The Russian Academy of Sciences forecasts that the proportion of gas in fuel and energy resources

Table 3.5: Strategic Gas Balance Indicators to 2030 (bcm)

	2007	*2010*	*2015*	*2020*	*2025*	*2030*
Resources	762	840	922–978	996–1082	1024–1120	1035–1132
Gas Production in Russia	654	717	781–845	850–941	871–974	876–981
in E. Siberia & Far East (East- 25)	12	23	44	77	87	89
Gas deliveries from C. Asia	63	69	70–82	70–82	70–86	70–87
Allocation	762	840	922–978	996–1082	1024–1120	1035–1132
National Economy	467	517	520–542	537–581	543–606	550–613
Asia-Pacific & USA	0	8.7	32–61	74–114	91–122	91–122
Mains gas to Asia-Pacific (Eastern Programme)			9	25–50	25–50	25–50
LNG for US, Europe, Asia-Pacific		8.7	23–52	49–89	66–97	66–97
Russia	417.9	464.8	465–485	480–523	485–548	491–555
of which E. Siberia & Far East	11.8	13.9	21	31	34	36
By Russian Region						
Siberia FD	15.5	17.6	18–22	19–24	19–29	20–30
Far East FD	9.2	9.8	10–12	17–19	18–22	19–23

Source: Russia & CIS OGW, 2–10 October, 2008, no. 38 (854), 5.

Table 3.6: Gas Production Forecast by Russian Region (bcm)

	2007	2010	2015	2020	2025	2030
Gas extraction	654	716.9	781–845	850–941	871–974	876–981
West Siberia	557	617	624–688	629–707	631–712	637–719
Yamal, onshore	-	-	78–116	124–177	187–236	250
Yamal, offshore						30–65
Barents Shelf			24	59–71	72–95	72–95
E. Siberia & Far East	11.8	22.7	44	77–108	87–118	89–121

Source: Russia & CIS OGW, 2–10 October, 2008, 6.

will drop from the current 51.5 per cent to about 45 per cent. The strategy assumes that gas prices in the domestic market will provide a rate of return equal to that of exports starting in 2011.

The draft Gas Industry Development Strategy to 2030 pays special attention to building what will be Russia's largest gas chemicals plant, which is in the Amur region, with a throughput of 40–50 bcm. The plant would produce 2.8–4.0 mt of polyolefins annually. Gazprom could commission the plant in 2016–24. There are also plans to build a total of seven gas chemicals plants, of which four will be located in East Siberia and the Far East. Besides the Amur plant, the plan calls for the building of gas chemical plants in the Irkutsk, Krasnoyarsk, and Khabarovsk regions, with gas throughput of 5.5, 12, and 30–40 bcm respectively and polyolefin production of 0.5, 1.0, and 2.7–3.3 mt. The Irkutsk and Krasnoyarsk plants are to come on line between 2014 and 2017, and the Khabarovsk plant is supposed to open between 2013 and 2020. However, the draft plan does not include the Yakutsk gas chemicals plant, the construction of which was planned under the EGP. Sibur-Vostok general director, Denis Solomatin, has said that the petrochemicals giant proposes to drop this project and ship liquid fractions in gas from the Chayanda gas field to the plant near Khabarovsk. He added that Sibur thinks it makes sense to build two gas chemicals plants in the East – in the Irkutsk region and the Khabarovsk territory. The company thinks that only a primary gas processing plant should be built in the Krasnoyarsk territory.[26]

The 2030 energy strategy

In November 2009 the Energy Minister, Sergei Shmatko, said that the Russian government had approved the latest draft version of the country's Energy Strategy for the period until 2030. The Energy Strategy contains three stages: from 2013 to 2015, Russia plans to move past the crisis in the energy industry and create conditions for growth; from

2015 to 2022, the country plans to raise overall energy efficiency on the basis of innovative development in the fuel and energy industry; and from 2022 to 2030 the sector will concentrate on the efficient use of energy resources and will start the transition to non-fuel types of energy.[27] (See Table 3.7.)

Table 3.7: Russia's Gas Balance until 2030 (million tonnes of fuel equivalent)

	2008	*First stage 2013–2015*	*Second Stage 2020–2022*	*Third Stage by 2030*
Gas Output	760.9	784–853	919–958	1015–1078
Gas Import	64	76–80	79–80	80–81
Domestic consumption	526	528–573	592–619	656–696
Gas export to CIS	91	101–103	100–105	90–106
Gas export to non-CIS	190	210–235	281–287	311–317

Source: Russia & CIS OGW, 26 November–2 December, 2009, 8.

As shown in Table 3.8, according to the final version of the Energy Strategy, released in late November 2009, production will reach 885–940 bcm by 2030, of which East Siberia will account for 45–65 bcm and the Far East some 85–87 bcm. That means that East Siberia and the Far East will account for 15 per cent of total production. Exports to Asia–Pacific countries will account for 11–12 per cent of

Table 3.8: Gas Production in Siberia and the Far East up to 2030 (bcm/y)

	2005	*2008*	*Phase I**	*Phase II***	*Phase III*
TOP	641	664	685–745	803–837	885–940
Tyumen R	585	600	580–592	584–586	608–637
N–P	582	592	531–559	462–468	317–323
O-TB			0–7	20–21	67–68
B V	3	8	9–10	24–25	30–32
Yamal			12–44	72–76	185–220
Tomsk R	3	4	6–7	5–6	4–5
European R	46	46	54–91	116–119	131–137
E. Siberia	4	4	9–13	26–55	45–65
Far East	3	9	34–40	65–67	85–87
of which Sakhalin	2	7	31–36	36–37	50–51

Note: TOP means total oil production; N–P = Nadym–Purtazovsky; OTB = Ob-Taz Bay; BV = Bolshekhetskaya Valley; R = region; I = Island.
*   By the end of the first implementation phase of the Strategy (2013–15);
** By the end of the second implementation phase of the strategy (2020–2).

Source: Ministry of Energy, Russian Federation, Energy Strategy of Russia.

exports in the first stage, 16–17 per cent in the second stage, and 19–20 per cent in the third stage. LNG will account for 4–5 per cent of exports in the first stage, 10–11 per cent in the second stage, and 14–15 per cent by 2030.[28]

In short, the revised strategy envisages gas exports to the East gaining a 20 per cent share of total exports, and oil shipments rising to a quarter, from the current 6 per cent. Total investment in the fuel and energy industry is estimated at 60 trillion rubles in the period. The forecast for gas production in 2030 was increased by 4 per cent from 870–880 bcm to 885–940 (bcm). Gas exports should amount to 349–368 bcm in 2030 instead of 353–380 bcm as previously expected. The strategy projects liquefied gas exports to rise to 14–15 per cent of total gas exports. The Energy Strategy calls for the share of gas in overall energy consumption to decline to 47 per cent by 2030 from 52 per cent, and the share of non-fuel energy to rise to 14 per cent from the current 10 per cent. The source said that the Energy Strategy is premised on a conservative forecast for oil prices of $70–80 per barrel in 2009 prices during the period until 2030.[29]

The Gasification Programme in East Siberia and Russia's Far East

Jonathan Stern and Michael Bradshaw explain the importance which was given to the Far East in general and the Altai Republic in particular:

> The Russian government has become increasingly concerned about the economic situation in the Russian Far East. During the 1990s the region suffered from significant population decline and its economy has not rebounded as fast as the European regions of the country. Thus gasification is part of a wider concern to maintain enough population and economic activity to ensure the effective occupation of the region. By 2004, there was little tangible evidence of gasification in East Siberia and the Far East: gas was used in only 6 out of 15 districts in Siberia and four out of 10 districts in the Far East. Since 2004, however, Gazprom has shown significantly greater interest in these regions. Its budget for regional gasification was substantially increased by 8 billion rubles to 60 billion rubles in 2007, Altai and Irkutsk being cited as two of five regions to benefit specifically from the increase.
>
> The Altai Republic was set to benefit significantly from an export pipeline to China and a regional gasification agreement was signed in 2006. Altai Krai is already minimally gasified (5.9 per cent in urban and 1.2 per cent in rural areas) and in 2006 eleven new settlements were added to the network via the Barnaul–Biysk pipeline. The first leg of the Barnaul–Biysk–Gorno-Altaisk pipeline was completed in late 2006 and branch lines will allow

gasification of nearly 120,000 apartments in 300 towns and villages; more than 100 million rubles was allocated for this project in Gazprom's 2007 budget.[30]

Gazprom started to pay attention to gasification in Kamchatka for the first time in 2007, and a press release suggested that it was a top priority for the socio-economic development of the region. The greater urgency may be connected with the fact that a consortium of Rosneft and South Korean companies made gas discoveries in 2006.[31] As part of the Kamchatka Krai gas supply project, Gazprom set about production drilling and pre-development of two fields – Kshukskoye and Nizhne-Kvakchikskoye. It planned to reach the design capacity of 175 mcm/y in the Kshukskoye field by 2011. The Nizhne-Kvakchikskoye field commissioning, with its design capacity of 575 mcm, was planned to allow total annual gas production capacity on the Peninsula to increase to 750 mcm in 2013. In September 2010 Gazprom commissioned the Sobolevo–Petropavlovsk-Kamchatsky gas trunk line and started gas deliveries to Petropavlovsk-Kamchatsky. CHPP-2 (combined heat and power plant) in Petropavlovsk-Kamchatsky will be the main gas consumer on the Peninsula at the initial stage. Gazprom aims at converting CHPP-1 in Petropavlovsk-Kamchatsky to gas and ensuring gasification of the Kamchatka Krai settlements located along the gas pipeline route.[32]

The initiative for the regional gasification scheme was continued. On 28 August 2008 Gorno-Altaisk hosted a joint meeting dedicated to the gasification of the Republic of Altai. It was co-chaired by Alexander Berdnikov, Head of the Republic of Altai, and Valery Golubev, deputy chairman of Gazprom's Management Committee. The meeting was attended by Viktor Ilyushin, member of the Management Committee and head of the Department for the Relationships with Russian Regions, together with heads and specialists from Gazprom's core business units and its subsidiaries (Mezhregiongaz and Gazprom Transgaz, Tomsk) as well as heads of the executive branch and of the municipalities of the Republic of Altai. Golubev announced that:

> … from 2008 to 2010 Gazprom is planning to invest 1.5 billion rubles in the Republic of Altai gasification. At the moment the gasification program is being implemented through the construction of the Barnaul–Biysk–Gorno-Altaisk gas main with the lateral to Belokurikha. The Altai Project contemplates supplying Russian natural gas to the People's Republic of China. The Altai pipeline will link the West Siberian fields with the Xinjiang–Uyghur Autonomous Region in western China.[33]

Reports on the Altai project confirmed the gasification plans for East

Siberia and the Far East, indicating that they were riding on the back of larger, export-oriented projects.

In Irkutsk gasification is a big issue, as would be expected given the problems of the Kovykta licenses in creating a market. Gazprom made regional gasification a very public issue, and had its own programme even before its takeover of the TNK–BP share in the ESGC, which started construction of a pipeline from Kovykta to Irkutsk in 2006 with the first gas being delivered in 2008. In addition to Kovykta, gas will be sourced from the much smaller Chikanskoye and Dulisminskoye fields. In 2008 Gazprom projected that gasification of Irkutsk Oblast would be over 82 per cent overall, and over 40 per cent even in rural areas.[34]

In April 2010, Gazprom and Transbaikal Krai signed an accord on gasification. In the first stage, regional gasification will rely on LPG and LNG utilization, while pipeline gas will be supplied in the long run. It is worth noting that the Transbaikal Krai was established on 1 March 2008 when the Chita Oblast and the Aginsk Buryat Autonomous Okrug were merged. Gazprom had signed an agreement of cooperation with the Chita Oblast in July 2006 and with the Aginsk Buryat Autonomous Okrug in April 2007. The general scheme for gas supply to, and gasification of, the Transbaikal Krai was approved in December 2009.[35]

In October 2010, Gazprom CEO Miller and Irkutsk Oblast governor Dmitry Mezentsev signed an Agreement of Cooperation, paying special attention to the development of gas production from the Chikanskoye field and to the elaboration of the synchronization schedule for preparation of the population centres located near the Chikanskoye gas and condensate field–Sayansk–Angarsk–Irkutsk gas trunk line, which was in the process of being designed. The construction of phase 2 of the Bratskoye gas and condensate field–Bratsk gas trunk line was also under consideration.[36]

In the same month, Gazprom signed the 2010–13 Prioritized Action Plan on gas supply to, and gasification of, the Republic of Buryatia. Gazprom was to apply various approaches, such as liquefied and compressed natural gases, or LPG. Gasification of Buryatia is to be divided into three stages. In the first stage Gazprom will focus on local gasification with LPG. The second stage specifies the use of liquefied natural gas in addition to LPG. Pipeline natural gas will be supplied to a number of settlements at the third stage, which will begin after the establishment of the required resource base.[37]

In Krasnoyarsk, the emphasis is on production, with gasification getting only a passing reference in Gazprom's agreement with the region. Gazprom intends to use associated gas for gasification, particularly in the Evenkia region where many of the fields are located. In Tomsk

Oblast, 6.3 per cent of the urban and 3.8 per cent of the rural population had access to some of the 1.3 bcm of gas supplied by Gazprom in 2006; the distribution network was due to be expanded in 2007. In April 2007, the first agreement to develop distribution infrastructure in the Aginsk Buryat Autonomous Region was signed. Gazprom also has significant aspirations for gas development in the Sakha Republic, a region where gas is already supplied to 18.5 per cent of the population, but not by Gazprom. Although the Sakha Republic is one of the four designated centres of gas activity in East Siberia and the Far East, few definite commitments have been made to expand gas development, despite the consortium with Rosneft and Surgutneftegaz, and substantial numbers of meetings between the company's senior management and key regional politicians in 2006. In mid-2007, however, Gazprom signed a major cooperation agreement with the region for all phases of gas development, including exports, although this did not appear to include any definite commitments.[38] In December 2010, Gazprom CEO Miller and Sakha (Yakutia) Republic President Borisov signed the Partnership Agreement for socio-economic development driven by implementing the Eastern Gas Programme.[39] The gasification development is one of the key agendas of the agreement, and Chayanda gas development will be the core of the Republic's socio-economic development.

In the case of the Primorskii Krai region, there is at present no gas supply, but this is planned to change when the Sakhalin–Komsomolsk-on-Amur–Khabarovsk pipeline is extended to Vladivostok as part of Gazprom's five year cooperation agreement with the region. In January 2006 Gazprom signed a regional gasification agreement with Khabarovsk, where 12 per cent of the population already had access to natural gas but (as in Sakha Republic) this was not supplied by Gazprom. The Khabarovsk pipeline would clearly provide the means for a substantial expansion of supply to the region.

From the perspective of the regions, there is a strong desire to see the gas projects move ahead quickly since they promise local employment, increased tax revenues, bring substantial foreign currency revenues (a proportion of which will find their way back to the local economy), and provide infrastructural development and secure and environmentally benign energy supplies. Stern and Bradshaw pointed out that the regions should see energy development as a potential source of economic development, although for various reasons Sakhalin's experience presents a cautionary tale. First, if a region's infrastructure is under-developed to start with this actually limits its ability to absorb the potential economic benefits generated by energy-related investment. Consequently much of the generated activity takes place outside the region. Second, since the majority of the impact is associated with the construction phase and the first generation Sakhalin projects are now either completed or nearing completion, local politicians on Sakhalin are worried that there will now be a lull in activity, and the economy will experience the classic boom and bust cycle often associated with oil and gas projects. Third, the Sakhalin-2 project, which is export-oriented, will make no contribution to regional gasification. This has been a major bone of contention for local politicians, and Gazprom has promised to address this problem. Fourth, the economic benefits of energy-related investments can be tempered by the nature of Russia's federal fiscal structure. The substantial increases in Sakhalin's Gross Regional Product generated by oil and gas and related construction activity actually resulted in a reduction in the level of federal transfers to the region. Finally, the Federal Government has reneged on an agreement to give the Sakhalin Oblast 50–60 per cent of the royalties generated by oil and gas production; instead it will receive only 5 per cent. In any event Stern and Bradshaw concluded that the 2007 Eastern Gas Programme, in combination with Gazprom's commitments to individual regions, would appear to create a political commitment to gasification from which it will be difficult to step back.[40]

Gazprom argued in 2010 that the gasification of the Russian regions, and achieving the maximum economically viable level of gasification throughout the Federation, was one of its top priorities. In 2009 Gazprom announced its intention to invest 18.5 billion rubles in the gasification of 69 regions. The Primorskii and Kamchatka Krais, and the Jewish Autonomous District were later added to the Gasification Programme. In 2010, the average gasification level in East Siberia and the Far East was no more than 7 per cent, while the figure for all Russia was 62 per cent. However, Gazprom has signed Agreements of Cooperation with 11 out of the 14 East Siberian and Far Eastern regions and Accords on Gasification with nine of them.[41]

The Sakhalin–Khabarovsk–Vladivostok (SKV) gas transmission system development and its implications

Gazprom had begun to show that the agreements and accords were not just words but did imply real action. The company started building the 1,830 km SKV pipeline in July 2009 and construction was completed in September 2011, well before the APEC summit in Vladivostok in 2012. The pipeline's initial capacity is 6 bcm/y, and the ultimate capacity 30 bcm/y.[42] On 15 June 2009 a Russian Government Directive was signed granting Gazprom the subsurface licenses for the three blocks of the

Sakhalin-3 project (the Kirinsky, Ayashsky, and Vostochno–Odoptinsky blocks) for an unlimited period.

According to *OGJ*:

> Russian officials have been canvassing Japanese investors to persuade them to join the SKV pipeline project. In early June [2009] Vladimir Putin had visited Tokyo and invited Japanese companies to participate in the construction of the pipeline as well as other facilities. In addition to the pipeline Gazprom also is said to be considering the construction of a gas liquefaction plant and a gas chemical facility on Sakhalin Island. During the Tokyo visit, Gazprom's Miller [held meetings] with a group consisting of Eizo Kobayashi, president of Itochu Corp., Yutaka Kase, president of the Sojitz Corporation, Toru Ishida, the Director General of the NREA of Japan, Hiroshi Watanabe, president of the JBIC, and Masami Iijima, president and chief executive officer of Mitsui. During the visit, an MOU was signed with Japan's NREA (under METI), the Itochu Corporation, and the Japex. The MOU prepares the way for the joint exploration for gas around Vladivostok, as well as cooperation in the transportation, marketing, and processing of the final products to potential customers in the Asia–Pacific, including Japan.[43]

A similar step was taken for South Korea, and in June 2009, Gazprom CEO Miller, and Kogas CEO Kangsoo Choo signed an agreement for the joint exploration of a gas supply project.[44]

The priority given to SKV development was indirectly confirmed by Gazprom deputy CEO Alexander Ananenkov's announcement in 2011 that Gazprom had already slowed down work on field construction at Bovanenkovskoye and on the Bovanenkovskoye–Ukhta gas pipeline. Some funds and contractors moved from the Bovanenkovskoye–Ukhta project to the SKV gas line, on which the rate of construction increased in order to be able to transport gas from Sakhalin to Vladivostok in 2011. Gazprom increased the 2009 spending on the SKV pipeline from 20 to 50 bn rubles.[45]

In fact, on 2 July 2009 in Yuzhno-Sakhalinsk, Gazprom's delegation took part in celebrations to mark the start of exploration drilling in the Sakhalin Peninsula's offshore Kirinskoye field. The start of geological exploration in the Kirinskoye field was another crucial step towards establishing Gazprom's resource base in the Far East and creating a new Sakhalin offshore gas production area as part of the state Eastern Gas Programme. This was the first project in the Sakhalin offshore area carried out solely by Russian companies. According to Gazprom's plan, starting from 2014 the Kirinskoye field will become a gas source for the SKV gas pipeline. The Kirinskoye gas and condensate field is located 28 km off Sakhalin Island (with sea depth of 90 metres) within the Kirinsky block of the Sakhalin-3 project. The C1 reserves of the Kirinskoye field amount to 137 bcm of gas and 15.9 mt of condensate.[46] In September 2010, Gazprom discovered the Yuzhno-Kirinskoye field with 260 bcm of C1+C2 gas reserves.[47]

Gazprom has also been able to take a step towards gas development in the Sakha Republic. On 10 March 2009 the Russian Government adopted the investment certificate for the comprehensive investment project permitting the elaboration of the documentation for the South Yakutia development project. In accordance with the Russian Government Directive of 16 April 2008, the license for the Chayanda oil and gas condensate field was granted to Gazprom as the owner of the Unified Gas Supply System. The Chayanda oil and gas condensate field is located in the Lensky district of the Republic of Sakha (Yakutia). The C1+C2 reserves of the field account for 1.24 tcm of gas and 68.4 mt of oil and condensate. Gazprom and the Sakha Republic (Yakutia) signed the Cooperation Agreement in July 2007 and the Gasification Accord in June 2008.[48]

On 5 June 2009 in Saint Petersburg, within the Thirteenth International Economic Forum, Valery Golubev, Deputy Chairman of the Gazprom Management Committee, agreed to implement a Comprehensive Investment Project which aimed to elaborate project documentation for the South Yakutia Comprehensive Development Investment Project, with the backing of the Russian Government and financed by budget allocations from the Russian Federation Investment Fund.[49] According to this document, Gazprom will be a co-investor in the South Yakutia project, with the objective of creating the Yakutsk Gas Production centre.

The Three Main Gas Supply Sources for the Eastern UGSS

Kovyktinskoye (or Kovykta) Gas[50]

Start-up and preparation work
The Kovyktinskoye deposit was officially opened in 1987. The Kovykta gas condensate field is 450 km from Irkutsk, in the Zhigalovo and Kazachinsko-Lensk Districts. The deposit is located within the Angara-Lena terrace abutting the southern part of the mid-Siberian plateau. In 1990, the Irkutsk administration initiated development of Kovykta by creating the Baikalekogaz regional consortium composed of the local energy companies Angarsknedteorgsintez, Varyengannefte-gaz, Irkutskenergo, Vostsibneftegazgeologiya, and Irkutskgeophysica.

In the same year the BP–Statoil consortium also became interested in the deposit. For two years this consortium and Baikalekogaz worked on a Kovykta feasibility report. In 1992, however, the consortium, concluding that developing the deposit would be unprofitable because of the absence of an export market, the low domestic prices for gas, and the meagre progress towards a production sharing agreement (PSA), ended its participation. This was also the year in which Rusia Petroleum was established; it was in fact an initiative of the Irkutsk administration and Baikalekogaz. Among other founders were Burgazgeoterm, Sayankhimprom, Usolyekhimprom, Irkutskbioprom, the Potential for Social Partnership Fund, and the Angarsk Bank. In 1993 the firm received a 25 year license for Kovykta. In 1994, 50 per cent of Rusia Petroleum shares were transferred to the control of the newly established Sidanco.[51]

With the beginning of rapid growth in the Chinese economy, Beijing made it clear to Moscow that it was ready to sit down at the negotiating table. In 1996, during the first visit of Russia's President Boris Yeltsin to China, a preliminary agreement was reached on the construction of a gas pipeline from the Irkutsk region to China. In November 1997 the Russian Federation and the People's Republic of China signed a Memorandum of Understanding on the main elements of a project feasibility study for a gas pipeline from East Siberia and other north-east Asian countries, and another feasibility study on of the development of Kovykta gas in Russia.

Until late 1997 lack of finance had held up the regional development programme covering the 1994–2001 period, although partial financing was found after the East Asia Gas Co. (EAGC), a subsidiary of Korean Hanbo Group (bankrupted in January 1997), invested $44 million in exchange for a 27.5 per cent equity share in Rusia Petroleum. The total investment required for the regional programme was estimated at $725 million. In November 1997 BP Exploration decided to form a strategic partnership with Sidanco by taking a 10 per cent equity stake at a cost of $571 million, at the same time acquiring 45 per cent of Sidanco's equity in Rusia Petroleum in exchange for covering $172 million of the costs of appraising the Kovykta gas field. These financial manoeuvres provided a major breakthrough in the financing of the programme.[52]

As to the international pipeline project, however, no progress was made. At the end of December 1997, a multilateral memorandum was signed by five countries (Russia, China, Mongolia, South Korea, and Japan) which met in Moscow. The memorandum calls for a coordinated approach to improving reserves estimates, further development of the field, a clearer identification of the market for the developed gas, and the carrying out of a feasibility study for the transnational pipeline project.[53] This multilateral approach, however, was suspended after one year of negotiation, and both Russia and China decided to restart the work based on a bilateral agreement. In late February 1999, during Chinese Prime Minister Zhu Rongji's visit to Moscow, agreement was reached on a feasibility study of the development of the Kovyktinskoye field and of the construction of a gas pipeline to China.

A joint Sino-Russian technical and economic study confirmed that a 3,400 km transnational pipeline connecting Irkutsk and Rizhao (a port city in Shandong province) and passing through Mongolia was feasible. The option of the Mongolian route, however, was sidelined by the major discovery in late 2000 of the Sulige-6 gas field in the Ordos basin. It was at this time that work began on the three year feasibility study on Kovykta gas exports to China and South Korea. Negotiations on the Kovyktinskoye project had been stagnating since the second half of 2001, but the negotiations were resumed in the spring of 2002 during a visit to China by the head of Rusia Petroleum. CNPC argued that the stagnation was largely due to Russia's inability to solve the key problems raised by the Chinese side, including the question of the route. China regarded the adoption of the eastern route, reaching north-east China, through Manzhouli, as a precondition for its participation. This differed from the Russian plan which favoured the western route through Mongolia to reach Beijing and the north China market. The western route would be shorter, especially the portion within Russia, would be less topographically challenging, and would allow Mongolia to collect a sizable transit fee.[54]

China preferred the eastern route for several reasons: first, it minimized the political risk and saved the transit fee by avoiding transit through Mongolia; second, the economic benefits brought by the pipeline could rapidly make Mongolians better off than people in China's inner Mongolia, a potentially negative influence on stability in the autonomous region; and third, if the eastern route was used the economic benefits of the pipeline would be concentrated in north-east China, a region desperately in need of them.

The three party feasibility study
In November 2000, Rusia Petroleum, CNPC, and Kogas signed an agreement to undertake a new feasibility study for an export gas pipeline with 30 bcm/y capacity. This study took three years to complete and during this time the Kovykta situation changed radically. Gazprom openly declared its interest in the project. The initial reaction of Rusia Petroleum shareholders to this was that they had no interest in

inviting Gazprom into the project. In the summer of 2001, Gazprom renewed work on the East Programme, the full name of which became a 'Programme to Create in Eastern Siberia and the Far East a Uniform System of Extraction, Transportation of Gas and Gas Supply with Possible Gas Export to the Markets of China and Other Countries of the Asian Pacific Region'.[55]

On 17 April 2002, at a meeting attended by Rusia Petroleum, TNK, Gazprom, Transneft, Sakhaneftegaz, Yukos, and all corporate shareholders of the Kovykta project, the Russian energy minister, Igor Yusufov, emphasized the need to accelerate work on a programme which he defined in the same words as the new name of the East Programme, mentioned above. Participants agreed that the minimum price at which 1,000 cubic metres of Kovykta gas would be supplied should be $100. They agreed to resume active negotiation with Beijing and Seoul regarding their readiness to receive an agreed volume of Russian gas, the price for the gas supplied, and the route of the gas pipeline. They also agreed to form a standing working group under the aegis of the Ministry of Energy to resolve the entire list of issues relating to development of Kovykta field, including the route of the export gas pipeline. [56]

A session of the coordinating committee for managing work on the feasibility study took place in July 2002, after more than a year's break. It decided that only the eastern route of the gas pipeline (bypassing Mongolia) would be considered from then on. In the course of three sessions of the coordinating committee – in July 2002, October 2002, and January 2003 – the parties succeeded in pushing the project forward substantially, defining the gas pipeline's route and seeking consensus on the gas price and the volumes to be supplied.[57] The October 2002 session set June 2003 as the date for finalizing the study. During the October session, Russia managed to obtain a promise from the other parties that gas supplies to China and then to South Korea would start in 2008. The proposal to transport 20 bcm per year of gas was dropped during the January 2003 meeting.[58]

The Gazprom factor
Regardless of these three party negotiations the Russian government in the summer of 2002 gave Gazprom the exclusive right to negotiate gas exports to Asia. Gazprom made it very clear that no Kovykta gas could be exported without Gazprom's full support. The deputy chairman of Gazprom, Ananenkov, outlined the reasons why the feasibility study of Kovykta, prepared without inviting Gazprom experts, was not legitimate. The most notable deficiencies of the document were:[59]

- the three signatories – TNK–BP's subsidiary Rusia Petroleum which holds the license to Kovykta, South Korea's KOGAS, and China's CNPC – quoted different prices of gas; each party was basing its assumption on its own idea of the final price;
- the study said nothing about the need to remove helium from the natural gas, yet helium is a strategic commodity which Russian law requires to be extracted on Russian soil; this operation would add greatly to the cost of the project;
- the study did not discuss domestic supply, and was therefore in contradiction to government plans for the development of the gas industry in the east of the country; and finally,
- the study was based on the idea that the export pipeline was to be built south of Baikal Lake, exactly where Yukos had been planning to build its Angarsk–Daqing oil pipeline before its plan had had to be abandoned after its rejection by environmental experts.

At the end of 2002, Gazprom recommended to the then Ministry of Energy a substantive Programme which the ministry approved on 13 March 2003. The government also approved the programme but sent it back for revisions. In 2004, Gazprom managed to lobby successfully for the creation of a single uniform seller of Russian gas in Asian markets.

The key finding of the feasibility study was that the project's development costs would be substantially higher than previously thought. It estimated that capital expenditure on developing the Kovykta field and building a gas pipeline to China and South Korea would total US$17 billion, compared with the previous estimate of US$10–12 billion. The feasibility study partners looked at both an operating tax regime and a production sharing agreement (PSA). The CNPC representative, Zhang Xin, argued that the PSA regime was preferable but the project could still be implemented without it. The real problem was the price, which, in drafting the feasibility study, the Russian side failed to resolve. Rusia Petroleum calculated that Kovykta gas would cost about US$5.0 per mmbtu while CNPC was only willing to pay $2.40 per mmbtu. In view of this difference it was decided to remove the question of the price from the feasibility framework.[60]

In November 2003, when work on the feasibility study was complete, CNPC was seriously considering an investment in the project. The CNPC vice president, Su Shulin, announced that 'we are interested in the Kovykta project. How we will participate in this venture is yet to be defined, but we are conducting negotiations and exploring the possibility of buying 25.82 per cent of Rusia Petroleum's shares'. This 25.82 per cent blocking share parcel of the Kovykta operator Rusia

Petroleum belonged to Interros Holding Company. In mid-December 2003, Interros announced that it had hired Merrill Lynch investment bank as a consultant to prepare a sale of the stake.[61] China, however, took no action.

In March 2004, the East Siberian Gas Company (VSGK), founded on a parity basis by TNK–BP and the Irkutsk Region Administration, began to implement the local gasification project.[62] Around the same time, Gazprom failed to reach an accord on acquiring a 25 per cent plus one share in the Rusia Petroleum project, even though Gazprom had agreed to hold a 25 per cent equity interest in the project to tap Kovykta as early as December 2001 for its Programme to develop the Gas Reserves of East Siberia and the Far East. The distance between TNK–BP and Gazprom was too great to bridge. In January 2004, Gazprom made a scathing criticism of the feasibility study completed in November 2003, and this was a strong signal that Gazprom's patience was running out. Finally, in September 2004, the TNK–BP CEO, Robert Dudley, said his company was ready to participate as a minority partner of Gazprom in developing the Kovyktinskoye field. This change came after the MNR minister, Trutnev, had made clear in mid-2004 that Rusia Petroleum's license was in danger of being withdrawn.[63]

The Helium question

Around this time, there was intense discussion on the issue of helium. In May 2004 Rusia Petroleum altered its stance at the initiative of the Gelimash Science and Production Association. On 24 May Sayanskkhimplast, Rusia Petroleum, and NPO JSC Gelimash signed a protocol of intent to build a plant to remove helium from Kovykta gas in Sayansk. The plant would be able to process up to 5.5 mcm/y of helium, a gas whose global turnover stood at over 160 mcm/y.[64] In early June 2004, Gazprom CEO, Alexey Miller, and TNK–BP Chief Operating Officer, Viktor Vekselberg, reached an accord on the need to use a single export channel in organizing East Siberian gas shipments to Asian-Pacific countries. Under the accord, Gazexport LLC, a 100 per cent subsidiary of Gazprom, would lead negotiations on the sale of gas from the Kovykta gas field to CNPC and Kogas. Gazprom and TNK–BP, however, interpreted the accord in different ways. Gazprom tried to fully monopolize export talks and insisted that TNK–BP and Rusia Petroleum should no longer participate in export negotiations. TNK–BP contended that Gazprom did not have full authority to monopolize the export negotiations, and that Gazexport had the authority to work on the terms of supply only.[65]

Also in June 2004, the MNR Commission on Subsoil Use demanded that Rusia Petroleum redress violations of the licensing agreement relating to initiating production at the Kovyktinskoye field. In December 2004, Rusia Petroleum proposed changes to the licensing agreement that would postpone the start of production by five years. In mid-May 2005 the Minister of Natural Resources, Yury Trutnev, announced that he could not exclude a range of measures being taken against the company for delays in developing the Kovyktinskoye field. In early June 2005 TNK–BP responded, presenting a new addendum to the licensing agreement for Kovyktinskoye. The company proposed that it would begin with 0.9 bcm/y of gas at the field in 2006 and then bring the production level up to 9 bcm/y by 2010. In late May 2005, the board of directors of TNK–BP allocated $136 million to finance the first phase of the Irkutsk gas project involving the construction of a gas pipeline to supply consumers in the Zhigalovsky district of the Irkutsk region. This early gas was the initial phase of a $1.1 billion project to gasify the Irkutsk region. The gas from Kovyktinskoye would be shipped to Sayansk, Usolye-Sibirskoye, Angarsk, and Irkutsk.[66]

Gazprom's stance on the helium question in both Kovykta gas and Chayanda gas can only be properly clarified once the development of the gas fields begins. One of the most frequently raised questions is whether the helium is a real issue or just a way for the Russian government and Gazprom to delay the development. According to data from the Siberian division of the Institute of Oil and Gas Geology of the Russian Academy of Sciences Kovykta contains around 37–42 per cent of Russia's A+B+C1 helium reserves.

Table 3.9: Natural Gas Composition of the East Siberian and Yakutia Gas Fields

	Methane	Nitrogen	Helium	Ethane	C3-C6
		Main Components (%)			
Sredne–Botuobinskoye	88.61	2.93	0.45	4.95	3.12
Chayandinskoye	85.48	6.44	0.58	4.57	2.58
Yurubcheno–Tokhomskoye	81.11	6.32	0.18	7.31	5.06
Sobinsko–Paiginskoye	67.73	26.29	0.58	3.43	1.55
Kovykta gas	91.39	1.52	0.28	4.91	1.78

Source: Gazprom.

Gazprom has argued that Kovykta reserves will not be needed before 2015. But, Gazprom has said, Kovykta contains large volumes of helium, a strategic product with military uses, and so Russia should first pass a law on helium before approving Kovykta's development.

Gazprom deputy CEO, Alexander Ananenkov, has said that Kovykta contains 50 per cent of Russia's helium reserves, and the feasibility study does not give an answer on how to deal with the problem. It is, therefore, unacceptable to start developing the biggest fields in East Siberia with high helium content by producing only simple gases, such as methane, for heating needs. According to Gazprom's estimates, Russia controls over one third of the world's helium reserves of 28 bcm.[67]

JOGMEC's Masumi Motomura pointed out that the level of the helium price, which hovered within the range $2,880–3,060 per 1,000 cubic metres in the USA in 2006, should not be ignored. It is true that helium is a very expensive energy source and it needs special management by gas producing companies. A big increase of helium production could cause a price collapse. To maintain the price level, controlled production, including construction of underground storage by Gazprom, would be necessary.[68]

TNK–BP agreed to sell its controlling stake to Gazprom
In August 2004, Gazprom acquired exploration licenses for two neighbouring blocks at Yuzhno-Kovyktinskaya with estimates of hypothetical gas reserves of 1 tcm. OOO Kubangazprom, a Gazprom subsidiary, is already drilling exploratory wells on the Chikansky block, just to the south of Kovyktinskoye.[69] The differences between Gazprom and BP seemed to diminish somewhat during negotiations between Alexey Miller and John Browne in Moscow in late May 2005. Miller proposed that Browne should examine the possibilities of collaboration on oil projects and on selling an interest in some oil assets to Gazprom. Browne responded that he was prepared to discuss selling a portion of the 50 per cent equity held by TNK–BP's Russian shareholders to Gazprom;[70] they could be persuaded to sell if the price was right. In September 2005, however, the Ministry of Natural Resources raised the issue of withdrawing the license, and then in late September 2006 the Irkutsk Region Public Prosecutors' Office demanded that the Federal Agency for Subsoil Use suspend TNK–BP's development license at the Kovykta gas condensate field.[71]

The sale of Udmurtneft to a consortium of SINOPEC and Rosneft in March 2006 did not improve relations between TNK–BP and Gazprom. Gazprom had regarded the acquisition of Udmurtneft as the main condition for reaching accord on Kovykta. It had offered $2.5 bn, but the winning bid was $3.5 billion. Having lost Udmurtneft, Gazprom went on the offensive and argued that the development of Kovykta could not take place without the existence of a federal programme for Russia's helium and the passing of federal laws on helium and on multi-component fields. Among other matters, such formalities would define the scientifically grounded criteria for considering fields containing helium as strategic.[72]

In mid-2007, however, there was a breakthrough in the tedious negotiations between Gazprom and TNK–BP. On 22 June Gazprom's deputy CEO, Alexander Medvedev, BP Group's executive vice president, James Dupree, TNK–BP's CEO, Robert Dudley, and TNK–BP's executive director of gas development, Victor Vekselberg, signed an agreement of mutual understanding creating a strategic alliance between TNK–BP and Gazprom. Its MOU with TNK–BP allowed Gazprom to take over the controlling 62.9 per cent stake in Rusia Petroleum and ESCG. There was no consensus among the three parties on the exact amount of money involved and they agreed to continue the negotiation. But the compromise amount would lie between \$600 and 800 million. Alexander Medvedev argued that market conditions would determine the final price. The TNK–BP financial report, completed according to International Financial Reporting Standards (IFRS), stated that the Kovyktinskoye project expenses up to 31 December 2006 totalled US\$405 million. But TNK–BP argued that actual spending from 1993 totalled \$800 million.[73]

In exchange for giving up Kovykta, TNK–BP received a long-term call option to buy a 25 per cent plus one share of Rusia Petroleum at the market price. According to Medvedev, Gazprom and BP would create a joint venture 50 per cent owned by Gazprom and 50 per cent owned by either TNK–BP or BP itself, depending on what assets were involved. BP CEO, Tony Hayward, said that BP would initially be looking for projects worth at least \$3 billion but the potential for further growth could be very significant. TNK–BP ceded control over the Kovykta gas deposit to Gazprom, under the threat of license withdrawal. The Commission of the Federal Agency of Subsoil Use (Rosnedra) at its session on 1 June 2007 considered a possible license withdrawal from Rusia Petroleum, TNK–BP's subsidiary. President Putin remarked, during a conversation with G8 member-state journalists at his residence in Novo-Ogarevo on 4 June, that:

> Kovykta co-owners have incurred obligations on development of this deposit and, unfortunately, have not satisfied the license conditions; they should have already begun production, extracting a specified volume of gas. Unfortunately they have not done this … it is possible to discuss many reasons for this, including the need to put in a pipeline system. But they knew that when they received the license. Nevertheless, they purchased it. I will not even now say how this license was obtained. We shall leave it to the conscience of those who did it then, at the beginning of the 1990s.

Putin drew attention to the fact that the gas reserves on this deposit are almost equivalent to the total reserves of Canada.[74]

TNK–BP's deal to sell its vast Siberian Kovykta gas project to Gazprom marks another large step in the Kremlin's takeover of the energy sector after months of government pressure. TNK–BP's development had long been held up by Gazprom's refusal to join the project on TNK–BP's terms. Without Gazprom, TNK–BP had no real market for the gas, due to the state-controlled gas champion's monopoly on export sales. Instead of withdrawing TNK–BP's license over production violations at the field, as the Russian government had threatened, Gazprom paid for it and the price was nearly double what TNK–BP had spent on developing the field.[75] The final transaction took place in the late summer of 2008.

During the fifth Baikal Economic Forum in September 2008, the Natural Resources Minister, Yury Trutnev, said that the failure of TNK–BP and Gazprom to close a deal on the giant Kovykta gas condensate field was delaying the project's commercial launch:

> … we have not forgotten about Kovykta but we have not received any proposal from Gazprom. Clearly there are some complications. The general Kovykta plan has not altered but unfortunately the timing seems to have changed. We are not going to adjust the volume of gas production in the license. Our task is to develop the field.[76]

On 14 October Russian environmental watchdog Rosprirodnadzor began inspecting the fulfilment of license agreements for the Kovykta gas condensate field in the Irkutsk region.[77]

On 16 October 2008 Yury Trutnev said that there were no grounds for renewing Rusia Petroleum's license to the giant Kovykta gas condensate field in the Irkutsk region:

> … the results of the inspection are not yet in because it is still in progress and will not finish before 20 October. The owners have been trying to convince us that they are honouring their license commitments in connection with the agreement reached with Gazprom. But in my view nothing has changed and nothing has been put right. And I don't see any grounds to renew their mineral developer's license.[78]

On 23 October Trutnev told the Kommersant that Rusia Petroleum's license to the giant Kovykta gas condensate field might be withdrawn immediately.[79]

Rosprirodnadzor completed an inspection of Rusia Petroleum, and sent the results to the subsurface management agency, Rosnedra. Violations of the license agreements were discovered during the inspection, the main one being that Rusia Petroleum was not fulfilling the level of gas production envisaged in the license agreement. Rosprirodnadzor recommended that the results of the inspection be considered by Rosnedra's early license recall commission. Rosnedra would make the final decision about whether to withdraw or continue the license.[80] However, no final decision was made.

Kovykta ownership switches to Gazprom
Meanwhile the ownership of Rusia Petroleum had changed. On 23 October 2008 OGK-3 confirmed that it had acquired 25 per cent minus one shares in Rusia Petroleum, which held the license to the giant Kovykta gas condensate field. The main shareholders in Rusia Petroleum prior to that deal were TNK–BP with 62.89 per cent, the Irkutsk regional administration with 10.78 per cent, and Interros Holding with 25.82 per cent. In mid-October it was reported that Interros would sell its stake to the wholesale generating company OGK-3, which is owned by MMC Norilsk Nickel (75 per cent) and United Company RUSAL (UC RUSAL) (25 per cent). OGK-3 had purchased the stake from Jarford Enterprises Inc. for $576 million. Interros is the ultimate beneficiary of the sale. Acquisition of the Rusia Petroleum stake will potentially make OGK-3 a partner of Gazprom in developing Kovykta. OGK-3 includes the Kostroma, Pechora, Cherepetsky, Kharanorsk, Gusinoozersky, and Yuzhnouralsky power plants, with installed capacity of 8,500 megawatts. The shareholders in OGK-3 are Norilsk, of which Interros owns 75 per cent, the remaining 25 per cent belonging to RUSAL.[81] It was reported in early 2009 that talks with Gazprom on Kovykta had been suspended.

On 2 September 2009, however, Gazprom deputy CEO, Valery Golubev, said that Gazprom was still interested in the Kovykta gas condensate field: 'we remain interested [in Kovykta]. A general plan for the deal is currently being drawn up'. But he declined to provide more details of this plan.[82] At the end of September 2009, Gazprom CEO, Miller, said that 'this is a large gas project and no matter what it will be carried out by Gazprom. The only question is the time frame within which it will be implemented'.[83] In March 2010, it was reported that TNK–BP was planning before the end of the year to sell its stake in Kovykta gas field to the State Oil firm Rosneft for between $700 and 900 million, but no following action was taken.[84] Three months later Rusia Petroleum, developer of the Kovykta field, filed for bankruptcy and TNK–BP's move (to sell its stake in Kovykta to Rosneft before the end of the year) was regarded as a way of putting pressure on the Moscow authorities to decide the fate of the Kovykta field.[85]

TNK–BP's initiative opened the door for an auction for the Kovykta field. In early March 2011, Gazprom won the rights to the giant Kovykta gas field. The starting price at the auction, held in the Siberian city of Irkutsk, was set at 15.1 billion rubles. Gazprom beat state holding company Rosneftegaz, owner of Russia's top oil company Rosneft, with its 22.3 billion ruble ($773 million) winning bid. TNK–BP Chief Financial Officer Jonathan Muir said that 'coincidentally it is the same price we agreed with Gazprom some years ago … and allows us to fully recover the investments on the balance sheets'. VTB Capital oil and gas analyst Lev Snykov commented that 'it is virtually for free … the stock market values Gazprom at $1.80 per barrel and global majors at $10'. The value to Gazprom would be crystallized as soon as Gazprom reached a clearer agreement with China on prices. But observers were sure that 'it will definitely be more than they paid for it'.[86] *Interfax* reported that, including VAT, Gazprom paid 25.8 billion rubles, of which 21.01 billion rubles was payment for Rusia Petroleum's property, 3.5 billion rubles for VAT, and 1.3 billion rubles for LLC Kovykta Neftegaz. It added that the Russian law of subsoil usage gives the property buyer the rights to readdress the license for the Kovykta field without introducing changes.[87] Gazprom aims to take over the license for the Kovykta development. This would put it in a position to juggle with both the Chayanda gas and the Kovykta gas project for exporting Russian gas to Asia. But its dilemma is that northern China and South Korean gas markets are not big enough to require the simultaneous development of both fields, so Gazprom has to prioritize one of them. On paper priority seems to go to Chayanda gas but Gazprom can change this if necessary, since the development cost of Chayanda gas is much higher than that of Kovykta gas

After all, it is not an exaggeration to say that the Kovykta project became a casualty of Gazprom's absence from the ownership structure of the project. Gazprom's strategy was to give the highest priority to both the Sakhalin-3 block and the Chayandinskoye project. Gazprom's decision to control the ownership of the Kovykta project in 2011 strengthened Gazprom's position as the sole exporter of natural gas from East Siberia and Russia's Far East, but at the same time this acquisition indicates that Gazprom could rethink the priority it had given to Chayanda gas and replace it with priority to fast-track development of the Kovykta project if the gas price negotiation between Russia and China is settled by 2012. Gazprom cannot ignore the fact that the production cost of Kovykta is clearly decidedly lower than that of Chayanda gas, as the price burden of pursuing Vladivostok LNG based on long-distance pipeline gas supply is very high. It remains to be seen how the balance between the development of Kovykta gas and that of Chayanda gas will be managed by Gazprom.

Sakhalin Offshore Gas

Sakhalin-1

The Sakhalin-1 project, which is composed of the Chayvo, Odoptu, and Arkutun-Dagi fields, boasts 2.3 billion barrels of oil and 17.1 tcf of recoverable gas reserves. Exxon Neftegaz Limited, a subsidiary of ExxonMobil Corp, is the operator of the project. Other consortium members are Russia's Sakhalinmorneftegaz-Shelf and RN-Astra, Japan's SODECO (Sakhalin Oil and Gas Development Co Ltd), and India's ONGC Videsh Ltd[88] Sakhalin-1 is one of the first generation of projects based on the initial exploration activity funded by the Japanese government in the 1970s and continued in the 1980s by Sakhalinmorneftegaz (SMNG), now a subsidiary of Rosneft. Development rights to these fields were awarded in 1991 and were subsequently the subject of production sharing agreements (PSAs) in 1994–5.[89]

In October 2001, the Sakhalin-1 partners declared Phase I of the project – which involved the start of oil production in 2005 with limited gas supplies available to help meet Russian domestic demand – to be commercially viable. The first oil deadline was met and in the autumn of 2006, despite problems with the Ministry of Natural Resources, the oil export terminal at De-Kastri on the Russian mainland commenced year-round oil exports. The project made considerable use of Extended Reach Drilling (EDR) technologies from the world's largest land rig, the Yastreb, thereby reducing the need to build offshore structures. By February 2007, the project had supplied 35 billion cubic feet (1 bcm) of gas to customers in Khabarovsk Krai.[90]

No real progress, however, was made on preparations for Sakhalin to export gas by pipeline or as LNG to Japan, China and South Korea for many years, indeed not until the end of 2008. In the late 1990s, a project by Japan Sakhalin Pipeline Feasibility Study Co. (JSPFSC) planned to deliver natural gas from Sakhalin-1 to Tokyo/Niigata taking an offshore route. Between April 1999 and the spring of 2002, Exxon Japan Pipeline Ltd and the Japan Sakhalin Pipeline Co. (JSPC) together prepared a $40 million feasibility study for Sakhalin gas supply to Japan. JSPC (composed of Japex 45 per cent, Itochu 23.1 per cent, Marubeni-Itochu Steel Inc. 18.7 per cent, and Marubeni Corporation 13.2 per cent) was the operator in attempts to develop the feasibility study. The study assumed that the pipe diameter was 26–28 inches (65–70 centimetres) and delivery capacity was 8 bcm/y. The distances from Sakhalin-1 to Tokyo and Niigata are 1,400 km and 1,120 km respectively. In August 2002, Japex announced that firms cooperating with Exxon Japan Pipeline Ltd had completed the feasibility study report for the section located on Japanese territory and had concluded that the project was technically and economically feasible. However, the Japanese government privatized JAPEX in December 2003 and consequently the owners of JSPC liquidated and closed the firm in October 2005.[91]

On 19 May 2003, the director general of the Resource and Energy Agency (under METI) officially discussed supporting the gas pipeline project before a Diet Committee. In particular, a loan arrangement by the Development Bank of Japan (DBJ) was mentioned. DBJ offered a financial programme for the construction of the natural gas pipeline project, terminating in the fiscal year 2006. The applicable interest rate would be the Policy Loan Interest Rate (1.7 per cent); the loan limitation would be 30 to 50 per cent of capital expenditure and the term of lending 15 years.[92]

The project of exporting Sakhalin-1 gas to Japan, however, was not followed up, and the focus moved towards gas supply to Khabarovsk Krai and then export to China by pipeline. On 11 June 2004, the Sakhalin-1 consortium signed protocols of intent with the Khabarovsk Governor, Viktor Ishayev, and potential consumers – Khabarovskenergo and Khabarovskkraigaz – for future supplies of natural gas from Sakhalin. According to the Sakhalin office of Exxon Neftegaz Limited, these documents set down the conditions to be included in the actual purchase and sale agreements. According to the documents, the first supplies of gas were to be made by the start of the 2005–6 winter season, and gas sales to Khabarovsk territory would reach 3 bcm per year by 2009. Exxon Neftegaz President, Stephen Terni, said at the signing ceremony that this would be the first gas to have been produced off the Sakhalin coast for consumers in the Russian Far East as part of a production sharing agreement, involving long-term commercial supplies of gas, and taking international prices and market conditions into consideration.[93]

Khabarovsk Krai governor Ishayev offered China the opportunity to help build a gas pipeline from Sakhalin to Khabarovsk, which could later be extended to China. Ishayev, who led the Russian delegation to the 37th session of the Pacific Economic Council in June 2004, stated in Beijing that:

> … we are building a gas pipeline from Sakhalin to Khabarovsk to start delivering gas to Khabarovsk in 2006. We are ready to give 1.5–2 bcm of gas to China. Let us extend the pipeline to China. We can build another pipeline if one pipeline does not cope with the amount of deliveries. During his visit to Khabarovsk, Mr Bango said that China's north-eastern regions want to buy up to 20 bcm of Russian gas a year.[94]

On 2 November 2004, ExxonMobil confirmed that it was in talks with CNPC on potential sales of natural gas from the Sakhalin-1 project. On the same day, Gazprom said it was considering cooperating with ExxonMobil in exporting gas to Asia. On 4 November the ExxonMobil chairman, Lee Raymond, said that an ExxonMobil-led consortium was considering several options, including converting gas to LNG or building a pipeline to Japan or China. Previously the company had expressed the view that a pipeline would be the most cost-effective way to deliver the gas.[95]

A source close to the Sakhalin-1 consortium revealed that the gas price would be indexed to the oil price in Japan, and would be roughly $55 per 1,000 cubic metres in the first years of the contract. This price is definitely higher than Russian domestic regulated prices, which in 2004 were hovering between $22 and $33 per 1,000 cubic metres. At preliminary negotiations in 2003–4, buyers in China proposed purchasing Russian gas from the Sakha Republic for $60 per 1,000 cubic metres at the border, and bids for Kovykta gas were in the $20–70 price range. Gazprom reckoned that a $50–55 gas price was most suitable for the unregulated sector of the Russian market. Starting on 1 October 2005, the gas monopoly was set to buy gas from Lukoil at the entrance to the Yamal gas transportation system for a minimum of $22.50 per 1,000 cubic metres and is selling gas on the border with Europe at an average price of $100.

Khabarovsk consumers welcomed gas even at relatively high prices, since the need to import fuel made electricity tariffs in the Far Eastern region on the whole significantly higher than the average elsewhere in Russia. Insiders at Rao Unified Electricity System (UES), the parent company of Khabarovskenergo, were pleased with the June 2004 supply agreement and pointed out that Sakhalin gas would be cheaper than coal, which the majority of local electric power stations had been using. UES calculated that Khabarovsk Territory, once it switched to gas, would save 600 million rubles per year.[96]

The Japanese government was not ready to give up the idea of distributing gas from Sakhalin to Japan rather than to China. Shoichi Nakagawa, Japan's trade minister, said ExxonMobil had been negotiating with Beijing to build a pipeline to north-east China via Russia's Pacific coast and Exxon had always viewed China as a potential customer. But the Sakhalin-1 project had long regarded Japan as the most natural and secure market. The US energy company wanted to build a pipeline from Sakhalin to Japan, but Japanese utilities, no longer under the control of the government, said they already had enough gas supplies to fuel their gas-fired power stations. Japan's utilities also had virtually no experience with piped gas, nor did Japan have an integrated gas network. The Japanese government was keen to pipe gas from Sakhalin because it would help its long-term policy goals of diversifying from Middle Eastern oil and shifting to cleaner fuel, as a way of meeting its Kyoto Protocol obligations.[97]

In August 2005 the JPDO,[98] a Japanese registered natural gas transportation and sales company, together with Stroytransgaz/Rosneft initiated a feasibility study for a natural gas pipeline from Sakhalin Island to Hokkaido/Honshu. Phase One of the study report, completed in December 2005, concluded that the project was technically, physically, and economically feasible. The companies submitted the study to the Russian Federation government in March 2006 for review and project approval.

Two planned new power plants would play the role of inaugural customers: a 300 MW combined cycle generator in Nayoro City and either a 2,000 MW combined cycle facility at the Mutsu-Ogawara Industrial Park or a 500 to 1,000 MW plant near Sapporo city. Large-scale power plants, those over 149 MW, require three and a half years of extensive and detailed environmental assessments. The overall construction period for such a power plant can last up to six years. Natural gas would be supplied by Gazprom. Estimates put initial 2011 annual gas demand at 0.5 bcm, mostly for the Nayoro Power Plant. That would rise to 3.5 bcm in 2013, after the commencement of commercial operation of the Mutsu-Ogawara Power Plant, with forecasts indicating an annual requirement of 8 bcm in 2019. The initial 0.5 bcm per year would come from Sakhalin-2. During President Putin's visit to Japan in November 2005, the Japanese and Russian governments agreed on the necessity for long-term cooperation, and the agreement stipulated that Russian natural gas, as gas or transformed to electric power, would have access to the Japanese market.[99]

By 2006, Khabarovsk Territory had already begun to receive Sakhalin-1 gas. According to Viktor Lyubushkin, general director of Rosneft–Sakhalinmorneftegaz, the region would receive 0.157 bcm of gas in 2006. Contracts relating to the supplies were signed by Sakhalin-1 project operator Exxon Neftegaz with Khabarovskenergo and Khabarovskkraigaz. The annual sale volumes would amount to about 1 bcm initially, rising in five years to as much as 3 bcm.[100]

In late September 2006, during a meeting with Victor Ishayev, governor of the Khabarovsk Territory, President Putin rejected the idea of building a new gas pipeline into China from a neighbouring Russian region. He insisted that 'we have laid a pipeline leading to Khabarovsk and it is designed for domestic gas supplies only. Under law, Russia has only one exporter – Gazprom'. The construction of the 445 km (277 mile) domestic pipeline project connecting Sakhalin and Khabarovsk via Komsomolsk-on-Amur was launched in March 2002 and completed ahead of schedule on 14 September 2006.[101]

In October 2006, Rosneft commissioned a 425 km section of the Komsomolsk-on-Amur–Khabarovsk gas pipeline (in 1987, it had built the first gas main in the region, the 557 km Okha–Komsomolsk-on-Amur pipeline with a capacity of 4.5 bcm/y). In 2006, Rosneft operated the entire Far East gas transport system.[102] On 16 October 2006, ExxonMobil (a Sakhalin shareholder) entered into a preliminary agreement to deliver 8 bcm of gas annually to China, and India's ONGC (another shareholder) has expressed interest in selling to India. The shareholders would be ready to build a gas main from Sakhalin to north-east China but Gazprom has opposed this.[103]

Daltransgaz has played a significant role in gas pipeline development in Russia's Far East. Construction of a gas pipeline to Khabarovsk is part of the programme to provide gas supply to the Sakhalin Region, Khabarovsk, and Primorsk Territory. The programme was drafted by Rosneft–Sakhalinmorneftegaz and approved by Russian Federation Government Resolution No. 852 in July 1999. To start with, the programme envisaged using the existing 140 km Okha–Komsomolsk-on-Amur gas pipeline, which had a pressure of 4 MPa and a capacity to move 4.5 bcm per year. This pipeline was pumping no more than 1.5 bcm of gas per year in 2004 because of the depletion of the Sakhalin fields run by Rosneft–Sakhalinmorneftegaz. The programme then called for building a new gas main SKV pipeline, of the same capacity and 1,587 km in length. The first stage of this gas line, the Komsomolsk-on-Amur–Khabarovsk section, composed of 375.2 km of trunk gas pipelines and 47.6 km of laterals was constructed during 2009–10, and the whole pipeline was completed in September 2011.[104]

Initially Rosneft refused to agree to Gazprom's taking a blocking shareholding in Daltransgaz. In 2003, Rosneft concluded that the project was unprofitable and sold 47.59 per cent of the shares to the Khabarovsk Region's Administration. Rosneft retained a blocking package of 25 per cent plus one share. In 2007, through an intermediary company, Gazprom purchased the 47.59 per cent shares from the regional government for 10.4 billion rubles (around US$420 million). Currently the share package is under the management of KIT Finance (a company close to Gazprom's Gazfond), while the Russian Federal Property Fund (the agency managing state property) owns 27.39 per cent.

In 2008, after the Russian president instructed Rosneft and Gazprom to end their dispute, they signed a complex agreement on cooperation in the Far East. Its essence was that, to start with, Rosneft would sell 25 per cent plus one share of Daltransgaz to Gazprom. It would then help Gazprom to negotiate purchases of gas from the Sakhalin-1 operator ENL. In exchange, Gazprom guaranteed Rosneft access to free capacity in the future SKV pipeline that would become a part of Daltransgaz. That agreement was due to be commissioned in 2011. By then, Gazprom also wanted an agreement on purchases from Sakhalin-1 for delivery to the Primorsk region. On 23 July 2008, Gazprom approved the purchase of the first stage of the SKV.

The Russian government, expecting resistance from foreign members of the Sakhalin-1 project, prepared at the end of June 2008 an order transferring to Gazprom the right to secure the state share of gas extracted at both Sakhalin-1 and Sakhalin-2. The Gazprom deputy CEO, Alexander Medvedev, had disclosed that according to Gazprom calculations, Sakhalin-2 would be profitable from 2014, when the state share would amount to nearly 2 bcm of gas, a volume large enough for the gasification of the Sakhalin Islands.[105]

In October 2008 Viktor Timoshilov, the coordinator for Gazprom's eastern projects, revealed that the capacity of the SKV gas pipeline may ultimately be as much as 30 bcm. In its first phase up to 2011, capacity would be roughly 10 bcm. The pipeline would be supplied with gas from the Sakhalin-1 project and later from the Kirinskoye, East Odoptu, and Ayashskoye blocks. The pipeline will be tied to the unified gas pipeline system via the Yakutia–Khabarovsk pipeline. Gazprom is drawing up the general scheme for gasification of the Sakhalin region, which will include the construction of a branch of the SKV gas pipeline to Yuzhno-Sakhalinsk.[106]

In the wake of the ceremony to mark the launch of the SKV gas pipeline on 31 July 2009 it was Gazprom rather than Sakhalin-1 which felt under pressure.[107] Gazprom and ExxonMobil were yet to agree on gas supplies. Prime Minister Putin asked Gazprom Deputy Chief Executive Alexander Ananenkov whether Gazprom could guarantee gas supplies for domestic consumers such as the new car assembly line built by car maker Sollers, which had just received a $158 million loan from the state bank VEB. Ananenkov responded that Gazprom did not have gas for Sollers because it could not contract enough for the new pipeline from the operators of the Sakhalin-1 consortium. He asked for state help. Ananenkov told Putin after the inauguration ceremony that 'we will need to work intensively with Exxon to obtain these 8 to 10 bcm of gas from Sakhalin-1'.[108]

The pressure on Gazprom to secure the initial 6 bcm/y of supply volume would be continuously intensified. It was not surprising to hear of a swap initiative by Gazprom. According to Gazprom's head of Foreign Relations, Stanislav Tsygankov, the state gas giant was considering a possible swap of LNG supplies to India in exchange for its 20 per cent stake in the Sakhalin-1 project. He explained that India would like to receive Russian LNG, so they were talking about swaps, but this plan was not, at that point, at the implementation stage.[109] This indication of the possible sell-off of its Sakhalin-1 stake shows that Gazprom was and is desperate to find supply sources for SKV. A stern reality is that it will take a long time to fill the SKV gas pipeline with either Sakhalin-3 gas or Chayanda gas.

Sakhalin-2
Sakhalin-2 is the world's biggest integrated oil and gas project. Phase 1 involved oil production from an offshore platform, Molikpaq, installed at the Piltun-Astokhskoye field in 1999. Phase 2 included the installation of two further platforms, 300 km of offshore pipelines connecting all three platforms to the shore, more than 800 km of onshore oil and gas pipelines, an onshore processing facility, an oil export terminal, and the construction of Russia's first liquefied natural gas plant.[110] The solid basis of this development was the massive proven reserves. The Russian Federation State Commission for Mineral Reserves lists Piltun-Astokhskoye's oil with condensate recoverable reserves (categories A B C1+C2) at 133.5 mt and the figure for Lunskoye is 35.5 mt. Furthermore, the Commission estimates Piltun-Astokhskoye's gas reserves (gas free and in gas caps, categories A B C1+C2) at 102.8 bcm and Lunskoye's at 530.8 bcm.[111]

The starting point of this giant project was the Sakhalin-2 production sharing agreement signed on 22 June 1994 between the Russian side (Russian Federation and Sakhalin Oblast Administration) and SEIC, which originally comprised the US companies Marathon Oil (30 per cent) and McDermott (20 per cent), together with Mitsui (20 per cent), Royal Dutch Shell (20 per cent), and Mitsubishi (10 per cent). Subsequently, McDermott sold its 20 per cent stake to the remaining shareholders on a pro rata basis. Of the four remaining companies, Marathon Oil with 37.5 per cent had the largest equity stake and remained the project operator. In December 2000, however, Marathon withdrew from the project leaving Shell with 62.5 per cent, Mitsui with 25 per cent and Mitsubishi with 12.5 per cent. Finally, only a few days later, Shell sold part of its holding to Mitsubishi leaving the final SEIC structure as Shell (55 per cent), Mitsui (25 per cent), and Mitsubishi (20 per cent).[112] In June 2001, Sakhalin-2's Supervisory Board approved the development plan for Phase 2 of the project, involving the construction of the infrastructure to enable year-round production and export of oil and gas as LNG.

To begin with, in 2003, Gazprom opposed the existence of an LNG project at the terminus of the eastern gas pipeline. It argued that the cost of gas delivery from East Siberia would be $100 per 1,000 kilometres and so its liquefaction was devoid of economic sense. However, this stance changed and, for two reasons, Gazprom came to support the idea of an LNG plant at the end of the gas pipeline from Sakhalin. First, China sought to buy gas at domestic prices and South Korea certainly had no plans to overpay. To make the Chinese and South Koreans more tractable, Gazprom could say that if they were not willing to obtain pipeline gas at market prices, it would sell it as LNG to Japan or the USA. Secondly, Gazprom needed to say something about its own plant in Primorskii Territory in case Shell refused to sweeten the terms of Gazprom's entry into that project and insisted on an increase in the project's cost.

In June 2006 at the annual general meeting of Gazprom sharehold-ers, Ananenkov announced that the company was developing feasibility studies for building gas liquefaction plants in the regions of the Far Eastern ports of Nakhodka (in Primorskii Territory) and Vanino (in Khabarovsk Territory). In late November 2006, Primorskii's governor, Darkin, signed an agreement with Gazprom to build a gas pipeline to the southern ports of the region. At that time, he said that 'we will either build an LNG plant or execute other gas sale options up to laying a gas pipeline under the Sea of Japan'.[113]

These plans provoked intense pressure from the Russian authorities on the Sakhalin-2 project. On 18 September 2006 the Russian govern-ment announced that it had withdrawn the environmental approval granted in 2003 for the Sakhalin-2 LNG project. The MNR minister, Yury Trutnev, commissioned a comprehensive survey of Sakhalin-2 compliance with project and environmental rules and requirements – this took place between 25 July and 20 August. Experts from the Rosprirodnadzor under MNR photographed the hazardous areas through which the pipeline would be passing, and this showed that engineering works to protect the pipe from mudflow were not in place. At the Agency's request (according to its deputy head, Oleg Mitvol), the Far Eastern division of the Russian Academy of Sciences conducted an analysis of this segment of the pipeline. Oleg Mitvol reported that the scientists had discovered that mudflow volumes in the region could reach 0.5 mcm while pipelines can be destroyed by as little as 70,000 cubic metres of mudflow.[114]

Rosprirodnadzor approved the environmental review of the adjust-
ments to the offshore pipeline route from the Piltun-Astokhskoye field
in Sakhalin-2. In 2004, Sakhalin Energy, the operator of the project,
had to suspend laying sub-sea pipelines at the Piltun-Astokhskoye field
for two seasons in order to conduct additional research. In 2005,
Sakhalin Energy decided to move the offshore pipeline route 20 km
further south than the original route, in order to protect the Okhotsk
and Korean populations of grey whales, which are an endangered
species. Rosprirodnadzor also approved the environmental review for
the onshore pipelines from the Chayvo field of Sakhalin-1 and the
3D offshore seismic survey programme for the Veninsky block of
Sakhalin-3.[115]

The withdrawal of environmental approval threatened the Sakhalin
government's plan to develop the island as a hub for broader-based
regional oil and gas projects. According to the Sakhalin Governor, Ivan
Malakhov (speaking at a round table discussion entitled 'The Strategy
for Developing the Sakhalin Region in the Period to 2020', held in
Moscow on 16 June 2006), the Sakhalin region may become a centre
for developing the continental shelf of the entire Far East. He added
that, at that moment, there were seven official proposals on how to
use Sakhalin gas.[116]

On 21 December 2006, Gazprom agreed to pay $7.45 bn to take
majority control in Sakhalin-2, the $20 bn oil and gas project led
by Royal Dutch Shell, thus strengthening the Kremlin's grip on the
country's energy resources. The price paid by Gazprom for its 50
per cent plus one controlling share exceeded analysts' expectations.[117]
Gazprom announced that it would take control on 22 December 2006,
following a Kremlin meeting between President Putin and the heads of
the four companies Mitsubishi, Mitsui, Shell, and Gazprom.[118] Moscow
was very concerned about the cost overrun of the project, which had
led to the environmental problems and equity changes. The scale of
cost overrun was exposed after the preliminary agreement between
Shell and Gazprom in July 2005 and, as a result, Shell was forced to
disclose the total expected cost to be $22 billion, compared with the
original figure of a little over $10 billion.[119] In other words, Shell had
completely lost control of the project and the additional $12 billion
of costs would mean that, under the terms of the PSA, the Russian
government would never have made any money from it. Gazprom saw
the urgency of stepping into the project as the leading shareholder.

During a conversation on 4 June 2007 with G8 member-state journal-
ists at his residence in Novo-Ogarevo, President Putin also commented
on the Sakhalin-2 transaction. He called the original project PSA

... a colonial agreement which has absolutely nothing in common with the interests of the Russian Federation ... I only regret that at the beginning of the 1990s Russian officials permitted such tricks for which they should be imprisoned. Execution of this agreement opened a long period during which Russian natural resource exploitation took place while nothing was received in return, or practically nothing. If our partners took their obligations seriously then we would have no reason to correct them. But they are guilty of breaking environmental laws. This is a recognized and obvious fact confirmed by objective environmental experts. Unlike others, Gazprom has not come in response to our pressure and then taken something away. Gazprom has paid a huge sum to participate in the project – US$8 billion. This is the market price.[120]

On 21 December 2007 Shell, Mitsui, and Mitsubishi announced that they had signed a protocol to bring Gazprom into Sakhalin-2 as the leading shareholder. In the spring of 2008 an official of Tokyo Gas and the Russian Minister, Viktor Khristenko, confirmed that Sakhalin-2 would postpone the beginning of LNG deliveries until 2009. The first stage of the Sakhalin LNG facility would deliver 4.8 mt of LNG from 2009 and the second stage would start to deliver the same amount from 2010. Gazprom has also declared its intention to obtain control over Sakhalin-1. If in that role it takes an approach similar to that used in Sakhalin-2, it is reasonable to expect re-orientation of gas deliveries from the domestic market to exports. It is unlikely that Gazprom will build a pipeline to China, preferring rather to re-orient Sakhalin-1-produced gas to the LNG facility at Sakhalin-2 after an expansion of its capacity. According to Gazexport, the extraction capacity of Sakhalin-1 amounts to approximately 10 bcm/year. In 2005, Gazprom had discussed the advantage of combining all Sakhalin projects so as to allow an expansion of Gazprom's production base. The most optimistic scenario involves the construction of a pipeline from the Chayvo, Odoptu, and Arkutun-Dagi deposits to an LNG facility in Primorsk, which could be expanded within two years. In 2008 it was assumed that about 50 per cent of gas extraction from Sakhalin-1 would go to the Russian domestic market.[121]

On 18 February 2009 Russia inaugurated the $22 billion Sakhalin-2 gas project. According to the Sakhalin Energy CEO, Ian Craig, in 2009 the project would ship around 50 tankers with 145,000 cubic metres of capacity and 50 cargoes of oil, each of 700,000 barrels (or 95,500 tonnes). In 2010, the number of LNG cargoes would reach 160 as the facility reached its full capacity.[122] On 29 March 2009 Sakhalin-2 sent its first LNG cargo (45,000 cubic metres for Tokyo Gas and Tokyo Electric) via the LNG carrier *Energy Frontier* for delivery at Tokyo Bay.

The first train at the 9.6 mt/y liquefaction complex has begun production and the second came on stream in 2009 as scheduled. According to Sakhalin Energy, 2009 and 2010 saw a gradual rise to full capacity production. The newly built Sakhalin-2 infrastructure includes three offshore platforms, an onshore processing facility, 300 km of offshore pipelines and 1,600 km of onshore pipelines, an oil export facility, and the LNG plant.[123] At the end of 2010, Sakhalin-2's LNG plant reached its full production capacity.[124]

Sakhalin-3

As a result of a tendering process held in 1993 by the Sakhalin Administration and the Ministry of Natural Resources, the right to explore and produce hydrocarbons within the Kirinsky block went to Mobil and Texaco, while Exxon Neftegaz won the rights for the Ayashsky and Vostochno–Odoptinsky blocks. In November 1997 Mobil and Texaco agreed to cede a third of their stake in the project to Rosneft and SMNG. The parties then formed a new operating company, PegaStar Neftegaz, which was intended to develop Kirinsky. In 1996 ExxonMobil and SMNG formed a consortium for the development of the Ayashsky and Vostochno–Odoptinsky blocks.[125] Kirinsky was put on a production sharing agreement list in 1999, while the preparatory work on the PSA (drawing up a feasibility study under production sharing terms) was carried out in 2000–2. In 2003, however, PegaStar Neftegaz began to consider the possibility of implementing the project within the framework of the then applicable taxation and licensing regulations.

In January 2004, Deputy Premier Viktor Khristenko declared the 1993 tender results null and void, because of subsequent amendments introduced into the second section of the Russian Federation Tax Code with regard to the PSA procedure. As a result, all three blocks – Kirinsky, Vostochno–Odoptinsky, and Ayashsky – were placed in the non-appropriated subsoil fund.[126] In 2005, hypothetical recoverable reserves of the Vostochno–Odoptinsky block were estimated at 70 mt of oil and 30 bcm of gas. Reserves at the Ayashsky block are 97 mt of oil and 37 bcm of gas; at Kirinsky 452.3 mt of oil, 700 bcm of gas and 53 mt of gas condensate; and at Veninsky 114 mt of oil and 315 bcm of gas.[127] The potential annual extraction capacity of Kirinsky, East Odoptinsky, and Ayashsky blocks was estimated at 15 bcm. It was expected that exploration works and commercial development preparation would take place in 2012–13. If Gazprom succeeds in merging all of the Sakhalin-3 gas, another 5.5 mt/y of LNG will be available from the Sakhalin Islands. The assumption remains that domestic Russian use would account for about 50 per cent of total

Sakhalin-3 gas extractions. Under the most optimistic forecast, total delivery volumes of LNG from Sakhalin by 2012–13 could amount to nearly 19 mt annually.[128]

Rosneft was given a five year exploration license for the Veninsky block in 2003. Its share in the venture was 74.9 per cent and the remaining 21.5 per cent was assigned to a regional state enterprise, the Sakhalin Oil Company. In August 2004, these two companies signed an agreement to conduct a geological study and subsequent exploitation from the Veninsky block, and in early 2005 they completed a study of the environmental impact and impact on the fishing industry.[129] In 2006, Rosneft drilled and tested the first exploration well at the South Ayashskaya structure. The well revealed promising oil and gas-bearing strata and tests confirmed the presence of hydrocarbons. The Kantan-3 submersible rig, owned by the Shanghai Offshore Drilling Co, conducted the drilling. In 2007, the company shot 680 square kilometres of 3D seismic survey, three times greater than the license requirements. Both companies started preparation for the drilling of an exploration well at the North Veninskaya structure, planned for 2008. They also obtained an extension of the Veninsky exploration license until the end of 2010.[130]

On 15 June 2007, however, Gazprom, at a session of the governmental Commission on the Fuel and Energy Complex (FEC), asked for licenses for deposits of the Sakhalin-3 project and Chayanda to be issued without any auctions or competitions. Prime Minister Mikhail Fradkov was ready to support this but the MNR minister, Trutnev, reacted sceptically. On 19 June 2007, at a session of the presidium of the Commission on Social and Economic Development of the Far East and the Trans Baikal, Ananenkov declared that Gazprom would not be able to deliver sufficient gas to the Far East until 2014. In his opinion, the requirements of the four Far East regions exceed 15 bcm of gas annually. Meanwhile, ExxonMobil, the operator of Sakhalin-1, was intending to deliver most of its gas to China. Ananenkov considered that it would be necessary to issue an instruction that ExxonMobil sell this gas to Gazprom rather than export it.[131] The Moscow authorities were sympathetic towards the Far East region's serious shortage of gas which Gazprom had highlighted.

On 6 May 2008 Viktor Zubkov, the former prime minister, approved within a month the documentation for the transfer of the Sakhalin-3 Kirinsky block to Gazprom. Hypothetical resources at Kirinsky totalled 930 bcm of gas and 453 mt of oil, which were much larger than those of Veninsky. Earlier, Gazprom had sent letters to the MNR and to Rosnedra requesting user rights for three Sakhalin-3 shelf blocks at once – East Odoptinsky, Ayashsky, and Kirinsky. Gazprom's eastern projects coordinator, Viktor Timoshilov, revealed at a Sakhalin oil and gas conference that Gazprom planned to launch production at the Kirinskoye gas field in the Sakhalin-3 project, in 2014. He said Gazprom would prepare documentation of design and estimates in 2009–10, carry out seismic work, and drill two wells. Gas reserves at the Kirinskoye field are 70 bcm and Gazprom planned to develop the field independently.[132] It is not surprising that JBIC had already expressed its interest in financing the Sakhalin-3 project, on condition that Japanese companies participated.[133]

According to Gazprom, the start of geological exploration in the Kirinskoye field was another crucial step towards establishing Gazprom's resource base in the Far East and creating a new offshore gas production area in Sakhalin as part of the state's Eastern Programme. This was to be the first offshore project in Sakhalin carried out solely by Russian companies.[134] On 16 October 2009 Gazprom's senior officer said that the firm intended to start production at the Sakhalin-3 project's Kirinsky field in 2011 or 2012 – two years earlier than previously planned – in order to achieve full use of its pipeline to the Pacific port of Vladivostok. Alexander Mandel, the head of Gazprom's subsidiary which oversees offshore projects, said that 'it had been planned that development of Kirinsky will start in 2014. We are thinking of beginning production there at the end of 2011 or the beginning of 2012'.[135]

In 2010 Gazprom announced the discovery of a new gas field in the Kirinsky block in Sakhalin-3. Gazprom is planning exploration activity through 2013 to evaluate the resources, whose estimates are as much as 1 to 1.4 tcm. The exploration programme includes over 3,000 square kilometres of 3D seismic surveys and more than 20 exploration wells. The reserves are forecast to grow by some 600 mt of fuel equivalent, including approximately 500 bcm of gas (in 2010, C1+C2 reserves of Kirinskoye and Yuzhno-Kirinskoye were, as mentioned earlier, estimated to be 137 bcm and 260 bcm respectively, definitely smaller than that of Sakhalin-2 reserves).[136] The scale of estimated gas reserves is clearly a very strong incentive for Gazprom to accelerate the exploration and consequently the produced gas supply to the SKV pipeline, but producing gas in Sakhalin-3, to start in 2014, looks very ambitious. According to the *Gas Industry Development Strategy to 2030*, Sakhalin-3 is expected to come on stream in 2017–20, with a total of 58 wells to be drilled.[137] This time frame of 2017–20 is more realistic than 2011–14, despite Gazprom's emphasis on its early production schedule.

Regardless of its failure with the Kirinsky block, Rosneft continues to develop the Veninsky block of Sakhalin-3 together with SINOPEC.

Rosneft controls 74.9 per cent of the joint operation and SINOPEC the remaining 25.1 per cent. ONGC has declared that it might purchase a 23 per cent share from Rosneft for US$300 million. If that transaction takes place, Rosneft will retain a controlling stake in the project with 51.9 per cent.[138]

Sakhalin-4, 5 and 6

The Astrakhanovskaya structure was considered one of Rosneft's top priority projects. The structure's hypothetical recoverable reserves total 110 bcm, and the maximum annual production rate has been projected at 4.3 bcm. The structure lies between 3 and 44 kilometres from the shore at a sea depth of 10 to 30 metres. However, no operations were conducted on the structure prior to 2000 since neither Rosneft nor Sakhalinmorneftegaz had enough funds even to pursue geological exploration. The first exploratory well, drilled in 2000, did not prove to be very successful.

In June 2001, Rosneft and BP signed a protocol of intent under which BP would participate in developing a field in the Sakhalin-4 project. Instead of a second exploratory drilling, in the first quarter of 2003 SakhalinNIPImorneft Institute and Sakhalinmorneftegaz, jointly with ZAO Sakhalinskie Projekty, performed an adjustment of the feasibility study for the Astrakhanovskaya structure drafted in 2002; this aimed to analyse project efficiency without conducting offshore drilling. Rosneft projected three deposits of the Daginsky and Nizhneokobykayaskaya horizons, which are accessible for onshore drilling and contain possible recoverable resources totalling 3 mt of condensate and 40 bcm of gas. The final calculations did not satisfy BP, which in May 2003 stated that it had lost interest in the Astrakhanovskaya structure because its reserves were too small.[139]

In early March 2004, BP agreed in principle with Rosneft's terms for the financing of investments in their joint project at Sakhalin. BP would fund not only its share of capital expenses on development, but also those of Rosneft. The two parties reached an accord that Rosneft would have a 51 per cent stake in the project and BP 49 per cent. According to Rosneft, overall investment would amount to around US$3 billion to 3.5 billion, of which US$150 million to 170 million would be spent on exploring the deposits.[140] Since 2004, Elvary Neftegaz (51 per cent of which belongs to Rosneft, 49 per cent to BP) has been the operating company for Sakhalin-4 (Zapadno-Shmitovsky block) and in 2007 it drilled two test wells on the Medved and Toiskaya structure.[141]

On 2 September 2009 the Gazprom deputy CEO, Valery Golubev, said that Gazprom was preparing a bid for the license to develop

Sakhalin-4 but, due to the existing levels of market demand, this was the only bid it was currently working on. He said that 'we currently have 4.1 tcm, but we're planning a lot of exploration before 2020. We aim to drill some 100 wells and add approximately 5.6 tcm to our reserves, in other words we will have around 10 tcm of C1 offshore gas reserves by 2020'. He added that an oil price of $50 per barrel was needed to make Sakhalin-3, to which Gazprom already held a license, profitable.[142]

The Sakhalin-5 project involves the development of four blocks – Kaigansko-Vasyukansky, Lopukhovsky, Vostochno-Shmitovsky, and Yelizavetinsky. The sea depth within the Lopukhovsky block is up to 100 metres. The Ministry of Natural Resources puts the project's ultimate recoverable resources at 325 mt of oil equivalent. According to TNK–BP, the exploration targets of the project contain 130 mt of oil and 500 bcm of gas in recoverable resources. The project's two largest blocks are Lopukhovskaya, with recoverable resources of 41 mt of oil and 135 bcm of gas and Vostochno-Shmitovsky, with 65 mt of oil and 215 bcm of gas in recoverable resources.[143] Elvary Neftegaz has carried out considerable 2D and 3D seismic exploration in order to meet the license requirements, within the framework of the Sakhalin-5 project (Vostochno-Shmitovsky block). Geological study at the project's Kaigansko-Vasyukansky block actually began in 2002 with the drilling of four exploratory wells. It received a certificate to open the deposit and in 2008 the company was continuing geological study of the block.[144]

Gazprom Neft purchased TNK–Sakhalin in 2005 from TNK–BP, which decided to sell the license after BP seismic data interpretation had concluded in 2003–4 that the company's book showed a low potential. On 31 May 2007 the license for geological study of the Sakhalin Lopukhovsky block (jointly located in Sakhalin-4 and Sakhalin-5) expired. TNK–Sakhalin (75 per cent owned by Gazprom Neft and 25 per cent by Sakhalin Oil Co), a subsidiary of the Sakhalin Administration, held the license.[145]

Sakhalin-6 is unique in having operations conducted by a single Russian company, Petrosakh. In 2000, the Russian Alfa Group bought 97 per cent of Petrosakh which has a geological license for the block and conducted a seismic survey in mid-2001. From 2001 to 2002, Petrosakh drilled an additional three producing wells. In 2001, Petrosakh was licensed for the geological study of the Pogranichnoye Block, which lies directly offshore and parallel to the existing Okruzhnoye field. In the summer of 2002, Petrosakh acquired 480 square kilometres of 3D seismic programme covering the central portion of the Pogranichnoye

Block. In 2004, it acquired an additional 65 square kilometres of 3D seismic data in two programmes covering the northern and southern areas of the near-coastline transition zone. In 2004, however, Petrosakh was sold to a strategic investor, the Urals Energy Company. Urals Energy agreed to pay Alfa a perpetual royalty payment of $0.25 per tonne ($0.03 per barrel) for any commercial quantities of oil produced and landed from the currently non-producing offshore areas covered by the license for the Pogranichnoye Block.[146] There has been no further sign of serious interest in comprehensive exploration in Sakhalin-5 and 6 at the time of writing.

Sakha Republic Gas

Sakha is the name of the principal Yakut ethnic group, and the Sakha Republic was previously known as the Yakut ASSR. The Republic, covering one-fifth of the Russian Federation (3.1 million square kilometres) is the country's largest autonomous republic but has a population of only 1.3 million. The total area regarded as being possible locations of oil and gas deposits covers 1.64 million square kilometres, of which only 0.2 square kilometres of seismic data has been shot. During the Soviet era, the development of Yakutian gas resources was largely ignored, due to the combination of geographical remoteness, the inhospitability of the climate, and the failure to attract foreign investment.[147]

In September 1990, the Yakut ASSR proclaimed the sovereignty of the Yakut-Sakha SSR within the Russian Federation. Demarcation of the powers of the Russian Federation and the Republic of Sakha (Yakutia) was stipulated by the Federal Treaty signed on 31 March 1992 and in an agreement 'On Mutual Relations between the Governments of the Russian Federation and the Republic of Sakha'.[148] These documents granted the Sakha Republic the right to develop, extract, and sell the natural resources in its territory. Sakha Republic was already exploring all the alternative ways of accelerating the development of its huge energy (especially gas) resources, including foreign investment.

Hydrocarbon exploration in the Yakut ASSR actually began in 1935, but the first serious exploration work began in the 1950s. During the 1970s and 1980s, all the major gas fields discovered (such as Verkhnevilyuchanskoye, Taas–Yuryakhskoye, Talakanskoye, and Chayandinskoye) were in the Botuobinsky geological region. Despite these discoveries, the Republic of Sakha's oil and gas reserves were barely tapped due to their remote location. In the Republic, the initial geological gas reserves are estimated at 9.6 tcm, which includes 8.3 tcm of recoverable. Almost three quarters of the initial geological gas reserves are predicted to be located at depths up to 4 km (4.8 tcm within depths of 1–3 km, 2.2 tcm within depths of 3–4 km). The gas reserves are concentrated in the Vilyuiskaya, Nepsko-Botuobinskaya, and Predpatomskaya oil and gas bearing regions.[149]

By the mid-1990s 30 deposits of hydrocarbons had been discovered, 19 of which are located in the south-west Nepsko-Botuobinskaya oil and gas bearing region, and 11 in the central Vilyuisk region (9 deposits) and the Predpatomsk region (2 deposits). As shown in the Appendix, the explored commercial gas reserves in Sakha Republic are concentrated in two oil and gas bearing regions: Vilyuiskaya and Nepsko-Botuobinskaya. According to Vasiliy Moiseyevich Efimov, the then president of Sakhaneftegaz, in 1998 the registered C1 category reserves were 1,000 bcm; in the Vilyuisk region there were 10 fields with 437.8 bcm and in the Botuobinsk region 21 fields with 586.3 bcm. In addition to this, the reserves of the Chayandinskoye field in the Botuobinsk region were estimated at 755 bcm (previously 208 bcm), of which 535 bcm were exploitable. This was a significant increase on the official figure of 165 bcm. A total of 64 wells had been drilled in the field.[150] This revised figure removed the uncertainty about the scale of the proven reserves. However, at this time Chayandinskoye had to compete against the giant Kovykta gas field with its 870 bcm of proven reserves.

Looking to be restored to its position as the main source of gas exports in the region, the new Sakha Republic took a number of initiatives. The first was in December 1997 when the Sakha Republic president, Nikolayev, and the Irkutsk regional governor, Boris Govorin, signed an agreement on social-economic, scientific-technical, and cultural cooperation, which stipulated the possibility of jointly exporting gas to the Asia–Pacific region. If their gas resources were combined, Sakha and Irkutsk could guarantee annual exports of 50 bcm of gas for the next 40 years.[151] In 1998, the Sakha Republic took a bold initiative. Sakhaneftegaz proposed the setting up of an East Siberian consortium based on the Irkutsk region, the Sakha Republic, and the Evenki Autonomous region of Krasnoyarsk Krai, and the proposal was supported by Rosneft, the Chita region Administration, JSC UES of Russia, the Administration of the Evenki Autonomous region, and Rusia Petroleum. Interestingly, Sakhaneftegaz signed an agreement with Rusia Petroleum for the joint development of the Kovyktinskoye and Chayandinskoye fields even though, in the first place, the priority would be given to Kovyktinskoye.[152]

At that time, this consortium proposal was the only way to remove suspicion about the scale of the proven reserves. Its significance lay in the fact that the combined development of the Kovyktinskoye and

Chayandinskoye fields would guarantee proven gas reserves that could justify a 4,000 km long distance pipeline development. It was not surprising to see this hybrid export scheme promoted at the fourth USA–China Oil and Gas Industry Forum by Xu Dingming, the then counsel of the Industrial Department of the State Development Planning Commission. The export scheme had two options, even though there was no difference between them regarding the section of pipeline within Chinese territory. The first option was a 4,961 km pipeline, of which the section in Russian territory was 1,960 km, plus the two pipelines from the Kovykta and Chayanda field which met at Bodaybo, adjacent to the northern tip of Lake Baikal. The second option was a 5,626 km pipeline, of which 2,625 km was located in Russian territory. This second option gave absolute priority to the Kovykta project, and the Chayandinskoye field was connected as a back-up supply source.[153]

It is true that up to the early 2000s the Chayanda gas project was not as advanced as Kovykta in terms of development preparation. However, significant work had been carried out to demonstrate the quality of proven reserves. First of all, on 26 July 2002, Sakhaneftegaz completed a preliminary feasibility survey for a pipeline exporting gas from Chayanda to Shenyang. This work began in April 1999 soon after the February 1999 agreement between CNPC and Sakhaneftegaz. The initial export volume would be 12–15 bcm/y and the figure could be increased to 20 bcm/y at a later stage. Secondly, Moscow's Central Commission for Reserves, of the Russian Federation's Ministry of Natural Resources, approved the 2002 revised figure for Chayanda gas proven gas reserves of 1,240 bcm. Thirdly, in October 2002, Gazprom and the Sakha Republic Government signed a framework agreement to form a joint venture to bid for the development license for the Chayandinskoye field and other fields in the Sakha Republic.[154] Following the preliminary feasibility work, Chayanda is estimated to have extractable gas reserves totalling 379.7 bcm of category C1, 861.2 bcm of C2 and 20 bcm of D1, and oil reserves totalling 42.5 mt of C1, 7.5 mt of C2, and 78.8 mt of D1. Besides this, the deposit also contains helium and other components.[155] The sizable reserves have attracted huge interest from Gazprom, which lost no time in making a strategic move towards the Sakha Republic.

Gazprom's strategic alliance with the Sakha Republic Government has special implications. In February 2002, the Sakha Government reported to the local legislative assembly that Yukos had secured a 47 per cent controlling stake in Sakhaneftegaz, which had previously been controlled by the Sakha Republic Government.[156] Yukos' initiative forced the Sakha Republic Government to become a minority shareholder in

Sakhaneftegaz. The Government wanted a strategic partnership with Gazprom in order to make sure that the Russian government would give the highest priority to the Sakha–Gazprom strategic project. In February 2003, Gazprom CEO Miller and Rosneft's President Bogdanchikov asked President Putin to instruct the Russian Federation's Ministry of Natural Resources and other relevant ministries to consider developing the Chayandinskoye, Talakanskoye, Sredne-Botuobinskoye, Kovyktinskoye, and Verkhnechonskoye oil and gas fields as a single project, and initiate an auction in accordance with the legislation in effect. Putin accepted this proposal.[157]

In March 2003 the Russian Government, at its first Cabinet meeting about the development of oil and gas reserves in East Siberia and the Far East, adopted the draft of the Programme to Establish a Unified System of Production, Transportation and Supply of Gas in East Siberia and the Far East. This was seen as a prelude to the development of gas exports for markets in China and other countries in the Asia–Pacific region as the basis for further work. In the same year, the Yakut authorities created a kind of local Gazprom. They transferred to STN, which is 100 per cent owned by the Republic, almost all the gas trunk lines (except the Alrosa-Gas local pipeline), all gas distribution pipelines, the rights to market gas (which Yakutgazprom produces), to provide gas supply, and to process gas (at the sole Yakut gas processing plant in the Republic).[158]

In late February 2004, the government of the Sakha Republic proposed a new route for the crude oil supply pipeline, laying it in a single corridor with the Chayanda–Nakhodka gas line, and in future attaching it onto a pipeline linking the Kovyktinskoye, Dulisminskoye, and Yaraktinskoye fields in the Irkutsk Region. The Sakha Republic's President Shtyrov, argued that the route proposed by the Sakha Republic was backed by the MNR, Gazprom, Surgutneftegaz, scientific centres, and the majority of ministries and governmental agencies. The Sakha Republic's prayer was not in vain. At a conference on the development of transportation infrastructure in the Far East held in Khabarovsk on 26 February 2004, President Putin asked Shtyrov to submit development studies on a new pipeline.[159] This is how Chayandinskoye's position as a major gas export source was reconfirmed in Gazprom's three scenarios for its Eastern UGSS, announced in the autumn of 2004.

In May 2007 Gazprom CEO Miller put to the government his request to receive Chayanda without either auction or competition. The deputy prime minister and Gazprom board chairman, Dmitry Medvedev, directed core government departments to consider the request. The MNR and MEDT gave its opinion that the law 'On

Subsoil' did not provide for licensing without competition, and the transfer would have to remain under consideration. Gazprom could not receive the license for the Chayandinskoye deposit earlier than the end of June 2008, owing to the need to calculate the compensatory payment to the budget for subsoil use. The MEDT, the MIE, and the Ministry of Finance with the participation of Rosnedra (the Federal Subsoil Use Agency) had to calculate the historic costs of the project. The main Rosnedra research institute, VNIGRI, had already calculated a minimum starting payment of 9.46 billion rubles.[160]

On 17 July 2007 in Yakutsk, Alexander Ananenkov, acting chair of Gazprom's management committee, and President Shtyrov of the Sakha Republic signed a cooperation agreement, stipulating the following major directions for their joint work as part of the implementation of the East Programme:[161]

- joint participation in the development and execution of gas supply and gasification projects and programmes;
- preparation and implementation of hydrocarbon field prospecting, exploration, and development projects;
- organization of regional processing and chemical production;
- energy saving technology development and implementation;
- preparation and coordination of regional fuel balances resulting from efficient local resource usage;
- development of a complex environmental monitoring programme.

GIV is a 100 per cent owned Gazprom subsidiary company set up pursuant to the board's 2007 resolution to implement Gazprom's investment projects in eastern Russia. The following projects have GIV as a customer:

- the Chikanskoye gas and condensate field;
- the Angarsk–Irkutsk gas main;
- gas supply to the Kamchatka region;[162]
- SKV gas main;
- YKV gas main;
- Phase 2 of gas supply to the Kamchatka region;
- gas processing facilities in the Republic of Sakha (Yakutia);
- gas processing facilities in the Irkutsk region;
- gas processing facilities in the Krasnoyarsk region;
- Phase 2 of the Bratskoye gas and condensate deposit and upgrades to the Bratsk gas main.

Gazprom's initiative in setting up GIV confirms that the YKV gas main and the gas processing facilities in the Sakha Republic were Gazprom's two main priorities in the Republic. According to Prime Minister Victor Zubkov's order of 28 November 2007, Chayanda and almost all those deposits for which Gazprom petitioned were included on the list of gas deposits having federal value. In addition, there was the law 'On Gas Supply' which provided that deposits of federal value could be transferred without competition to the organization controlling the uniform system of gas supply.[163]

On 25 January 2008, President Shtyrov and Gazprom CEO Miller met in Moscow to discuss the status of the Cooperation Agreement between Gazprom and Sakha. The meeting focused on issues relating to the development of gas production and processing, and gas chemical centres in the East Programme. On 29 January Shtyrov declared that Gazprom should receive subsoil use rights at Chayanda in 2008. On 6 February Dmitry Medvedev said 'as chairman of Gazprom's board of directors and first deputy prime minister I have instructed the Ministry of Industry and Energy to accelerate the preparation of operations'.[164]

In February 2008 Medvedev, then a candidate for the Russian Presidency, decided to give Chayanda to Gazprom and charged the MIE and the MNR, together with Gazprom, to prepare offers on licensing by 12 March. Even though the Minister of Natural Resources insisted on carrying out an auction, on 16 April, the Russian government issued an order whereby Gazprom received the Chayanda license. The license, however, was not totally free: Gazprom must pay the state an indemnity of 8–10 billion rubles ($339–423 million).[165]

To accelerate progress, a series of working group meetings was held. In early October 2007 the Gazprom head office hosted the first session of the Joint Working Group for cooperation between Gazprom and the Government of the Sakha Republic.[166] On 18 March 2008 the Gazprom headquarters hosted an extended meeting of the Joint Working Group co-chaired by Gennady Alekseev, First Deputy Chairman of the Government of the Sakha Republic, and Viktor Timoshilov, Head of the Eastern Projects Coordination Directorate of Gazprom. The meeting considered issues of interaction between Gazprom and the Sakha Republic concerning the development of gas processing and chemical facilities in the region. The meeting scrutinized the progress made in preparations for bilateral accord on the gasification of the Republic. Focus was also placed on the synchronization of activities carried out by Gazprom and the Government of the Republic as part of the Eastern Programme.[167] This was followed on 12 May by a similar meeting between Gazprom's Deputy CEO, Ananenkov, and the first vice president of the government of the Sakha Republic,

Gennady Alekseev.[168] These working group meetings continued during 2008 and 2009.

In 2008 the Sakha Republic government's plan was to increase its current gas production capacity of 1.6 bcm/y to 34 bcm/y by 2020, largely through the development of the Chayandinsky deposit. According to the strategy of social and economic development for Sakha through 2020, the total regional product would grow 2.8 times, the average income of the population would grow 3.6 times, and the region would become completely unsubsidized. Social, infrastructural, and industrial projects required 2 trillion rubles (about $809 billion) of private investment and 500 billion rubles of state infrastructure development investments which Sakha did not, at that time, have. Possible sources of finance were the State Investment Fund or the Federal Target Programme for Far Eastern development (at that time providing 566 billion rubles for projects throughout the Far East).[169]

In order to meet the challenges set out in the EGP, Gazprom's Board of Directors charged the Management Committee with continuing the work aimed at developing the company's resource base with the following objectives: i) to secure the loading of the SKV gas transmission system with gas from the Kirinskoye field beginning in 2014, ii) to organize gas supply to the Kamchatka Krai, with the first gas to be delivered to Petropavlovsk-Kamchatsky in the fourth quarter of 2010, and iii) to prepare the hydrocarbon reserves of the Chayanda field for commercial development, with gas to be fed to the gas transmission system in 2016.[170]

In August 2009, Alexey Miller reported on the work done by Gazprom to create the Yakutsk gas production centre as part of the Eastern Gas Programme. At that date, two exploratory wells were being drilled, and 2D and 3D seismic surveys, scheduled for the winter of 2009, were at the initial preparatory stage. Engineering geological studies and a geodesic survey had been done. Development of the oil rim and gas deposit was to be started in 2014 and 2016 respectively. However, gas from the Chayanda field has a complicated composition with high helium content. In 2010 Gazprom should finalize an investment plan for a gas chemical facilities development in East Siberia and the Far East, as well as the Reservoir Management Plan for the Chayanda oil and gas condensate field development.

To ensure the injection of Yakutia's gas into the gas transportation system by 2016, it is first necessary to construct the YKV gas trunk line, and Gazprom will start constructing this pipeline in 2012 following completion of the SKV gas transmission system. Miller has pointed out that:

the gas production forecast has been updated according to the follow-up data obtained during operations at the Chayanda field. In this regard, it is necessary to resolve the issue of granting Gazprom in 2009 the subsurface rights for Yakutia's Srednetyungskoye, Tass-Yuryakhskoye, Sobolokh–Nedzhelinskoye and Verkhnevilyuchanskoye fields in order to ensure an optimal loading of the YKV gas transportation system. This is especially significant as helium is only found in some of the fields and Yakutia's gas may be delivered to the market simultaneously.[171]

On 12 March 2010, after signing the general plan for gas supply and gasification for the republic, Miller announced that Gazprom was ready

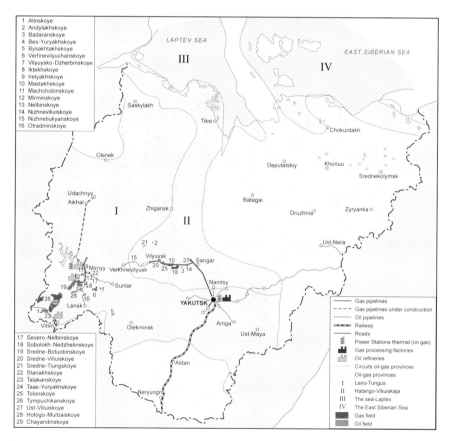

Map 3.2: Gas and Oil Fields in the Sakha Republic

Source: Sakhatransneftegaz, Sakha Republic.

The author is grateful to I. K. Makarov, Director General of Sakhatransneftegaz, and Svetlana Yegorova-Johnstone, advisor to the Ministry of Nature Protection of the Sakha Republic for help in the production of this map.

to begin gasification of the Republic of Sakha as part of the EGP. He elaborated on the tough deadline set by the government for bringing on stream the prioritized facilities in Yakutia, launching the YKV gas transportation (GTS) construction in 2012, and starting production from the Chayandinskoye field in 2014 (for oil) and 2016 (for gas). The plan required the company to put the top-priority gas processing and gas chemical facilities into operation in 2016. The construction of the 3,500 km YKV gas main, with a delivery capacity of 36 bcm/y and 13 compressor stations, would be another major element of a new powerful GTS in the Russian East. This gas line was planned to run parallel to the ESPO line, and its construction would allow Gazprom to build an LNG plant and petrochemical complex in the Primorye territory.[172]

In July 2010 the Gazprom Gas Industry Commission for Field Development and Subsurface Use approved the development plan for the Chayanda gas. The plan was submitted to the Central Commission for hydrocarbon field development under the Federal Sub-surface Use Agency.[173] As discussed earlier (in the section on the Gasification Programme in East Siberia and Russia's Far East), in December 2010, Gazprom CEO Miller and Sakha Republic president Borisov signed the Partnership Agreement for socio-economic development driven by the EGP implementation. Chayanda gas development is the cornerstone of this initiative.

Conclusion

During the 2000s there was much talk, but not much real progress, regarding gas field development in East Siberia. It is not an exaggeration to say that virtually the whole decade was dedicated to preparing development for the coming decades. The approval of EGP during the second half of the 2000s was the one tangible achievement which accelerated movement towards the development of the massive proven gas reserves confirmed in remote East Siberia and the Far East. If there was any positive news, it was that in 2010 Gazprom took a significant step by approving the development scheme of the Chayanda gas field in the Sakha Republic, which envisages the development of the YKV gas pipeline network; Gazprom was also hugely encouraged by the new gas discovery in the Kirinsky block in Sakhalin-3 in the autumn of 2010. On top of this, Gazprom's unexpected decision to buy out the Kovykta gas asset, by joining in the spring 2011 auction, added a new possible source of supply for gas exports from the Eastern Gas Programme. This is assuredly a positive change, not only for the

Gazprom management, but also for potential gas importers in north-east Asia. Will the decade of 2010 be really different from the 2000s? The preparation work during the 2000s provided the Russian planners with a number of options.

The first option is not really an export-oriented one. If the Russian planners are just gasifying the Far East region, then the SKV gas pipeline plus some more production will be good enough, but it will not result in exports in the near future. The second and third options are export-oriented ones. The second option is the self-financed one. If the planners want to place more emphasis on export, then they will need to make a critical decision based on their priorities about which gas fields – Sakhalin-3 or Chayanda gas – will be developed first, and to finance it themselves or perhaps with non-Chinese partners. The third option is to fall in with what the Chinese negotiators want in terms of fields, pipelines, and investments.

Gazprom is very well-positioned to accelerate the development of Chayanda or Kovykta gas, as soon as a price deal with China material-izes. The summer of 2011 failed to witness the long-awaited settlement on gas price negotiations. Hard evidence that Russia is taking the steps necessary for the full development of its East Siberian gas resources will not exist until the leaders of both Russia and China are prepared to reconsider and change their stance towards the price negotiations. Failure to reach a gas deal right after Russia's leadership change in 2012 will mean a substantial reduction in the prospects of large-scale Russian gas exports to Asia in the foreseeable future.

Appendix to Chapter 3

Table A.3.1: Russian Gas Industry Indicators (bcm)

	GO	GC	Gas Exports	Kazakh	Gas Imports Turkmen	Uzbek
1995	596	378	190.6	3.2	0.3	0.0
1996	602	380	196.5	2.0	1.9	0.0
1997	571	350	198.4	2.7	1.9	0.0
1998	591	365	202.5	2.3	0.0	0.0
1999	591	364	204.5	3.6	0.0	0.0
2000	584	377	217.1	5.3	29.1	2.4
2001	582	373	200.1	3.8	15.5	0.0
2002	595	389	201.2	7.2	14.0	4.5
2003	620	393	203.7	7.1	4.3	1.3
2004	634	402	217.9	6.6	7.0	7.1
2005	641	405	222.3	12.4	3.4	8.2
2006	656	419	262.5	14.4	38.5	9.6
2007	654	426	237.4	15.2	42.6	9.7
2008	665	420	248.1	17.4	42.3	10.3
2009	596	390	202.8	17.7	11.3	15.4

Note: i) GO = Gas Output; ii) GC = Gas Consumption.

Source: GO and GC figures RC (2010), 18; Gas exports and imports figures RC (2010), 123.

Table A.3.2: Russia's Gas and Condensate Production by Company (bcm)

	2003	2005	2007	2008	2009
Russia	620.326	641.015	654.136	664.852	596.443
Oil Companies	40.489	50.833	60.327	59.705	
Rosneft	7.012	13.045	15.467	12.162	12.214
Surgutneftegaz	13.883	14.361	14.139	14.123	13.592
Lukoil	4.769	5.795	13.725	14.234	12.408
TNK–BP	6.809	10.517	10.156	12.800	13.592
Gazprom Neft	1.985	1.994	1.759	3.026	3.110
Russneft	0.665	1.058	1.555	1.343	1.325
Yukos	3.448	1.970	1.519	-	-
Slavneft	0.823	0.994	0.928	0.899	0.905
Tatneft	0.728	0.737	0.738	0.762	0.761
Bashneft	0.369	0.363	0.341	0.356	0.374
Novatek	20.134	25.369	28.516	30.812	32.675
Gazprom	540.180	547.058	550.143	550.911	455.165
IPSAO	19.523	17.754	15.150	23.424	50.323

Note: IPSAO = Independent and PSA Operators.

Source: RC (2010), 31.

Table A.3.3: East Siberia's Gas Production by Company, 2008

Company	Gas mcm	%
Lenaneftegaz (supervised by Surgutneftegaz)	45.6	0.9
Ust-Kut Neftegaz (supervised by Irkutsk Oil Co)	56.1	1.1
Verkhnechonskneftegaz(supervised by TNK–BP and Rosneft)	1.7	0.0
Yakutgazprom	1543.0	28.9
Irelyakhneft		
Dulisma (supervised by Urals Energy)	29.6	0.6
Taimyrgas (supervised by Norilsk Nickel)	1145.6	21.4
Vostsibneftegaz (supervised by Rosneft)	1.1	0.0
Danilova (supervised by Irkutsk Oil Co)		0.0
Taas-Yuryakh Neftegazodobycha		0.0
Vankorneft (supervised by Rosneft)	67.6	1.3
Alrosa-Gaz (supervised by Alrosa)	227.1	4.3
Norilskgazprom (supervised by Norilsk Nickel)	2161.0	40.5
Sakhatransneftegaz	5.3	0.1
Suzun (supervised by Gazprom and TNK–BP)		0.0
Branches vertically integrated oil and gas mining and smelting companies	3649.7	69.1
East Siberia, Total	5283.7	100.0
Russia, Total	**664852.0**	
Share of East Siberia in the Russian Federation (per cent)	0.8	

Source: Kontorovich and Eder (2009).

Table A.3.4: Gas Production in Siberia and the Far East up to 2030 (bcm/y)

	2010	2015	2020	2025	2030
W.S	574	632	682	699	744
E.S	6	35	109	142	152
F.E	22	30	45	52	72
S1	8	11	12	12	12
S2	14	18	22	22	22
S3–9	0	1	11	18	38
ST	1	1	1	0	0
WKS	0	0	1	8	12
SFE	602	697	845	893	967
RT	644	732	920	992	1076

Note: W.S. = West Siberia; E.S .= East Siberia; F.E .= Far East; S1 = Sakhalin 1; S2 = Sakhalin 2; S3–9 = Sakhalin 3–9; ST = Sakhalin Terrestrial; WKS = West Kamchatka Shelf; SFE = Siberia and Far East; RT = Russia Total.
Source: Kontorovich and Eder (2009)[174].

Table A.3.5: Fields in the Sakha Republic, 2010[175]

Status of fields	Reserves & Resources	Subsoil user
Federal fields		
Chayandinskoye gas condensate field	Gas: B+C1, 379.7 bcm, C2, 861.2 bcm; Oil (recoverable): C1, 42.5 mt, C2, 7.5 mt; Condensate (recoverable): C1, 5.7 mt, C2, 12.7 mt	Gazprom
Srednetyungskoye gas condensate field	Gas: B+C1, 153.2 bcm, C2, 9.2 bcm; Condensate (recoverable): C1, 8.0 mt, C2, 0.64 mt.	Will be transferred to Gazprom without a tender
Taas-Yuryakhskoye oil and gas field	Gas: B+C1, 102.7 bcm, C2, 11.3 bcm; Oil (recoverable): C1, 2.0 mt, C2, 5.3 mt.	Will be transferred to Gazprom without a tender
Sobolokh–Nedzhelinskoye gas condensate field	Gas: B+C1, 64.0 bcm, C2, 0.7 bcm; Condensate (recoverable): C1, 3.0 mt, C2, 0.04 mt	Will be transferred to Gazprom without a tender
Verkhnevilyuchanskoye oil and gas field	Gas: B+C1, 139.6 bcm, C2, 69.7 bcm; Oil (recoverable): C1, 1.5 mt, C2, 32.3 mt; Condensate (recoverable): C1, 2.7 mt, C2, 1.3 mt.	Will be transferred to Gazprom without a tender
Distributed fields		
Alinskoye gas and oil field	Gas:B+C1, 0.7 bcm, C2, 1.7 bcm; Oil (recoverable): C1, 0.5 mt, C2, 4.6 mt.	Surgutneftegaz
Verkhnepeleduyskoye gas and condensate field	Gas: B+C1, 1.0 bcm; Condensate (recoverable): C1, 0.03 mt	Surgutneftegaz
Vostochno-Alinskoye oil field	Oil (recoverable): C1+C2, 8.0 mt	Surgutneftegaz
Irelyakhskoye oil gas field	Oil (recoverable): C1, 9.6 mt, C2, 0.3 mt; Gas: B+C1, 4.7 mcm; Condensate (recoverable): C1, 0.1 mt	Irelyakhneft
Mastakhskoye oil and gas field	Gas: B+C1, 18.3 bcm, C2, 6.5 bcm; Condensate (recoverable): C1, 0.4 mt, C2, 0.3 mt.	Yakutgazprom
Machobinskoye oil and gas field	Gas: B+C1, 3.6 bcm, C2, 2.1 bcm; Oil (recoverable): C1, 2.3 mt, C2, 2.8 mt.	Yakutgazprom
Miminskoye oil and gas field	Gas: B+C1, 1.4 bcm Oil (recoverable): C1, 0.67 mt	Yakutgazprom
Nelbinskoye gas field	Gas: B+C1, 4.3 bcm, C2, 2.3 bcm	Yakutgazprom
Otradninskoye gas field	Gas: B+C1, 0.9 bcm, C2, 5.4 bcm	Lensk-Gaz
Severo-Nelbinskoye gas field	Gas: B+C1, 0.54 bcm; Condensate (recoverable): C1, 0.02 mt,	Yakutgazprom
Severo–Talakanskoye oil field	Oil (recoverable): C1 + C2, 34.9 mt	Surgutneftegaz
Sredne–Botuobinskoye oil gas field, northern block	Gas: B+C1, 19.1 bcm; Condensate (recoverable): C1, 0.35 mt	Alrosa-Gaz

Status of fields	Reserves & Resources	Subsoil user
Sredne–Botuobinskoye oil gas field, central block	Gas: B+C1, 134.2 bcm, C2, 15.5 bcm; Oil (recoverable): C1 + C2, 123.4 mt; Condensate (recoverable): C1, 2.45 mt.	Taas-Yuryakh Neftegazdobycha
Sredne–Botuobinskoye oil gas field, eastern block	Gas: B+C1, 2.2 bcm, C2, 1.8 bcm; Oil (recoverable): C1, 2.9 mt; C2, 3.6 mt; Condensate (recoverable): C1, 0.04 mt, C2, 0.03 mt	Rosneftegaz
Sredne-Vilyuiskoye gas condensate field	Gas: B+C1, 126.1 bcm; Condensate (recoverable): C1, 5.5 mt, C2, 0.03 mt	Yakutgazprom
Stanakhskoye oil and gas field	Gas: B+C1, 5.5 bcm, C2, 14.5 bcm; Oil (recoverable): C1, 0.06 mt, C2, 5.8 mt;	Surgutneftegaz
Talakanskoye oil and gas field, central block	Gas: B+C1, 18.9 bcm, C2, 3.8 bcm; Oil (recoverable): C1, 99.5 mt, C2 13.0 mt; Condensate (recoverable): C1, 0.2 mt, C2, 0.04 mt.	Surgutneftegaz
Talakanskoye oil and gas field, eastern and Taransky blocks	Gas: B+C1, 16.6 bcm, C2, 14.8 bcm; Oil (recoverable): C1, 47.5 mt, C2 5.2 mt; Condensate (recoverable): C1, 0.2 mt, C2, 0.1 mt.	Surgutneftegaz
Tympuchikansky gas and oil field	Gas: B+C1, 2.2 bcm, C2, 11.2 bcm; Oil (recoverable): C1, 0.4 mt, C2 16.6 mt	Gazprom Neft-Angara
Khotogo-Murbayskoe gas field	Gas: B+C1, 1.0 bcm, C2, 9.6 bcm	Gazprom Neft-Angara

Undistributed fields

Andylakhskoye gas condensate field	Gas:B+C1, 7.8 bcm; Condensate (recoverable): C1, 0.6 mt.	
Badaranskoye gas field	Gas: B+C1, 6.1 bcm	
Besyuryakhskoye gas condensate field (Iktekhskoye block)	Gas: B+C1, 1.2 bcm, C2, 9.2 bcm	
Bysyhlakhskoye gas condensate field	Gas: B+C1, 5.5 bcm, C2, 9.7 bcm; Condensate (recoverable): C1, 0.14 mt, C2, 0.25 mt	
Tolonskoye gas condensate field	Gas: B+C1, 33.5 bcm, C2, 10.6 bcm; Condensate (recoverable): C1, 1.6 mt, C2, 0.5 mt	
Vilyui–Dzherbinskoye gas field	Gas: B+C1, 19.0 bcm, C2, 16.4 bcm	
Ust–Vilyuiskoye gas field	Gas: B+C1, 0.8 bcm	

Source: Glazkov (2010), 12.

CHAPTER 4

THE OIL INDUSTRY IN CHINA

From the very beginning of China's rapid economic growth in the late 1970s it was obvious that it could only continue if the bulk of energy supplies came from outside the country. Nonetheless, through exploration and reorganization, the government attempted to derive maximum interim benefit from China's own relatively small oil endowment before the emphasis shifted to the search for supplies from abroad.

This chapter aims at understanding the achievements of China's oil industry during the 2000s. It will start with a brief review of China's energy balance and the related energy policy. It will then cover the issues of oil industry restructuring, exploration and production, the national oil and gas reserves survey, oil imports, pipeline development, the country's refining sector expansion, and strategic oil reserves. The chapter's main focus is to appreciate China's capacity for oil production based on a full knowledge of national reserves. This will explain why the rapid expansion of oil imports – and consequently China's Out-Going policy for overseas oil production – are inevitable. And the rapid expansion of the country's strategic oil reserve will demonstrate China's strong concern for security in its energy supply. This general focus will also yield some hints about a more specific, but central, question of China's stance towards buying oil from Russia.

China's Energy Balance

During the early decades of its rapid economic development China was able to meet its rapidly growing energy needs entirely from domestic resources, so its direct impact on global markets was minimal. From 1980 to 2000, primary energy consumption more than doubled, from about 603 to 1,386 mtce, implying an average annual rate of growth of 4.2 per cent. Over the same period, GDP quadrupled, growing at an average annual rate of 9.7 per cent, indicating a low elasticity of energy demand relative to GDP of 0.43. But then, between 2000 and 2005 (the period of the Tenth Five Year Plan), annual growth in primary energy consumption accelerated and averaged 9.9 per cent; the energy/GDP elasticity was on average greater than 1.0 – more

142

than double the 0.5 elasticity which energy projections implicitly assumed for the period 2000 to 2020. The data in Table 4.1 comes from the Chinese government's *Statistical Yearbook* and is different from that used by the World Bank in its widely quoted estimates.[1] In 2007 the World Bank concluded that if the energy consumption trends of the Tenth Five Year Plan (which covered the period 2006–10) continued for the following 15 years, energy consumption in 2020 would be as much as twice the level of the then current high projections.[2] Without prompt application of comprehensive energy policies to rein in energy consumption, China could have been embarking on a self-defeating energy-intensive voyage that would erode its energy supply and have unsustainable impacts on the environment.

According to the IEA's *World Energy Outlook 2010*, China's primary energy demand stood at 2,131 million tonnes of oil equivalent (mtoe) in 2008, equal to 93 per cent of the energy demand of the USA in

Table 4.1: Primary Energy Production and Consumption in China (mtsce and %)

	Total (mtsce)	Coal	Crude Oil	Natural Gas	Electricity
Production					
1980	637.4	69.4	23.8	3.0	3.8
1985	855.5	72.8	20.9	2.0	4.3
1990	1039.2	74.2	19.0	2.0	4.8
1995	1290.3	75.3	16.6	1.9	6.2
2000	1289.8	72.0	18.1	2.8	7.2
2005	2058.8	76.5	12.6	3.2	7.7
2006	2210.6	76.7	11.9	3.5	7.9
2007	2354.5	76.6	11.3	3.9	8.2
2008	2612.1	76.6	10.7	4.1	8.6
2009	2750.0	77.3	9.9	4.1	8.7
Consumption					
1980	602.8	72.2	20.7	3.1	4.0
1985	766.8	75.8	17.1	2.2	4.9
1990	987.0	76.2	16.6	2.1	5.1
1995	1311.8	74.6	17.5	1.8	6.1
2000	1385.5	67.8	23.2	2.4	6.7
2005	2246.8	69.1	21.0	2.8	7.1
2006	2462.7	69.4	20.4	3.0	7.2
2007	2655.8	69.5	19.7	3.5	7.3
2008	2914.5	70.3	18.3	3.7	7.7
2009	3066.5	70.4	17.9	3.9	7.8

Note: electricity includes hydro-power, nuclear power and wind power.

Source: *China Statistical Yearbook*. 2010.

that year. Despite the global financial and economic crisis, China's economy had remained resilient, growing at 9.1 per cent in 2009.[3] A year later, IEA announced that, based on preliminary data, China had overtaken the USA to become the world's largest energy user. IEA added that since 2000 China's energy demand had doubled, yet on a per capita basis it was still only around one-third of the OECD average. Prospects for further growth were very strong considering the country's low per-capita consumption level and the fact that it is the most populous nation on the planet, with more than 1.3 billion people.[4]

Data from the National Bureau of Statistics showed that China's energy consumption increased by 5.9 per cent in 2010 as factories stepped up production to fuel the world's fastest-growing major economy. The Bureau said that energy use climbed to 3.25 billion mtsce) in 2010, while energy consumption per unit of GDP fell by 4.01 per cent.[5] Despite the fact that China continued to be the largest coal producer in the world, its coal imports were increasing rapidly. In 2010 China imported 165 mt, up 31 per cent on the previous year. Net coal imports in 2010 were 145.8 mt, up 42.4 mt or 29 per cent from the previous year.[6] *China Daily* reported that net coal imports may reach as much as 233 mt in 2011.[7] The growth of coal imports provides a further problem for China's energy planners, whose main priority is oil supply security. Primary oil demand under the IEA's new policy scenario will be 11.7 mb/d in 2020 and 14.3 mb/d in 2030.[8] Assuming that China maintains an oil production level of 4 mb/d in the coming decades, the level of oil imports will be 7.7 mb/d in 2020 and 9.3 mb/d in 2030.

In early January 2011 Zhang Ping, chairman of China's National Development and Reform Commission, said that China had cut coal use by nearly 2 bt over the previous five years by replacing outdated thermal power plants with environment-friendly ones and by pushing clean energy. It had already been announced that 5 trillion yuan ($755 billion) would be spent on clean energy over the next decade to lift the non-fossil fuel component from 8 per cent of primary energy demand in 2009 to 15 per cent by 2020.[9] However, this non-fossil fuel initiative is not enough to slow down galloping oil demand or to reduce the share of coal from its present 70 per cent to the hoped-for 50 per cent in China's energy balance.

The scale of China's energy demand in the coming decades is gigantic. IEA's chief economist, Fatih Birol, has said that China requires total energy investments of some $4 trillion over the next 20 years to keep feeding its economy and avoid power blackouts and fuel shortages. He added that China is expected to build some 1,000 GW of new power-generation capacity over the next 15 years. That is about equal to the current total electricity generation capacity in the USA – a level achieved over several decades of construction.[10] A brief overview of China's energy policy will give some hints as to whether the Chinese planners are well placed to handle such an enormous expansion of energy demand.

China's Energy Policy

This section aims at reviewing the change of priorities of China's energy policies, in particular the White Paper issued at the end of 2007, and then understanding how the government's bureaucracy has restructured both energy policy in general, and oil policy in particular.

A change of priorities

The two main central preoccupations of the Chinese government's energy policies are: growing demand for oil as a whole and reliance on oil imports.[11] As shown in Figure A.4.5, the National Development and Reform Commission (NDRC) is the primary policymaking and regulatory authority in the energy sector, while four other ministries and commissions oversee various components of oil policy. Since the early 1990s, the government has recognized the vital role of energy in underpinning its desired rate of economic growth; hence, obtaining sufficient energy has become a top priority of economic policy.

According to the Department of Communications and Energy's 1997 *Energy Report*[12] this priority was translated into four specific energy objectives: security of supply, social equity, economic efficiency, and environmental protection. During the period 1998–2002, secure access to international oil supplies became a particular concern, while environmental protection appeared to have been the least important of the energy sector objectives.[13]

The time of energy surpluses had come to an end, and by the end of 2002 shortages and blackouts became commonplace. By 2003 they had become a serious threat to the Chinese economy. This combination of energy supply shortages, rising oil imports, and growing global concerns about the environment showed the pressing need for a comprehensive solution to the country's energy problems. A new approach to energy policy making was unavoidable. In 2003, the Energy Bureau was created, followed by the Energy Leading Group (ELG) with its own State Energy Office (SEO) in 2005. Consultations with energy think tanks also became more common. This more urgent approach to energy

policy culminated in the government's announcement in 2004 that
the sustainable use of energy was from then on a key priority for the
whole nation.[14]

In mid-May 2005 the establishment of the ELG, headed by Prime
Minister Wen Jiabao and consisting of two vice prime ministers and 13
ministers, including those of foreign affairs and defence, was approved
by the State Council. In addition a 24 member SEO was to assist the
work of the ELG. The tasks of the ELG and the SEO were to secure
overseas oil and gas reserves, to ease the chronic electricity shortage,
to stabilize the supply of coal, to promote energy efficiency in industry,
and to promote nuclear power and renewable energy resources. One
senior official in the Research Office of the State Council said that
energy security was the ELG's key task, and the search for foreign
cooperation in accessing overseas oil and gas reserves would be the
pivotal security issue.

Another challenge facing this new institutional environment will
be to avoid overlaps of authority in the energy industry, which exist
even in the case of the earlier established Energy Bureau under the
NDRC. For example, the pricing department of the NDRC is still in
charge of setting energy prices; the Ministry of Trade oversees oil and
coal imports and exports as well as the domestic oil market; while the
Ministry of Land and Resources is responsible for resource exploration.
The result is that energy policy is not always consistent and is sometimes
even contradictory.[15]

On 18 October 2005 China issued the full text of the Eleventh
Five-Year Plan (2006–10), stressing the importance of the development
of the energy sector over this period. Its proposals were aimed at 20
per cent reduction of energy consumption per unit of GDP by 2010,
notwithstanding the fact that the country's GDP would then be more
than twice the size in 2000. To achieve this target, the central government called for the development of energy-efficient products and the
optimization of the energy consumption structure. Proposals included
the establishment of large coal production bases, the renovation of
medium-sized and small coal mines, and the development and utilization of coal-bed methane. The central government would pay equal
attention to the development of oil and gas resources, sparing no efforts
in the exploration and development of oil and gas, and would promote
cooperation with foreign countries to enhance oil stockpiling capacity.
The government would vigorously develop nuclear power, build more
power grids, transmit more electricity from the west of the country to
the east, and step up the pace of developing renewable energy sources
such as wind and solar power.

According to Xu Dingming, then director of the Bureau of Energy of the NDRC, the central government's new energy policy would do the following: i) establish an energy-saving society using measures such as removing restrictions on low-emission vehicles and initiating oil-saving management methods; ii) improve capacity, to guarantee oil supplies; iii) reform oil pricing and the tax system; iv) establish an oil stockpiling system; v) develop new energy and renewable energy; and vi) strengthen international cooperation.

The central government has promulgated Energy Laws which embody the guiding principles of the development of different sectors of the energy industry. The Power and Coal Laws have been in existence since 1966, while the Renewable Energy Law has existed since 1 January 2006 and was last amended in September 2011. Regulations on Sino-foreign joint exploration and development of onshore oil resources were adopted in 1993 and were last amended in September 2011.

China does not have a Petroleum Law, at least not under that name. Instead, there are two regulations, for onshore and for offshore resources. Regulations on the exploitation of offshore petroleum resources in cooperation with foreign enterprises were adopted in 1982, and were last amended in September 2011. An onshore equivalent set of regulations was adopted in 1993 and was also last amended in September 2011.

During the Eleventh Five Year Plan, the Chinese government expected its known hydrocarbon reserves to grow by 0.9–1.0 bt annually for oil and 400–450 bcm for natural gas. The authorities claimed that the country had proven oil reserves of 24.8 bt, proven natural gas reserves of 4.4 tcm, and proven coal reserves of 1 trillion tonnes; consequently China still believed, overoptimistically, that it would be able to continue to rely fundamentally on its domestic resources.[16] Some changes were proposed to accelerate the construction of oil/gas pipelines during the five years of the eleventh plan, and to improve the national pipeline system, by establishing a pipeline grid to transport oil and gas from the west to the east, and from the north to the south. The second west–east gas pipeline and pipelines to import oil from abroad were also scheduled in the programme.[17]

On 26 December 2007, the Information Office of the State Council published a White Paper summing up the state of the energy situation and announcing some changes to future policies. The While Paper drew attention to three main achievements. The first was a remarkable improvement in aggregate energy supply. China had brought into operation a group of extra-large coal mines each with an annual output of over ten million tonnes. Several large oil fields had been opened as oil production bases, and crude oil production had increased steadily to 185 mt, making China the world's fifth-largest oil producer by 2006. Natural gas output increased from 14.3 bcm in 1980 to 58.6 bcm in 2006. At the same time the proportion of commercial renewable energy in the structure of primary energy was rising.

The second notable achievement was that the structure of consumption had improved. In 2006, China's total consumption of primary energy was 2.46 billion tonnes of standard coal equivalent, and the proportion supplied by coal was 69.4 per cent, down from 72.2 per cent in 1980. Consumption of other forms of energy had risen correspondingly and the share of renewable and nuclear energy had nearly doubled, rising from 4.0 to 7.2 per cent.

Thirdly, progress had taken place in environmental protection. In 2006, almost 100 per cent of coal-fuelled generation units had dust-cleaning facilities and nearly 100 per cent of the discharge of waste water met the relevant standards. The total installed capacity of thermal power units with flue gas desulphurization (FGD) built and put into operation grew quickly during 1998–2007: new capacity installed during the year 2006 was 104 million kw, exceeding the combined total of the previous ten years. Such thermal power units accounted for 30 per cent of total power in 2006 compared with 2 per cent in 2000.[18] More recently, 'China has emerged in the past two years as the world's leading builder of more efficient, less polluting coal power plants, mastering the technology and driving down the cost'.[19] And, more recently still, 'work is underway on China's first clean coal-based power plant in the northern city of Tianjin. The $1bn project, called GreenGen, will be the country's first commercial-scale plant to use carbon capture and storage'.[20]

The White Paper emphasized three major challenges. The first is the unbalanced geographical distribution of energy resources. Coal is found mainly in the north and north-west, hydropower in the south-west, and oil and natural gas in the eastern, central, and western regions, and along the coast. The consumers of energy resources, however, are mainly in the south-east coastal areas, where the economy is most developed. Such great distances between producers and consumers has meant that enormous expenditures are incurred on large-scale, long-distance transportation of coal and oil from the north to the south and on the transmission of natural gas and electricity from the west to the east.

Second, it pointed out the physical difficulties of developing energy resources. There are severe geological problems in accessing coal resources, and most of China's coal requires underground mining; only a small amount can be mined by opencast methods. Oil and gas resources are located in areas with complex geological conditions and at great depths, requiring the use of advanced and expensive prospecting and drilling techniques. Untapped hydropower resources are mostly located in the high mountains and deep valleys of the south-west, far from the centres of consumption, the consequent distribution issues entailing technical difficulties and high costs.

Third, there are problems of organization. So far, the energy market system is not complete and prices fail to fully reflect the scarcity of resources, the balance of supply and demand, and environmental costs. In addition, the safety of coal production is far from satisfactory, the structure of the power grid is not rational, the oil reserves are not sufficient, and a much more effective emergency warning system needs to be established to deal with energy supply breakdowns and other major unexpected emergencies.[21]

The White Paper argued that the basic themes of China's energy strategy were: giving priority to thrift, reliance on domestic resources, encouraging diverse patterns of development, relying on science and technology, protecting the environment, and increasing international cooperation for mutual benefit. The country needs to strive to build a stable, economical, clean, and safe energy supply system, so that sustained economic and social development can be supported by the development of sustainable energy.

The Seventeenth National Congress of the Communist Party of China, held in October 2007, was another forum at which energy policy was defined. In addition to the goals of accelerated transformation of the development pattern and the quadrupling of per capita GDP between 2000 and 2020, the Congress emphasized that this should be done through optimizing the economic structure and improving economic returns, while at the same time reducing the consumption of energy resources and protecting the environment. As already mentioned, the Eleventh Five Year Plan for National Economic and Social Development had already set the goals of a 20 per cent reduction between 2005 and 2010 in energy use per unit of GDP, and a 10 per cent aggregate reduction in the emission of major pollutants.[22]

The White Paper stressed the need to expand energy supply by the active and orderly development of coal, oil, and gas production, electric power generation, and renewable energy. It especially advocated greater efforts in prospecting for and exploiting oil and natural gas, concentrating on major oil and gas basins, including those of Bohai Bay, Songliao, Tarim, and Ordos, and on prospecting new areas, fields, and strata on land and in major offshore areas so as to increase the amount of recoverable reserves. It pointed to the need to tap the full potential of major oil producing areas, to maintain stable yields, to increase recovery ratios, and to slow down the trend of decreasing yields in older oil fields.[23]

Restructuring the Government's Energy Bureaucracy

The energy crisis of 2003–4 was an indictment of China's then-existing energy bureaucracy. The widespread energy bottlenecks and shortages that plagued the country prompted the Chinese leadership to embark on yet another round of restructuring the energy institutions. The first attempt at reform took the form of the bureaucratic restructuring of 2003, in particular the establishment of the NDRC's Energy Bureau.

The electricity shortages that began to emerge in late 2002 and the growing dependence on imported oil strengthened the voices of experts who viewed the country's fractured energy bureaucracy as being ill-equipped to manage the challenges of a rapidly growing, increasingly market-oriented, and internationalized energy sector. The Energy Bureau, however, was the outcome of a compromise between key stakeholders in China's energy sector. Both the NDRC and the Chinese NOCs were strongly against the establishment of a ministerial-level agency, and the Energy Bureau, under the NDRC did not have the political power or the financial and human resources to manage the energy sector effectively.

By the end of 2004, the leadership had reached a consensus on the creation of a new energy authority. The formal announcement of the establishment of the ELG and the SEO was made in May 2005 with the publication of document 2005-14 by the State Council, although in late April an NDRC official had publicly confirmed that the SEO had already been established, under the leadership of director Ma Kai and the deputy directors Ma Fucai and Xu Dingming. The ELG under the State Council was headed by Prime Minister Wen Jiabao, and the SEO reports directly to the prime minister.

Erica Downs has described how energy concerns have moved to centre-stage in Chinese politics:

> The widespread energy shortages, which coincided with the replacement of the third generation of leaders with the fourth generation, appear to have an important influence on the approach of President Hu Jintao and Premier Wen Jiabao to energy governance. Not only have Hu and Wen, by necessity, paid more attention to energy than their predecessors Jiang Zemin and Zhu Rongji, but they have also sought to redress the supply-side bias in the energy [sector]. The energy crisis exposed the limits of the 'growth at any cost' model of economic development associated with Jiang and Zhu

and undoubtedly helped to forge a consensus among the current leadership that demand moderation – essentially a domestic policy issue – is necessary for sustainable economic development.[24]

The ELG and SEO set-up failed to silence the calls for further institutional reform in the energy sector. At the end of 2005, PetroChina had 424,175 employees and SINOPEC 389,451, while the Energy Bureau had a staff of only 57. In other words, the government was heavily reliant on the energy companies for manpower and expertise. There was a serious debate about whether to restore a Ministry of Energy. Xia Yishan, a member of the expert team under the Executive Office of the Energy Leading Group considered that China did not need an administrative bureau so much as an integrated energy administrative system, and he admitted that defects were bound to appear under the existing energy management system. But he added that the government was cautious about the reestablishment of the Energy Ministry in the near future[25] (for the restructuring during the 1976–2007 period, please see Figures A.4.1–A.4.5).

On 11 March 2008 the Beijing authorities announced that they were setting up a National Energy Commission (NEC) to develop national strategy and security, and a National Energy Bureau (NEB) to administer the sector under the powerful National Development and Reform Commission (NDRC). These were to replace the Energy Office and the current Energy Bureau – also under the NDRC – and several other small departments, including one which managed civilian nuclear energy. The NEC would be responsible for studying and developing national energy strategy and assessing major issues in energy security and development, and the NEB would be mainly in charge of formulating and organizing the implementation of energy sector plans, industry policy and standards, developing new energy sources, and encouraging energy conservation.[26]

On 24 March 2008 the NEB was established and Zhang Guobao,[27] vice minister of the NDRC, became its director. Contrary to the general expectation that the NEB would take full control of energy issues relating to macro-control, project approval, and pricing system reform, the new Bureau's role may be more limited. Hua Jianmin, former Secretary General of the State Council, said that the Bureau should mainly engage in framing plans, policies, and standards relating to the energy industry. At the China Development Summit Forum, Zhang Guobao said that the NEB did not expect to participate in setting energy prices but rather to act as a suggestion-maker. He admitted that it was inconceivable that less than 100 officials in the governmental energy sector could preside over all China's energy issues. The NEB is composed of 50 members from the NDRC's former Bureau of Energy and 24 from the former Energy Leading Group of the State Council. The NEB directly manages China's energy sector.[28]

The NEB seemed likely to start as a macro-regulator, or perhaps as a provider of the improved strategic energy research needed by the policy-makers. Many experts hoped the new Bureau might resume the energy research that was initiated by the ELG but then postponed, pending the governmental reform. On 25 June 2008 China's State Council approved the details of the infrastructure of the newly established NEB. It was to be divided into nine departments and to have approximately 112 employees.[29] The NEB is under the direct administration of the NDRC. Zhang saw the establishment of the NEB as a step on the path towards setting up an independent energy ministry under the State Council.

According to a statement released by the NDRC on 29 July 2008, China's state council had specified the functions and obligations of the newly established NEB in March 2008. The functions of the NEB are a combination of those exercised by the former office of the ELG, the NDRC, and the former China Commission of Science, Technology and Industry for National Defence (COSTIND). The NEB will design energy development strategy, draft plans and policies, make proposals on energy industry reform, manage the country's oil, natural gas, coal, and power industries, manage national oil reserves, design policies on renewable energy development and energy conservation policies, and carry out international energy cooperation. According to this statement the NEB has a different focus in relation to energy from that of the NDRC and the Ministry of Industry and Information Technology (MIIT) since it will specifically manage the refining, coal-based fuel, and fuel ethanol sectors. According to Pang Changwei, a senior researcher at the China University of Petroleum, the NEB has the rank of a vice ministry.[30]

On 27 January 2010, the state council decided to set up a National Energy Council (NEC), with Prime Minister Wen Jiabao as head, to improve strategic policy-making and coordination. Vice Prime Minister Li Keqiang will act as NEC's deputy head, and Zhang Ping, head of the NDRC, will be head of the general affairs office of NEC. Zhang Guobao, the head of NEB, will serve as the office's deputy head. The NEC committee is composed of 21 members, most of them ministers of various government agencies.[31] The new NEC occupies a higher position in the hierarchy of administration than the NEC, which was set up as part of the 2008 reform, and this confirms the increasing importance of energy issues to the sustainable development of China's economy.

Oil Industry Restructuring

Between 1994 and 1998, the Chinese government reorganized most state-owned oil and gas assets into two vertically integrated firms: the China National Petroleum (Group) Corporation (CNPC) and the China Petrochemical (Group) Corporation (SINOPEC). These two conglomerates operate a range of local subsidiaries, and together dominate China's upstream and downstream oil markets. On 27 July 1998, at the grand inauguration of CNPC and SINOPEC at the Great Hall, Vice Prime Minister Wu Bangguo described the industry reshuffle as a pivotal part of the ongoing government restructuring. Under the supervision of the State Council, the two group companies report directly to the State Economic and Trade Commission (SETC). The 1998 reform introduced two vertically integrated NOCs: CNPC and the SINOPEC Group. The two corporations do not assume government functions and they are responsible for their own profits and losses.

In accordance with a geographic division, CNPC will own crude production capacity of 106 mt and refining capacity of 100.3 mt/y. It controls oil and gas fields, refineries, and petrochemical plants in 12 provinces in north and west China – Inner Mongolia, Tibet, Ningxia, Xinjiang, Liaoning, Jilin, Heilongjiang, Sichuan, Shaanxi, Gansu, Qinghai, and Chongqing. In 1997, CNPC's sales revenue was 249.4 billion yuan ($30.05 billion) making it ninety-third in *Fortune Magazine*'s list of the world's top 500 companies, and it had fixed assets of 483.05 billion yuan ($58.2 bn). CNPC remains much the largest and most influential NOC, and is the leading upstream player in China. CNPC, along with its publicly-listed arm PetroChina, account for roughly 60 per cent and 80 per cent of China's total oil and gas output, respectively.

SINOPEC has crude oil production capacity of 36 mt and refining capacity of 117.9 mt/y. It covers oil fields and refineries in 19 provinces in east and south China – Beijing, Shanghai, Tianjin, Hebei, Shanxi, Jiangsu, Zhejiang, Anhui, Henan, Jiangxi, Shandong, Guizhou, Yunnan, Guangxi, Guangdong, Hubei, Hunan, Hainan, and Fujian. In 1997 SINOPEC's sales revenue was 331.5 billion yuan ($37.53 billion), giving it sixty-second place in the *Fortune* list. SINOPEC has 89 subsidiary companies, with total assets of 380.6 billion yuan ($45.9 bn).[32] It has traditionally focused on downstream activities such as refining and distribution; these sectors made up 76 per cent of the company's revenues in 2007.[33]

Three Chinese NOCs were established between 1982 and 1988 (see Table 4.2). In 1982, the China National Offshore Oil Corporation (CNOOC) was set up under the direction of the Ministry of Petroleum

Table 4.2: Chinese NOCs, 2010

	CNPC (PetroChina)	CNOOC Ltd.	SINOPEC Corp.
Percentage owned by the state, in 2005/2010	90.0/86.29	70.6/	71.23/75.84
Annual revenues (Million yuan)	1,465,415	183,053	1,913,182
Net profit (bn yuan)	139.9	54.4	70.7
E&P revenue (Million yuan)	525,895	-	-
Refining revenue (Million yuan)	657,728	-	-
Year-end proven crude oil reserves (Million barrels), as of 2005/2010	14,187/ 11,278	1,578/ 1,915	2,841/ 2,888
Year-end proven natural gas reserves (bcf))	65,503	6,458	6,447
Crude oil output (Million barrels), 2010	857.7	263.4	327.9
Gas output (bcf), 2010	2221.2	379.6	441.4
Number of employees, as of 2005/2010	439,220/ 552,698	3,584/ 4,650	358,304/ 373,375
Parent company employees, as of 2005/2010	1,133,985/ 1,587,900	37,000/	730,800/

Sources: CNPC/PetroChina, CNOOC, and SINOPEC annual reports, and the three NOCs' official websites.

Industry (MPI) to facilitate offshore foreign cooperation. In 1983, the downstream activities were separated from the MPI, and these, together with other refinery and petrochemical activities of other industries and local governments, were transferred to a new NOC named Sinopec Corporation under the direct supervision of the State Council. The most notable change occurred in 1988 when the government transformed the MPI into a new NOC named China National Petroleum Corporation (CNPC). Thus during the period 1982–8 three NOCs had been born. Most of the regulatory and administrative functions of the previous ministry were retained in CNPC, although the Ministry of Energy coordinated the regulation of the whole energy sector.[34]

Additional state-owned oil firms have emerged in the more competitive landscape in China over recent years. CNOOC, which is responsible for offshore oil exploration and production, has seen its role expand as a result of growing attention to offshore zones. Also, the company has proved to be a growing competitor to CNPC and SINOPEC, not only by increasing its E&P expenditures in the South China Sea, but also by extending its reach into the downstream sector, particularly in the southern Guangdong Province, through its recent 300 billion yuan investment plan. In October 2007 CNOOC announced that it planned, within three years, to open 1,000 service stations in the Pearl River Delta, Yangtze River Delta, and Bohai Rim areas.[35] The Sinochem Corporation and CITIC Group have also expanded their presence in China's oil sector, although their involvement remains dwarfed by CNPC, SINOPEC, and CNOOC. The government intends to use the stimulus plan to enhance energy security and to strengthen the Chinese NOCs' global position by offering various incentives to both upstream and downstream investment.[36]

In addition to the three main NOCs, a new player has entered the oil industry. Shaanxi Yanchang Petroleum (Group) Co. Ltd (YPCL), China's fourth largest oil company, is a local state-owned oil company. Unlike CNPC (PetroChina), SINOPEC and CNOOC, which are directly owned by the state-owned Asset Supervision and Administration Commission, YPCL is a local oil company owned by the Shaanxi provincial government. On 14 September 2005, YPCL was formed by merging 21 exploration and development companies and three refineries. Its predecessor, Yanchang Oil Mining, a local-owned oil company, had been involved in a dispute with PetroChina over their overlapping blocks in Shaanxi province, in particular those located in Changqing oilfield. Originally the central government intended to merge Yanchang into PetroChina to tighten central control over oil and gas resources. But the Shaanxi provincial authority and Yanchang Oil Mining managed to block the merger, and finally Yanchang Oil Mining was transformed into a share-holding company in 2005. In response to fierce competition from the domestic giants CNPC, SINOPEC, and CNOOC, YPCL also has its eyes on overseas assets and access to finance, as a result of being listed on a stock exchange.[37]

Up to April 2004, China had four state oil trading companies: UNIPEC (China International United Petroleum & Chemicals Co. Ltd), a wholly owned affiliate of SINOPEC,[38] the PetroChina International Co. Ltd (China Oil) a subsidiary of PetroChina Company Ltd (PetroChina),[39] Sinochem,[40] and Zhuhai Zhenrong[41] (China's exclusive trader of Iranian crude). A fifth, a joint venture between CNOOC[42] and SINOPEC, was established in 2004, and a sixth, China Zhenhua Oil Co. Ltd, established in 2003, was later allowed to trade in oil.[43]

Exploration and Production

In 2005, Che Changbo (deputy director of Oil and Gas Resources Strategic Research Centre, Ministry of Land Resources) estimated that

China's crude oil and natural gas production would reach 183 mt and 50 bcm respectively. He added that crude oil production would probably reach 200 mt in 2010 and that that production level would be maintained for the following 15 years. According to the third round of appraisal of oil and gas resources (to be discussed in Chapter 5), 0.8–1.0 bt of proven oil reserves is expected to be added annually during the years 2005–20. Changbo predicted that China's oil imports in 2010 would reach 170 mt, or 44.7 per cent of the national total demand of 380 mt. Oil imports would probably rise to 200 mt by 2020, representing 44.4 per cent of the total demand of 450 mt. Another estimate, however, is significantly different: Cao Xianghong (senior vice president of SINOPEC) gives a much higher figure: oil imports in 2020 would be 270 mt or 60 per cent of total demand of 450 mt, and annual oil output could reach 180 mt before 2020. He added that China's total demand could reach as much as 600 mt in 2020 if a projection is made based on the growth of the oil demand in the two years 2004 and 2005.[44] So far this prediction has not been wide of mark.

Total oil production reached 190 mt (3.8 mb/d) in 2008, similar to the 2007 level. China's largest and oldest oil fields are located in the north-east region of the country. CNPC's Daqing field produced about 40.2 mt (0.804 mb/d) of crude oil in 2008, as shown in Table 4.3. SINOPEC's Shengli oil field produced about 27.7 mt (0.580 mb/d) of crude oil during 2008, making it China's second-largest oil field.[45] However, Daqing, Shengli, and other ageing fields have been heavily tapped since the 1960s and are expected to produce less in the coming years. Recent exploration and production (E&P) activity has focused on the offshore areas of Bohai Bay and the South China Sea, as well as onshore oil and natural gas fields in western interior provinces such as Xinjiang, Sichuan, Gansu, and Inner Mongolia.

The Third National Oil and Gas Reserves Survey

In 2006, the Ministry of Land & Resources (MLR) disclosed that the first government-initiated national oil and gas survey was being undertaken. Strictly speaking, this is the third national survey of oil and gas reserves. But according to Che Changbo (deputy director of the Strategic Research Centre of Oil and Gas Resources of MLR) this was the first government-led oil and gas survey, the previous ones having been conducted by the state oil companies. The MLR, the NDRC, and the Ministry of Finance (MOF) jointly organized the survey. A total of 17 units, including governmental research centres, three Chinese

Table 4.3: China's Crude Oil Production (mt)

	2002	2005	2008	2010
Daqing	50.131	44.951	40.200	39.871
Huabei	4.380	4.351	4.452	4.260
Liaohe	13.512	12.255	11.998	9.500
Xinjiang	10.050	11.664	12.230	10.891
Dagang	3.939	4.993	5.112	4.780
Jilin	4.298	4.585	6.622	6.100
Changqing	6.101	9.399	13.791	18.250
Yumen	0.597	0.770	0.702	0.482
Qinghai	2.140	2.215	2.212	1.860
Sichuan	0.138	0.138	0.141	0.138
Yanchang	3.667	8.123	11.230	13.725
Jidong	0.653	1.250	2.009	1.731
Tarim	5.020	6.001	6.537	5.541
Tuha	2.496	2.098	2.095	1.630
CNPC total	**107.122**	**112.792**	**119.333**	**118.758**
Shengli	26.715	26.945	27.740	27.340
Henan	1.880	1.871	1.805	2.270
Zhongyuan	3.800	3.200	3.003	2.725
Jianghan	0.965	0.955	0.965	0.965
Jiangsu/Anhui	1.570	1.647	1.710	1.710
New Star	-	4.560	6.523	7.525
SINOPEC Star	2.931	-	-	
SINOPEC Total	**38.036**	**39.286**	**41.802**	**42.561**
CNOOC & others	25.125	28.784	28.889	42.512
China Total	**170.284**	**180.861**	**190.024**	**203.831**

Source: *China OGP* (2002, 2005, 2008, and 2010).

state oil companies, and the Petroleum Universities were involved in the survey. The intention was to produce a more rigorous survey than had been achieved previously.

Yang Hulin (director of the Development Strategy Division of the Strategic Research Centre of Oil & Gas Resource, an affiliated research organization of MLR) pointed out two new features of this survey: i) it would fill several gaps that had never been filled in the previous rounds, for example, the recoverable amounts of China's oil and gas resources, and the initial survey results for the Tibetan regions and the southern part of the South China Sea; ii) more non-conventional oil and gas resources were to be included in the survey, oil sands and oil shale being the newly added items in this round. As of 2006, 10 sub-items of conventional oil and gas resources had passed the acceptance inspection conducted by state-designated experts, and so could be published. The acceptance inspection report of non-conventional oil and gas resources, namely coal-bed methane, oil sands, and oil shale, was ready to be released in June 2005. The survey, started in November 2003, reported its final results after nearly three years of work, in June 2006.[46]

According to this third national survey, as shown in Table 4.4, China has a total of 106.8 bt of oil reserves. The onshore oil reserves are 82.2 bt, of which 41.97 bt are the in east, 37.24 bt in central and west China, and 2.5 bt in the south. The remaining 24.6 bt are offshore oil reserves, of which 15.0 billion tonnes are recoverable.[47] The second survey, conducted in 1994, had evaluated 150 sedimentary basins across the country and concluded that there were 94 bt of oil in place.[48]

Table 4.4: China's Oil Reserves, Evaluation Results

	Oil Reserves Billion tonnes
1987	78.7
1994	94.0
2006 *	106.8

*   The Third Survey was initiated in 2003.

Source: Yang Liu (2006c), 31.

The third resource evaluation was not complete. It was conducted in only 150 out of 429 deposit basins. According to Chen Mingshuang of PetroChina, oil resources in the Qinghai–Tibet Plateau, as well as parts of the South China Sea, are quite rich, but those areas are not included in the estimate of the national total oil reserves. A little over half are conventional resources and the rest are divided almost equally between heavy oil and low penetrate resources, along with a much smaller quantity of bitumen and oil sands. This, as already mentioned, was the first time that an evaluation had been made of unconventional resources such as CBM, oil shale, and oil sands.

China's known oil reserves have shown an upward growth trend. From 1949 to 1959 proven reserves grew slowly, but between 1959 and 1988 there were two booms in the identification of oil reserves, in 1977 and 1987. From 1985 to 2004, an average 767 mt was added to proven oil reserves annually. Between 1991 and 1995, the annual increment in proven and recoverable oil reserves averaged 123 mt, from 1996 to 2000 158 mt, from 2001 to 2005 186 mt, and from 2006 to 2010 it is expected to have been 180 mt.[49]

The survey identified four target exploration areas where increased oil reserves might be found. The first is Sag basin in east China. There are 55 basins in east China, covering 1.02 million square kilometres. Oil reserves still to be proven amount to 12.213 bt. A large number of discoveries have been made in the Nanbao sag since 2004, and predictable oil reserves amount to 250 mt. The second area is west China, the location of 25 basins covering 1.587 million square kilometres. Ten basins have been proven and 13 basins already have oil and gas flowing. The west and central areas have 11.423 bt of oil reserves to be proven. Several large oilfields such as Zhijing, Ansai, and Xifeng have been found in the Ordos basin in recent years. The third target area is offshore China. The offing areas close to the shore have 8.631 bt of oil reserves to be proven. The fourth area is south China with estimated oil reserves of 2.5 bt.[50] (See Table 4.5).

Table 4.5: Forecast of Oil Reserves by Area[51]

| | Oil reserves waiting to be proven (billion tonnes) | |
	Reserves	*Recoverable reserves (recovery rate: per cent)*
East China	12.213	2.482 (20.0)
Central China	1.890	0.340 (18.0)
West China	9.533	2.097 (22.0)
Other areas	0.008	0.001 (15.0)
Land Total	23.644	4.921 (20.8)
Offshore Total	8.631	1.899 (22.0)
National Total	32.275	6.820 (21.1)

Source: Yang Liu (2006c).

In 2007, China invested 66 billion yuan in geological prospecting, of which 50 billion yuan were for the prospecting of oil and gas resources. PetroChina's investment in geological prospecting exceeded 300 million yuan in 2007, three times the amount in 2000. The company discovered more than 500 mt of geological reserves of oil in the 2004–7 period, and reportedly a further 700 mt in 2008. SINOPEC found new geological oil reserves amounting to 2.35 bt during the years 1998–2007 and new natural gas reserves of 1.07 tcm.[52]

In August 2010 the Ministry of Land and Resources also revealed that it had completed geological studies of oil and gas in south-east Asia, the North Pole area, and the Central Asia–Caspian Sea Region. The ministry said it had a four year project (2008–11) to assess global oil and gas resources and to provide data and materials for national oil and gas strategy. The project studied 25 programmes in eight large regions including the Asia–Pacific, the Middle East, the former Soviet Union, Africa, South America, North America, Europe, and the North Pole region. The ministry added that it has analysed oil and gas distribution and the potential of 20 oilfields in south-east Asia, but revealed no details.[53]

Onshore

Roughly 85 per cent of Chinese oil production capacity is located onshore. The NOCs are also investing significantly in technologies to increase oil recovery rates at the country's mature oil fields. Increasingly, CNPC is utilizing natural gas supplies from the Daqing field for rein-jection purposes to fuel enhanced oil recovery (EOR) projects. CNPC hopes that EOR techniques can help stabilize Daqing's oil output in the years ahead. However, China's domestic demand for natural gas supplies is also increasing, which may place a competing claim on natural gas output from Daqing.

Daqing field

Daqing field, discovered in 1959, contains 40 individual fields all located in the Songliao basin, and covers an area of 160 km from north to south and between 6 and 30 km east to west. The seven main fields are situated in a large anticlinal basin called the Changyuan structural formations. In 1995, the estimated geological reserves of the field totalled 4.67 bt, which was later increased to something in excess of 5 bt. However, by the end of 2005, 1.9 bt had already been taken, leaving only 610 mt of recoverable reserves in place. Only if the ultimate recovery rate can be raised to 50 per cent, could the remaining 2.5 bt be extracted. Daqing also produces significant volumes of natural gas. Up to now, this gas has only been associated gas. More recent exploration suggests that in the southern Daqing fields there may be significant reserves of non-associated gas trapped at much deeper levels than that found associated in the existing oil fields.

Tatsu Kambara and Christopher Howe explain that:

> the character of the Daqing fields means that water injection methods be used to raise pressures for recovery from the outset. The water ratio rose from 60 per cent in 1980 to 80 per cent in 1995 and 85 per cent in 2000. During the 1990s the main recovery methods changed fundamentally. Not only have pumping systems been installed, but infill wells have been drilled, reducing the space between wells from around 500 metres to as little as 100 metres in some cases. The total number of oil producing and water servicing wells in Daqing now exceeds 50,000. Another technological innovation of

importance has been the use of the polymer flooding method. This is a tertiary recovery procedure that involves injecting water-displacing polymers into the wells. Up to 17 per cent of the total oil recovered from Daqing is estimated to have been produced by this system.[54]

Daqing field gained combined proven oil reserves of about 250 mt in four oil rich blocks in the first half of 2006. The four blocks are located in Heilongjiang province and Inner Mongolia Autonomous region. Two blocks, one holding around 73 mt and the other 60 mt, were found in the Mongolian Autonomous County of Dorbod in Heilongjiang. The other two blocks are one with 96 mt in Zhaoyuan county of Heilongjiang, and another with 30 mt in the Hailaer basin in Inner Mongolia.[55] These discoveries are not good enough, however, to permit the field to return to its peak 50 mt/y capacity.

In 2003, Daqing oilfield announced that it planned to reduce production by 7 per cent per annum in the succeeding seven years. At this rate of decline Daqing oil production would have shrunk to 30 mt in 2010, although this turned out to be a little pessimistic as Table 4.6 shows.[56] In 2003 Daqing had produced 48.4 mt, or 28.6 per cent of China's total production. PetroChina hopes to extend Daqing's service life by diminishing the planned production. Besides, Daqing is undergoing a strategic switch to become a petrochemical stronghold. Daqing plans to expand its ethylene cracking capacity to 1.5 mt/y and propane capacity to 1 mt/y on the assumption that crude throughput is maintained at around 20 mt/y.[57]

Table 4.6: Projected Oil Production of China's North-Eastern oilfields, 2005–2015 (mt/y)

		Daqing	Jilin	Liaohe	Total
2005	(Projection)	45	5.5	12.2	62.7
	(Actual)	44.95	4.58	12.26	61.79
2010	(Projection)	34.71	5.89	10.57	51.17
	(Actual)	39.87	6.10	9.50	55.47
2015		30	6.0	9.35	45.35

Source: Li Yuling (2003c), 2; *China OGP* (2002, 2005, 2008, and 2010).

Daqing's actual production decline was not as severe as the earlier prediction (shown in Table 4.6), but the 10 mt/y decline from the peak level is still a heavy blow to Daqing field. Daqing still aims to produce 400 mt of crude over the period 2008–17. According to Wang Yongchun, Party secretary of Daqing oilfield, the field will maintain its annual crude oil output at 40 mt over the next 10 years. In March 2009,

CNPC announced that crude oil output at Daqing had cumulatively exceeded 2 bt. In 2008, 40.2 mt of crude oil and 2.76 bcm of gas were produced from Daqing.[58]

Changqing oil field, located near the Ordos basin, exceeded an annual output of 30 mt for the first time on 19 December 2009, reaching 30.06 mt, making it China's second largest oil–gas field after Daqing. In 2010, the figure exceeded 35 mt.[59] According to *China Petroleum Daily*, a newspaper run by CNPC, Changqing produced a record 10.83 mt, or around 878,400 barrels per day of oil equivalent in the first three months of 2011. This figure is definitely higher than that of Daqing.[60] If it is maintained, Changqing will become the biggest field in China. The oil and gas resource in Changqing is mostly characterized by 'three lows' (low permeability, low pressure, and low yield) and is not easy to exploit. PetroChina Changqing Oilfield Company has developed special techniques to effectively exploit this oil–gas field. In 2007, the field produced over 20 mt of oil and gas in equivalent weight (the field's gas production was 11 bcm, accounting for 16 per cent of China's total, just below Tarim's 15.4 bcm). In July 2008, CNPC announced that Changqing oil and gas field under CNPC planned to raise its annual production of oil and gas in equivalent weight to 50 mt by 2015. In 2015 the field should be capable of producing 25 mt of crude oil.[61]

Tarim as an alternative to Daqing

China's interior provinces, particularly Xinjiang Uygur Autonomous Region in the north-west, have also received significant attention. According to an EIA report:

> the onshore Junggar, Turpan–Hami, and Ordos Basins have all been the site of increasing E&P work, although the Tarim Basin in north-western China's Xinjiang Uygur Autonomous Region has been the main focus of new onshore oil prospects. Reserve estimates for Tarim vary widely, with IHS Energy reporting that some estimates are as high as 78 billion barrels of total in-place oil reserves. The basin is home to Sinopec's Tahe oil field, with an estimated 996 million tons of in-place oil and gas reserves after a recent addition of 135 million tons in 2008. Since 2005, hydrocarbon production from Tarim has doubled, and the NOCs are taking advantage of tax breaks and other incentives to develop the region and offset declines in mature basins.[62]

Exploration for oil and gas started in China's western regions as early as the 1950s, at which time the Karamay field was discovered. The Tarim (which means the place where the waters gather) is a sedimentary basin located in the Xingjian Uygur Autonomous Region. It extends over 560,000 square kilometres (maximum length east–west 1,820 km, and north–south 510 km). The basin is bounded by the Tianshan Mountains in the north, the Kunlun range in the south, and the Karakorums in the west. Its height above sea level is in the order of 1,000 to 1,500 metres and its central area is occupied by the Taklamakan Desert, which extends to 330,000 square kilometres.

According to Kambara and Howe:

> after the discovery by the Ministry of Geology and Mining and the Ministry of the Petroleum Industry, of two important oil wells – Shasan 2 in 1984 and Lunnan in 1988 – the Tarim Petroleum Exploration and Development Command became the leading institution in surveying and testing in the area. Its first big success was the discovery of the Tazhong field. So important was this field considered that in 1994 a high speed motorway was constructed through the dunes.[63]

Table 4.7: Oil Fields in the Tarim Basin

	Area Size (sq. km)	Proven reserves (in place) Crude Oil (mt)
Tazhong 4 Oil	35.7	81.370
Yaha Oil/Gas	48.9	44.429
Kekeya Oil/Gas	27.5	30.655
Lunnan Oil	36.6	51.130
Yingmaili Oil/Gas	48.3	19.501
Donghetang Oil	16.5	32.927
Hade 4 Oil	66.6	30.680
Yangtake Oil/Gas	18.3	5.675
Jilake Oil/Gas	52.5	7.820
Jiefangjudong Oil	14.0	15.322
Sangtamu Oil	18.6	15.010
Tazhong 16 Oil	24.2	9.760
Tazhong 6 Gas	58.0	0.734
Yudong 2 Gas	10.2	1.425

Source: Kambara and Howe (2007), 87.

In April 2009, CNPC announced that Tarim oil field in Xinjiang province was expected to see its combined output of crude oil and natural gas reach 50 mt in 2020. A year later CNPC announced that the company planned to turn Xinjiang province into China's largest oil and gas production base by 2020, by boosting the oil and gas production capacity to over 60 million tonnes of oil equivalent.[64] Recorded output in 2008 was 20.3 mt, of which 6.45 mt was crude oil and 17.3 bcm natural gas.[65] According to Jiao Fangzheng, vice president of SINOPEC, the Tarim basin is becoming the latest strategic development base for SINOPEC. The group has already designated the north-west region as an important base for its upstream oil industry, with its focus on Tarim. Tahe oil field in Tarim basin will be the major battlefield on which SINOPEC will pursue the growth of oil and gas output in the long term. The field produced six mt of crude oil and 1.27 bcm of gas in 2008. Tahe oilfield's production is expected to reach 15 million tonnes of oil equivalent in 2015, including 10 mt of crude and 5 bcm of natural gas, and 20–25 mt in 2020, including 15 mt of crude and 5–10 bcm of gas.[66] In 2010, Tahe field completed construction of a 44 km crude oil pipeline to connect the oil field with Kuche, Uygur Autonomous Region. This pipeline, with a capacity of 3 mt/y, will add up to 5 mt/y to SINOPEC Northwest Oilfield Company's total oil transportation capacity.[67]

Strictly speaking, as shown in Table 4.3, Tarim's oil production level is not good enough to provide the alternative for Daqing. In 2010, oil production from Tarim was recorded at only 5.5 mt, while Shengli and Changqing recorded 27.3 mt and 18.3 mt respectively. Even if Xinjiang's 10.9 mt is combined with Tarim, the total is only 16.4 mt. However, when oil and gas production is combined, Tarim's role as an alternative to Daqing will make sense.

Offshore

The EIA reports that:

> about 15 per cent of overall Chinese oil production is from offshore re-
> serves, and most of China's net oil production growth will probably come
> from offshore fields ... current offshore production is 680,000 b/d, and is
> expected to rise to 980,000 b/d by 2014. These volumes will offset some
> of the declines from the more mature onshore fields in eastern China.
> Offshore E&P activities have focused on the Bohai Bay region, Pearl River
> Delta, South China Sea, and, to a lesser extent, the East China Sea. The
> Bohai Bay Basin, located in north-eastern China offshore from Beijing, is
> the oldest oil-producing offshore zone and holds the bulk of proven offshore
> reserves in China.[68]

On 3 May 2007, CNPC announced the discovery of the massive Jidong Nanpu oil field on the rim of Bohai Bay. Wen Jiabao said that he felt too excited to sleep. However, the so-called unprecedented or milestone discovery could not reverse the trends of rocketing domestic oil and natural gas demand and of home crude oil production falling well behind demand. The Jidong Nanpu oil field, which contains 1.02 bt of oil reserves, including 405.07 mt of proven reserves, is scheduled to complete its first phase construction by 2012, when crude oil out-put of the oil field should reach 10 mt per year. The field's peak oil production will be 25 mt per year. Only Daqing and Shengli oil fields have annual oil production above 30 mt. Both fields, located in east China and the flagship oil fields for CNPC and SINOPEC respectively, are facing the fate of declining oil production and rising development costs. PetroChina's vice president, Hu Wenrui, said the discovery cost of the Jidong Nanpu oil field is quite low (about $0.59/barrel), much lower than the domestic average cost of about $3.5/barrel, and the development cost of the oil field should also remain at the level of 2.3–2.5 billion yuan for one mt of crude oil production. He predicted that the primary recovery rate of the oil field could reach 40 per cent and the stable production term would last 10–15 years.[69]

On 14 August 2007, the Ministry of Land and Resources certified the discovery of 445 mt of proven reserves in PetroChina's Jidong Nanpu oil field, as well as 86.59 mt of economically recoverable oil reserves (a recovery rate of 19 per cent). The company's estimate was that the field contained proven reserves of 405.07 million tonnes of oil equivalent or nearly three billion barrels, and it aimed to extract 40 per cent or more of the oil it has proven in the field off north China, that is, some 1.2 billion barrels.[70] The 40 per cent recovery rate is above the national average of 20–30 per cent, and Jiang Jiemin, the general manager of PetroChina, predicted that recovery could even reach 40 per cent during the second phase development. Nanpu field also has some 52 bcm of associated gas.[71]

PetroChina aimed to develop the Nanpu oilfield as soon as possible, and its average annual investment between 2007 and 2011 was expected to be eight billion yuan, about 4.3 per cent of PetroChina's 2007 budgeted expenditure. The company expected annual oil production at Nanpu to reach seven mt by 2009, from about two mt in 2007, and it further aims to produce 10 million tonnes of oil equivalent per year by 2012, when the first phase of the Nanpu oil field development project is completed.[72]

It is sometimes argued that Nanpu field can play an important role in building China's strategic reserves. To store imported oil as strategic oil reserves in oil tanks is important, and so is maintaining oil fields with large undeveloped oil reserves. The arguments for using Nanpu as a strategic reserve are as follows:

• Bohai bay, where the Nanpu is located, has an ideal geographical position for storing oil reserves;
• delaying the development of a big oil field is a universal choice in developed countries because it endows a nation with power in the international market and enhances its flexibility;

- China's reserves of foreign exchange had risen to over 1.3 trillion dollars by the middle of 2007. Such a huge amount of foreign exchange may do harm to healthy economic development and so purchasing oil is a good way to release foreign exchange;
- China is working on producing a commercial oil reserves plan, in parallel with its initiative to build up the national strategic reserves. The discovery of the Nanpu oil field provides it with a favourable pilot scheme to improve the strategic oil reserves system.[73]

The Bohai Rim has been home to a group of large scale refineries and petrochemical projects. CNPC's Dalian Petrochemical and Fushun Petrochemical and SINOPEC's Tianjin Petrochemical, Yanshan Petrochemical, and Qingdao Refinery, are located there. CNPC is planning to build a large-scale greenfield refinery in the region, while continuing its effort to expand the capacity of the existing refineries. The natural gas to be produced in Nanpu oil field will also satisfy some of the gas demand in Beijing and Tianjin, and cushion the gas supply pressure of the Shaanxi–Beijing gas pipeline.

Strictly speaking, CNPC and SINOPEC are of almost equal weight in the Bohai Rim in the refining and petrochemical sectors. Each of them has plans to build a 200,000 b/d greenfield refinery and a one mt per year ethylene complex in Caofeidian, a massive port and industrial zone in Hebei province. The Bohai Rim, however, is also an important area for CNOOC, which has built the Penglai 19-3 oil field offshore Daqing.

Even before the Nanpu discovery was logged, offshore areas were expected to account for much of China's growth in oil production. CNOOC has made eight new discoveries of offshore reserves, increasing the company's proven oil reserves to 1.6 billion barrels. The company intends to double oil production in Bohai Bay, where over half of the NOC's production is expected to originate by 2015. ConocoPhillips, the largest foreign company acreage holder in Bohai Bay, is expanding production at the company's Penglai field, China's largest producing offshore field. On 16 September 2008 CNOOC Ltd, the listed arm of CNOOC, announced that a new platform on the Penglai 19-3 oil field, China's largest offshore oil field, had come online earlier than expected. The new platform is the third to be put into operation under Penglai 19-3's second development phase.[74] In 2011 ConocoPhillips had expected Penglai to reach its peak production of 150,000.[75] Unfortunately, in July 2011 oil and mud leaked from two platforms in the Penglai 19-3 oilfield, and so some reduction in production was inevitable in order to clean up the oil spill.[76] In 2008, the figure was a

mere 45,000 b/d. ConocoPhillips and CNOOC were aiming to complete Phase II of the field development, which includes five platforms, by May 2010. CNOOC brought one block in the company's Bozhong oil field online in March 2009, pumping 4,000 bbl/d. Total production from the Bozhong fields was expected to reach 25,000 b/d in 2011.[77]
    EIA reports that:

> in 2008, CNOOC, along with its partner Husky Energy of Canada, began commercial production at the Wenchang oil fields at an initial rate of 14,000 b/d. Wen 19-1 is expected to produce nearly 19,000 b/d from other platforms under development. CNOOC also brought the Xijiang 23-1 field on stream in 2008 and this is expected to produce 40,000 b/d of crude oil. CNOOC, ConocoPhillips, and Devon Energy have been developing the Panyu oilfields with the aim of reaching a peak output of 60,000 bbl/d. In 2009, CNOOC's total hydrocarbon production in the South China Sea (SCS) was 245, 000 boe/d – 191,000 boe/d in oil and 54,000 boe/d (324 MMcf/d) in natural gas. Also, according to PFC Energy, CNOOC's proven hydrocarbon reserves in 2009 in the SCS were 957 million boe, up 28 per cent from a decade ago. In the same year, CNOOC offered 17 blocks in the deepwater offshore parts of the South China Sea in order to encourage more exploration in these more technically challenging areas. In 2010, CNOOC made another significant discovery of the Enping Trough in the shallow waters of the SCS, and the area could generate up to 30,000 b/d.[78]

Oil Imports

Starting from 1988, China's strategy has been to import growing amounts of crude oil to feed the surging home demand fuelled by the high economic growth rate.[79] During the first decade of high growth (the 1980s), the country was able to maintain quasi-autarky in energy supply. However, by 1993 it had become a net oil importer and by 1995 its dependence on oil imports had increased to about 5.3 per cent. This level of dependence was still no cause for alarm, but the situation was soon to change completely. In 2000, the growth of demand propelled China's dependence on oil imports to nearly 33 per cent. Moreover, net oil imports grew at 14.2 per cent annually in the subsequent five years, to reach 143.6 mt in 2005 (although the *Xinhua News Agency* gives a figure of only 127 mt). BP's annual statistical data suggests China's oil consumption in 2005 was 328 mt, giving a dependency rate of around 44 per cent. If this is even close to the real level of dependence on foreign oil supplies, it is a good reason why the country's energy security became a major concern for decision makers.[80]

In fact, as shown in Table, 4.8, crude imports to China rose over six times during the 1998–2008 period. The Middle East remained the largest source of China's oil imports, although African countries also contributed a significant amount. In 2008, Saudi Arabia and Angola were the two largest sources of oil imports, together accounting for over one-third of the total. Imports from Russia and Kazakhstan increased quite significantly from 2.4 mt in 2001 to 25.3 mt in 2010 (Table 4.9).

While oil pipelines are dealt with in general in the next section, it is relevant here to mention China's initiative for the building of

Table 4.8: China's Crude Oil Imports by Region, 1998–2010 (mt/y)

	Middle East	Africa	Asia–Pacific	Europe	Total
1998	16.37	2.19	5.47	3.00	27.32
1999	16.90	7.25	6.83	5.63	36.61
2000	37.65	16.95	10.61	5.05	70.27
2001	33.86	13.55	8.68	4.17	60.26
2002	34.39	15.80	11.85	7.37	69.41
2003	46.37	22.18	13.85	8.73	91.13
2004	55.79	35.30	14.16	17.57	122.82
2005	59.99	38.47	9.68	18.94	127.08
2006	65.60	45.79	5.16	18.98	145.18
2007	72.76	53.04	5.73	20.84	163.18
2008	89.62	53.96	5.06	17.44	178.89
2009	97.46	61.42	9.62	21.68	203.79
2010	112.76	70.85	8.80	25.86	239.31

Source: CPCC (2003) and *China OGP* (2005–11).

Table 4.9: China's Crude Oil Imports from Russia and Kazakhstan, 2001–10 (mt/yr)

	Russia	Kazakhstan
2001	1.77	0.65
2002	3.03	1.00
2003	5.25	1.20
2004	10.78	1.29
2005	12.78	1.29
2006	15.97	2.68
2007	14.53	6.00
2008	11.64	5.67
2009	15.30	6.01
2010	15.25	10.05

Source: *China OGP*, various issues.

a China–Burma (Myanmar) oil pipeline. China revived its plans to construct this pipeline from Myanmar in an agreement signed in March 2009. As Myanmar is not a significant oil producer, the pipeline is seen as an alternative transport route for crude oil from the Middle East and Africa that would bypass the potential choke point of the Malacca Strait.[81] CNPC and Myanmar's Ministry of Energy, building on the March 2009 agreement, have signed a memorandum of understanding to build a 442,000 b/d oil pipeline. CNPC will be responsible for the line's design, construction, operation, and management. CNPC said that the project includes the 1,100 km pipeline that will start at Myanmar's west coast port of Kyaukryu, enter China at Ruili, and then extend to its terminus at Kunming, capital of Yunnan province. The oil line will shorten the shipping distance of China's Middle Eastern and African oil imports by 1,200 km, reducing transport time and boosting China's energy security by avoiding the pirate-infested Malacca Strait.[82] In combination with the pipelines from Central Asian Republics and Russia to China, the China–Burma (Myanmar) oil pipeline would help China to diversify its oil supply routes. Even if the combined capacity of oil supply by pipelines only reaches 1.14 mb/d, China's heavy dependence on oil supply through the sea lanes would be lessened, and this is its major energy supply security concern.

CNPC's website estimated that in 2010 China's overseas oil and gas production was equal to more than 25 per cent of national output (50 mt per year).[83] Xinhua's *China OGP* had estimated in 2007 that by the year 2015, China's oil demand was projected to have an annual growth rate of 5.3 per cent against a mere 0.9 per cent annual growth rate of domestic crude oil production. Demand is projected to reach 546 mt in 2015. Due to the limits of domestic oil resources, domestic crude oil production is likely to be around 199 mt in the same year. In other words, net crude oil imports will be 347 mt in 2015, and the oil import dependency rate will have climbed to 64 per cent.[84]

According to DRC (Development Research Centre under the State Council) estimates, China's remaining exploitable oil reserves amount to about 2.4 bt. Estimates of annual domestic oil output show production peaking at about 200 mt in 2015 and thereafter ranging between 180 and 200 mt up to 2020. The DRC and ERI's projections, however, indicate that by 2020 China's oil demand will be double or triple that amount, between 450 and 610 mt according to the low and high scenarios. This would mean that China's incremental oil demand between 2000 and 2020 would range from between 12 and 28 per cent of the incremental global oil demand projected by the IEA. The consequence will be that the country would increasingly rely on oil imports, and the share of imports could rise to between 50 and 60 per cent of oil consumption by 2020.[85] However, research carried out by the Chinese Academy of Social Sciences projected that 64.5 per cent of oil consumption is likely to be met by imports in 2020,[86] and ERI also projected that the dependence rate will be over 65 per cent by 2020.[87] Such a high level of Chinese demand will result in increasing pressure on world oil markets, and the cost to China's own economy would be considerable.

In the face of the security implications of growing import dependency, a small group of Chinese specialists has been encouraging the Chinese government to actively prepare for the commercial and strategic opportunities presented by a melting Arctic. In June 2009, Gati Al-Jebouri, the general director of Litasco (Lukoil global trader), and Dai Zhaoming, the president of Unipec Asia Company Ltd (Sinopec Trader), signed a framework agreement on oil supplies. The agreement calls for Russian export of 3 mt of oil blend and/or YK blend of oil produced from Yuzhno Khylchuyu Field, Nenets Autonomous Okrug to China. In early September 2010 CNOOC received its first consignment of stable gas condensate from a Russian independent gas producer, Novatek, via the Arctic Northern Sea route.[88] Arctic route oil trading[89] will take time to develop meaningful dimensions, but it could deliver a sizable volume around the mid-2020s.

Pipelines

In a summary of the importance of pipelines to China, the US Energy Information Administration (EIA) reports that the country:

> has actively sought to improve the integration of the country's domestic oil pipeline network as well as to establish international oil pipeline connections with neighbouring countries to diversify oil import routes. In March 2007, CNPC spearheaded the Beijing Oil & Gas Pipeline Control Center that monitors all long-distance pipelines and performs data collection to enhance the efficiency of the system. According to CNPC, China has about 18,100 km of total crude oil pipelines (69 per cent managed by CNPC) and nearly 4,900 km of oil products pipelines in its domestic network. The total length of oil liquids and natural gas pipelines is increasing at about 6 per cent per year. At present, the bulk of China's oil pipeline infrastructure serves the more industrialized coastal markets. However, several long-distance pipeline links have been built or are under construction to deliver oil supplies from newer oil-producing regions or from downstream centres to more remote markets.[90]

On 20 October 2006, the Western China Refined Oil Pipeline came into operation. The 1,842 km link with annual handling capacity of 10 mt delivers petroleum products from Urumqi in Xinjiang Province to Lanzhou in Gansu Province. PetroChina's three large scale refineries, Urumqi Petrochemical, Karamay Petrochemical, and Dushanzi Petrochemical, provide their oil products for the new pipeline. The capacity of the three refineries is about 20 mt/y.[91] Gradually, this pipeline will connect with other regional spurs to deliver supplies to the eastern coast, as well as to accommodate additional oil imports from Kazakhstan. Previously, most oil supplies from Xinjiang were delivered by rail. In addition, the Western Pipeline consists of a crude oil line from Xinjiang to the Lanzhou refinery, which came online in 2007.[92]

The west oil product pipeline has now been connected with the existing 1,250 km Lanzhou–Chengdu–Chongqing oil product pipeline which was also put into operation in December 2002 and is run by PetroChina. The LCC pipeline carried about five mt of oil products in 2005. PetroChina controls about 75 per cent of oil supply in Sichuan province due to the LCC pipeline. In addition, PetroChina started to construct the Lanzhou–Zhengzhou–Changsha oil pipeline in 2007.[93]

When the Lanzhou–Zhengzhou–Changsha and Jinzhou–Zhengzhou pipelines are completed, PetroChina will be able to kill three birds with one stone. First, it will be able to save a lot of its costs by piping oil products from north-east and north-west China to south China. At present PetroChina supplies the south China market mainly through railways and shipping. Second, PetroChina's expectation of fully exploring the central and south China markets will be realized. And third, when the two new pipelines are operative, it will increase its oil product sales.

Lanzhou will have reliable crude supply backup from Xinjiang and Kazakhstan. Oil products produced in Gansu and Xinjiang can be directly piped by the Urumqi–Lanzhou pipeline and the future Lanzhou–Zhengzhou–Changsha pipeline. And Jinzhou will shift to processing Russian crude and become a hub from which to direct oil products to central and south China through the future Jinzhou–Changsha pipeline. Currently Lanzhou and Jinzhou Petrochemical have a crude processing capacity of 12.5 mt/y and 10 mt/y respectively.[94]

PetroChina's market position in the south-west, however, faces challenges from SINOPEC, which in December 2005 completed the Maoming–Kunming oil product pipeline with an annual planned capacity of 10 mt. In October 2006, SINOPEC's 1,135 km Pearl River Delta oil product pipeline was put into operation. The pipeline starts from Zhanjiang and ends at Shenzhen, with Maoming as a hub on the line, covering 11 cities in Guangdong. The pipeline is designed to handle 9.5 mt of oil products annually.[95]

In addition, in 2005 a new crude oil pipeline was added to SI-NOPEC's oil transportation network. This runs from the 250,000 DWT oil terminal in Ningbo, Zhejiang province, to Nanjing, Jiangsu province, via Shanghai. The pipeline is designed to transport 40 mt of imported crude oil to feed SINOPEC's Shanghai Petrochemical, Gaoqiao Petrochemical, Jinling Petrochemical, and Yangzi Petrochemical. This pipeline will reduce SINOPEC's oil transportation costs by between 1.2 and 1.3 billion yuan annually. SINOPEC is also constructing an oil product pipeline in south-west China.[96]

On 7 January 2008, at the State Council meeting, vice Prime Minister Zeng Peiyan argued that the increase in the construction of pipelines would play a significant role in exploration, development, transportation, and storage of oil and gas resources. According to the NDRC, China had a total of close to 60,000 km of oil and gas pipelines at the end of 2007, but its target was to build a 100,000 km nationwide pipeline network before 2010.[97] As of 2010, China's oil and gas trunk pipeline reached 68,000 km, far short of the target distance of 100,000 km. China aims to build 210,000 km oil and gas trunk pipeline by 2020.

In 2007, CNPC put three main pipelines into operation, giving it, by the end of that year, 33,000 km of long distance oil and gas pipelines, 70 per cent of the total. As discussed earlier, the first of these three is the West Crude Pipeline that is capable of transporting 20 mt/y of crude oil from Xinjiang to Lanzhou (1,562 km). The whole West Pipeline spans 4,000 km, including the Sino-Kazakh crude pipeline. The second is the 647 km Dagang–Zaozhuang oil products pipeline from CNPC's Dagang Petrochemical in Tianjin to Zaozhuang, on its competitor's turf, Shandong province. The third is the Lanzhou–Yinchuan natural gas pipeline connecting CNPC's gas fields in Tarim, Qaidam, and Changqing, three of China's largest gas fields. In 2007, three other pipelines were being built by CNPC: the Lanzhou–Zhengzhou–Changsha oil products pipeline, the Daqing–Harbin segment of the northeast natural gas pipeline grid, and the Sino-Russian crude pipeline. CNPC had also signed agreements to build towngas grids in Zhuhai (Guangdong province) and Kunshan (Jiangsu province).[98] As shown in Table 4.10, the 3,000 km Lanzhou–Zhengzhou–Changsha oil product pipeline was put into operation in 2009.

The pipeline development achievement during the Eleventh Five Year Plan (2006–10) was impressive, and the total distance reached 27,000 km. Table 4.10 shows the performance during the years 2007–10.[99] In 2011, CNPC planned to build 7,100 kilometres of pipelines involving

Table 4.10: China's Crude Oil and Oil Product Pipelines, 2007–10

Crude Oil pipeline, 2007–2010	Length km	Operational Year	Owner	C/D/P mt per year/mm/MPa
Urumqi–Lanzhou	1,852	June 2007	Western pipeline Co, CNPC	20/-/-
Mohe–Daqing	965	2010	PetroChina	15/813/8.0
Daqing–Tieling	210	2010	PetroChina	27/813/6.3
Rizhao–Dongming	462	2011	SINOPEC	10-20/711–610/8.0
Shikong–Lanzhou	359	2010	PetroChina	5/457/-
Hui'anbao–Yinchuan	141	2010	PetroChina	6/508 -
Hejian–Shijiazhuang	147.5	2009	SINOPEC	8/-/-
Tianjin–Canzhou upgrading	168	2009	SINOPEC	-/-/-
Daqing–Skovorodino	927 + 70	August 2010	CNPC	15/1220/-

Oil Product pipeline, 2007–2010	Length km	Operational year	Owner	C/D/P mt/mm/MPa
Dagang–Zaozhuang	654	2007	PetroChina	3/-/-
Lanzhou–Zhengzhou–Changsha	3,007	2009	PetroChina	10-15/ 508–990/ 8-14
Yan'an Refinery–Xian	200	2009	PetroChina	5/-/6.3–12
Kunming–Dali	323	2009	SINOPEC	2/323.9–273/10
Liuzhou–Guilin	190	2010	SINOPEC	1.5/ 273.1/10
Liaoyang–Bayuquan	200	2009	PetroChina	-/406.4/-
Fujian refinery integrated pipeline	345	2009	SINOPEC	-/-/-
Second phase of Shandong–Anhui	905	2010	SINOPEC	9.7/-/10
Southern Jiangsu	393	2010	SINOPEC	-/406.4/-
Second phase of Pearl River Delta	498	2010	SINOPEC	3.75/-/-

Note: C/D/P = Capacity in mt/y / Diameter / Pressure

Sources: Lin Fanjing (2008a); www.cnpc.com.cn/eng/company/businesses/Pipe-linesTransportation/PipelineTransportation.htm. (Last accessed autumn 2011.)

35 projects. The company has 51,000 km of pipelines including 30,000 km of gas pipelines (accounting for 90 per cent of the country's total), 13,000 kilometres of crude pipelines (accounting for 70 per cent), and 8,000 kilometres of oil products pipelines, (accounting for 50 per cent). Between 2011 and 2015, CNPC plans to build 54,000 kilometres of pipelines,[100] of which gas pipelines are planned to account for 30,000 km. This construction would double the length of its current network.[101]

As shown in Table 4.11, CNPC now has seven subsidiary companies engaged in the pipeline business. Petroleum Pipeline Bureau (CPPLB)

Table 4.11: CNPC and PetroChina's Companies Involved in Pipeline Business

Company	Main Task
China Petroleum Pipeline Bureau (CPPLB),[102] under CNPC	Mainly engaging in construction of pipeline and oil & gas tanks
China Petroleum West Pipeline Co Ltd, under CNPC	The company is responsible for constructing and operating West crude and oil products pipelines. It is co-established by CNPC, CNPC Daqing Petroleum, CNPC Xinjiang Petroleum, CPPLB, CNPC Changqing Petroleum and Tuha Oil E&D Headquarters, and Tarim Oil E&D Headquarters.
PetroChina Petroleum Pipeline Company (CPPE)	CPPE mainly undertakes operation of a majority of inland pipelines with a total length of 8,900 km. CPPE was also the constructor of the Sebei–Ningxia–Lanzhou natural gas pipeline, Lanzhou–Chengdu–Chongqing oil products pipeline, and Zhongxian–Wuhan natural gas pipeline.
PetroChina West East Gas Pipeline Company	The company is now mainly engaged in operating and maintaining the first West–East pipeline.
Beijing Huayou Natural Gas Co Ltd, under PetroChina	Beijing Huayou was the constructor of the first and second Shaanxi–Beijing natural gas pipeline. The company is now the operator of Shaanxi–Beijing pipelines.
PetroChina North China Gas Marketing Company	Besides the responsibility of natural gas sales in Beijing, Tianjin, Shanxi, Hebei, Shandong, and Liaoning, the company also engages in selling natural gas from Shaanxi and North China oil field to foreign countries and also undertakes PetroChina's LNG business.
PetroChina Central China Gas Marketing Company	The company deals with natural gas sales along the Zhongxian–Wuhan pipeline and the W–E pipeline.

Source: Lin Fanjing (2008a), 4.

and China Petroleum West Pipeline Co. Ltd are under the control of their parent company CNPC, while another five companies, including PetroChina Petroleum Pipeline Company (CPPE), PetroChina West East Gas Pipeline Company and Beijing Huayou Natural Gas Co Ltd, are under the control of PetroChina.

CNPC has designed its pipeline business so that it is composed of three kinds of companies – pipeline constructors (currently under the control of CNPC), pipeline operators (under the control of PetroChina), and pipeline sellers (under the control of PetroChina). At the end of 2007, CNPC transferred CPPLB's ten cities where pipelines have been developed (Changji (Jilin), Daqing (Heilongjiang), Changchun (Jilin), Shenyang (Liaoning), Jinzhou (Liaoning), Dalian (Liaoning), Qinhuangdao (Shandong), Beijing, Zhongyuan (Henan), and Changqing (Shaanxi)) to PetroChina's China Petroleum Pipeline Engineering Corporation (CPPE), a dedicated pipeline operating company. In January 2008 CNPC transferred its Changqing–Yinchuan natural gas pipeline to PetroChina.

SINOPEC has several oil products pipelines, including two long-distance oil products pipelines: the Luoyang–Zhengzhou–Zhumadian pipeline went into operation in June 2007 and the Shijiazhuang–Taiyuan pipeline which was completed in 2008 (See Table 4.12 – neither of these long-distance pipelines is included in this table).[103] The 425 km Luoyang–Zhengzhou–Zhumadian pipeline, stretching from SINOPEC's Luoyang Petrochemical via Zhengzhou, Xuchang, Luohe, and ending at Zhumadian (all in Henan province), will transport 3.9 mt/y. The Shijiazhuang (Hebei)–Taiyuan (Shanxi) oil products pipeline is 316 km in length and passes through Shijiazhuang (Hebei), Yangquan (Shanxi), Jizhong (Shanxi), Taiyuan (Shanxi), and 13 other counties.

SINOPEC's short-distance oil products pipelines enjoy an advantage.

Table 4.12: SINOPEC's Oil Products Pipelines

in Service	*under Construction*
South-Western oil products pipelines	Beijing–Tianjin pipeline
Oil products pipeline around Beijing	Jinshan–Jiaxing–Huzhou pipeline
Jinshan–Shanghai pipeline	Quanzhou–Fuzhou pipeline
Shandong–Anhui-Jiangsu pipeline	Quanzhou–Xiamen pipeline
Jingmen–Jingzhou pipeline	Jiujiang–Nanchang pipeline
Luoyang–Zhumadian pipeline	Yueyang–Changsha–Xiangtan–Zhuzhou pipeline

Source: Lin Fanjing (2008a), 5.

Its 2,890 km oil products pipeline grids in south China and the Pearl River Delta play a significant part in encouraging local oil products supply, as well as guaranteeing SINOPEC's dominant position in the region. SINOPEC has six oil products pipelines in service and six short distance pipelines under construction, most of them are provincial city-to-city pipelines.

The fundamental reason behind the zeal for pipeline transportation is cost saving; costs of pipeline transport of oil products are only 62 per cent those of rail transport, and savings are even greater on natural gas and crude oil. Compared with other means of transportation, including railways and highways, pipeline transportation is safer and more convenient, with lower wastage rates. The wastage rate of rail transportation is 0.5 per cent, while that of pipeline transport is less than one half of this (0.225 per cent). The management costs of pipeline transportation are also lower than for rail. For all these reasons pipeline transportation is more economic: the cost of rail transport is 0.1237 yuan per tonne of oil products per kilometre, compared with the pipeline cost of only 0.0772 yuan.[104]

Summing up recent developments, EIA reports that:

> in order to push the supply from the Lanzhou refinery to market centres in the east and south, CNPC recently commissioned various oil product pipelines. The company launched the Lanzhou–Chengdu–Chongqing pipeline in 2008 and the 300 thousand bbl/d Lanzhou–Zhengzhou–Changsha pipeline in 2009. The Zhengzhou–Changsha segment is expected to be completed by 2010. PetroChina also has plans to build at least two additional spurs from Zhengzhou, which would help deliver crude oil supplies eastward. One is the Zhengzhou–Jinzhou pipeline, which would deliver oil north-eastward to Hubei Province. The other is the Zhengzhou–Changsha link, which would terminate in Hunan Province near the industrial south-east. Parts of these links came online in 2009, and together will form the country's largest oil product pipeline network.[105]

Refining

In May 2009 the NDRC announced that during the three years to 2011 they would phase out outdated production capacities in petrochemical, non-ferrous metals, iron and steel, textiles, and light industries. In particular, China planned to eliminate refining facilities with a capacity less than one mt and renovate those with a capacity between one and two mt. It is estimated that roughly 50–60 mt of refining facilities consists of single capacity units of less than one mt. Currently there is about 80 mt of crude oil primary capacity including 45 mt in East China's Shandong province.

More than 10 mt of outdated refining capacity had been phased out by the end of 2008. In particular, CNPC phased out 212 refining facilities with low efficiency and high emissions at seven under-performing refineries, including Jiangnan Refinery, Jilin Petrochemical, Jilin Oilfield Refinery, and Anshan Refinery. One hundred and eleven refineries with less than one mt of capacity had already been closed down between 2001 and 2005.[106]

The following is the breakdown of the projected demand for oil products by 2015:

i)   Diesel demand for diesel-fuelled vehicles, agricultural machinery and equipment, and farming vehicles is projected to reach 90.41 mt, 38.23 mt, and 17.49 mt respectively. Diesel demand in 2015 is projected to be 194.35 mt with an annual growth rate of 5.8 per cent.

ii) According to the Eleventh Five Year Plan for the ethylene industry, national ethylene production will reach 18 mt in 2010. So the demand for chemical light oil is expected to rise to 120 mt by 2015.

iii) As the number of gasoline-fuelled vehicles in service will increase rapidly, gasoline demand for gasoline-fuelled vehicles is predicted to reach 54.18 mt in 2015. In addition, the number of motor cycles in service will also grow rapidly and the gasoline demand for these is expected to be 20.27 mt in 2015. So total gasoline demand will be 78.56 mt.

iv) Fuel oil demand will maintain a constant growth rate to reach 52.27 mt in 2015. Fuel oil demand from the manufacturing sector will decline to 11.75 mt in 2015, but demand from the power generation sector will amount to 13.88 mt and that from the shipping transportation sector will rise to 25.14 mt.

v)   It is predicted that jet fuel demand will grow at an annual rate of 6.5 per cent to reach 16.49 mt in 2015 (double the 2005 level).[107]

The expanding refining sector has undergone modernization and consolidation in recent years. Dozens of small refineries (teapots), accounting for about 20 per cent of total fuel output, have been shut down, and larger refineries have expanded and upgraded their existing production systems. Domestic price regulations for finished petroleum products have hurt Chinese refiners, particularly smaller ones, because of the large gap between international oil prices and the relatively low domestic rates. In 2008, SINOPEC and CNPC (PetroChina) reportedly incurred refining losses of nearly $29 billion, which were partially covered by direct government subsidies.[108] Despite the occasional modest increases in consumer prices, China's oil companies have been losing money on their refining activities for several years. The country's major oil refiner, SINOPEC, has been in a powerful position to negotiate compensation from the government for its financial losses; subsidy payments of 10 billion yuan were made at the end of 2005, 5 billion yuan at the end of 2006, and 12 billion yuan in March 2008. Despite operating losses relating to refining of more than 20 billion yuan, PetroChina has not received any subsidy.[109]

China has achieved its fastest and largest growth of refining capacity during the 2000s, when it jumped 72.8 per cent, from 276 mt/y in 2000 to 477 mt/y in 2009, with an average annual growth rate at 6.3 per cent. In 2010, the newly added refining capacity is estimated to be 30.5 mt, after Qinzhou refinery in Guangxi province and Huajin refinery in Liaoning province had come into operation, together with the expansion of some existing refineries. China's total refining capacity achieved 500mt in 2010, and the figure will hit 750 mt/y by 2015 if the planned and under-construction projects go into operation as scheduled. The annual growth rate during 2011–15 is estimated at 6–7 per cent. In addition to this, overseas investment in refining capacity in China by 2015 was expected to reach 31.5 mt/y, from 10.5 mt/y in 2010.[110]

To date, a refinery industry with focus on the eastern area and supplemented by central and western areas has formed in China, driven by the principle of close attachment to resources and markets and proximity to the coast. In 2010, China's refining capacity was mainly concentrated in the east, north-east, and south of the country, each representing 32 per cent, 21 per cent, and 15 per cent of the national total respectively. According to the NDRC's mid-and long-term plan for the refining industry, China expected to increase the number of 10 mt/y refining bases from 17 to 20 by 2010,[111] with combined capacity expanded from 50 per cent to 65 per cent of the national total, which was expected to raise the average refining capacity of domestic refineries to 5.7 mt/y[112] (see Table 4.13).

The latest revitalizing plan for China's petrochemical industry specified 10 large future refining bases, including Ningbo, Shanghai, Nanjing, and Dalian with a refining capacity exceeding 30 mt/y each, and Maoming, Guangzhou, Huizhou, Quanzhou, Tianjin, and Caofeidian with a refining capacity of over 20 mt/y each. By 2015, however, the country is likely to have a 20 mt/y supply shortage in central and south-west China, and north-east and north-west China will remain important resources of product oil. The expansion of refining projects in east, south, and north areas is expected to achieve fast growth in

Table 4.13: China's Refining Bases with Capacity Exceeding 10 mt/y (mt/y)

Refinery	Ownership	Capacity in 2005	Capacity in 2010	Newly added capacity
Dalian Petrochemical	PetroChina	10.50	20.50	10.0
Fushun Petrochemical	PetroChina	10.0	10.0	0.0
Yanshan Petrochemical	SINOPEC	8.0	10.0	2.0
Shanghai Petrochemical	SINOPEC	14.0	14.0	0.0
Gaoqiao Petrochemical	SINOPEC	11.0	11.30	0.3
Jinling Petrochemical	SINOPEC	13.0	13.50	0.5
Zhenhai Petrochemical	SINOPEC	20.0	20.0	0.0
Qilu Petrochemical	SINOPEC	10.0	10.0	0.0
Guangzhou Petrochemical	SINOPEC	7.7	13.0	5.3
Maoming Petrochemical	SINOPEC	13.5	13.5	0.0
Lanzhou Petrochemical	PetroChina	10.5	10.5	0.0
Dalian WEPEC	PetroChina	10.0	10.0	0.0
Tianjin Petrochemical	SINOPEC	5.5	15.0	9.5
Fujian Refinery	SINOPEC	4.0	12.0	8.0
Dushanzi Petrochemical	PetroChina	5.5	10.0	4.5
Qingdao Refinery	SINOPEC	-	10.0	10.0
Huizhou Refinery	CNOOC	-	12.0	12.0
Guangxi Petrochemical	PetroChina	-	10.0	10.0
Yangzi Petrochemical	SINOPEC	8.0	9.5	1.5
Hainan Refinery	SINOPEC	-	8.0	8.0
Total		**161.2**	**242.8**	**81.6**

Source: Lin Fanjing (2010d), 3.

refining capacity. By then, four refining industrial zones in the circum-Hangzhou Bay, Pearl River delta, Bohai rim, and north-west China will have been formed, in accordance with the regional economic development plan.[113]

The refining capacity of local refineries, or teapots, represents almost half of China's total. At the end of 2008, the refining capacity of teapots was 88.05 mt (including 14 mt at the Shaanxi-based Yanchang Petrochemical Group). By region, Shandong province in East China has 37 refineries, Liaoning province 15, and Guangdong province 14. But the operational rates of local refineries have remained low, in particular after the consumption levy on fuel oil was raised to 0.8 yuan/litre from January 2009 (an eight-fold increase), since local refineries had traditionally taken fuel oil as feedstock due to lack of crude oil supply.[114]

Since the end of 2009, China has been a net gasoline and diesel exporter, as producers attempt to escape the losses imposed upon them by the NDRC's price capping. Even though the state offers subsidies, selling abroad can be more profitable. However, this outlet may not last long, given that refining oversupply also exists in the Asia–Pacific countries. SINOPEC is seeking large-scale compensation to cover its accumulated losses. Meanwhile, China is attempting to phase out small-sized or backward refining operations in a step towards streamlining its refining industry, and it is expected that about 50 mt/y of capacity will be shut down before 2015.[115]

Strategic Oil Reserves

According to the Eleventh Five Year Plan (2006–10), the second phase of the country's strategic oil reserves planning is already on the agenda. The first phase was focused on locations close to the consuming market, as well as those having favourable infrastructure facilities, such as oil pipelines, crude terminals, or railways. The second phase will not be confined to coastal areas. Preliminary studies were made of Heilongjiang, Inner Mongolia and Xinjiang, among other places, most of which have important railway ports and oil pipelines for the import of crude from Russia and Central Asian countries. In the second phase, underground storage is likely to be adopted. Locations with layers of salt rock and granite are both suitable for the building of underground oil deposits, although storage spaces in salt rock are much cheaper than those in the granite layer. Geographic studies have shown that Guangdong, Huai'an in Jiangsu province, Yulin in Guangxi province, and Jianghan basin in Hubei province have good granite layers for this purpose.[116]

From as early as 2006, the Beijing authorities began to import Russian crude for the strategic petroleum reserve. After 11 August 2006, about 3 million barrels of Russian crude oil were put into the Zhenhai base. According to *Shanghai Securities News* (a branch of *Xinhua News Agency*), this batch of crude came from oilfields in Russia's Ural region. The Zhenhai oil reserves base, 18 km from Ningbo, is the site of 52 oil depots with a designed capacity of 100,000 cubic metres each, adding up to total storage capacity of 5.2 mcm (32.7 million barrels), equal to 4.6 days of China's current total oil consumption.[117]

On 18 December 2007, the State Council approved the establishment of the National Petroleum Reserve Centre (NPRC). The NPRC is designed as an executor of China's petroleum reserves policies, which aim to maintain a petroleum reserve to safeguard national economic security. The NPRC's duty is to contribute to capital needs and to control the construction and management of national petroleum reserve bases. The

NPRC also controls the procurement, takeover, rotation, and utilization of strategic reserves (SPR), and monitors the domestic and international oil market. The NPRC is headed by Yang Liansong, former president of the *China Economic and Trade Herald*, which is owned by the Academy of Macroeconomic Research, a branch of the NDRC.[118] China's oil reserves issues are jointly administered by the National Energy Leading Group, the Energy Bureau under the NDRC, the National Energy Reserve Office, and the office of the National Energy Leading Group (based in the NDRC). The NDRC established the national petroleum reserve office inside the Energy Bureau in 2003, and the Bureau's director always heads this office.[119]

On 25 September 2009, construction of the Dushanzi (Xinjiang province) strategic petroleum reserve base, with a storage capacity of 3 mcm, began. Zhang Guobao, head of the National Energy Administration and vice minister of the NDRC, said that it marked the start of the second phase of SPR base construction (Table 4.14). It was the first time government officials had specified the location of one of the second phase SPR bases. Construction of the Dushanzi SPR was completed in July 2011, and in September 2011 it was reported to have come into operation. Its 30 oil tanks, each having a capacity of 100,000 cubic metres, will be filled up mainly with crude oil from Kazakhstan and, in order to use more of the pipeline's capacity, Russian oil imported via the China–Kazakhstan crude oil pipeline. It is part of the effort to build four SPR bases with total storage capacity of 26.8 mcm or

Table 4.14: First and Second Phase of SPR in China

	Capacity	*Operator*
First Phase		
Zhenhai, Zhejiang province	5.2 mcm	SINOPEC
Aoshan, Zhejiang province	5.0 mcm	SINOCHEM
Huangdao, Shandong province	3.0 mcm	SINOPEC
Dalian, Liaoning province	3.0 mcm	PetroChina
Second Phase		
Jinzhou, Liaoning province	3.0 mcm	PetroChina
Qingdao, Shandong province	3.0 mcm	SINOPEC
Jintan, Jiangsu province	2.5 mcm	SINOPEC
Zhoushan, Zhejiang province	3.0 mcm	SINOCHEM
Huizhou, Guangdong province	2.0 mcm	CNOOC
Zhanjiang, Guangdong province	7.0 mcm	SINOPEC
Dushanzi, Xijiang province	3.0 mcm	PetroChina
Lanzhou, Gansu province	2.0 mcm	PetroChina

Source: *China OGP*, 1 October 2009, 14.

21.44 mt. The interim goal is to achieve storage capacity for about 40 days of normal consumption, rising to the equivalent of 100 days of consumption by 2020, after the second and third phases of SPR bases are completed.[120] In 2009, China imported crude oil at an average cost of $58 a barrel for the SPR.[121]

In addition to the SPR, Beijing is also encouraging the state-run oil firms to build their own commercial reserves. PetroChina Dushanzi Petrochemical Co has built a 1.4 mcm commercial oil reserve. At present there are commercial crude reserves in Wangjiagou district in the city of Urumqi, Dushanzi in Karamay, and Shanshan in Turpan, all in Xinjiang province. Xinjiang province is expected to accommodate 13 mcm of total oil reserves in the future, including the SPR built by the central government, oil reserves stocked by local governments, commercial reserves kept by corporations, and reserves built up by small and medium sized companies.[122] The government reported that, in addition to the crude reserves, it also planned to create a strategic stockpile of refined oil products to be operated by a subsidiary of the NDRC, and aimed to boost stocks to 80 million barrels by 2011. In addition, there are plans to increase commercial oil product storage to 252 million barrels by 2013.[123]

In July 2010 PetroChina announced it had completed the construction of four crude oil storage tanks, each with a capacity of 100,000 cubic metres, in Qinzhou in the Guangxi Autonomous Region. These tanks are built as the first phase of PetroChina's international oil reserves project, in which a total of 42 such tanks will be built. PetroChina is test-running its 12 mt/y Qinzhou Refinery in Guangxi, and the storage tanks will be used partly as the company's commercial reserves.[124]

According to Zhao Youshan, director of China's private oil firms' organization, the Petroleum Flow Committee of China General Chamber of Commerce (PFCGCC), SPR will be open to private oil firms. Zhao said that the PFCGCC's appeal to have 36 mt of crude oil storage capacity held by private oil firms to serve as part of the SPR has won approval from high officials, and he indicated that related policies may be announced soon. In June 2010, China issued 36 new clauses which aimed to encourage private firms to cooperate with state-controlled industries such as the petroleum industry.[125]

Conclusion

China's ravenous appetite for fossil fuels is driven by its shifting economic base – away from light export industries like garment and shoe production, and towards energy-intensive heavy industries like steel, cement, car manufacturing, and construction for the domestic market.[126] In particular the explosion of car ownership, in line with nationwide road network development, will accelerate the rapid growth of oil demand in the coming decades. The headache for energy planners is that domestic oil production is not big enough to cover the galloping oil demand, and a massive quantity of oil imports in the coming years is inevitable. In 2010, crude oil imports reached 239.3 mt, of which 47.1 per cent (112.8 mt) came from the Middle East and 29.6 per cent (70.8 mt) from Africa. Oil from these two sources alone reached almost 184 mt. The percentages from Russia and Kazakhstan were 6.4 (15.2 mt) and 4.2 (10.1 mt) respectively. By 2020, the total figure is projected to reach roughly 385 mt, a figure which assumes that in 2020 the total demand will be 585 mt (or 11.7 mb/d) and domestic production 200 mt. Net imports are very likely to reach 10 mb/d during the 2020s, and security of oil supply thus becomes one of the Chinese planners' highest priorities.

The intensified activities of Chinese NOCs' Going-Out policy during the recent past has been driven by the necessity of securing overseas supply sources, and this momentum is almost certain to be maintained during the 2010s. In late March 2010, PetroChina's CEO, Jiang Jiemin, estimated that a total investment of no less than $60 billion was needed to form five regions of global oil and gas cooperation by 2020, and added that the company wanted half its oil and gas to come from abroad by 2020. As discussed earlier, the total investment in overseas oil and gas related projects during the years 1992–2009 (Spring) was US$44.4 bn, carried out by a combination of three Chinese NOCs plus other investors. The figure of US$60 bn investment in the coming ten years by PetroChina alone is a very ambitious target, but it is not impossible considering that in 2009 alone Chinese companies spent a record $32 billion on mining and energy acquisitions.[127]

There is no limit to the global expansion of Chinese NOCs. China's rapidly expanding need for energy promises to have major geopolitical implications as the country hunts for ways to satisfy its energy hunger. The Middle East and Africa will continue to occupy pivotal roles as the main supply sources. In particular, once the two large oil fields in Iraq return to normal operation, the volume of supply from the Middle East will increase significantly. That means that China's heavy dependence on sea transportation will not be easily reduced in the coming decades, despite the massive investment in pipeline development for oil supply from the Central Asian Republics, Russia, and Myanmar.[128]

Despite the sharp increase in PetroChina's spending budget, Russian oil and gas assets are very likely to be exempted from CNPC's and PetroChina's mergers and acquisitions list. Sino-Russian oil cooperation will have to find another improvised vehicle, such as loans for oil, to increase the scale of oil and products trading between two countries. It is inconceivable that the trading of crude and oil products between the two parties will diminish in the coming decades. Additional crude supply could be added when the joint development of upstream assets in East Siberia and Sakhalin offshore starts to produce some tangible results, presumably during the second half of the 2010s.

Appendix to Chapter 4

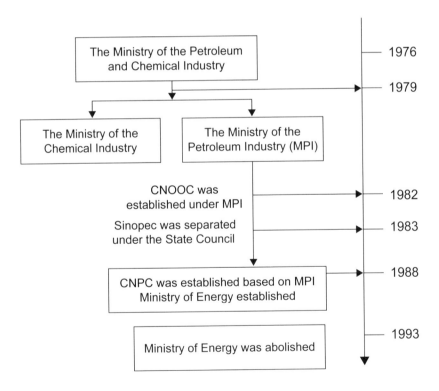

Figure A.4.1: The Evolution of the Regulatory and Industrial Framework (1976–93)

Source: Ma Xin (2008)

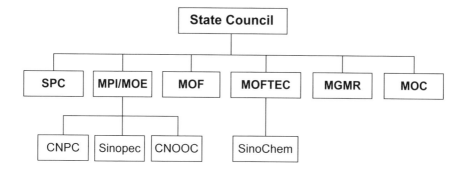

SPC: State Planning Commission
MPI: The Ministry of Petroleum Industry
MOE: The Ministry of Energy (1988–92)
MOF: Ministry of Finance
MOFTEC: Ministry of Foreign Trade and Economic Co-operation
MGMR: Ministry of Geology and Mineral Resources
MOC: Ministry of Commerce
CNPC: China National Petroleum Corporation (from 1988)
Sinopec: China Petrochemical Corporation (from 1983)
CNOOC: China National Offshore Oil Corporation (from 1982)
SinoChem: China National Chemical Import and Export Corporation

Figure A.4.2: The Regulatory and Industrial Framework (1978–93)

Source: Ma Xin (2008)

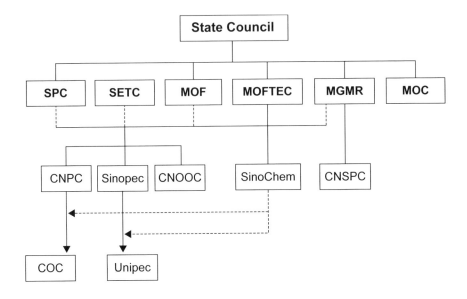

SPC: State Planning Commission (1993–2003)
SETC: State Economic and Trade Commission
MOF: Ministry of Finance
MOFTEC: Ministry of Foreign Trade and Economic Co-operation
MGMR: Ministry of Geology and Mineral Resources
MOC: Ministry of Commerce
CNPC: China National Petroleum Corporation (from 1988)
Sinopec: China Petrochemical Corporation (from 1983)
CNOOC: China National Offshore Oil Corporation (from 1982)
SinoChem: China National Chemical Import and Export Corporation
COC: China Oil Corporation
Unipec: China International United Petroleum and Chemical Company

Figure A.4.3: The New Regulatory Framework and Industrial Structure

Source: Ma Xin (2008)

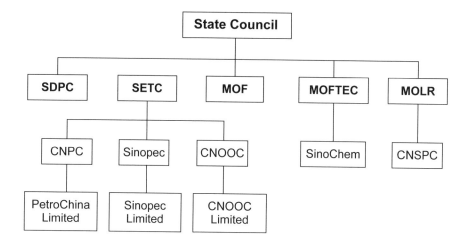

SDPC: State Development and Planning Commission (1998–2003)
SETC: State Economic and Trade Commission (1998–2003)
MOF: Ministry of Finance
MOFTEC: Ministry of Foreign Trade and Economic Co-operation (1998–2003)
MOLR: Ministry of Land and Resources
CNPC: China National Petroleum Corporation
PetroChina Limited: PetroChina Company Limited
Sinopec: China Petrochemical Corporation
Sinopec Limited: Sinopec Company Limited
CNOOC: China National Offshore Oil Corporation (from 1982)
CNOOC Limited: China National Offshore Oil Company limited
CNSPC: China National Star Petroleum Corporation (before 2001);
                   it was acquired by Sinopec after 2001
SinoChem: China National Chemical Import and Export Corporation

Figure A.4.4: The Regulatory Framework (1998–2003)

Source: Ma Xin (2008)

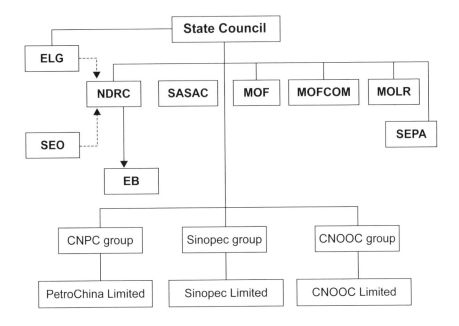

ELG: Energy Leading Group
SEO: State Energy Office (of Energy Leading Group)
NDRC: National Development and Reform Commission
EB: Energy Bureau (of NDRC)
SASAC: State Asset Supervision and Administration Commission
MOF: Ministry of Finance
MOFCOM: Ministry of Commerce
MOLR: Ministry of Land and Resources
SEPA: State Environmental Protection Administration
CNPC: China National Petroleum Corporation
PetroChina Limited: PetroChina Company Limited
Sinopec: China Petrochemical Corporation
Sinopec Limited: Sinopec Company Limited
CNOOC: China National Offshore Oil Corporation (from 1982)
CNOOC Limited: China National Offshore Oil Company limited

Figure A.4.5: The Regulatory Framework (2007)

Source: Ma Xin (2008)

CHAPTER 5

THE GAS INDUSTRY IN CHINA

In gas, as in oil, China's policy emphasis was to start by accelerating domestic gas discoveries and then increasingly to integrate imports (via pipelines and LNG) into the national energy plan. But there was no fixed policy about where the gas should come from and, so far, to the surprise of many observers, it has not come from Russia.

During the 2000s China's gas industry expanded vigorously. This chapter aims to explain the nature and consequences of this expansion. It starts with a brief review of China's domestic gas resources, focusing on domestic production basins, moving on to give a picture of coalbed methane gas and shale gas options. It then touches on China's domestic gas demand, paying special attention to the reform of gas prices – the main stumbling block to the long delayed Sino-Russian gas supply negotiation. The focus then moves to the growth of city gas, the main feature of China's gas expansion in the coming decades. After a detailed picture of China's nationwide gas pipeline network and its important trans-national pipeline connections, the chapter concludes with a discussion of the various balances and imbalances between the role of domestic production on the one hand and imports (both by pipeline and in the form of LNG) on the other in meeting the country's vast and growing demand for gas.

Domestic Gas Resources

According to BP's annual Statistical Review of World Energy, at the end of 2009 China's total proven gas reserves were 2.9 tcm, and the reserves-to-production (R/P) ratio was 29.0.[1] This somewhat conservative estimate, however, has done little to dent Chinese confidence in the country's ability to expand its domestic capacity.

In November 2004, an authoritative report on China's energy future was prepared by the State Council's Development Research Centre (DRC) under the title *Research on National Energy Strategy and Policy in China*.[2] This advocated the greater use of natural gas as a clean alternative to coal, in particular in the power and residential sector. It also stressed the importance of raising natural gas to 10 per cent of the energy mix by 2020.

In 2000 China's total energy consumption was 1.39 billion tonnes of coal equivalent. If this level of consumption were to double by 2020 (a common projection), demand would have risen to 2.78 btsce.[3] However, the actual figure in 2010 was already 3.2 btsce, and that for 2015 is projected by Zhang Guobao, former head of NEA, to be as much as 4 btsce.[4] Ten per cent of this is 278 mtsce, equivalent to 209 bcm of natural gas.[5] It is far from obvious that the country's natural gas resources are sufficiently abundant to provide for the production of over 200 bcm of natural gas annually in 2020, and the Beijing authorities have, therefore, been very anxious to understand the true scale of natural gas resources.

According to the second national natural gas resources survey (conducted in 1994 in China's 69 sedimentary basins, excluding the Spratly Islands, by the China National Petroleum Corporation (CNPC) and the China National Offshore Oil Corporation (CNOOC)), China's geological resources of conventional natural gas are 38.04 tcm, of which 79 per cent or 29.9 tcm are in onshore deposits and the remaining 21 per cent or 8.14 tcm are offshore. Approximately 89 per cent of the geological resources of these 38 tcm reserves were found in 13 basins – Songliao, Bohai Bay, Ordos, Sichuan, Tarim, Junggar, Turpan-Hami, Qaidam, Middle Yangtze River Reaches, East China Sea, Yinggehai, Qiongdongnan, and Pearl River Mouth.[6]

The third national resource survey was initiated in 2003,[7] and nearly 130 basins were evaluated. The preliminary results pointed to a higher estimate of China's natural gas base: the total was 52.7 tcm, a significant increase from the 1994 estimate of 38 tcm (see Tables 5.1 and 5.2). The 2003 survey also estimated remaining undiscovered resources to be 17.4 tcm, of which 4.1 tcm are located in the Ordos basin, 3.5 tcm in the Tarim basin, and 2.7 tcm in the Sichuan basin.[8]

The majority of the gas fields found between 1949 and 1976 were in the Sichuan basin. From 1977 to 1988, six large gas blocks took

Table 5.1: China's Gas Reserves Evaluation Results

	Natural Gas Reserves tcm
1987	33.6
1994	38.0
2006 *	52.7

* The Third survey was initiated in 2003 and completed in 2006.

Source: Yang Liu (2006c), 31.

Table 5.2: China's Natural Gas Resources according to the Third Survey, 2003 (tcm)

	Resources	*Undiscovered*
Onshore	38.8	14.5
east	4.4	1.8
central & west	31.3	11.9
south	3.2	0.8
Offshore	13.8	2.9
Total	**52.7**	**17.4**

Source: Chen Mingshuang (2006).

shape, including the basins of Sichuan, Tarim, and Ordos. Since 1989, a group of large gas fields has been discovered. By the end of 2005, the national total proven and recoverable gas reserves reached 3.5 tcm, up 25 per cent from the 2004 estimate. From 2000 to 2005, China discovered and proved eight large gas fields, each with reserves of over 100 bcm. The newly added recoverable gas reserves between 1991 and 2003 reached 2.06 tcm, an annual average of 158.5 bcm. In the latter part of that period (1999–2003) a total of 1.616 tcm of proven and recoverable gas reserves was added, an annual average of 201.9 bcm. In addition, it is projected that between 2004 and 2020 newly added recoverable reserves will be 3.13 tcm, an annual average of 183.9 bcm. Recoverable gas reserves will reach 5.59 tcm by 2020.

The third national survey was completed in 2006 and China was found to have a total of 52.7 tcm of natural gas. The onshore natural gas reserves are 38.82 tcm, of which 4.36 tcm are in east China, 31.26 tcm in central and west China, and 3.2 tcm in south China. The offshore gas reserves are 13.8 tcm. Proven but undeveloped gas reserves amount to 17.4 tcm, of which 6.78 tcm (39 per cent) are located in central China, 5.14 tcm (29.5 per cent) in the west, and 2.9 tcm (18.5 per cent) offshore. It is worth noting that the first ever evaluation, in 47 basins, of unconventional resources such as CBM, oil shale, and oil sands took place as part of the third survey.[9]

Since figures on the breakdown of this 52.7 tcm resource base by individual basins are not available, it is reasonable to use CNPC's 2005 data as a rough guide to the relative capacity of different basins. This is done in Table 5.3, which shows that China's natural gas resources, distributed in 115 gas-bearing onshore and offshore fields, totalled 35.03 tcm, including recoverable reserves of 22 tcm. These reserves are distributed mainly in the central, western, and offshore regions which, respectively, have geological reserves of 10.11 tcm (28.86 per cent), 11.60 tcm (33.12 per cent), and 8.10 tcm (23.13 per cent). Geologically, natural gas resources are mostly located in Cenozoic and Mesozoic rock formations, which contain proven resources of 13.25 tcm (37.82 per cent of the total) and 11.31 tcm (32.29 per cent of the total) respectively. In terms of depth, the gas resources are located at almost all levels geographically, that is, in the shallow layers there is 8.3 tcm (23.69 per cent), in the middle-deep layers 10.21 tcm (29.15 per cent), in the deep layers 10.94 tcm (31.23 per cent), and in the ultra-deep layer 5.58 tcm (15.93 per cent). Natural gas resources can be characterized as shallow reserves in the eastern and offshore regions, and deep reserves in the western and high seas zones. Conventional natural gas resources stand at 26.66 tcm, accounting for 76.11 per cent of the country's total; these resources are mainly scattered throughout central and western China.[10]

Table 5.3: Natural Gas Resources by Basin (tcm)

	Prospective	*Geological*	*Recoverable*
Tarim	11.3	8.9	5.9
Ordos	10.7	4.7	2.9
Sichuan	7.2	5.4	3.4
East China Sea	5.1	3.6	2.5
Qaidam	2.6	1.6	0.9
Yinggehai	2.3	1.3	0.8
Bohai Bay	2.1	1.1	0.6
Qiong south-east	1.9	1.1	0.7
Songliao	1.8	1.4	0.8
Others	10.8	6.0	3.6
Total	**55.9**	**35.03**	**22.0**

Source: CNPC, the 2005 (latest) resources investigation, quoted in Higashi (2009).

Despite the massive potential of the gas reserves identified by the third national survey, Beijing's energy planners had never, up to 2008, appeared confident that much more than 150 bcm/y of conventional gas could be produced by 2020. However in 2009, for the first time, the figure of 150 bcm/y was mentioned and in 2010 CNPC projected that gas production would reach to 210 bcm in 2020 and 300 bcm in 2030.[11] At the time of writing, however, it seems a more realistic prediction that the maximum production of conventional gas by 2020 will be around 150 bcm. During the 2010s, the fourth national survey will tell whether China's proven reserves can really push conventional gas production over the 200 bcm/y level by 2030.

China's Main Gas Production Bases

China's major gas production regions are located in nine basins, namely Tarim, Ordos, Sichuan, Qaidam, Songliao, Bohai Bay, Yinggehai, Qiongdongnan, and the East China Sea. In China, gas fields with gas reserves of more than 30 bcm are called large gas fields. Eleven gas fields have reserves exceeding 100 bcm, including Sulige, Jingbian, Kela-2, Yulin, Puguang, Daniudi, Wushenqi, Zizhou, Dina-2, and Kelameli. Gas development is to focus on four key regions – the Tarim basin, the Sichuan basin, the Ordos basin, and the South China Sea basin. At their peak, production in these regions will reach, respectively, 75–80 bcm, 55–65 bcm, 40–45 bcm, and 40–50 bcm,[12] a combined total of 210–240 bcm, although the projection did not state when peak

Table 5.4: Natural Gas Production (bcm)

	2002	*2005*	*2008*	*2010*
Daqing	2.021	2.443	2.796	2.995
Huabei	0.533	0.573	0.565	0.821
Liaohe	1.132	0.921	0.870	0.801
Xinjiang	2.019	2.895	3.434	3.806
Dagang	0.394	0.332	0.448	0.369
Jilin	0.217	0.273	0.545	1.408
Changqing	3.913	7.531	14.360	21.113
Yumen	0.061	0.079	0.053	0.022
Qinghai	1.150	2.121	4.415	5.614
Sichuan	8.751	11.629	14.834	15.364
Yanchang				
Jidong	0.041	0.077	0.306	0.432
Tarim	1.088	5.677	17.384	18.362
Tuha	1.143	1.532	1.522	1.257
CNPC total	**22.463**	**36.082**	**61.537**	**72.363**
Shengli	0.750	0.879	0.770	0.508
Henan	0.110	0.101	0.061	0.059
Zhongyuan	1.614	1.661	1.061	4.709
Jianghan	0.127	0.121	0.135	0.160
Jiangsu/Anhui	0.023	0.064	0.058	0.056
New Star		3.196	6.042	6.860
SINOPEC Star	2.296			
SINOPEC Total	5.095	6.285	8.308	12.493
CNOOC & others	5.307	8.124	10.668	8.744
China Total	**32.865**	**50.492**	**80.513**	**93.600**

Source: *China OGP* (2002, 2005, 2008, and 2010). Empty cells denote zero production.

production would be reached. In 2010, China's gas production was recorded at 93.6 bcm (Table 5.4).

As shown in Table 5.5, in 2007 it was projected that China's domestic gas production would reach 120–150 bcm by 2020, but CNPC's most recent projection suggests that total production could even reach 202 bcm, of which 62 bcm would be unconventional gas production (30 bcm of tight gas, 20 bcm of coalbed methane, and 12 bcm of shale gas).[13]

Table 5.5: China's Natural Gas Production Projections (bcm)

		2005	*2010*	*2015*	*2020*
CNPC	2005	35.3	65.0	70.0–75.0	80.0–90.0
	2007	44.2	73.0	85.0–90.0	90.0–105.0
SINOPEC	2005	6.3	7.0–10.0	12.0–14.0	18.0–20.0
	2007	7.7	9.5	10.0–13.0	10.0–20.0
CNOOC	2005	7.2	8.0–10.0	12.0–13.0	14.0–17.0
	2007	6.7	7.5	10.0–12.0	10.0–15.0
CUCBM	2005	0	2.0	4.0	8.0
	2007	0	2.0	5.0	10.0
Total	2005	48.8	82.0–87.0	98.0–106.0	120–135
	2007	58.6	92.0	110.0–120.0	120–150
	2009				over 150

Source: Asia Gas & Pipeline Cooperation Research Center of China, quoted in NAGPF (2005, 2007, and 2009).

The Beijing energy planners have high expectations of non-conventional gas production in the coming decades. The figure of 62 bcm/y by 2020 is a very ambitious one. If it can be achieved, despite the shortage of water supply for non-conventional development, it will produce a major change in production perspectives.

Coalbed Methane (CBM)

In China, some 15 bcm of CBM is released freely into the atmosphere every year, and if all of this were utilized, it would reduce emissions by approximately 0.76 mt of CO_2 equivalent and 1.86 mt of dust.[14] *Xinhua News Agency* reported that:

the reserves of CBM in China are around 36.7 tcm, ranking China as the world's third largest country in CBM reserves, after Russia and Canada. 330 CBM wells were drilled during 2005 in China, and CBM reserves were found to be almost equal in quantity to the expected reserves of onshore natural gas, which total some 38 tcm. The Qinshui and Ordos basins hold the biggest reserves with more than 10 tcm of resources between them. CBM reserves in Xinjiang are estimated at up to 6.8 tcm, accounting for 19 per cent of the national total. Xinjiang has three large CBM fields out of nine with reserves over 1 tcm.[15]

The development of CBM in China is mainly conducted in the Qinshui basin, Shanxi province, where the world's largest CBM power plant and a large scale liquefaction project have been built.[16] There were five main Chinese players engaged in the CBM development business: PetroChina, China CBM, SINOPEC, Jincheng Anthracite Mining Group, and Fuxin Mining Group. The US company Chevron was the first foreign oil major to sign an agreement with China CBM for exploration and development in China, and a number of CBM projects have been developed in association with foreign investors, which include Sino Gas & Energy (SGE), Greka Energy International, Fortune Liulin Gas Company (a subsidiary of Fortune Oil), Far East Energy, Verona Development, Terra West Energy, Canada Energy, and Ivana Ventures.

Commercial production and utilization of CBM started when, on 1 November 2005, the first phase of the Panhe CBM project, located in the Qinshui county of Jincheng, in Shanxi province and operated by China United Coalbed Methane (CUCBM), was formally completed. The project's target was to drill 100 wells in 2005 (up to October 2005, 81 wells had been drilled, of which 15 had begun to produce CBM with a total daily production of 2,000 cubic metres). Another 40 wells were due to start production in November 2005, when the construction of basic facilities for transporting and selling CBM was to have been completed.[17]

In mid-2006, the NDRC approved the Eleventh Five Year Programme for CBM development and utilization. According to Guo Benguang, assistant general manager of CUCBM, the five main points in this programme were: i) China was expected to produce 10 bcm of CBM in 2010, half of which would be produced through surface development, and two CBM production bases were to be built, one in the Qinshui basin and the other in the Ordos basin; ii) in order to reduce coalmine exploration along with its dangers, the state aims to give CBM priority over coal; iii) in remaining coal exploration, priority should be given to safer coal mines; iv) CBM would be planned as a complete chain, upstream to downstream; v) for the first time, consideration would be given to building CBM pipelines, of which 10 would be built, with a total length of 1,390 km and an annual handling capacity of 0.8–1.0 bcm each. Accordingly, the Eleventh Five Year Plan for CBM committed 3.09 billion yuan for building long distance CBM pipelines.[18]

In 2008, a *Feasibility Study of Coal Bed Methane Production in China*, undertaken by the EU–China Energy and Environment Programme, concluded that China's CBM industry should be treated more favourably and should receive more investment. Luo Dongkun, of the China University of Petroleum in Beijing, who led the study, recommended a 40 per cent increase on the existing subsidy for some CBM E&P activities, as well as an increase in government spending, in order to foster basic R&D in the production and marketing of planned CBM. According to Luo's calculations, the current subsidy to some projects, amounting 0.2 yuan/cubic metre of CBM sold, is lower than the benefits which CBM brings to society. The subsidy to CBM E&D related to residential use of CBM, as in Fengfeng (Hebei province), Zhina (Guizhou province), Liupanshui (Guizhou province), and Hongmao (Guangxi province), should be increased by 40 per cent to 0.28 yuan/cubic metre.[19]

PetroChina, a shareholder of CUCBM, had independently signed an MOU with the Australian gas supplier Arrow Energy to develop methane in Xinjiang, ahead of potential rivals such as SINOPEC and Shenhua Group. In November 2007 the NDRC and MLR decided to expand cooperation in the CBM sector, after a rule was released on 24 October that weakened the dominance of CUCBM in the CBM sector and allowed other domestic companies to forge alliances with foreign investors to exploit CBM In 2008 CNPC decided to leave CUCBM to become an independent CBM developer. PetroChina planned to finish Phase 1 of a CBM processing plant with an annual capacity of one bcm in 2008.[20] In June 2008, CNPC announced that it had started the construction of China's first CBM pipeline in Shanxi province. The 35 km pipeline in Qinshui County of Shanxi province feeds in to CNPC's West–East Gas Pipeline.[21]

In 2008, China's total CBM production was 0.5 bcm, and this rose to 1 bcm in 2009. CNPC's share of this was only 0.19 bcm.[22] Even though *Xinhua News Agency*'s report suggested that by 2010 China's CBM production would reach 10 bcm, including 5 bcm from surface and 5 bcm from underground,[23] the target figure seemed unrealistic as the feasibility of increasing the production scale so rapidly was questionable. According to Wood Mackenzie, production in 2010 reached only 1.25 bcm/y, but ambitious targets of 10 bcm/y by 2015 and 20 bcm/y by 2020 are certain to be set.[24] The actual result of CBM production during the 2000s, however, was a big disappointment, even though the potential is still huge.

A pipeline to pump CBM from Shanxi province to the WEP II pipeline came into operation in 2009. The 35 km CBM pipeline from

Qinshui county to the town of Duanshi, with a transportation capacity of three bcm, is set to pump CBM produced in the Qinshui basin, Shanxi province, to eastern China via CNPC's WEP II.[25] According to Yang Jianhong, vice director of the PetroChina Planning Institute, CBM and Synthetic Natural Gas (SNG) will become important supplementary gas resources for China's nationwide gas supply system. The supply of CBM and SNG is projected to reach a combined volume of 30 bcm/y by 2020. As a modern coal and chemical programme encouraged by the government, the CBM project in China contributed 0.7 bcm/y to supplies in 2009, and that figure is expected to reach 10 bcm/y by 2020. Fifteen SNG programmes have been initiated in China, with a combined production capacity of 25 bcm/y, although some are still at the planning stage or under construction.[26]

Preferential policies and government subsidies will help greatly in maintaining the momentum of CBM development in coming years. Among such policies are: i) a 0.2 yuan/cubic metre subsidy from the government, ii) giving surplus electricity from CBM power plants priority in supplying the grid at a 'de-sulphur' preferential price, iii) relieving power plants from responsibility for peak-shaving, iv) entitling CBM enterprises to enjoy VAT rebates, v) exempting surface drainage companies from resources tax, vi) entitling Sino-foreign joint ventures to preferential tax scales, vii) exempting imports of CBM equipment from import duty and VAT, and viii) the compensation fee rate for CBM mining being set at 1 per cent.[27]

According to NEA's development plan for the CBM industry in the 2011–15 period, China's CBM output is projected to reach 20–24 bcm by the end of 2015, of which 10–11 bcm will be from surface wells and 11–13 bcm from underground sources.[28] The target figures are big, and it remains to be seen whether performance during the first half of the 2010s can be more impressive than during the 2000s.

Shale Gas

According to researchers at PetroChina's New Energy Institute, based in Langfang:

> China's total shale gas resource is estimated at 21.5–45 tcm. China is modelling its nationwide shale research project on basins that have geologic characteristics similar to the US hydrocarbon provinces. The Chinese have selected four large provinces for study and development: the South China basin with a maturity similar to the Appalachian, the Junggar and Tuha basins similar to the Rocky Mountains and the Qaidam and East China basins similar to the Michigan basin.[29]

In 2008, Petromin Resources announced that the 'Terrawest Energy Corporation (TWE) has tapped into what may be a massive new unconventional natural gas resource in China'.[30] A year later, the Chinese NOCs decided to team up with international oil companies. On 10 November 2009, the first joint development project in shale gas got underway as Royal Dutch Shell and PetroChina signed an agreement to develop the Fushun-Yongchuan shale gas block project in Sichuan province. According to Dow Jones Deutschland:

> this marks the nation's latest effort to tap shale gas resources after the launch of a Sino-US Shale Gas Resource Cooperation Initiative ... during U.S. President Barack Obama's first state visit to China. The initiative announced on 17 November 2009 in Beijing is expected to assess China's shale gas potential through joint technical studies with reference to American experience with shale gas.[31]

CNPC announced on August 20 2010 that the Langfang branch of its PetroChina Exploration and Development Research Institute had set up a national shale gas laboratory and has been conducting shale gas research since 2007.[32]

In January 2010 it was reported that SINOPEC was in talks with BP over potential collaboration in the exploration and development of shale gas. The move underlines growing international interest in China's shale gas fields.[33] BP was already involved in extracting methane from coal in China. For oil and gas multinationals, keen to invest more in China's rapidly evolving but relatively closed energy sector, shale gas presented an attractive opportunity.[34] In the same month Norway's state oil company Statoil denied that it had an agreement with CNPC to study shale gas or conduct test drilling in China, despite a report from CNPC's research unit that the two companies had begun study and test-drilling in a Chinese shale gas block.[35] In November 2010, Statoil was reportedly on the verge of a deal to explore shale gas reserves in China.[36]

The NDRC is reviewing a plan to encourage the development and utilization of this unconventional gas source in an effort to meet rising energy demand without excessively increasing greenhouse gas emissions. According to the Ministry of Land and Resources (MLR), China aims to raise the annual production capacity of shale gas to 15–30 bcm by 2020, compared with the MLR's (possibly over-optimistic) projection for total gas production of between 187.5 and 250 bcm in that year. In other words, China aims to achieve recoverable shale gas reserves of 1 tcm by 2020 after identifying between 20 and 30 main exploration and development blocks, and between 50 and 80 potential target blocks.[37] It expects to find commercially recoverable reserves in Sichuan amounting to as much as 100 bcm within two years. The MLR said that the Sichuan basin might have shale gas resources which are between 1.5 and 2.5 times larger than its conventional gas reserves.[38] In late October 2010, the MLR invited China's four NOCs to participate, for the first time, in an acreage tender for shale gas exploration. There are six target acreages, of which three are in Guizhou province, and one in each of the provinces Chongqing, Anhui, and Zhejiang.[39] The Energy Research Institute's deputy director, Jiang Xinmin, said that the central government plans to give shale gas companies subsidies of no less than 0.33 yuan ($0.049) per cubic metre (the current rate given to seam gas producers) in order to achieve the ambitious targets. He added that although it has 3 tcm of shale gas resources, China currently lacked both the key technologies and extensive pipeline network necessary to promote the sector's healthy development.[40] Besides this, tapping the shale gas will be expensive and difficult for a country that is desperately short of water.[41]

The *Financial Times* has reported that 'China's shale gas potential is estimated at 50–100 billion barrels of oil equivalent, on top of its 75 billion boe of coal-bed methane reserves',[42] and argued that 'in the long run, China and the US agreeing that natural gas is the fuel of the future may well prove to be more important than their failure to come up with binding emissions targets at Copenhagen in December 2009'.[43]

China's Domestic Gas Demand

During the decade 2000–2009, China's natural gas consumption increased from 24.5 bcm to 88.7 bcm, an annual growth rate of 15.4 per cent. In volume terms, annual average growth was 4.5 bcm per annum between 2000 and 2005, and 10.5 bcm per annum between 2005 and 2009. The share of natural gas in China's energy consumption mix increased from 2.4 per cent in 2000 to 3.8 per cent in 2009.[44] The gas market is forecast to reach its peak during the next 20 years. Between 2005 and 2030 natural gas demand will rise by about 10 per cent (more than 15 bcm) annually. During the period 2020–30, when China's natural gas production is expected to exceed 200 bcm, natural gas demand, also according to very imprecise projections, might reach 280–530 bcm per annum, of which (depending on what is assumed about production) anywhere between 30 and 60 per cent would have to be satisfied by imports.[45]

Until 2008, most of the projections on China's gas demand in 2020 hovered around the broad range of 200–240 bcm. In 2009, however, the prevailing estimate underwent a considerable increase (Table 5.6).

Table 5.6: Gas Demand Projections (bcm/y)

	2005	2010	2015	2020
CNPC	63.7	106.8	153.4	210.7
CNPC (2006)		120.0		200.0
CNPC (2009 & 2010)				300.0
SINOPEC (2006)		140.0		240.0
NAGPF (2009)				230.0
NAGPF (2007)		100.0		210.0
NAGPF (2004)		106.8		210.7
ERI/NDRC	64.5	120.0	160.0	200.0
ERI/NDRC (2010)			200.0	300.0
ERI/NDRC (2011)			230–240	
NEA/NDRC (2010)			260.0	
Xinhua News Agency (2010)		110.0		270–300
CNOOC	61.0	100.0	150.0	200.0
BP	42.0	74.0	135.0	177.0
EIA/DOE	51.0	79.0	127.0	181.0
IEA (2002)		61.0		109.0
IEA (2009)				176.0
IEA (2011)			247.0	335.0
UBS (2009)				212.0
Actual		110.0		

Note: NAGPF = Northeast Asian Gas & Pipeline Forum.

Sources: IEA (2002); CNPC/SINOPEC (2006); NAGPF (2004), NAGPF (2007), and NAGPF (2009); *China Securities Journal* (2010); IEA (2009); *People's Daily* 30 January 2011, 'China to see gas demand soar by 20% in 2011', http://english.peopledaily.com.cn/90001/90778/7276466.html; 'The conference of the Bureau of Energy introduced the energy economic situation in the first half of the second half of the trend', www.gov.cn/xwfb/2010-07/20/content_1659303.htm; IEA (2011b), 23.

For the first time CNPC mentioned a projected figure as high as 300 bcm. In March 2010, the NDRC's Energy Research Institute also supported the higher figure.[46] At the very least, CNPC was confident that China's gas production figure could exceed 150 bcm by 2020. This is a huge change considering that only a decade before, Ma Fucai, the then president of CNPC, predicted that the gas industry's major boost would come within two decades, suggesting that annual gas production would be 70–80 bcm (nearly four times the then current level of 22 bcm), and by 2020, it would increase to 100–110 bcm, boosting the gas share in China's primary energy consumption to 8 per cent from the current 2 per cent.[47] Further changes of expectation occurred in the subsequent ten years, and the predictions rose even higher. In 2008 PetroChina's vice president, Jia Chengzao, saw China's natural gas production at least doubling in the next decade to reach 150–200 bcm by the end of the 2010s.[48] In fact, during the Tenth Five Year Plan period (2001–5), eight gas fields, each with more than 100 bcm of proven reserves, had been discovered.[49] As mentioned earlier, in 2010, Jie Mingxun, president of CBM, under PetroChina, projected that China's conventional and unconventional gas production by 2020 would be 140 bcm and 62 bcm respectively.[50]

Despite the rapid rise of domestic capacity, the level of demand outpaced it. According to the National Energy Administration, China's gas demand in 2010 was 110 bcm, 20 per cent up on 2009, while production was only 94.5 bcm, 12 per cent up on 2009. The NEA projected that the demand figure in 2011 would be 130 bcm, with production at only 110 bcm.[51] In 2010, the NEA signalled that China would witness an unprecedented increase of gas use, and projected that demand in 2015 would reach 260 bcm,[52] 8.3 per cent of China's primary energy mix.[53] Immediately after the Twelfth Five Year Plan (2011–15) announcement, ERI projected that China's natural gas supply by 2015 would be as high as 230–240 bcm/y, of which 150 bcm would be domestic production, 30 bcm imports in the form of LNG, and 50 bcm imports by pipeline.[54] CNPC projected that demand in 2030 would reach 392 bcm.[55] Wood Mackenzie went so far as to project that total gas demand would rise from 93 bcm in 2009 to 444 bcm in 2030, a compound annual growth rate of 7.5 per cent, most of the growth coming before 2020.[56]

Changes in Consumption Structure

David Fridley summarizes the present structure of consumption in this way:

> Natural gas remains primarily a fuel of the industrial sector. At the time that economic reforms started in the early 1980s, industry accounted for over 90 per cent of total consumption, and nearly half of that amount was used in the oil and gas sector itself. As the value of commercial gas has grown, natural gas increasingly penetrated other sectors, including residential, power generation, and transportation. As of 2008, the share of industrial use is in the production of chemicals, including chemical fertilizers, as both fuel and feedstock. Fertilizer plants remain priority recipients of natural gas, and their supply is guaranteed through both allocation and pricing controls.
>
> ...
>
> Since 2000, natural gas consumption has risen on average 14 per cent per year, with residential use soaring at a 20 per cent annual rate of growth, and power sector use by 16 per cent per year.[57]

Since 2004, in particular, the annual growth rate has accelerated to more than 20 per cent, far above China's GDP growth rate. Fridley argues that:

> two fundamental factors explain this shift. The first is the response to the electricity crisis of 2003–2006, and the second is the doubling of the size of the national economy since 2000 ... The boom in economic growth led primarily by industry resulted in a scramble for all forms of energy. Coal demand rose by 10 per cent per year from 2000 to 2005, while oil went up 8 per cent a year. But Electricity – hydro and nuclear power – soared 13 per cent a year. From an average annual capacity addition of 30 GW per year in the late 1990s, a boom in construction of new power plants added 60 GW of new capacity in 2005, 102 GW in 2006, with 90 GW additional slated for 2007. Nearly all of these were coal-fired, but China added 15.6 GW of gas-fired generation during this period, with 19 GW additional planned'.[58]

Faced with increasing demand for natural gas, the government issued a new sectoral priority sector policy in August 2007, specifying that city residential use and combined systems for heat and power should have the highest priority. Since late 2008, the increase in natural gas consumption has slowed, due to falling oil prices and the effect of the world economic recession, but it continues to grow.[59] As discussed earlier, natural gas consumption grew in 2009 to reach 89 bcm.

During the 2000s China's gas consumption market has shifted from peripheral areas of gas fields to eastern economic developed areas. Except for Tibet, all of China's 30 major provinces and municipalities have access to natural gas, but the rate of penetration is quite variable.[60] In parallel with the development of China's nationwide pipeline network and a string of major LNG terminals, four major consumption centres were established; i) Bohai Rim area, ii) Yangtze River Delta, iii) Pearl River Delta, iv) Sichuan and Chongqing region. Due to governmental restrictions relating to the development of a natural gas petrochemical industry, however, the south-west and north-west regions (where the gas petrochemical industry is major gas consumer) have seen slower growth in the consumption of natural gas.[61]

As shown in Table 5.7, natural gas consumption by region shows that the share of the south-west, where Sichuan basin is located, has declined significantly from 43.8 per cent to 20.5 per cent during the 1990 to 2008 period, and this relative decline is projected to continue until 2030. The main demand areas by 2030 will be Yangtze River Delta, the south-east coast, the Bohai area, and the central south areas, which are projected to account for 66 per cent of China's total demand. The central south area will have the fastest growth of natural

Table 5.7:   Natural Gas Consumption by Region (%)

	1990	2000	2005	2008	2015 (P)	2030 (P)
Yangtze River delta	0.2	1.2	7.4	13.5	18.9	18.0
south-east coast	0.0	2.9	5.4	10.1	16.5	16.7
central south	7.4	5.2	6.8	8.8	12.4	15.2
Bohai area	15.6	12.1	14.7	15.8	14.0	16.0
south-west	43.8	43.7	29.4	20.5	14.2	12.2
north-west	3.7	12.0	19.0	12.8	9.7	7.4
mid-west	0.5	3.4	7.5	12.2	9.6	9.2
north-east	28.8	19.6	9.8	6.3	4.6	5.3
Total	100.0	100.0	100.0	100.0	100.0	100.0

Note: (P) = projection.
The regions:
> Yangtze River delta: Shanghai, Jiangsu, and Zhejiang
> south-east coast: Fujian, Guangxi, Guangdong, and Hainan
> central south: Hubei, Hunan, Anhui, Henan, and Jiangxi
> Bohai area: Beijing, Tianjin, Hebei, and Shandong
> south-west: Sichuan, Chongqing, Yunnan, and Guizhou
> north-west: Xinjiang, Gansu, and Qinghai
> mid-west: Inner Mongolia, Shaanxi, Shanxi, and Ningxia
> north-east: Heilongjiang, Jilin, and Liaoning

Source: Duan Zhaofang (2010).

gas demand – from 5.2 bcm in 2007 to 57.6 bcm in 2030, an annual average growth of 11 per cent.

As shown in Table 5.8, by 2030 the Bohai area, mid-west, and central south will form a consumption market dominated by city gas, and the north-west will form the industrial fuel-oriented market. The south-east and the Yangtze River Delta will form the urban market for what is seen as a cleaner way of producing power, and the resources

Table 5.8: Natural Gas Consumption Mix by Region in 2030 (%)

	City Gas	Power Generation	Industrial Fuel	Chemical Sector
Yangtze River delta	39	34	26	1
south-east coast	33	41	18	9
central south	50	12	35	3
Bohai area	55	19	24	2
south-west	40	8	18	35
north-west	29	14	45	12
mid-west	51	5	21	22
north-east	31	11	29	29

Source: Duan Zhaofang (2010).

areas such as the south-west, north-east, and central west (or mid-west) areas will provide a large part of the market for gas to be used in the chemical industry.

Table 5.9 shows the shift of balance in China's natural gas consumption mix between 1995 and 2009 and the trend up to 2030. Before 2000, China's gas consumption was dominated by industrial fuel and chemical sector use. Following the introduction of long-distance pipelines like WEP I, the gas consumption pattern has changed greatly; by 2009, city gas was up from 12 to 43 per cent, power generation up from 5 to 12 per cent, while industrial fuel was down from 61 to 26 per cent and the chemical sector use was down from 22 (having risen to 39 per cent in 2003) to 20 per cent. The trend towards an increased role for city gas and power generation is reflected in the projected shares of demand from these two sectors of 34 and 21 per cent in 2020, and 42 and 21 per cent in 2030.

Table 5.9: China's Natural Gas Consumption Mix, 1995–2030 (%)

	City Gas	Power Generation	Industrial Fuel	Chemical Sector
1995	12	5	61	22
2000	18	4	41	37
2001	21	4	40	35
2002	22	4	39	35
2003	21	3	36	39
2004	26	4	36	33
2005	24	5	37	34
2006	26	6	33	35
2007	27	11	30	33
2008	34	15	28	23
2009 E	43	12	26	20
2010 E	31	18	30	22
2015 E	31	21	32	17
2020 E	34	21	30	15
2025 E	39	21	27	12
2030 E	42	21	25	11

Note: E = estimate.

Source: Duan Zhaofang (2010).

In 2000, total residential gas consumption was 3.2 bcm and it rose to 13.3 bcm in 2007, an annual average growth rate of 22 per cent. In the coming decades, residential gas consumption will grow rapidly to reach 64 bcm in 2020 and 105.4 bcm in 2030. Per capita consumption will keep (or be maintained) at its level of around 14 bcm in 2020 and 2030 respectively. In the case of commercial gas consumption, during the

2000–7 period it grew from 0.34 bcm to 1.71 bcm, with annual average growth rate of 25.8 per cent. In parallel with city gas network expansion and the increase of the gasification rate, commercial gas consumption will continue to grow fast to reach to 16.2 bcm by 2030. During 2000 and 2008 gas consumption for the transportation sector rose from 0.58 bcm to 2.7 bcm, and this figure is expected to reach to 31.6 bcm by 2030.[62]

Between 2000 and 2008, China's installed power capacity grew rapidly at an annual growth rate of 12 per cent, reaching 792.93 GW. Gas-fired units amounted to 25.14 GW, accounting for 3.2 per cent of the total. The south-east coast, Yangtze River delta, central south, and Bohai area are the key areas for gas-fired generation in China. Natural gas consumption for power generation grew to 11.9 bcm in 2008, accounting for 14.7 per cent of China's total gas consumption. As shown in Table 5.9, gas consumption for power generation is projected to expand by 2015, with its share in national total consumption up to 21 per cent. This level will be maintained until 2030.[63]

Gas consumption for industrial fuel rose during the 2000–8 period from 10.1 bcm to 20.8 bcm a year. The consumption of industrial fuel is at present concentrated in the four sectors of oil and gas production, petrochemical industry, construction materials, and metallurgy. By 2030, the figure will reach 99 bcm, and its application will be extended to multiple sectors. In the case of natural gas consumption for the chemical industry, China's synthetic ammonia and methanol output, with natural gas as the raw material, has continued to grow in recent years. In 2000, the volume was 9.1 bcm and it rose to 18.6 bcm in 2008. Natural gas consumption for the chemical industry will grow and then maintain a stable level after 2020, but its share in China's total gas consumption will decline to 11.4 per cent by 2030.[64]

In short, natural gas usage will become more commonplace in urban areas as it replaces petroleum and coal. Compared to the current consumption structure, the proportion of natural gas for the city gas sector will rise but its use as an industrial fuel will maintain the current level until 2020 and then will witness a minor scale decline; usage for power generation will remain flat, and feedstock usage for the petro-chemical industry will fall. Government policies such as the 'Natural Gas Utilization Policy' are guiding the gas market in this direction.[65]

Gas Price Reform[66]

The debate about the future of the demand for, and supply of, gas amounts to little more than guesswork as long as the question of the price is not included. Markets have been playing a growing role in the Chinese economy, and even though many of the uses of gas were in areas of economic activity still dominated by the state, the price of energy products has had a growing effect even on the decisions of planners. That was most emphatically the case with imports of gas, since the world market, especially for LNG, is a complex one, in which Chinese planners have had to learn to move. Prices began to be seen not only as essential elements in the determination of demand and supply, but also as influences on the whole energy sector, on the behaviour of foreign companies, and in international relations. By the early 2000s the serious discussion of prices and price reform had arrived at the question of gas.

After a long delay, in late May 2010 Beijing announced that it would raise the domestic onshore natural gas producer benchmark by 0.23 yuan ($0.034)/cubic metre (See Table 5.22). This price reform offers a big incentive to China's main gas producers, like PetroChina and SINOPEC. This section will review the process of gas price reform during the second half of the 2000s, and its implications. In late 2005 the *Xinhua News Agency* reported that China's low natural gas price has resulted in a vicious circle of production and demand. The government-pegged price had deviated far from the market value. The direct consequence was undersupply, because the low price had boosted demand but inhibited the zeal for production on the part of producers. The culprit of this vicious circle was the government-pegged price.[67]

Even Zhao Xiaoping, director of the Price Department of the NDRC admitted, at a seminar on resources price reform in October 2005, that the natural gas price should be increased. Wang Guoliang, CFO of CNPC, proposed a natural gas price formula very close to the international pricing system. For the eastern part of China, he argued, the retail price should be the LNG CIF international price plus the cost of storage, re-gasification, and transport, and in the western part, the ex-factory price should be the retail price in the eastern part minus the cost of storage and transport.[68]

The 2006 Gas Price Reform

On 26 December 2005, the NDRC summarized the features of its price reform proposals as follows:[69]

i)   the classification of the price of natural gas for industry, towngas, and fertilizer production will be simplified;
ii)   the factory price of natural gas will be classified into two grades.

The gas produced by the Sichuan–Chongqing gas field, Changqing gas field, Qinghai gas field, all gas fields in Xinjiang province (except for WEP I gas which will not follow the new pricing mechanism), and some gas fields in Dagang, Liaohe, and Zhongyuan will be classified as Grade I, which is set at a lower price. The gas produced by the remaining gas fields will be classified as Grade II, which is priced at 980 yuan per 1,000 cubic metres. The benchmark price for Grade I natural gas will be gradually adjusted to that of Grade II natural gas within the next 3–5 years;

iii) the government will gradually change its role in natural gas pricing. It will only issue a guide price for natural gas. Suppliers and buyers will be allowed to negotiate and decide a specific price based on the government-set guide price. The actual price will be allowed to fluctuate in a range of 10 per cent above or below the guide price;

iv) the benchmark price of natural gas will be adjusted once a year in line with the changes of the average price of crude oil, LPG, and coal with weights of 40, 20, and 40 per cent respectively. The crude oil price will be set according to the average of the FOB prices of WTI, Brent, and Minas. The LPG price is based on the LPG FOB price in the Singapore market, and the coal price is the average price of high-quality Shanxi coal blends, high-quality Datong coal blends, and ordinary Shanxi coal blends at the Qinhuangdao port;

v) the benchmark price of natural gas will be adjusted each year according to the prices of other resources such as petroleum; natural gas producers may vary the factory price of natural gas by a maximum of 8 per cent in either direction.

Based on these guidelines, the nationwide factory price of natural gas would be adjusted upwards by 50–150 yuan per 1,000 cubic metres for the industry and towngas sectors, and by 50–100 yuan per 1,000 cubic metres for the fertilizer production sector (Table 5.10). An NDRC official said that the upward adjustment would have only limited influence on residents and enterprises.

By 1 August 2006, Xinjiang had adjusted the retail price of natural gas by 1.3 per cent to 1.366–2.076 yuan per cubic metre, and from July 2006 Sichuan raised the price by 10 per cent. In Shanghai, prices of 2.1 yuan/cubic metre of natural gas and 1.05 yuan/cubic metre coal gas had been set in 1999 and 2003 respectively. Since 1 September 2006, Beijing raised the price of natural gas for non-household use by 0.15 yuan/cubic metre and the city planned to hold a public hearing on raising the price of natural gas for household use in October 2006.

Table 5.10: Benchmark Price of Grade I Natural Gas, 1 January 2006
(yuan/1000 m³)

Gas field	Sector	Benchmark price
Sichuan–Chongqing	Fertilizer production	690
	Industry	875
	Towngas	920
Changqing	Fertilizer production	710
	Industry	725
	Towngas	770
Qinghai	Fertilizer production	660
	Industry	660
	Towngas	660
Gas fields in Xinjiang	Fertilizer production	560
	Industry	585
	Towngas	560
Others	Fertilizer production	660
	Industry	920
	Towngas	830

Note: Suppliers and buyers are allowed to negotiate and decide specific prices which
are not more than 10 per cent above or below the above benchmark factory
price. The benchmark factory price of natural gas from the Zhongxian–
Wuhan pipeline is set at 911 yuan per 1,000 cubic metres.

Source: Lin Fanjing and Mo Lin (2006).

On 12 September 2006, Shanghai held a public hearing on setting up
a natural gas pricing mechanism and on gas price adjustment. The city
had two alternatives: the gradual growth plan, with a price increase
in the range 19–38 per cent, and the unified pricing plan with a price
increase of 23.8 per cent. Under the gradual growth plan, Shanghai
set three prices of 1.25, 1.35, and 1.45 yuan/cubic metre for coal gas,
and 2.5, 2.7, and 2.9 yuan/cubic metre for natural gas according to
the different grades of gas consumed by Shanghai residents. Under
the unified pricing plan, the price is 1.3 yuan/cubic metre for coal gas
and 2.6 yuan/cubic metre for natural gas. For non-resident coal gas
consumers, the base price was raised from 1.45 yuan/cubic metre to
1.8 yuan and a seasonal fluctuation mechanism introduced.[70]

At the end of October 2006 even Ma Kai, chairman of the NDRC,
entered the debate, confirming the need for price reform, given that the
domestic gas price was too low and should be gradually raised,[71]while
implicitly acknowledging that the NERC was the ultimate authority on
price reform in the sector. In late autumn 2006 the NDRC reportedly
conducted a cost investigation in Sulige gas field, which belongs to

PetroChina's Changqing Oilfield Exploration Bureau. The aim was to investigate the feasibility of raising the domestic natural gas price and to suggest how to do so. By 2020, electricity from gas-fuelled power plants will probably account for 6.7 per cent of total national electricity production. The absence of a natural gas pricing mechanism and of a policy to ensure long-term gas supply were still, however, two major obstacles to the rapid development of gas-fuelled power generation.[72]

It was also reported that Beijing was planning a rise of nearly 8 per cent in the household natural gas price. In other words, the price of household gas would rise by 0.15 yuan to 2.05 yuan. About 2.88 million Beijing households were connected to natural gas at the end of 2005. The city purchased 3 bcm of gas in 2005 at a price for households which was lower than that in Tianjin and Shanghai, where gas prices stood at 2.20 yuan and 2.10 yuan per cubic metre respectively.[73]

At the end of April 2007, the relevant government departments and the domestic oil giants gathered in Chengdu, capital of Sichuan province, to discuss steps towards reforming the natural gas pricing system. It was not the first time that the state had done this, nor was it the first time that, after the conclave, the system failed to make significant progress. The meeting sent out a signal that natural gas should be, first of all, considered as a commodity and, second, it should become market-oriented, and this is the accepted direction for ongoing reform. Insiders, however, said it would take a long time to reach this objective, although they agreed it was a correct and unavoidable direction. An NDRC official said that the government should guarantee social stability before taking action to reform the natural gas pricing system. One thing was certain, the gas price would constantly rise, and monopoly enterprises such as CNPC and SINOPEC would continue to benefit. In June 2007, the existing import license systems set for natural gas and LNG were removed, and this allowed the state to impose stricter import control through quotas or licences. The three oil giants which have imported LNG, or been in talks with overseas LNG suppliers, will benefit from this policy change since some local state-owned enterprises will be stopped from directly purchasing LNG from overseas.[74] In order to guarantee the existence of a market for relatively expensive natural gas – such as imported LNG – the state can adjust the price by changing taxation. For example, less or even no VAT could be levied on the imported LNG. Meanwhile, no steps were taken.

An observer has described the next part of the story:

> In November 2007, the Beijing authority decided to drastically increase ex-plant prices. For the industrial sector, in particular, the average prices rose by 50 per cent from around 800 yuan/1,000 cubic metre ($3.04 /

mmbtu) to 1,200 yuan/1,000 cubic metre ($4.57/mmbtu). Current tariffs are fixed for each individual pipeline, thus reflecting the construction cost. For example, the tariffs of the Ordos–Beijing pipeline are 310 yuan/1,000 cubic metre ($1.18/mmbtu) for Shanxi province, and 480 yuan/1,000 cubic metre ($1.82/mmbtu) for Tianjin. The tariffs of the WEP I are 680 yuan/1,000 cubic metre ($2.58/mmbtu) for Henan province and 980 yuan/1,000 cubic metre ($3.72/mmbtu) for Shanghai.[75]

Tables 5.11–5.14 show gas prices in 2008. When, in 2008, PetroChina conducted studies to decide on the terminal gas cost of the WEP II, an official of PetroChina predicted that it would be higher than the current price. Zhang Guobao, vice-minister of the NDRC, said in

Table 5.11: Ex-Plant Prices by Gas Field and by Sector, 2008

Gas Field	Sector	Ex-Plant RMB/1000m³	Ex-Plant $/mmbtu
Chunyu	Fertilizer	690	2.62
	Industry	1275	4.84
	Residential	920	3.50
Changqing	Fertilizer	710	2.70
	Industry	1125	4.27
	Residential	770	2.93
Qinghai	Fertilizer	660	2.51
	Industry	1060	4.03
	Residential	660	2.51
Xinjiang	Fertilizer	560	2.13
	Industry	985	3.74
	Residential	560	2.13
Dagang Liaohe Zhongyuan (category 1)	Fertilizer	660	2.51
	Industry	1320	5.02
	Residential	830	3.15
Dagang Liaohe Zhongyuan (category 2)	Fertilizer	980	3.72
	Industry	1380	5.24
	Residential	980	3.72
Other fields	Industry / Other	1380 / 980	5.24 / 3.72
WEP I	Industry / Other	960 / 560	3.64 / 2.13
Zhongwu pipeline	Industry / Other	1281 / 881	4.86 / 3.34
Shaanjing pipeline	Industry / Other	1230 / 830	4.67 / 3.15

Source: CNPC RIE & T. (2008).

Table 5.12: Ordos–Beijing Pipeline Gas, 2008

Destination	Ex-Plant	Pipeline Tariff	City Gate	$/mmbtu City Gate
Shaanxi	830	120	950	3.61
Shanxi	830	310	1140	4.33
Shandong	830	400	1230	4.67
Hebei	830	420	1250	4.75
Beijing	830	450	1280	4.86
Tianjin	830	480	1310	4.55

Source: CNPC RIE & T. (2008).

Table 5.13: WEP I Gas, 2008

Destination	Sector	Ex-Plant	Pipeline Tariff	City Gate	$/mmbtu City Gate
Henan	Industry	960	640	1600	6.08
	Residential	560	680	1240	4.71
Anhui	Industry	960	750	1710	6.50
	Residential	560	750	1310	4.98
Jiangsu	Industry	960	790	1750	6.65
	Residential	560	940	1500	5.70
	Power	560	620	1180	4.48
Zhejiang	Industry	960	980	1940	7.37
	Residential	560	980	1540	5.85
	Power	560	720	1280	4.86
Shanghai	Industry	960	800	1760	6.69
	Residential	560	980	1540	5.85
	Power	560	670	1230	4.67

Source: CNPC RIE & T. (2008).

February 2009 that the government would calculate the terminal price of the second pipeline by the end of 2009.[76] The discussion on gas price reform has been at the top of the agenda, but no easy solution has been found and no radical gas price reform has been carried out. Industry insiders were sure that China would raise gas prices before the end of 2009, but no action was taken. In reality, state-capped natural gas prices remained at less than half the level of international prices.

Table 5.14: Natural Gas Price in Major Chinese Cities, 2008

	Natural gas as Towngas		*Natural gas for commercial use*	
	yuan/m³	*$/mmbtu*	*yuan/m³*	*$/mmbtu*
South China				
Shenzhen	3.50	14.22	3.70–3.95	15.04–16.05
Guangzhou	3.45	14.02	3.70	15.04
Foshan	3.85	15.65	4.62	18.78
Dongguan	3.85	15.65	4.00	16.26
Zhongshan	3.85–4.80	15.64–19.50	5.50	22.35
Shantou	5.40–6.00	21.95–24.39	-	-
Zhanjiang	3.50	14.23	5.60	22.76
Shaoguan	5.98	24.30	7.18	29.18
Haikou	2.60	10.57	3.30–3.73	13.41–15.16
Nanning	4.19	17.03	-	-
East China				
Nanjing	2.20	8.94	2.35	9.55
Yangzhou	-	-	-	-
Shanghai	2.10	8.54	3.30	13.41
Hangzhou	2.40	9.75	2.50	10.16
Quanzhou	3.80	15.44	-	-
Central China				
Changsha	3.80	15.44	-	-
Wuhan	2.30	9.35	2.43	9.88
Hefei	2.10	8.54	2.48–3.20	10.08–10.13
Nanchang	6.50	26.42	-	-
south-west China				
Chengdu	1.43	5.81	1.66	6.75
Chongqing	1.40	5.69	1.67	6.79
Kunming	-	-	-	-
Guiyang	-	-	-	-
North China				
Beijing	2.05	8.33	1.95–2.55	7.83–10.36
Tianjin	2.20	8.94	2.40–2.80	9.75–11.38
Taiyuan	2.10	8.54	2.00–2.25	8.13–9.14
Zhengzhou	1.60	6.50	2.10–2.40	8.54–9.75
Luoyang	2.60	10.57	-	-
Hohhot	1.57	6.38	-	-
Jinan	2.40	9.75	2.82	11.46
Qingdao	-	-	-	-
north-east China				
Changchun	2.00	8.13	-	-
Shenyang	2.40	9.75	-	-
Heilongjiang	-	-	-	-

Table 5.14: continued

	Natural gas as Towngas		Natural gas for commercial use	
	yuan/m³	*$/mmbtu*	*yuan/m³*	*$/mmbtu*
north-west China				
Xian	1.75	7.11	1.95	7.93
Lanzhou	1.45	5.89	1.00–1.25	4.06–5.08
Xining	-	-	-	-
Yinchuan	1.40	5.69	-	-
Urumqi	1.37	5.57	1.85	7.52

Source: China OGP, 15 August 2008, 5.

The Unrealized 2009 Gas Price Reform and Gas Price Setting for the Sichuan-to-East Pipeline (SEP)

The National Energy Administration (NEA) had confidently announced that reforming the gas pricing system would be its 'key task' in 2009 and Zhang Guobao, head of NEA, shared this view: 'China' he said, 'should introduce a reasonable natural gas pricing mechanism as soon as possible'.[77] Chi Guojing, secretary general, China Gas Association, gave a strong indication that China would indeed reform the natural gas pricing mechanism before the end of 2009. The expectation of the market was that the reform scheme might appear before the WEP II, carrying Turkmenistan gas to Guangdong province, went into operation. Chi said that it was certain that the towngas and industrial use natural gas prices would increase eventually, whatever the plan was.[78]

Li Wei, vice director of the Natural Gas Market Office of the Oil & Gas Pipeline Planning Research Institute, under PetroChina's Planning & Engineering Institute, said that the NDRC had endorsed a plan to distribute the 30 bcm per year of gas to nine provinces along the WEP II trunk pipeline. In other words, PetroChina and local governments would begin to negotiate the gas supply price. Under the current pricing system, PetroChina's supply, based on Turkmenistan gas, would be uncompetitive. According to Li, the CIF gas price at Horgus was 1.8 yuan/cubic metre, higher than the price in some inland cities. If transmission fees were taken into account, PetroChina's gas supply to Guangdong would be around 3–4 yuan/cubic metre – much higher than the price Guangdong currently pays for gas supply from CNOOC's Dapeng LNG terminal, which is supplied by LNG from Australia at about $170/tonne or 1.60 yuan/cubic metre. Li also said that Guangdong province might take 10 bcm/y of gas, which is one third of the total supply of the WEP II pipeline.[79]

Chi suggested that one option was to follow a 'one source, one price' method to encourage natural gas suppliers to produce or import natural gas at a low price. But this method cannot solve PetroChina's dilemma. The other option is to set a natural gas pricing formula for the whole country pegged to the international oil price. Chi, however, ruled out the possibility that China would replace the current payment method by 'volume with payment by calorific value', as it is impossible to replace gauging meters quickly enough throughout the country. PetroChina is considering the option of mixing Xinjiang gas into the WEP II to dilute the high cost, while waiting for the new pricing system. The gas-rich regions must supply gas to other regions as a priority, even if that means their own industries suffer shortages of gas (fertilizer plants in Sichuan and methanol companies in Xinjiang have had to stop production in order to guarantee gas supply to other regions). This is the reason for building the main trunk gas pipeline from Xinjiang to the coastal areas. That is why they call it 'west gas to the east coast' (where the main gas consumption areas are located). The local producing areas have no say about this central policy. So, the gas rich regions have a strong interest in reforming the gas pricing system.[80]

The announcement of the Sichuan-to-East pipeline (SEP) gas price in the summer of 2009 gave a strong hint about the nature of the forth-coming gas price reform, even though there was no gas price reform in 2009. The NDRC announced that the producer price of Puguang gas to be pumped into the SEP would be 1.28 yuan/cubic metre, including VAT (which is $0.187), but it would be allowed to fluctuate within a ±10 per cent band after negotiations between buyers and sellers.[81] The average transportation fee is set at 0.55 yuan/cubic metre, with different fees to the provinces en route, depending on distance. Specifically, the fee for Sichuan is 0.06 yuan ($0.0088), for Chongqing 0.16 yuan, for Hubei province 0.32 yuan, for Jiangxi province 0.54 yuan, for Anhui province 0.65 yuan, for Jiangsu province 0.76 yuan, for Zhejiang 0.81 yuan, and for Shanghai 0.84 yuan ($0.123). Consequently, the maximum price SINOPEC can set for SEP gas is 2.248 yuan ($0.33) per cubic metre, which would be the benchmark plus 10 per cent, or 1.41 yuan ($0.20) per cubic metre, plus the maximum transportation fee of 0.84 yuan ($0.12) per cubic metre. This transportation fee takes wear-and-tear costs into account. The supply price to any customer (both producer price and transportation fees) will be uniform, a major change from the former arrangement of pricing by sectors, such as towngas, fertilizer production, and power generation.[82]

The pricing structure for the SEP has three novel aspects. First of all, the wear-and-tear costs (that is, the cost to the producers associated with gas losses in filling the natural gas into its pressurization stations) have been excluded. Traditionally, the cost of wear-and-tear is about 0.3 per cent of the ex-factory price. Secondly, it is the first time that a fluctuation within a band around the producer price has been sanctioned. Thirdly, the standardization of the natural gas price gives a strong indication that natural gas consumption by fertilizer producers will no longer be favoured, and that the priority has shifted towards residential towngas use. In short, this price reform is characterized by the flexibility of the producer price and the principle of 'one gas source, one price'.

Zhang Kang, vice director of the Consultation Commission of SI-NOPEC Oil Exploration and Development Research Institute, pointed out that the high Puguang gas price of 1.28 yuan/cubic metre against the WEP I price of about 0.48 yuan/cubic metre reflects the higher cost of exploration and development in the Puguang gas field, which is located in the mountainous eastern Sichuan region.[83] The ex-factory price of gas supplied by SEP is 0.792 yuan (for fertilizer makers and non-industrial towngas) or (for direct customers and industrial towngas) 0.352 yuan per cubic metre higher than that supplied by PetroChina's WEP I.[84] The price difference between WEP I and SEP is a reminder of the necessity of gas price reform as soon as possible (Tables 5.15 to 5.17).

A study by UBS predicted that the Chinese government was likely to introduce a pricing mechanism that linked the price of natural gas to imported gas, oil, and alternative fuels. The study concluded that such a mechanism would probably lead to domestic long run gas prices close to parity with, or at a discount from, imported natural gas, but made clear that the timing and magnitude of any gas price increase

Table 5.15: Producer and Gate Prices of the Sichuan-to-East Pipeline (SEP), 2009 (yuan/m^3)

	Producer Price	Transportation Fee	Average Gate Price	Residential Towngas Price
Sichuan	1.28	0.06	1.34	1.43
Chongqing		0.16	1.44	1.4
Hubei		0.32	1.60	2.3
Jiangxi		0.54	1.82	3.8
Anhui		0.65	1.93	2.1
Jiangsu		0.76	2.04	2.2
Zhejiang		0.81	2.09	2.4
Shanghai		0.84	2.12	2.5

Source: Lin Fanjing (2009e), 3.

Table 5.16: Gate Prices of the WEP I Pipeline (yuan/m³)

	Average Gate price (P+T)	P	T	Towngas	Fertilizer Production	Power Generation
Henan	1.14	0.48	0.66	1.16	1.12	-
Anhui	1.23		0.75	1.23	-	-
Jiangsu	1.27		0.79	1.42	1.27	1.10
Zhejiang	1.31		0.83	1.46	-	1.20
Shanghai	1.32		0.84	1.46	1.28	1.15

Note: P = producer price & T = transportation price

Source: Lin Fanjing (2009e), 3.

Table 5.17: Residential Towngas Price for the Major Cities, May 2009.

City & Provinces	Price (yuan/m³)
Beijing	2.05
Tianjin	2.20
Shanghai	2.50
Jinan, Shandong	2.40
Qingdao, Shandong	2.40
Chengdu, Sichuan	1.43
Guangzhou, Guangdong	3.45
Urumqi, Xinjiang	1.37
Bayingolin Mongol Autonomous prefecture, Xinjiang	1.30

Source: Lin Fanjing (2009e), 3.

was not certain.[85] As shown in Table 5.18 the UBS study used $7.0 / mmbtu as the border price (including border VAT) for gas imports from Turkmenistan. The city gate price at Shanghai and Shenzhen is $11.5/ mmbtu and $11.6/mmbtu respectively. The Shanghai price is in line with the figure suggested by IEA's China gas study, but the Shenzhen price shows some difference.

According to the *China OGP*, however, the CIF price of natural gas imported from Turkmenistan by PetroChina stands at 2 yuan/ cubic metre ($8.13/mmbtu), implying a city gate price of more than 3 yuan/cubic metre ($12.19/mmbtu) to Guangzhou after pumping via the WEP II pipeline.[86] *China OGP*'s figure of $12.19/mmbtu is in line with the figure suggested by IEA study (as shown in Table 5.19 and 5.20). A similar figure was suggested by an article in *Petromin Pipeliner*, which argued that the price should be $9.5/mmbtu at the Chinese border (inclusive of a 13 per cent VAT) at $80/barrel, assuming that

Table 5.18: Estimated Imported Gas Price ($/mmbtu)

Gas Source–Arrival City	Border price	Border VAT	Pipeline/ Re-gas	City Gate
	– – – – – *Price composition* – – – – –			
Turkmenistan–Shanghai	6.2	0.8	4.4	11.5
Turkmenistan–Shenzhen	6.2	0.8	4.6	11.6
Malaysia LNG–Shanghai	7.7	1.0	0.8	9.5
Dapeng LNG–Shenzhen	3.2	0.4	0.8	4.4
LNG (Qatar Gas III)–Ningbo	13.0	1.7	0.8	15.5

Source: UBS (2009), 21–2.

Table 5.19: Gas Price Estimation in Shanghai, 2009 ($/mmbtu)

Domestic Natural Gas

	Ex-Plant Price	Pipeline Tariff	City Gate Price
Offshore (Pinghu)			4.8
1st West-East Pipeline			
Industry	3.7	3.0	6.7
Residential	2.1	3.7	5.8
Power	2.1	2.6	4.7
Sichuan-East Pipeline (planning)			6.5–7.5

Natural Gas Imports

	Import (Border) Price	Tariff* /Re-gas Cost	City Gate Price
Malaysia Tiga LNG			
Low case (Oil: $50/bbl)	6	0.8	6.8
High case (Oil: $80/bbl)	8	0.8	8.8
Qatar Gas IV LNG			
Low case (Oil: $50/bbl)	8	0.8	8.8
High case (Oil: $80/bbl)	12.8	0.8	13.6
Turkmenistan Pipeline			
Turkmen border: $145/1000 m^3	5.1	3.8	8.9
Turkmen border: $195/1000 m^3	6.4	3.8	10.2
Turkmen border: $230/1000 m^3	7.4	3.8	11.2

* The pipeline tariff for 4,000km of the WEP II is estimated at 1000 yuan/1000 m^3 (USD144/1000 m^3), and the unit tariff per 100 km is calculated to $3.6/1000 m^3.

Source: Higashi (2009).

Table 5.20: Gas Price Estimation in Guangdong, 2009 ($/mmbtu)

Domestic Natural Gas

	Ex-Plant Price	*Pipeline Tariff*	*City Gate Price*
Offshore (Yacheng)			3.0
Sichuan-south Pipeline (planning)			6.5–7.5

Natural Gas Imports

	Import (Border) Price	*Tariff* /Re-gas Cost*	*City Gate Price*
NSM LNG			
Low case (Oil: $50/bbl)	3.16	0.8	3.96
High case (Oil: $80/bbl)	3.16	0.8	3.96
Qatar Gas II, IV LNG			
Low case (Oil: $50/bbl)	8.00	0.8	8.80
High case (Oil: $80/bbl)	12.8	0.8	13.6
Turkmenistan Pipeline			
Turkmen border: $145/1000 m³	5.10	4.6	9.70
Turkmen border: $195/1000 m³	6.40	4.6	11.00
Turkmen border: $230/1000 m³	7.40	4.6	12.00

*   The pipeline tariff for the 4.800 km of the WEP II is estimated at 1,200 yuan/1000 m³ ($173/1000 m³), and the unit tariff per 100 km is $3.6/1000 m³.

Source: Higashi (2009).

in the future China will pay the European netback. Adding the WEP II tariff, the Shanghai city gate price would be about $14/mmbtu, which is similar to the expensive Qatari contracts which CNOOC and PetroChina signed in 2008. The Turkmen gas price to China is linked to oil prices and it is currently set at about 85–90 per cent of the equivalent of the European netback price. At the Brent price of $80/barrel, the Chinese border price of Turkmenistan (pipeline tariff and VAT inclusive) is 2.2 yuan/cubic metres ($8.4/mmbtu). As shown in Table 5.18–21 and 5.23, China has set 1.1 yuan ($4.4/mmbtu) as the pipeline tariff for domestic WEP II to Shanghai, that is, a city gate price of around 3.3 yuan ($12.7/cubic metres).[87]

The UBS study shows that the city gate price of Turkmenistan gas is much lower than the LNG price (presumably from Qatar Gas III). The price burden of Myanmar gas imports is also high. A 2,806 km natural gas pipeline with an annual transportation capacity of 1.2 bcm is due to pump its first unit of gas to Kunming city in 2012 and

Table 5.21: Pipeline Transmission Tariffs

	$/mmbtu	*yuan/m³*
Sichuan–Shanghai	3.4	0.84
Tarim–Shanghai	4.0	0.98
Ordos–Shanghai	2.0	0.49
Tarim–Ordos	2.0	0.49
Ordos–Beijing	1.8	0.45
Tarim–Beijing	3.1	0.77
Turkmen border–Shanghai	4.4	1.10
Turkmen border–Guangdong	4.6	1.14
Turkmen border–Ordos	2.2	0.55
Sichuan–Ningbo	3.3	0.81

Source: UBS (2009), 21.

the supply price (CIF price) will not exceed 3.5 yuan per cubic metre ($14.23/mmbtu). This price is clearly higher than the terminal price of 1.1 yuan per cubic metre in Kunming, a fact which encouraged market speculations about further natural gas pricing reform.[88] One thing which is very clear at this stage is that the projected price of gas from the Myanmar–Kunming gas pipeline, and the LNG price based on Qatar Gas III, are far outside the Chinese gas consumers' budget, and such high gas prices would be a major obstacle to China's gas expansion. To satisfy the demand for 300 bcm/y of gas by 2020, it is unavoidable that the NDRC undertakes a gas price reform to relieve the price burden imposed by gas imports, either as LNG or as pipeline gas.

It is very difficult to ascertain an accurate figure for the Turkmenistan–China border gas price. PetroChina did not disclose the price of the gas exported from Turkmenistan to China, but a PetroChina insider has reported that the price is certainly lower than the price offered by Russia to China for the pending Sino-Russian gas pipelines.[89] Tables 5.19 and 5.20 based on IEA data confirm that the price burden of Turkmenistan gas is not inconsiderable.

The 2010 Gas Price Increase as the First Step in Price Reform

On 31 May 2010 the NDRC announced that it would raise the domestic onshore natural gas producer benchmark by 0.23 yuan ($0.034)/cubic metre (see Table 5.22).[90] As discussed earlier, two increases in the ex-factory price had already taken place, in December 2005 (50–150 yuan/1,000 cubic metre) and December 2007 (400 yuan/1,000 cubic metre). The new rise came after a long period of anticipation about the reform of the gas pricing system, and is regarded as the first step of that reform. The average price rise will be 24.9 per cent, far beyond the market expectations of 10–20 per cent. According to data from the NDRC, for Xinjiang oil fields and gas for WEP I and II, the ex-factory price of the gas provided to the chemical fertilizer industries and urban non-industrial sectors will be increased by between 25.2 and 41 per cent, while other gas prices will rise by about 20 per cent. The NDRC stated that this rise would only add 4.6 yuan ($0.67) to monthly household spending, and low-income families would receive subsidies to cover rising natural gas costs.

PetroChina will be the biggest beneficiary of the price increase since it will gain an extra 14 billion yuan in revenue. This would offset a considerable part of PetroChina's payment of an extra 3.6 bn yuan via the new resources tax levied in the north-west region of Xinjiang by the Ministry of Finance. SINOPEC would receive 1.95 bn yuan more in revenues by selling 8.47 bcm of gas, but this would be roughly offset by its extra payment of 1.4 bn yuan in the form of the new resources tax in Xinjiang.[91]

The price rise means a big increase in costs for chemical fertilizer producers. China's nitrogen fertilizer industry consumes about 11 bcm of natural gas annually, of which 8 bcm are provided at a preferential price. This price increase was one of a coordinated set of efforts designed to promote energy saving and rein in the six high-energy consuming industries (including power, steel, non-ferrous, and chemical industries) which had been responsible for increasing energy consumption per unit of GDP by 3.2 per cent year on year during the first quarter of 2010.[92] After the 31 May price increase, several regions, including Shanghai and Zhejiang, raised their own local retail gas prices accordingly. On 28 September 2010 the Beijing Municipality Commission of Development and Reform announced that it would raise local retail commercial gas prices by 0.33 yuan ($0.049) per cubic metre, with immediate effect.[93]

The Price Burden of Gas for Power

In China, gas prices are significantly higher than coal prices, and this inhibits switching from coal to gas. In 1999, the average price of natural gas in seven major cities was $241 per tonne of oil equivalent (toe) while that of steam coal was only $59/toe. As a comparison, in the USA in the same year the natural gas price was $83/toe and that of steam coal delivered to power plants $42.5/toe. That is, power plant coal was on average about 25 per cent of the price of natural gas in

Table 5.22: China's Ex-factory Benchmark Price by Major Gas Fields: 2007 and 2010 (yuan/1,000 m³)

Gas field	Category	2007			2010
Sichuan–Chongqing Gas field	Fertilizer production	690			920
	Industrial use (direct client)	1275			1505
	Industrial use (via urban supply)	1320			1550
	Urban supply	920			1150
Changqing Oil field	Fertilizer production	710			940
	Industrial use (direct client)	1125			1355
	Industrial use (via urban supply)	1170			1400
	Urban supply	770			1000
Qinghai Oil field	Fertilizer production	660			890
	Industrial use (direct client)	1060			1290
	Industrial use (via urban supply)	1060			1290
	Urban supply	660			890
Xinjiang Oil field	Fertilizer production	560			790
	Industrial use (direct client)	985			1215
	Industrial use (via urban supply)	960			1190
	Urban supply	560			790
Dagang, Liaohe, Zhongyuan Oil fields		1st*	2nd*	Ave*	
	Fertilizer production	660	980	710	940
	Industrial use (direct client)	1320	1380	1340	1570
	Industrial use (via urban supply)	1320	1380	1340	1570
	Urban supply	830	980	940	1170
Other fields	Fertilizer production	980			1210
	Industrial use (direct client)	1380			1610
	Industrial use (via urban supply)	1380			1610
	Urban supply	980			1210
West-to-East pipe	Fertilizer production	560			790
	Industrial use (direct client)	960			1190
	Industrial use (via urban supply)	060			1190
	Urban supply	560			790
Zhongxian-Wuhan Pipe	Fertilizer production	911			1141
	Industrial use (direct client)	1311			1541
	Industrial use (via urban supply)	1311			1541
	Urban supply	911			1141

Table 5.22: continued

Gas field	Category	2007	2010
Shaanxi-Beijing Pipe	Fertilizer production	830	1060
	Industrial use (direct client)	1230	1460
	Industrial use (via urban supply)	1230	1460
	Urban supply	830	1060
Sichuan–East Pipe		1280	1510

Note: Urban supply means towngas here.
*   The first category stands for the preferential price and the second one is the normal price. The dual-track natural gas pricing mechanism was cancelled in May 2010. Ave stands for average price.

Source: China Securities Journal (2010), 82–3.

Table 5.23: Gate Prices by Source to Shanghai, 2010 (yuan/m^3)

	Gate price
WEP I	1.6–1.7*
Shanghai LNG	1.8–2.0
SEP	2.35
WEP II	3.2–3.4

*   = price to be raised due to adjustment of ex-factory prices in May 2010.

Source: China Securities Journal (2010), 84, 106.

China, compared to about 50 per cent in the USA.[94] Gas pricing was a major concern to power producers, and it was not certain whether the pricing proposed by PetroChina would encourage enough power producers to convert to gas or to build new gas-fired plants in the wake of the WEP I development.

David Fridley analyses the details of the choice:

> If power producers in Shanghai region buy gas at the city gate price, then the fuel cost of their power will range from 0.29 yuan ($0.035) to 0.41 yuan ($0.05) per kWh, depending on whether they use combined-cycle or conventional natural gas turbines. By comparison, the average retail price of power in Shanghai is about 0.70 yuan ($0.085) per kWh, some of which accrues to the local power bureau and not the generator. The State Power Corporation has requested that prices for power generation be no higher than 0.90 yuan/cubic metre ($3.10/mmbtu), reducing average fuel costs per kWh generated to about 0.20 yuan ($0.024) to 0.28 yuan ($0.034).

This compares to the average sales price of 0.34 yuan ($0.042) per kWh recorded by the National Power Corporation on their total national power sales in 1999.[95]

...

[Five years later] a study by the State Grid Corporation reconfirmed that natural gas is not competitive with coal when the price of natural gas is 2.5 times that of coal on a Btu basis. In 2004, when coal delivered to a power plant in north China averaged 230 yuan per tonne ($1.3/mmbtu), power producers were being offered gas at $3.80 per mmbtu or a 2.9:1 ratio between gas and coal. In mid-2007, FOB steam coal prices at Qinhuangdao in Hebei province for 5500 kcal Shanxi coal reached 470 yuan/tonne, more than double the price in 2004. If such high coal prices continue, this implies natural gas may be competitive even in northern China at as much as $6.0–6.590/mmbtu.[96]

It could be added that reforms in electricity pricing have helped increase the competitiveness of gas-fired generation, in particular south of the Yangtze River.

Nonetheless, it would be safe to say that the price burden was and will remain a major obstacle to the rapid expansion of the gas-for-power sector, even though Chinese energy planners have seen the power and residential sectors as the driving force of China's natural gas expansion. A decade ago the authorities had a very ambitious plan for increased gas use in the power sector. The State Power Corporation (SPC) had prepared tentative plans for gas-fired generation up to 2010. To use gas from the WEP, the SPC had initially planned 5.4 GW of gas-fired generating capacity in Henan province and the Yangtze Delta area. According to IEA's 2002 gas study, the Chinese government aimed at building a total of 20.1 GW of gas-fired power generation capacity by 2010.[97] At the time, the target looked very ambitious. In 2007 the IEEJ (Institute of Energy Economics of Japan)'s study provided updated figures showing that China's gas fired power capacity had reached 10.6 GW in 2006, accounting for 1.7 per cent of total installed capacity and 2.2 per cent of total installed thermal capacity. It added that the China Electricity Council projected gas-fired power generation capacity would reach 60 GW by 2020, accounting for 5–6 per cent of total installed capacity.[98] If installed capacity in 2020 is 1,885 GW, 36.3 per cent of which will be from non-fossil fuels,[99] the achieving of 60–70 GW in gas for power capacity will represent only 3.2–3.7 per cent of the total. To become 5–6 per cent, the gas for power figure would have to be 94–113 GW. Without the reform of electricity pricing, however, the chance of success in achieving this target is not high.

According to the China Electricity Council, total installed power capacity in 2010 was 962.2 GW, of which, 650.1GW was coal-fired power, 26.7 GW gas-fired power, 10.8 GW nuclear power, 31.1 GW wind power, and 213.4 GW hydropower.[100] Zhang Guochang, of Dongying Gas Generation Consultancy, said that:

> in terms of cost, natural gas-fuelled power cannot at present compete with other energy sources. Each kilowatt hour of natural gas-derived electricity costs roughly 2.5 yuan ($0.39), while coal based power costs approximately 0.35 yuan ($0.05) and wind power costs nearly 0.55 yuan ($0.08).[101]

Under this price regime, it is a real achievement that the gas for power capacity was almost 27 GW. But, under the current gas price regime, further expansion will be a real struggle.

Fridley has warned of the effects of this distortion of relative prices:

> Without reforms in electricity pricing, including the adoption of peak and off-peak pricing systems in certain areas and the linkage of electricity prices to the coal price, the ambitious target of installing up to 70 GW of natural gas-fired power plants by 2020 will be a difficult one to achieve. In China the coal price is determined by the market while the price of electricity is controlled by the state. Under the mechanism that links the two, if the price of coal rises 5 per cent or more in one six month period, the electricity price is adjusted accordingly, the electricity companies bearing 30 per cent of the cost above a 5 per cent rise. The remaining 70 per cent is reflected in the electricity price. If the coal price rises less than 5 per cent, the change is calculated in the next period or when the total change reaches or surpasses 5 per cent.[102]

A Greenpeace document on the true cost of coal elaborates:

> From 2000 to 2007, power generation in China accounted for, on average 49.1 per cent of coal consumption. The figure during 2004 and 2007 recorded 51.9 per cent. As coal for power generation accounts for over 50 per cent of the total coal demand, the growth of demand for coal in general will also be reduced due to decreasing power demand. Other conditions being unchanged, the growth of the coal price in the market will decline in the end. However, with coal demand rising rapidly for other uses, the declining trend of the coal price will be offset. The coal price for the whole market will then rise. This is why the growth rate of the coal price for electricity generation never exceeds the average growth rate of the overall coal price since it was opened to the market in 2002. Major factors that distort the coal price can be corrected gradually by market forces if the relevant property, administrative and judicial systems are optimized. A conservative conclusion would therefore be that the coal price is undervalued by 17.73 per cent in China due to the cost and price distortion of government control over coal production.[103]

Even so, despite the price burden, Beijing's power plants are shifting from coal to natural gas. In late February 2010, CNPC's newsletter confirmed that the Gaojing Thermal Power Plant in Beijing Municipality had shifted from burning coal to natural gas.[104] CNPC will deliver 1 bcm of natural gas to the power plant in 2010 via its Shaanxi–Beijing gas pipeline, fed by the Changqing oilfield. Datang International Power Generation Company is operating the power plant in Beijing. According to CNPC, Beijing plans to refit the Gaobeidian and Shijingshan thermal power plants to use natural gas, and it also plans to demolish the downtown Dongjiao Thermal Power Plant and replace it with a new gas-fired power plant. Beijing consumed 5.9 bcm in 2009 and that quantity increased to 7 bcm in 2010. By 2015, Beijing's natural gas consumption is projected to be 18 bcm and, to meet this demand, CNPC completed the third phase of the Shaanxi–Beijing pipeline in December 2010.[105] This initiative on gas for power, seen in the Beijing area, was not applied to everywhere in China, but that could change during the 2010s.

The NEA forecasts that by 2020 total gas-fired power generation capacity in small and distributed units will rise from 5GW in 2011 to 50GW (or 3 per cent of installed power generation).[106] China's industry also sees the potential. Zhang Guochang, of Dongying Gas Generation Consultancy, has argued that China's gas for power expansion could be achieved through small-scale power plants in unconventional gas producing regions.[107]

City Gas (Towngas) Expansion[108]

For a long time, China's central authority prohibited private capital from entering the gas market, but the ban was in effect lifted in 2002 when the government began to implement a franchised operation system in the towngas market. Since the Foreign Investment Guide was published in March 2002, foreign and private firms have, for the first time, been allowed to distribute towngas. Qualified private companies were permitted to enter into the towngas business. This private players' operation in the sector was in full swing by 2003.[109] At that time, a number of private companies, such as Hong Kong and China Gas (Towngas), China Gas Holdings (China Gas), Xinao Group, Panva Gas Holdings, Wah Sang Gas, Zhengzhou Gas, the Guanghui Industrial Investment Corporation (GIIC), Baijiang Gas, and the Minglun Group, were focusing on the penetration of the earliest stages of the city gas business in China. By 2006, the active players in the towngas sector were confined to Towngas, China Gas, Xinao Group, and Zhengzhou Gas. Until then, Chinese NOCs had shown no appetite for entering the towngas business, but CNPC began to penetrate it by setting up a new entity. According to a Goldman Sachs research report, the top 10 downstream gas distributors in China in 2010 had about 32 per cent of the market (Beijing Enterprises 6.0 per cent, China Resources gas 5.2, Shenergy Company 4.2, ENN Energy 3.9, Towngas China 3.8, China Gas 3.2, Kunlun Energy 2.0, and Shaanxi Provincial Natural Gas 2.0 per cent).[110]

Table 5.24: Natural Gas Sales Volume by End Use in 2010 (%)

	Household	*Commercial/ Industrial*	*CNG*	*Power Generation*
Towngas	27	73	0	0
China Gas	13	71	10	0
ENN Energy	15	71	13	0
CR Gas	30	61	9	0
Beijing Enterprises	13	61	0	26

Source: GSGIR (2011).

A brief review of the towngas players will reveal their possibilities for expansion in the coming years.[111]

1. Hong Kong and China Gas Co. Ltd (Towngas): The front runner has been the Hong Kong and China Gas Co. Ltd, also known as Towngas, which has dominated the HK city gas sector, and decided to expand its business in China. By September 2003 the firm had investments in 16 city piped gas projects. Towngas was also awarded a 3 per cent minority equity stake in China's first LNG terminal project in Guangdong Province. By 2006, Towngas has secured 31 joint projects in mainland China, with more than three million clients, 12,000 km of gas grids, and annual sales of 1.4 bcm of gas equivalent. On 3 December 2006, Towngas signed an agreement to acquire some gas and grid assets and related business from the HK-listed towngas supplier Panva Gas Holdings[112] for a total payment of 3.23 billion Hong Kong dollars. Towngas became the largest shareholder of Panva Gas by taking a 45 per cent stake. It renamed the company Towngas China Company Limited and focused on piped city-gas projects.[113] In 2008, the sales volume was over 5.8 bcm. Hong Kong & China Gas Investment Limited, an investment holding company in Shenzhen, manages the Group's investments in mainland China. By the end of 2009 the Towngas Group had more than 100 projects in 19 provinces, municipalities, and autonomous regions. Over 30,000 employees work in the Group's diverse joint ventures on the mainland.[114]

2. China Gas Holdings Ltd (China Gas): China Gas owns a total of 121 natural gas projects, including exclusive piped gas development rights in 110 cities and regions, eight natural gas pipeline transmission projects, as well as the license to import and export LNG and other fuel products in China, and 35 LPG distribution projects. The Group has secured exclusive gas operations in Hubei, Hunan, Guangxi, Guangdong, An-hui, Jiangsu, Zhejiang, Hebei, Shaanxi, Inner Mongolia, and Fujian, which are the major cities along the West–East gas pipeline and the Chongqing–Wuhan and Shaanxi–Beijing gas pipelines. China Gas is the only city gas operator owning long-distance gas pipelines in China.[115] In November 2006, Zhongran Investment Ltd, a subsidiary of China Gas was granted by the Ministry of Commerce the right to import and export gas-related products including natural gas, LNG, methanol, etc., and also the wholesale and retail rights for such products.[116] In 2007, China Gas established a joint venture with South Korea's SK Energy and Oman Oil respectively.[117] Joining hands with SINOPEC was a wise option for China Gas's future survival.[118] In October 2009 China Gas signed a cooperation and non-competitive agreement on domestic gas sales and sourcing with CNPC Kunlun Natural Gas Co., covering the sale and sourcing of piped gas, liquefied petroleum gas, and liquefied natural gas.[119]

3. Xinao Gas Holdings Limited (Xinao Gas): Xinao Gas Holdings is ENN Group's publicly-traded subsidiary and distribution arm. It was listed on the Hong Kong Stock Exchange in 2001. The Xinao Group began to invest in city gas in 1992 and Xinao Gas Holdings (Xinao Gas) is part of Xinao group. In 2003 the firm claimed to be the largest non-state-owned city gas operator in Mainland China, and has operations in some 30 cities. In early June 2006, Xinao successfully acquired the right to import and export gas from the Ministry of Commerce. By 2006 Xinao has been testing the ground in about 60 cities, especially targeting overseas LNG supply sources.[120] ENN Energy Distribution, under its Chinese name Xinao Gas, primarily distributes natural gas, liquefied petroleum gas, methanol, and dimethyl ether, a clean-burning hydrocarbon fuel. It has developed a 10,000+ kilometre network of gas pipelines. Xinao Gas has brought clean heating and cooking fuels to 4.2 million households and 12,000 industrial and commercial customers. It operates gas fuel projects in 79 cities.[121]

4. Zhengzhou Gas: Zhengzhou Gas, based in Henan Province, expanded its natural gas sales very rapidly in the first nine months of 2003. Their sales rose to 135.5 million yuan ($16.4 million), of which the delivery to residential users in Zhengzhou accounted for over 70 per cent. The company has connected the WEP to Zhengzhou, the capital of Henan Province. In July 2009, Zhengzhou Gas Group, the parent of Zhengzhou Gas Co. Ltd signed a framework agreement with China Resources Gas (CR Gas) Group Ltd to set up a joint venture. Zhengzhou Gas Group will keep a 20 per cent stake, while CR Gas will acquire the remaining 80 per cent. After the purchase, CR Gas will indirectly hold a 34.5 per cent stake in Zhengzhou Gas Company.[122] In 2010, CR Gas was operating 48 city gas projects in 15 Provinces (Sichuan, Jiangsu, Hubei, Shandong, Shanxi, Hebei, Jiangxi, Yunnan, Anhui, Zhejiang, Fujian, Henan, Liaoning, Guangdong, and Inner Mongolia).[123] At the end of June 2010, CNPC signed a strategic cooperation agreement with CR Gas to jointly develop an urban gas business, but no details of the agreement were released.[124]

5. Beijing Enterprises: Supplies gas to the cities and is a core business operation of Beijing Enterprises Holdings Ltd, which has 3.4 million subscribed gas users and 7,500 km of gas pipes, and an annual gas sales volume of 3.6 billion cubic metres. Its principal business unit is the Beijing Gas Group Ltd. Operational since the 1950s, it is currently the country's largest natural gas supplier and service provider.[125]

6. Guanghui Industrial Investment Group: Based on the feasibility study done in early April 2002, the Xinjiang-based Guanghui Industrial Investment Corporation (GIIC) undertook to bring the gas project into operation by mid-September 2003.[126] In 2004, GIIC and Shenzhen Tianmin Ltd signed an agreement to transport some 0.25 mcm/d of LNG in tanks from Xinjiang to the southern coastal city starting in June 2004.[127] As the new National Gas Utilization Policy clearly banned the establishment of LNG projects using the natural gas produced by gas fields in China, however, GIIC as the largest LNG producer in China had to halt the construction of several LNG projects in Xinjiang.[128] To cope with the problem of natural gas supply, GIIC began to explore the possibility of gas supply from the Central Asian Republics, and GIIC has received approval from the Chinese and Kazakh government to build a cross-border pipeline to transmit 1.5 mcm/d of gas. In September 2009, GIIC acquired a 49 per cent stake in Kazakh Tarbagatay Munay (TBM) to develop an oil and gas block in the Zayan region.[129] (For China's domestic LNG, see the Appendix to this chapter.)

CNPC's Entry to the Towngas Sector

CNPC started its urban gas distribution business in 2000 and set up its first professional gas distribution subsidiary in 2004 in Beijing. By 2008, it had invested a total of 740 million yuan ($107.9 million) in gas retail projects in 46 cities in 14 provinces.[130] Some of CNPC's activities in 2006 signalled a serious interest in towngas business expansion. PetroChina has cooperated with Aptus Holdings to jointly establish and operate the Huayou Company which has branches in Hunan and Changde. The Hunan branch is mainly engaged in establishing the trunk pipeline from Changsha to Changde, which will provide natural gas to the cities on its route. The Changde branch is responsible for managing two natural gas projects in Changde, and also supplying Changsha–Changde gas to urban residents and other end users in the city.[131]

On 7 December 2006, CNPC's subsidiary China Petroleum Pipeline Bureau (CPPB) signed a framework cooperation agreement with Zhuhai Pipeline Gas Company (ZPGC) whereby it acquired an 85 per cent stake in ZPGC.[132] In order to enter into the domestic towngas sector, CNPC established a specialized towngas gas company named CNPC Pipeline Gas Investment Co. Ltd. The firm has obtained exclusive towngas marketing rights in 46 cities in 14 provinces, with a total investment of 740 million yuan and gas supply reaching three bcm.[133] On 20 August 2007 the NDRC announced the Natural Gas Utilization Policy, which prioritized the towngas sector in utilizing natural gas. The Policy calls for high utilization efficiency of natural gas, and the towngas sector is able to satisfy this requirement.[134] CNPC began to take steps towards city gas business development. At the end of 2007 CNPC started to build natural gas-using projects in Jiangsu province, Guangdong province, and Hebei province.[135]

In 2009, PetroChina Natural Gas Bureau had transferred the assets of nine towngas companies to China Huayou Group Corporation, a subsidy of CNPC, and in this way CNPC began to expand its towngas business. Kunlun Towngas, a unit established in August 2008 under the CNPC flagship, was the company set up to derive competitive advantage from its parent's natural gas supplies. Kunlun Towngas has acquired towngas assets from CNPC's China Huayou Group Corp, China Petroleum Pipeline Bureau, Sichuan Petroleum Bureau, CNPC Jilin Petroleum, and CNPC Shenzhen Industrial Company, including 100 towngas projects in 23 cities. In April 2009, CNPC Pipeline Bureau listed 100 per cent of its CNPC Natural Gas Pipeline Towngas Investment Co. Ltd for sale on the Beijing Equity Exchange market.[136] In late November 2009, CNPC took another step towards integrating its downstream towngas assets across China. CNPC put several towngas assets up for sale on the China Beijing Equity Exchange (CBEX) and the Shanghai United Assets and Equity Exchange (SUAEE). This is all part of CNPC's effort to integrate its downstream natural gas distribution system. Kunlun Gas was set to acquire the assets on sale, whose value was estimated at a total of 2 billion yuan[137] (for details see Table 5.25).

Table 5.25: Kunlun Gas Acquisitions in 2009

Acquired company	*Equity stake (%)*
CNPC Natural Gas Pipeline Towngas Investment Co	100.0
Yongqing Huayou Gas Co Ltd	66.11
Zhouzhou Huayou Gas Co Ltd	89.99
Bazhou Huayou Gas Co Ltd	51.0
Changde Huayou Gas Co Ltd	51.0
Hunan Huayou Natural Gas Distribution Co Ltd	43.55
Hainan Huayou Ganghua Gas Co Ltd	51.0
Zoucheng Huayou Gas Co Ltd	51.0

Source: Lin Fanjing (2009g).

Kunlun Gas under PetroChina and Hong Kong-listed CNPC Hong Kong, will be the key force in the development of CNPC's towngas sector. Kunlun Gas will be the major operator, while CNPC Hong Kong will serve as the provider of capital.[138] Kunlun Gas signed a framework cooperation agreement with CNPC Pipeline Bureau in a bid to expand towngas business for Kunlun Gas.[139] LNG terminals in Rudong, Dalian, Shenzhen, and Tangshan will be brought under the ownership of Kunlun Energy.[140]

In late July 2010, Kunlun Energy won approval to team up with Sichuan Petroleum Administration to acquire Sichuan Shizhong Petroleum Gas Transportation Technology (Shizhong Gas), a towngas company linked to CNPC China Natural Gas in Sichuan. This move is designed to take advantage of PetroChina's presence in Sichuan and develop business in the province. In early August, Kunlun Energy signed a joint venture agreement with the Tianjin Gas Group to start a new natural gas business, and engage in towngas pipeline construction and gas supplies to Tianjin. In September, Kunlun Energy won a bid to acquire a 75 per cent stake of PetroChina Dalian LNG for 2.21 bn yuan and the following month, Kunlun Gas Co Ltd paid 500 million yuan to acquire a 50 per cent stake in Lanzhou Gas and Chemical Group.[141]

One certainty is that China's city gas sector will be the biggest beneficiary of China's ambitious gas expansion scheme during 2010s, and at the end of this decade a major restructuring of China's city gas sector will be inevitable.

The Nationwide Gas Pipeline Network Development

China's natural gas transport infrastructure developed primarily on a regional basis, serving mainly the producing oil and gas fields and local industries. In 1999 the length of natural gas pipeline in China was only 11,630 km with 14.1 bcm per year capacity. By 2005 it had more than doubled (28,000 km, with 45 bcm/y handling capacity) and by the end of 2008 it had reached 35,000 km with transportation capacity of over 80 bcm/y. According to Yang Jianhong, vice director of PetroChina Planning Institute, the total length of China's natural gas pipelines is expected to reach 100,000 km by 2015. By that date the main and branch natural gas pipelines will have been extended by about 25,000–30,000 km but an even greater effort will be devoted to the construction of 35,000–40,000 km of sub-branches, to meet the galloping demand for gas across the country.[142] In parallel with this expansion of pipeline networks, CNPC plans to construct 10 gas storage tanks in the 2011–15 period to expand the total capacity to 22.4 bcm.[143]

Table 5.26 shows the longest and highest capacity pipelines constructed between 2001 and 2009. The most important of these, according to these two criteria, are the Chongqing–Wuhan pipeline, Changqing E–W pipeline II, Sichuan-to-East pipeline, and Yongqing–Tangshan–Qinhuangdao pipeline.

The first Shaanxi–Beijing pipeline was completed in September 1997. This 864 km pipeline was designed to transport Changqing gas at a maximum annual delivery rate of 3 bcm. The pipeline was designed by the Survey and Design Institute of the Sichuan Petroleum Administration and the German PLE company. CNPC and the Beijing Municipal government jointly invested 3.94 billion yuan.[144] The second Shaanxi–Beijing pipeline came into operation in July 2005 and was designed to be connected with WEP I.[145]

To meet the exploding demand in the Beijing area, CNPC undertook the third phase development of the Shaanxi–Beijing gas pipeline in order to increase gas supply to the city. The National Energy Administration announced in October 2010 that the third Shaanxi–Beijing gas pipeline was scheduled to come into operation at the end of October, but the work was not completed until the end of the year.[146] This third pipeline will provide some 2 bcm of additional gas supply to north

Table 5.26: PetroChina's Gas Pipelines in Operation between 2001 and 2009

	L km	D Mm	P MPa	C bcm/y	O
Zhouzhou–Shijiazhuang (Hebei)	202	508	6.4	2.0	2001
Sebei (Qinghai)–Lanzhou (Gansu)	953	660	6.4	3.0	2001
Sebei-Xining–Lanzhou	915			6.0	2009
Shijiazhuang–Xintai–Handan (Hebei)	161	508	6.4	2.5	2002
Cangzhou (Hebei)–Zibo (Shandong)	210	508	6.4	2.5	2002
Zibo–Weifang (Shandong)	125	610	6.4	1.0	2002
Lunnan (Xinjiang)–Shanghai	3,900	1,016	10.0	12.0	2004
Changzhou (Jiangsu)–Hangzhou (Zhejiang)	200	711	6.4	2.0	2003
Dingyuan (Anhui)–Hefei (Anhui)	84	406	6.4	0.5	2003
Nanjing (Jiangsu)–Wuhu (Anhui)	135	508	6.4	2.0	2003
Dongfang 1-1–Dongfang City(Hainan)	116	711	10.0	5.0	2003
Dongfang–Yangpu (Hainan)	135	914	10.0	5.0	2003
Jiaozuo–Anyang (Henan)	200	273	6.3	0.64	2003
Zhengzhou–Gushi (Henan)	501	406/355	6.4	1.0	2003
Weifang–Qingdao (Shandong)	165	611	6.4	2.0	2003
Bozhong 28-1–Longkou (Shandong)	100	355.6	6.4	1.0	2004
Qingshan (Hubei)–Nantong (Jiangsu)	195	529	6.4	1.7	2004
Chuzhou (Anhui)–Suqian (Jiangsu)	260	529/329	6.4	1.74	2004
Lixin (Anhui)–Xuzhou (Jiangsu)	170	529/426	6.4	1.2	2004
Wushenqi (Inner Mongolia) – Hohhot (Inner Mongolia)	497	457/377	6.1	1.2	2004
Zhongxian (Sichuan)–Wuhan (Hubei)	738	711	6.4	1.2	2005
Shanghai–Hangzhou (Zhejiang)	150	813	6.4	2.0	2004
Changqing E–W pipeline II	923	914/813 /711/508	10/8.4 /6.4	8.0	2004
Yongqing–Tangshan– Qinhuangdao	320	1,016		9.0	2009
Nanpu–Tangshan	51.6	660		2.5	2008
Daqing–Qiqihar (Heilongjiang)	155.7	406.4		0.82	2008
Yingxian (Shanxi)–Zhangjiakou (Jiangsu)	283	508		1.2	2009
Changling–Changchun– Jilin Petrochemical Pipeline	221			2.8	2008
Huaibei–Huainan (Anhui)	190	406.4	4.0	0.4	2005
Zhengzhou–Luoyang (Henan)	130	323.5	4.0	0.4	2005
Longkou–Weihai (Shandong)	185	406.4	4.0	0.5	2005
Lishui–Wenzhou (Zhejiang)	148	355.6	10.0	1.0	2005
Jiannan (Sichuan)–Lichuan (Hubei)	60	219	4.0	0.14– 0.21	2005
Jiannan (Sichuan)–Shizhu (Chongqing)	100	219	4.0	0.3	2005
Yancheng–Nantong (Jiangsu)	190	457	6.4	0.6	2005
Qinnan (Shanxi)–Handan (Hebei)	242	711/559	4.0	4.3	2005

Note: L = length, D = Diameter, P = Pressure, C = capacity, O = operation year.

Source: Lin Fanjing (2006c); *China OGP*, 15 April 2009, 6–7; www.cnpc.com.cn/ Resource/english/images1/pdf/08AnnualReportEn/08-Annual%20Business%20Review.pdf (last accessed spring 2011)

China during the winter. This 1,026 km pipeline, with a delivery capacity of 15 bcm, runs from Yulin in Shaanxi province and ends at the Changping district, Beijing Municipality. The pipeline project involves a total investment of 14.48 billion yuan, including 73.735 million yuan specified for environmental protection (0.5 per cent of total investment in the pipeline).[147]

A separate 3 bcm 950 km gas pipeline from Liangbei field in Qaidam basin to Xining, capital of Qinghai and Lanzhou, capital of Gansu, was completed in 2001, and a parallel line with 3.3 bcm capacity came online in early 2010. The Sebei–Xining–Lanzhou Pipeline, securing gas supply to Qinghai, Ningxia, and Gansu, is China's third most important pipeline preceded only by the WEP (I and II) and the Shaanxi–Beijing pipelines. In south-west China, the first major pipeline out of Sichuan was the 780 km, 3 bcm Zhongxin–Wuhan line from Chongqing to Wuhan, with spurs to Changsha and other cities in Hubei province.[148]

In north-east China, Daqing Oilfield Co. Ltd was responsible for building an 87 km gas pipeline with an annual handling capacity of 5 bcm. After this project was completed in October 2007, it was able to inject natural gas into Harbin's towngas grids. The construction of the Daqing–Harbin gas pipeline is a key further step towards CNPC's goal of transporting natural gas southwards to major gas consuming markets. The natural gas pipeline network in north-east China, of which the Daqing–Harbin pipeline is a part, will cover major cities in the region such as Daqing, Qiqihar, Harbin, Changchun, Shenyang, and Dalian, and then the line will be extended to Beijing.[149]

Another venture, started in 2009, was the PetroChina Natural Gas Pipeline Company, a joint venture between PetroChina and Beijing Gas, which:

> decided to invest 20 billion yuan in laying the third Shaanxi–Beijing pipeline, with a 60:40 ratio equity stake between PetroChina and Beijing Gas. The pipeline was on stream at the beginning of 2011 and is expected to provide Beijing and Tianjin with 15 bcm of natural gas, almost equal to the combined supplies of the first and second pipelines.[150]

By 2015 the Shaanxi–Beijing pipeline, operated by CNPC, will double the natural gas supply to Beijing, raising it to 12 bcm. This target follows a blueprint for Beijing which envisages expanding residential towngas supplies to all of its suburbs before 2015, including the Miyuan, Huairou, and Yanqing Districts, about 60–90 km away from central Beijing.[151] The above-mentioned gas pipelines could not have any nationwide links without the WEP I, II, and III pipelines. In short, the WEP I project at the beginning of the 2000s, and the subsequent decisions to go ahead with the WEP II and III projects, laid a solid foundation for natural gas expansion in China during the 2010s.

The West–East Gas Pipeline (WEP I) Project[152]

It was CNPC which initiated the planning and study of the WEP I Project in 1996, two years before the preliminary feasibility study was begun in August 1998. The overall plan of the project had been more or less settled by 2000. According to the West–East Natural Gas Transportation Project Management Organization, PetroChina, the WEP I aimed at transporting natural gas from both the Tarim and Changqing gas fields in western China to Shanghai in eastern China through a pipeline with a 12 bcm/y capacity. WEP I was set to pass through eight provinces and one municipality, crossing large rivers, like the Yangtze River and the Yellow River, six times and medium-sized rivers more than 500 times, crossing highways over 500 times and trunk railways 46 times. The pipeline would pass through areas with a seismic intensity of 6 or below for about 2,500 km, of 7 for about 700 km, and 8 for about 700 km.

The State Council requested the State Development Planning Commission (SDPC) now known as the National Development and Reform Commission (NDRC), to set up a working group before the end of

Table 5.27: SDPC Working Group for the WEP I Project

State Development Planning Commission (SDPC)	Xinjiang Province	People's Bank of China
State Economic and Trade Commission (SETC)	Gansu Province	Bank of China
Ministry of Finance	Ningxia Province	State Development Bank
Ministry of Land and Natural Resources	Shaanxi Province	Industrial and Commercial Bank
State Administration of Petroleum and Chemical Industries	Shanxi Province	of China
State Power Corp (SPC)	Anhui Province	China Construction Bank
China National Petroleum Corp (CNPC)	Jiangsu Province	
China Petrochemical Corp (SINOPEC)	Henan Province	
China National Offshore Oil Corp (CNOOC)	Shanghai Municipality	
China International Engineering Consulting Corp (CIECC)		

Source: compiled by author, based on Quan Lan (2000a), 1.

March 2000. Many central and local authorities were included in this SDPC working group. A working conference entitled 'West Gas Transport East' was organized by SDPC on 24–26 March.

The WEP I construction plan[153]

The WEP I was constructed and put into operation in phases. The entire project was divided into two sections, East and West. The preliminary feasibility study and the overall design of the project were completed in 2000. In 2001, construction of the East section project began, followed by the West section in 2002. At the end of 2003, the East section project was completed and in 2004 the West section was completed. The entire project, involving pipeline construction, city gas grid build-up, and gas downstream projects in the Yangtze River Delta, required an investment of 120 billion yuan in the first phase from 2000 to 2004. The total capital requirement was expected to be 300 billion yuan. Fixed asset investments for the pipeline construction project were pegged at 38.4 billion yuan.

Shortly before WEP construction started, Chinese steel manufacturers could not produce the X-70 steel tube with a diameter of 1,118 mm which was preferred for the trunk line. But in the end X-70 pipes with a diameter of 1,016 millimetres were used. According to *China OGP*, CNPC's Planning and Engineering Institute expected the entire line to require 1.74 mt of steel. The pressure of the trunk line was designed to be 8.4 MPa, higher than in any other Chinese operating gas pipeline. The trunk pipeline required 1.36 mt of spiral submersed arc-welded (SSAW) and 0.3 mt of longitudinal submersed arc-welded (LSAW) pipes. The trunk pipeline required 16 compressor stations along the 4,167 km pipeline. But overseas business monopolized the supply of compressors, and the bill for these came to 7.7 billion yuan, or 20 per cent of the total fixed investment. Centrifugal compressors and gas turbines were the first products which had to be imported. Besides this, more than 200 valves and an automatic control system also had to be imported. The project called for 49.5 billion yuan for construction, of which 45.6 billion yuan was for the trunk line, 1.8 billion yuan for the branch line, and 2.1 billion yuan for the underground depot.[154]

WEP I expansion and the adjustor pipeline

In order to meet the explosion of demand, the Beijing authorities decided to increase the planned capacity of WEP I from 12 bcm/y to 17 bcm/y, as a result of which an additional 22 gas compression stations and 24 compressors units had to be installed, involving an additional investment of 4.3 billion yuan. By August 2006, eight stations had been built and four compressor units installed. The upgrading project was scheduled to be complete by the end of 2007.[155] In 2009, Tarim's gas production reached 18.1 bcm, of which 16.6 bcm went to WEP I. According to CNPC, by July 2010, the Tarim oil field had supplied a total of 70 bcm to WEP I over the previous five and a half years.[156] By 2010, WEP I and its branch pipelines supplied 110 cities, 3,000 companies, and about 300 million residents with natural gas.[157] In 2010, CNPC started on the construction of underground natural gas storage tanks along WEP I. The tanks, which will be mainly used for peak-shaving stockpiles, include Jintan tank in Jiangsu province, Liuzhuang tank in north Jiangsu province, Pingdingshan tank in Henan province, and two tanks in Yingcheng in Hubei province.[158]

Collapse of foreign partnership for WEP I

It is well known that from the very early stages, CNPC did not wish to form partnerships with western companies for the WEP I project, because it saw no benefit of giving away the sizable returns from WEP I when the gas supply sources had all been discovered by CNPC itself. The first western company to give a negative verdict on the WEP I project was BP. In September 2001, BP attributed its decision to withdraw from the WEP I project mainly to its low rate of return on investment. PetroChina has affirmed that the IRR will be no less than 12 per cent, although that of the upstream sector will exceed 15 per cent. But many investors found an IRR of 12 per cent insufficiently attractive.[159] The real reason, however, was that BP was more interested in selling gas to China than in merely building gas pipelines in China. In this context, BP's preference for the Sino-Russian gas pipeline over WEP I was understandable.[160]

Despite BP's withdrawal, Beijing continued to pursue international cooperation for WEP I. On 4 July 2002 PetroChina got as far as signing a Joint Venture Framework Agreement with three international consortia. The project's capital was split in the ratio 55:45 between the Chinese and foreign participants. On the Chinese side, PetroChina had a 50 per cent stake and SINOPEC 5 per cent. The three international investing consortia, (Shell with Hong Kong & China gas, ExxonMobil with Hong Kong CLP Holdings, and Gazprom with Stroytransgaz[161]) each held 15 per cent. The total proven recoverable reserves of the six gas fields involved add up to 304 bcm. PetroChina's total proposed investment, therefore, upstream and midstream, amounted to 22.6 bn yuan. The total investment in the upstream sector and pipeline construction sector was estimated at 27.3 bn and 43.5 bn yuan respectively.[162]

However, the initiative for the joint development of WEP by CNPC and the western consortia collapsed within two years. On 2 August 2004, PetroChina made an official announcement that terminated the WEP I joint venture negotiations with the Shell-led consortium. The joint venture had been a crucial test of China's ability to work with western majors, and attracted much interest in the press. China's withdrawal gave some strong hints about how the government's and the NOC's decision making systems were working, something which continued to puzzle the western players and companies. The day after the joint venture fell through, CNPC's WEP I was successfully being completed. A trial transmission took place on 1 October 2004 and regular commercial supply began at the beginning of 2005.[163] The collapse of western participation in WEP I was a fatal blow to Gazprom's strategy to link WEP I with its Altai gas export scheme.

WEP II

It was in early January 2006 that reports first appeared that China was contemplating building WEP II, starting from the west and ending in south China's Guangdong province. The relevant government departments and companies were at a preliminary stage, studying the feasibility of building such a pipeline. PetroChina was drawing up the plan. The WEP II proposal was officially presented in the draft of Guidelines for the Eleventh Five-Year Development Plan for National Economic and Social Development (2006–10) submitted for examination and approval by the ongoing Fourth Session of the Tenth National People's Congress. The costs for the new pipeline project were expected to exceed $5.2 billion, and it was expected to have an annual handling capacity of 26 bcm. The logic for this pipeline was based on the scale of gas demand in Guangdong province. The main trunk of the pipeline has a length of 4,859 km and the total length is more than 7,000 km, if all the branch lines are included. WEP II travels through 13 provinces, autonomous regions, and municipalities (Xinjiang, Gansu, Ningxia, Shaanxi, Henan, Anhui, Hubei, Hunan, Jiangxi, Guangxi, Guangdong, Zhejiang, and Shanghai).

Opinions about WEP II were divided. Song Dongyu, an expert from PetroChina's Planning Institute, said the project should not be endorsed both because the cost was too high, and because Guangdong's natural gas demand could be met by LNG import and gas production from the South China Sea. However, Zhou Zhaohua, an expert from PetroChina's Langfang Research Institute of Petroleum Exploration and Development, argued that the pipeline was necessary in order to transport natural gas from Xinjiang to Guangdong, the Yangtze River

Delta, the Pearl River Delta, and north China – areas of very high natural gas demand. It would be hard to satisfy demand in the Pearl River Delta with imported LNG, and in the South China Sea natural gas exploration had only just begun.[164]

In September 2007, *Xinhua News Agency* reported that PetroChina was prepared to develop the WEP II project as soon as WEP I went into operation.[165] In late 2007 CNPC signed a deal with the Taiyuan Iron & Steel Group (TISCO) in Shanxi province to purchase steel pipelines for the WEP II project.[166] On 22 February 2008 CNPC began work at several sites simultaneously: in Shanshan (Xinjiang), Wuwei (Gansu), Wuzhong (Ningxia), and Dingbian (Shaanxi). WEP II was to be built in two phases, with the western branch line from Horgos (Xinjiang) to Jingbian (Shaanxi) via Zhongwei (Ningxia) due to come on stream first before the end of 2009, and the eastern branch line that runs from Zhongwei to Guangzhou (the construction of which started in December 2008), plus the feeder line from Weiyuan (Guangdong) to Shenzhen (Guangdong), to start operation by 2011.[167] In fact, WEP II went into operation on 30 June 2011. The project consists of eight sub-lines and one major line which will extend 4,865 km and run from Khorgos in Xinjiang to Guangzhou, Guangdong Province. The main trunk line and three sub-lines have been completed, and the other five branch lines will be finished in 2012.[168]

China OGP argued that the Russian government has been patently uneasy about the forthcoming construction of the WEP II project. An insider from PetroChina said that the Russian government had sent an urgent invitation to PetroChina asking for further negotiations. It is worth noting that PetroChina was starting to study the feasibility of WEP III immediately after the company had announced the route of WEP II on 27 August 2008. When WEP III goes into operation, PetroChina will have connected all the large gas pipelines, and so a nationwide natural gas pipeline grid will finally exist.[169]

WEP II's characteristics can be summarized as follows:[170]

- Location: it runs from Horgos to Guangzhou, covers 14 provinces, and has a total length of 8,700 km, including eight branches.
- Planned pressure: 12 MPa for the west section and 10 MPa for the east section.
- Pipeline grade and details: X80, 18.4 mm hot rolled coils, 22 mm hot rolled plates; diameter 1219 x 18.4 mm SSAW pipes, diameter 1219 x 22.0 mm LSAW pipes, hot bending bends and fittings.
- Annual Delivery Capacity: 30 bcm/y.
- Construction period: 2008–11.
- In operation: 2011 onwards.

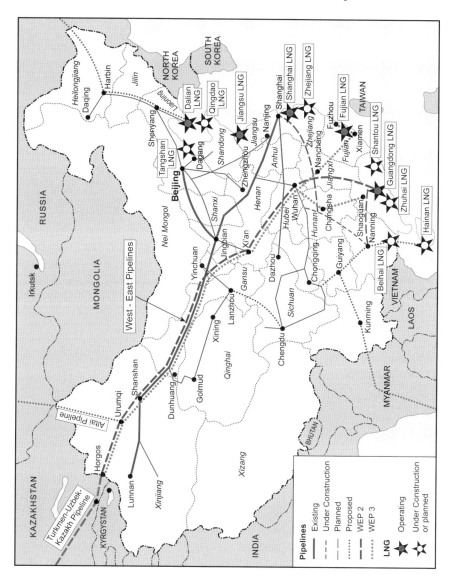

Map 5.1: WEP Corridor

Source: Compiled by author, based on maps from CNPC and ERI.

According to the CNPC Pipeline Bureau, China has upgraded the steel pipes used in three steps: in 1996 X60 steel pipe was first used for the Shaan–Jing Gas Pipelines; in 2001 X70 steel pipe was first used for WEP I; and in 2005, X80 steel pipe was first used for the Ji-Ning branch line and is now in general use in WEP II.

OGJ reported:

> PetroChina has agreed a contract with GE Oil & Gas to supply PGT25+
> gas turbines and PCL 800 compressors for seven pipeline compression
> stations in the western section of WEP II. The contract also includes
> electric-motor-driven units for another three stations. The estimated value of
> this deal is $300 million. WEP II requires 1,240 miles of branch lines. The
> Chinese section will use 1.1 million tonnes of X80 48-in. OD welded pipe
> and 3.2 million tonnes of X80 18-in. OD spiral pipe. GE had previously
> won orders for compression equipment for the expansion of the WEP I and
> for two stations, including the head station, on WEP II. GE also supplied
> the equipment for Sinopec's Sichuan-to-East China natural gas pipeline.[171]

The Central Asia–China Gas Pipeline and the western segment of WEP
II were put into operation on 14 and 31 December 2009 respectively.
Gulf Oil and Gas described the event and its importance:

> On January 20, 2010, after a 20-day preparation for pressure boosting, the
> 2,745.9 km long western segment of the WEP II was ready for gas delivery
> and began to distribute gas to the Second Shaan–Jing Gas Pipeline. The
> Zhongwei–Jingbian branch is a cross-link line of the WEP pipelines and the
> Shaan–Jing Gas Pipelines. As an important hub of gas pipelines in China,
> Jingbian station dispatches imported Central Asian gas to Eastern China
> through the WEP I, and to Beijing and surrounding regions through the
> Shaan–Jing Gas Pipelines.[172]

In harmony with the WEP II development, Guangdong province was
planning to build a pipeline network, with a total investment of 26
billion yuan, for the distribution of natural gas. The first phase of this
project aimed at transmitting natural gas from the Trans-Asia pipeline
to Guangdong by joining the WEP II in 2011. The construction of the
entire pipeline network is expected to be complete in 2015. A total of 10
bcm will be supplied to 21 cities in Guangdong and traded at an equal
price throughout the province. The first phase project of this pipeline
network will finish construction simultaneously with the WEP II. The
Guangdong Natural Gas Pipeline Network Company, co-founded by
CNOOC, SINOPEC, and Guangdong Yudean Group in 2008, will
oversee construction, operation, and management.[173] In August 2010 the
Shenzhen Gas Corporation announced that the company had signed an
agreement with PetroChina to purchase 4 bcm/y of natural gas from
WEP II. This contract will last for 28 years after coming into effect in
the second half of 2011, but the financial terms of the agreement, as
of August 2010, had still to be finalized by the NDRC.[174]

By June 2011 construction of the infrastructure for the major line
of WEP II, with a total distance of 8,653 km and a total budget of
142.2 billion yuan ($21.88 billion), was complete.[175] The intention of

the project is to improve China's energy consumption structure by increasing natural gas use. The project is expected to save 76.8 million tonnes of coal from being burned, so helping to cut emissions, reducing the discharge of 130 mt of CO_2, and 1.44 mt of SO_2.[176]

WEP III, IV, and V

As early as November 2007, it was reported that PetroChina was planning to construct WEP III, possibly originating in Altai in Xinjiang Uygur Autonomous region and terminating somewhere in the Bohai Gulf.[177] In September 2008 *Xinhua* reported that CNPC was planning that WEP III would run from the Xinjiang region to Fujian province as its ultimate destination, feeding on either domestic gas or gas imported from Russia.[178] WEP IV was mentioned for the first time in 2009[179] but CNPC has neither confirmed whether the Altai line will be connected with WEP IV, nor whether the line would lead to Bohai Bay. Finally in the spring of 2011 WEP V was mentioned for the first time, but no details were given.[180]

The details of WEP routes were revealed via the Northeast Asian Natural Gas and Pipeline conference that had been held, usually biennially, since 1995. For example, Chen Qingxun of China Petroleum Pipeline Bureau, a subsidiary of CNPC, had given a very detailed presentation on WEP I and II pipeline development at the Tenth Forum held in Novosibirsk on 18–19 September 2007. At that time, he confirmed that CNPC was reviewing the option of a WEP III pipeline by revealing the possible route. Two years later, the pipeline route map prepared by Zhang Xuezeng, of the China Petroleum Pipeline Bureau, CNPC, clearly showed that CNPC was studying the option of WEP IV as well.[181] According to *China OGP*, the rough blueprint of WEP IV shows that it will run from western Gansu, in Sichuan, to central Shaanxi and south to Guangdong, with the objective of diverting central China's natural gas to the eastern region.[182]

Even after WEP II goes into operation, the gas supply will still lag far behind the rapidly growing domestic demand. PetroChina indicated that WEP III would basically run in parallel with WEP II, and that the Yangtze River Delta and the Pearl River Delta will still be the targeted markets. PetroChina decided to increase the handling capacity of WEP I from 12 bcm to 17 bcm. Since WEP II is designed to carry Central Asian gas, however, PetroChina will reduce natural gas production to some extent in Xinjiang. After WEP III is built, PetroChina will restore gas production in Xinjiang to a normal level, and at that time Russian gas will become an important part of WEP III.[183] This *China OGP* report reflected how fully the WEP III project had been discussed internally by the autumn of 2007.

According to Yang Jianhong, deputy director of the Oil & Gas Pipeline Engineering Department at PetroChina's Planning and Engineering Institute, China in 2008 only had 40,000 km of natural gas pipelines, but would need at least 100,000 km to form a sound downstream gas market. He said that if China finally obtained Russian gas from west Siberia, WEP III would be built to transmit the Russian gas to China's eastern and southern regions. WEP III would probably pass through south-eastern Fujian province, which is not covered by WEP I and II, and would then extend to Shanghai and Guangdong.[184] Mr Yang's assumption was that the gas supply source would be Russia rather than a Central Asian republic.

In April 2010 the Chinese government clarified a number of points. WEP III will extend from Xinjiang province to Guangdong province, via Gansu, Ningxia, Shaanxi, Henan, Hubei, and Hunan. The pipeline, with a capacity of 30 bcm/y, will deliver imported gas from central Asia. WEP III will consist of two sections, a western section from Xinjiang to Zhongwei City, Ningxia Hui Autonomous Region, and an eastern section extending from Zhongwei City to Shaoguan City, in Guangdong province. The western section is scheduled to start operation in 2012 and the eastern one in 2014. The central government announced that work on WEP III would begin before the end of the year,[185] and in July it was reported by *China Business Post* that the route of PetroChina's WEP III had been worked out and would probably go from Xinjiang to Guiyang, capital city of Guizhou province.[186]

In the near to medium term future, therefore, the backbone of PetroChina's nationwide gas pipeline network would be formed by the existing first and second, and the planned third Shaanxi–Beijing gas pipelines, in addition to the existing WEP I and II and the prospective WEP III, and the existing Zhongxian (Sichuan)–Wuhan (Hubei) gas pipeline.[187] If, however, WEP III construction starts without any agreement between Russia and China on the Altai route gas supply and the export price, the development would indirectly confirm Beijing's intention to import more gas from the Central Asian Republics rather than Russia. If the import volume rises from 40 to 60 bcm/y, a figure officially mooted by both the Chinese and Turkmenistan authorities, WEP III will be the conduit for the extra gas imported, and will reinforce the WEP corridor as the core of the nationwide gas pipeline network. As discussed in Chapter 7, the gas supply source for WEP III and IV pipeline development is a critically important issue for both Sino-Russian gas cooperation and Sino-Central Asian Republics gas cooperation.

From the Beijing planners' viewpoint, however, securing the supply source of gas for WEP III and IV is a central part of the WEP (I–V) corridor development, and requires very large-scale pipeline gas imports from the Central Asian Republics, and possibly from Russia as well.

PetroChina will not consider linking its own gas pipelines with those of SINOPEC unless the Chinese central government decides to expropriate gas pipelines from their owners and establish a state-owned and independent gas pipeline company.[188] Until that happens, PetroChina and SINOPEC will continue to build gas pipelines independently.

International Gas Pipelines

The Beijing authorities had given CNPC full authority to promote the introduction of international gas pipelines, and since the mid-1990s CNPC had focused on the development of trans-border pipelines from Russia, the Central Asian Republics, and Myanmar.

Feasibility Studies on Two Gas Pipelines

The visit to Russia of the Chinese Prime Minister Zhu Rongji at the end of February 1999 injected new life into the Sino-Russian transnational oil and gas pipeline programmes. Among the 11 agreements signed at the end of the fourth regular meeting between Zhu and his Russian counterpart Primakov, three concern trans-boundary oil and gas pipelines, one for oil and two for gas.[189] The second accord on gas featured a feasibility study on a project which had long been on negotiating agendas, a possible natural gas pipeline from Irkutsk to Shenyang, Liaoning, and further south to Dalian. This gas pipeline, expected to run parallel to the crude oil pipeline, could move 20 bcm of gas to China annually.

The original plan envisaged that China would import 10 bcm, and the rest would be for South Korea. The route of the pipeline proved controversial. The so-called west line plan involved moving Russian gas to Rizhao in Shandong province via Mongolia and Beijing. This pipeline was designed with a total length of 3,364 km (1,027 km in Russia, 1,017 km in Mongolia, and 1,320 km in China) and a diameter of 1,420 mm. The required investment for the project was $6.85 billion. Apart from the route issue, CNPC has also worried about the insufficiency of reserves at the Kovyktinskoye field, saying that another 700 bcm of proven gas reserves needed to be added to its current 800 bcm to make the construction of such a long-distance pipeline economically viable. The State Council of the Chinese Government allowed CNPC to participate in the construction of the pipeline, but has not yet consented to CNPC involvement in upstream exploration and development in Russia.

To help keep the project on track, CNPC and Russia's Ministry of Fuel and Energy (MFE) proposed that Gazprom join the project but the gas giant was hesitant. CNPC believed that in the long run Kovyktinskoye's reserves were insufficient to sustain exports, as well as to supply East Siberia and the Far East, even if the reserves in Yakutia were added in. This explains why CNPC signed an agreement with Sakhaneftegaz in March 1998.[190]

Another gas accord on a preliminary feasibility study of a natural gas pipeline from west Siberia to Shanghai was mooted. It was the longest of the three pipeline projects, stretching over 6,800 km, of which 2,430 km would be within Russia and the rest in China. Designed to move 20–30 bcm annually, it would fortify China's domestic development of the West–East pipeline.[191] Gazprom expressed keen interest in constructing both the pipelines and underground gas depots for China's longest natural gas pipeline project. The Chinese side, however, rejected Russia's offer on the grounds that its technology was too underdeveloped and its financial resources too limited.[192]

In August 2001 Gazprom decided to join in the Shell-led consortium to bid for the construction of the WEP I pipeline. A Gazprom executive told the press that Gazprom was negotiating with the Chinese authorities on using Chinese debts to Russia, including debts for armament supplies, to finance the gas pipeline. The message signalled Russia's inclusion in the joint venture for the gas pipeline. The question was how to phase in Russia's role, and how big the role was to be.[193] In early August, 2002 the Beijing authorities took their first major decision on LNG supply.[194] This remarkable breakthrough was not matched by any turnaround on pipeline gas development. China regarded acceptance of the east route, which designated Manzhouli as the only way to reach north-east China, as a pre-requisite for its participation in the development. This differed from the Russian plan, which favoured the west route through Mongolia reaching Beijing and the north China market. The west route would be shorter, especially the portion within Russia, and less topographically challenging. The route would allow Mongolia to collect a sizable transit fee.

WEP and the Russian and Central Asian Gas Pipeline

Development of WEP I and II during the 2000s provided a solid basis for the introduction of trans-national pipeline gas from both Russia and the Central Asian Republics. (The details of Sino-Russian gas supply negotiation will be discussed in Chapter 7). Due to the expensive lesson of the Angarsk–Daqing pipeline project, China learned that, to avoid weakness when negotiating with Russia, it needed an alternative supply pipeline. That is why China, in the case of crude, initiated its pipeline connection with Kazakhstan. This lesson was applied to the natural gas pipeline project as well. While still pursuing the supply of Russian gas to China, Beijing decided to take full advantage also of the gas supply option from the Central Asian republics, in particular Turkmenistan and Kazakhstan. (The details of Central Asian gas supply to China will be reviewed in Chapter 6.)

Myanmar Gas Pipeline

Another important initiative towards developing a trans-national pipeline network, from Myanmar to south-west China's Yunnan province, would not have been possible without sustained pressure by the Chinese on the Myanmar government. On 14 February 2008, the Yunnan provincial government confirmed that preparatory work was underway for the Sino-Myanmar pipeline project and that the construction of the pipeline would begin in 2008. According to a framework agreement between CNPC and the Yunnan government signed in December 2007, CNPC planned to build a 20 mt/y oil and gas pipeline, and a refining complex with a 10 mt/y crude run and 0.8 mt/y of ethylene production capacity. The oil and gas pipelines, scheduled for construction to start in 2010 and to be completed by 2013, run from Kyauk Phyu in the western state of Rakhine and arrive at Mandalay, the city opposite the Chinese border town of Ruili, Yunnan province, from where it turns south, finishing at Chongqing.[195]

In June 2008 the South Korea-based Daewoo International Corporation signed an agreement to sell natural gas produced in Myanmar to CNPC. Daewoo, which holds a 51 per cent stake in the A-1 (Shwe and Shwe Phyu fields) and A-3 (Mya) gas fields in Myanmar, signed a memorandum of understanding with CNPC over the sale and transportation of natural gas from the project. The natural gas price would be set quarterly in accordance with global natural gas prices. Daewoo estimated the two gas fields' reserves at 5.4–9.1 tcf, of which 4.5–7.7 tcf could be commercialized.[196]

According to *Reuters*:

on 25 August 2009, it was finally announced that a consortium led by South Korea's Daewoo International will invest about $5.6 billion to develop

Myanmar gas fields as part of a 30-year natural gas supply deal with China. The investment came just a week after China signed a $41 billion liquefied natural gas import deal with Australia. The Myanmar gas development plan, which has been mooted since 2004, will allow the consortium to supply natural gas to CNPC, with a peak daily production of 500 million cubic feet, or about 3.8 million tonnes annually. The supply, due from 2013 from the Shwe and Shwe Phyu fields in Myanmar's A-1 offshore block and Mya field in A-3 offshore block, amounts to around 7 per cent of China's current (2009) gas consumption of 7.3 billion cubic feet per day. The consortium will undertake production and offshore pipeline transportation, while land transportation to China will be jointly managed with China National United Oil Corporation (CNUOC).[197]

In June 2010 construction of the Myanmar section of the Sino-Myanmar Oil and Gas Pipeline with a capacity of 12 bcm/y began. This section is 771 km long, while the natural gas section runs for 793 km, according to CNPC. Construction of the domestic section of the Sino-Myanmar Oil and Gas Pipeline started on 10 September 2010.[198]

LNG Expansion

In 2010, China's total LNG imports reached 9.4 mt. To meet the galloping LNG demand, China is accelerating the development of LNG receiving terminals. According to the three Chinese NOCs' LNG expansion plans, all of the coastal provinces will have receiving terminals. CNOOC's nine LNG receiving terminals alone will have a combined receiving capacity of 48.4 mt after new projects are completed.[199]

Review of the Guangdong and Fujian LNG Developments

Three years after China initiated pre-feasibility studies on LNG importation at the end of 1995, the State Council finally endorsed this ambitious scheme at the end of October 1998. Initially, the SDPC authorized CNOOC jointly with China Electric Power Corporation (former Ministry of Power Industry), the Ministries of Communications and of Construction, and eight international oil companies to conduct an LNG importation study.[200] In 1998 the State Council decided to carry out the LNG pilot work in China, and CNOOC acted as pioneer. As CNPC, SINOPEC, and CNOOC plunged into the establishment of LNG projects, the NDRC set a principle – one province, one LNG project. However, the actual situation has violated the state's original intention. First, LNG project construction occupied a large amount of land, while China suffers from limited land resources. Sometimes central government has no choice but to grant approval for land occupation by LNG project construction because of pressure from companies and local governments. Second, at that time China was expected to import 30 mt/y of LNG by 2010 and 60 mt/y by 2020. At this rate of growth, a major and reasonable concern would be to find sufficient LNG sources.[201] In mid-June 1999, an overall appraisal on Guangdong LNG was presided over by the CIECC, which submitted a detailed appraisal report to the SDPC, whose own views were sent to the central government for approval in July.[202] This study supplied the policy principles which would maintain adequate and secure sources of LNG in the long run. They enabled the authorities, in August 2002, to make one of the most important decisions for natural gas expansion, namely to choose the supply source for Guangdong LNG and Fujian LNG which would allow the epoch of LNG imports to begin.

Guangdong LNG:[203] In 1996 CNOOC submitted a 'Planning Report on the LNG utilization Project in the south-eastern Coastal Area' to the State Council. In the wake of the report, the Guangdong provincial government and CNOOC carried out the pre-feasibility study and the site selection. In May 1998, they jointly submitted the 'Proposal on Guangdong LNG Terminal and Trunk line Project' to the SDPC. At the end of 1998, the State Council approved the pilot LNG project in Guangdong, and in April 1999, the project proposal was submitted to the SDPC and won approval at the end of that year. International bidding for foreign investors started in August 2000. A total of 27 foreign applicants purchased the data package and by 8 September 2000, 23 firms in ten consortia had handed in bids. Four consortia were shortlisted by the Chinese side for the second round of bidding. In March 2001, it was announced that BP had been chosen as the foreign partner for the construction of the terminal and trunk pipeline. The Executive Office of the Guangdong project submitted the bidding result to the SDPC for approval on 24 April 2002 according to timetable, and the result was initially supposed to be announced in June 2002. After a delay of a few weeks, the decision was made. On 8 August 2002 the Chinese government made the historic announcement that Australia and Indonesia had won the gas supply contracts for the two terminals in the Guangdong and Fujian provinces respectively. Australian LNG, the marketing arm of the North West Shelf Project, won a 25-year contract to supply 3 mt/y of LNG to the Guangdong terminal. But the Tangguh project won a consolation prize with a contract to supply 2.5 mt/y of LNG to the Fujian terminal.

In May 2003, CNOOC signed an agreement to acquire a 25 per cent stake in the China LNG joint venture within the Australian North West Shelf (NWS) project. In addition to the interest acquired in the NWS gas production, the agreement also granted CNOOC a 5.3 per cent (approximately) interest in certain production licenses, retention leases, an exploration permit of the NWS project, and the right to participate in future exploration undertaken over and above the proven reserves. The agreement stipulated that CNOOC's share in the LNG joint venture would increase as the LNG quantity supplied to Guangdong terminal increased.

CNOOC confirmed that the acquisition price was $1.52 per boe. Including anticipated development costs of $1 per boe, the acquisition price was 37 per cent lower than CNOOC's historical exploration and development costs, and 48 per cent lower than the implied price paid by Australia's Woodside Petroleum Ltd, the operator of the North West Shelf. It confirms that the co-owners of the North West Shelf LNG project made a significant sacrifice to secure the Chinese market.

The Guangdong LNG Import and Development scheme was divided into two phases: the first incorporates a 3.7 mt/y LNG receiving terminal and a trunk gas grid extending 215.4 km, with a capacity of 4 bcm/y. The total investment (for receiving terminal and gas grid only) was 5.1 bn yuan ($600 million). The second phase added another 2.5 mt/y of receiving capacity to the terminal. A new gas grid from Zhuhai to Foshan, extending 181.7 km, aimed at moving 8.2 bcm of gas. Construction of two oil-to-gas power plants in Foshan (Desheng plant and Shakou plant) was included in the second phase work. The total investment (for receiving terminal and gas grid only) was 2.1 bn yuan.

On 26 May 2006 the first LNG shipment, after its ten day voyage from Australia, arrived at Shenzhen Dapeng LNG terminal wharf, heralding a new era for LNG in China. The LNG price finally signed in December 2004 was still relatively low, equal to an international oil price of around $25 per barrel. On 28 November 2006, an LNG-fired power plant saw its first generator put into operation in Shenzhen.[204] At the end of 2008, the Guangdong Dapeng LNG terminal had three storage tanks, approximately 400 km of high pressure gas pipeline, and 19 clients; it supplied gas to the cities of Shenzhen, Guangzhou, Dongguan, Foshan, and Hong Kong, and also provided gas to 12 power plants in Guangdong. The Dapeng LNG selling price to power plants fluctuates between 1.5 and 1.64 yuan/cubic metre, the grid-price is uniformly 0.581 yuan/Kwh (inclusive of tax).[205] A year later the capacity of the terminal had been expanded to 6.7 mt/y from a previous capacity of 3.7 mt/y, and over the longer term the expansion of the Guangdong terminal could also help to provide supplies for the province's new scheme – a $6.97 billion gas transmission grid set to supply 60 bcm of gas by 2020.[206]

Fujian LNG:[207] The central government decided to approve the Fujian LNG project. The Chinese government announced on 8 August 2002 that BP had won the 2.5 m tonne/year gas supply contract for the Fujian LNG terminal. CNOOC aimed at developing two gas-fired power stations in Nanpu and Songyu, and pipeline distribution networks in five cities (Fuzhou, Putian, Quanzhou, Xiamen, and Zhangzhou). CNOOC held 60 per cent of the equity of the LNG terminal and the trunk line project, and its partner, Fujian Investment & Development Co. Ltd (FIDC), owned by the Fujian government, retained the remaining 40 per cent equity.

On 27 September 2002, CNOOC announced that it had signed a HOA for the acquisition from BP of a 12.5 per cent stake in the Indonesian Tangguh LNG project, by payment of $275 million. BP retains a 49.66 per cent stake in the project. CNOOC's acquisition price for the Tangguh project was around $0.89 per barrel of oil equivalent, well below CNOOC's average historical exploration and development costs of around $4/boe. The Tangguh price was also lower than the company's earlier acquisition of a 5 per cent stake in Australia's North West Shelf Gas Project reserves, which cost CNOOC $320 million and translates to about $1.52/boe.[208] The HOA signed in Jakarta confirmed a 25-year, $8.5bn Sales and Purchasing Agreement for supplying 2.6m tonnes/year of LNG to the Fujian LNG project.[209]

A noticeable fact with the HOA was that the 12.5 per cent stake was larger than the 5 per cent stake which CNOOC received in the North West Shelf through leveraging a 3m tonnes/year gas contract for the Guangdong LNG terminal. The gas-supply volume was also raised from the earlier reported 2.5m to 2.6m tonne/year. In early 2003, the overall project proposal of the Fujian LNG terminal was approved by the State Council. On 23 August 2003 Fujian LNG construction started with land reclamation from the sea in preparation for the terminal building in Putian, Fujian province. The total investment for the project was around 25 billion yuan, of which 6 billion yuan was for the terminal and trunk line, 15 billion yuan for power plants, and 4 billion yuan for the towngas grids.[210]

> In late December 2005 [reports a Chinese source] the NDRC approved the feasibility studies of three gas power plants in Fujian by CNOOC Gas & Power Ltd. The three power plants located in Jinjiang, Putian, and

Xiamen would have a combined generating capacity of 3.5 GW and an annual output of 14 billion Kwh. The projects would use two mt of LNG, over 70 per cent of Fujian's total supply.[211]

The Fujian LNG installation 'received its first test run cargo in May 2008'.[212] On 29 July 2009 a Fujian-based LNG and chemicals project in which CNOOC had invested received the first shipment of LNG from the Tangguh LNG Project in Indonesia. CNOOC planned to boost capacity of Fujian LNG terminal to 5.2 mt/y before 2011.[213]

The Second Wave of LNG Expansion

Besides the Guangdong and Fujian LNG projects, a number of other LNG terminals were developed in the coastal areas of China. In 2006, the Chinese government approved about ten LNG terminals, belonging to PetroChina, SINOPEC, and CNOOC, along the energy-intensive eastern coast.[214] CNOOC intended to develop a comprehensive coastal pipeline network by connecting all the LNG terminals in each province along the coast. In fact, in late 2004, CNOOC reached cooperative outline agreements with local governments to build three LNG receiving terminals in Liaoning province, Jiangsu province, and Guangdong province.[215] However, CNOOC's initiative was partially realized when the NDRC decided to allow PetroChina to develop LNG terminals in Liaoning, Hebei, and Guangxi provinces, and SINOPEC to develop the terminals in Shandong province and Tianjin Municipality.[216]

Shanghai LNG: Shanghai LNG project was financed by CNOOC and Shanghai Shenergy, with 45 and 55 per cent stakes respectively. The first phase entered operation in 2009 with a 3 mt/y receiving capacity. The investment for the receiving terminal and undersea pipelines totalled 7 billion yuan. The second phase, that will be completed in 2020 was designed to receive 6 mt/y of LNG.[217] 'The initial LNG supply contract was signed between Shanghai Shenergy and Petronas in late 2006, for deliveries of 1.1 mt starting in 2009 and increased to 3 mt by 2012'.[218] According to *Xinhua News Agency*, Shanghai LNG terminal was under a contract with Petronas at a price of $6–7/mmbtu (or a CIF price of 1.8 yuan/cubic metre). At the end of 2005, Fu Chengyu, chairman of CNOOC, said that the company had completed LNG layout along the south-east coast, and Shanghai LNG terminal would boost its LNG strategy layout in China's coastal area.[219] On 11 October 2009 Shanghai LNG terminal received its first LNG cargo of 45,000 cubic metres aboard the 88,000 cubic metre *Arctic Spirit* LNG carrier from Bintulu, Malaysia.[220]

Shanghai Shenergy built the Wuhaogou LNG emergency station, which has 120,000 cubic metres storage capacity (fully occupied in March 2009). The Wuhaogou was one of the most modern LNG facilities in China, equipped with liquefaction and gasification facilities, tankers, and LNG vessels.[221] According to the *Shanghai Morning Post*, Shanghai's natural gas consumption was projected to reach 12 bcm by 2015 and 15–18 bcm by 2020, and the municipality's natural gas supply volume would be able to sustain the city for 30 days. By 2020, the city is expecting to run exclusively on natural gas. Shanghai has been aiming to have five sources of natural gas by 2011, namely the West–East Gas Pipeline project with 2.27 bcm/y of gas, the East China Sea with 0.5 bcm/y of gas, the Sichuan–Shanghai gas pipeline built by SINOPEC, with 1.9 bcm/y of gas, the LNG import project from 2009 with 4 bcm of gas, and the second West–East Gas Pipeline project, with 2 bcm/y of gas.[222]

Zhejiang LNG: CNOOC holds a 51 per cent stake and the remaining 49 per cent stake is shared by Zhejiang Energy Group (29 per cent) and Ningbo Power Development Company (20 per cent). The first phase, with 3 mt/y of LNG receiving capacity, is scheduled to come on stream in 2012, and a second phase is designed to expand capacity to 6 mt/y. The terminal is expected to play a peak-demand role in supplying Zhejiang's natural gas pipeline network and also in providing emergency supplies to Zhejiang province. In 2003, CNOOC signed a framework agreement with Chevron to buy 80–100 mt of LNG from Gorgon project within the coming 25 years, with a fixed price of $4/mmbtu. CNOOC also intended to secure 12.5 per cent equity in the Gorgon project but the agreement collapsed due to the price issue.[223] Due to the difficulty in securing the supply source, in early 2007 CNOOC announced a delay in completion until mid-2009. David Fridley explained that the special nature of Zhejiang's LNG market arose from the fact that it was:

> largely LPG based, with far less coal gas and coke oven gas consumption (0.33 bcm natural gas equivalent) compared to Shanghai. LPG consumption is 75 per cent residential based, and gas use in the power sector has accounted for less than 2 per cent of the total. About half of the initial LNG supply will be used for power generation.[224]

In May 2009 CNOOC and BG signed a contract to jointly develop the Queensland Curtis LNG project in Australia, with further agreement to allow CNOOC to buy 3.6 mt/y of LNG for 20 years after the project starts. The LNG from Queensland may be used to supply the Zhejiang terminal.[225]

Jiangsu LNG: Both PetroChina and SINOPEC have tried to win the Jiangsu provincial government's support for their LNG proposals. In 2005 Jiangsu provincial government signed a framework agreement with SINOPEC for the Lianyungang LNG terminal, and also finalized another deal with PetroChina over its Rudong LNG terminal in Nantong city. Through this deal, PetroChina was expected to build a gas-fuelled power plant with six 390 MW power units.[226] In March 2007, PetroChina was given the go-ahead from the NDRC to start construction on the Jiangsu LNG terminal. The 3.5 mt/y Rudong LNG terminal in Jiangsu province is under construction and the first phase, estimated to have cost 6 billion yuan, was completed in 2011. PetroChina holds 55 per cent equity, and the remaining 45 per cent is shared by Pacific Oil and Gas Co Ltd (35 per cent) and Jiangsu CITIC Asset Management Group (10 per cent). Jiangsu LNG will be fed by LNG imported from Qatar and will supply the gas through the WEP II and Hebei–Ningbo pipelines.[227] In June 2009, an accident occurred during the construction of the No. 1 storage tank at Rudong, but it started its operations as scheduled in 2011.[228]

Dalian LNG: In May 2009, PetroChina and Dalian Port (PDA) agreed to form a joint venture for a 6 billion yuan ($877 million) LNG terminal project in Xingang, Dalian. PetroChina will own 75 per cent of the venture, Dalian Port 20 per cent, and Dalian Construction Investment Corporation, a local government investment arm, the remaining 5 per cent. With the start of operations in 2011, the LNG terminal in Xingang will mainly serve users in Liaoning Province. Construction on the terminal, which will have an initial annual LNG receiving capacity of 3 mt, started in April 2008. CNPC finished constructing an LNG receiving dock on 29 August 2010. The dock is part of a larger LNG re-gasification and storage terminal, which will handle 10.5 bcm/y of natural gas supply. Australia's Gorgon LNG project is expected to be the main supply source for this terminal.[229]

Shandong Qingdao LNG: Construction has commenced on an auxiliary project to SINOPEC's planned LNG project in Qingdao, in Shandong province; this is the gas pipeline project between Jiaozhou and Rizhao. Jinan was selected as the site for the planned LNG terminal project. The Jiaozhou–Rizhao gas pipeline will stretch 193 km and have a designed gas transport capacity of 1.7 bcm. The Jiaozhou–Rizhao pipeline will feed on Daniudi (Ordos) gas from west China before the Qingdao LNG terminal is put into operation, which is expected to be in 2013. The project cost is 580 million yuan. The construction of the pipeline means that the LNG project has achieved a breakthrough. The Shandong LNG project was initially scheduled to be finished in 2007, but it was delayed due to lack of progress in securing overseas LNG supply. In September 2010, SINOPEC announced the start of construction of the Qingdao LNG project which will take place in two phases. The first, with total investment of 9.66 bn ($1.43 bn), is expected to be completed by September 2013 and put into operation in November 2013. SINOPEC initially expected to receive the 10 mt/y of LNG from Iran, but later began to explore LNG supply from Russia, Australia, Indonesia, and Papua New Guinea. In 2009, SINOPEC signed a long-term supply contract with ExxonMobil to receive 2 mt/y from Papua New Guinea.[230]

Tangshan LNG: In November 2010, the NDRC approved PetroChina's Tangshan LNG terminal project, which is expected to be put into operation in 2013 with annual receiving capacity of 3.5 mt/y. The gas will mainly be supplied to Beijing, Tianjin, and Hebei Province. According to a report from *China Energy News*, the company will invest about 8.41 billion yuan. The first phase of the LNG terminal is designed to receive 3.5 mt, and three 160,000 cubic metre storage tanks will be built. A dock for LNG carriers in the size range 80,000–270,000 cubic metres will also be built. The transportation project includes a 129 kilometre pipeline and a 28 kilometre branch pipeline, with transportation capacity of 9 bcm and 3.5 bcm, respectively. The Hebei LNG project is PetroChina's third receiving terminal approved by the NDRC. PetroChina has already transferred its controlling stakes in the projects to its listed unit Kunlun Energy. Construction actually began before final approval, because delaying projects can result in a critical loss of market share to rivals.[231]

Zhuhai LNG: The first phase of Zhuhai (Nanjing Bay of the Gaolan Port Economic Zone) LNG terminal, approved by the NDRC in March 2010, will be able to receive 3.5 mt/y of LNG when it is put into operation in 2013. The schedule is one year behind that originally planned. The first phase of the project encompasses an LNG receiving station, an LNG appropriated wharf for dropships up to 420m long with 80,000–270,000 cubic metres of gas storage, and 291km of gas transmission trunk lines. In addition, three 160,000 cubic metre storage tanks will be built onshore. The first phase investment cost will be 11.3 bn yuan ($1.65 bn). Zhuhai LNG will be able to handle 7 mt/y of LNG when its second phase is also completed, and ultimately it will have a 12 mt/y capacity. The terminal, first proposed in 2005, is jointly owned by CNOOC and a number of local energy companies. It will provide gas supplies to the cities of Guangzhou, Foshan, Zhongshan, and Jiangmen in the Pearl River Delta area.[232]

Hainan LNG: CNOOC and Hainan province plan to invest 5.6 billion yuan in building the Hainan LNG terminal, in which CNOOC holds 65 per cent equity and Hainan Development Co. Ltd the remaining 35 per cent. The first phase of the Hainan LNG installation will cost 3.8 billion yuan, includes the building of a special LNG port, and an LNG terminal and pipeline. The first phase is designed to have a receiving capacity of 2 mt/y and can be expanded to receive up to 3 mt/y.[233]

As shown in Table 5.28, the required LNG import for all the terminals being operated, under construction, or approved is at least 28.8 mt, and the figure will be 54.9 mt when the second stage volume is added. If the volume of several planned terminals is also included, the required LNG import for the first stage LNG terminals could easily be more than 35 mt; and that figure will rise to almost 60 mt when the second stage volume is added.[234] Table 5.29 shows that the contracted LNG supply volumes by three Chinese NOCs are far short of the total import volume. According to the SASAC, CNOOC plans to build nine LNG receiving terminals along the Chinese coast. Among the nine terminals, the following have come into operation: Guangdong in 2006; Fujian, 2009; Shanghai, 2009; Dalian, 2011; and Rudong, 2011. CNOOC also plans to set up 15 CNG (compressed natural gas) service stations.[235] In line with the large scale LNG imports, more LNG carriers will be needed to take care of the LNG transportation.[236]

Chinese NOCs must secure supply before they can receive government approval to build a regasification terminal, and by December 2009 CNOOC, PetroChina, and SINOPEC had signed several long-term supply contracts with a supply volume of 28.45 mt. These contracts were primarily with Asian firms sourcing LNG from Indonesia, Malaysia, and Australia; however, Chinese NOCs have also signed long-term contracts with other sources such as Qatar Gas and from global upstream developers which can supply LNG from various international liquefaction assets. But the volumes were not big enough to supply the operating, unfinished, and planned LNG terminals along the coastal regions of China.

Difficulty in Securing LNG Supply Sources

A number of the LNG supply contracts which Chinese NOCs have signed have given a strong hint about the kind of obstacles which lie ahead for LNG supply contracts. During the 2004–6 period, Zhuhai

Table 5.28: China's LNG Import Projects

Project	Phase I/II (mtpa)	Operation date	Company (%)	Progress
GuangdongDapeng LNG	3.7 / 2.6 (2011)	2006	CNOOC (33), BP (30)	Full operation
Fujian Putian LNG	2.6 / 3.0 (2011)	2009	CNOOC (60), FIDCL (40)	Full operation
Shanghai LNG	3.0 / 3.0 (2012)	2009	CNOOC (45), Shenergy (55)	Full operation
Dalian LNG	3.0 / 3.0 (TBD)	2011	CNPC (75), Dalian Port (20)	Operation started
Jiangsu Rudong LNG	3.5 /3.0 (TBD)	2011	CNPC (55), Pacific Oil & Gas Ltd (35)	Operation started
Hebei Tangshan LNG	3.5 / 3.0 (TBD)	2013	CNPC	under construction
Zhejiang Ningbo LNG	3.0 /3.0 (TBD)	2012	CNOOC (51), Zhejiang Energy (29)Ningbo Power (20)	under construction
Zhuhai Jinwan LNG	3.5 / 3.5 (TBD)	2013	CNOOC (30), Guangdong Yudean Group Co Ltd (25)	under construction
Shandong Qingdao LNG	3.0 / 2.0 (TBD)	2013	SINOPEC	under construction
Sub-Total	**28.8–54.9**			
Hainan Yangpu LNG	3.0 /3.0 (TBD)	2012	CNOOC	proposed
Tieshan Guangxi LNG	3.0	2015	SINOPEC	planned
Total	**34.8–57.9**			

Note: FIDCL = Fujian Investment & Development Co Ltd
TBD = to be determined.

Source[237]: Lin Fanjing (2008g); CNPC RIE & T. (2008); Higashi (2009); China Securities Journal (2010), 36–41; *China EW*, 9–15 September 2010, 6.

Table 5.29: Existing Long-term LNG Sales and Purchase Agreements

Buyer	Supply Source	Volume (mtpa)	Term (years)	Signing Date	Destination Terminal / Capacity (mtpa)	First Cargo
CNOOC	Australia NWS	3.7		2003	Guangdong (3.7)	2006
	Indonesia Tangguh	2.6	25	2006 Sep	Fujian (2.6)	2009
	Malaysia Tiga	3.0		2006	Shanghai (3.0)	2009
	Qatar Gas II	2.0	25	2008 Jun	Multi-destination?	2009
	Total (Portfolio)	1.0	15	2009 Jan	Multi-destination?	2010
	BG, Queensland, Australia	3.6	20	2009 May	Zhejiang	2012
CNPC	Qatar Gas IV	3.0	25	2008 Apr	Jiangsu province	2011
	Shell Gorgon+	2.0	20	2008 Nov	Dalian	2011
	Shell Browse	3.3	20	2007 Sep	Cancelled at the end of Dec 2009	
	ExxonMobil Gorgon	2.25	20	2009 Sep	Dalian	2011
SINOPEC	Exxon PNG	2.0	20	2009 Dec	Qingdao	
Total		**28.45**				

Source: Higashi (2009); China Securities Journal (2010), 36–42; *China ERW*, 12–18 June, 2008, 5; *China OGP*, 1 July 2008, 16; Crooks (2009c); 'Sinopec, Exxonmobil ink Papua New Guinea LNG deal', http://www2.china-sd.com/News/2009-12/4_3926.html.

Zhenrong, CNOOC, SINOPEC, and PetroChina signed an MOU with NIOC involving combined volumes of at least 45.5 mt, including the 30 mt/y LNG supply agreement between CNPC and NIGEC.[238] No serious progress, however, was being made regarding the delivery of LNG on schedule. On 4 September 2007 CNPC signed an LNG purchase and sale agreement with Shell in Perth, Australia, under which Shell will sell 1 mt/y of LNG from their Gorgon project to CNPC for 20 years. On 6 September 2007, CNPC signed an LNG supply contract with Australia-based Woodside Energy Ltd in Sydney, witnessed by President Hu Jintao and Australia's Prime Minister John Howard. The agreement confirmed the potential sale of 2 mt/y of LNG from Browse LNG to CNPC for a period of 15 years starting from 2013–15. It is the first LNG purchase agreement signed by CNPC with foreign partners. An insider revealed that the ceiling CIF price of the CNPC–Shell deal would be $10/mmbtu or 2.7 yuan/cubic metres. At that time *China OGP* argued it was wise for Chinese oil companies to secure long term LNG import as a first priority, even if the import price seemed higher than expected.[239]

However, CNPC presented other Chinese LNG buyers with a major problem when the supposed $10/mmbtu purchasing price looked as if it might be used as a reference price for other potential buyers. A CNOOC insider said:

> it is still impossible for us to accept such an import price as high as $10/ mmbtu, though end-users are keen on using natural gas as fuel and tend to cope with increasingly high prices. Unlike CNOOC, CNPC uses LNG in a mix of energies and the high LNG cost can be partly diluted by CNPC's low-cost natural gas piped from its own gas fields.[240]

Chinese experts like Hu Jianyi, of the Chinese Academy of Engineering, believe that LNG demand and supply will reach a basic global balance in the long term and that as projects come into production after 2012, the tightness of the market may diminish.[241] Both Japan and South Korea have advocated the formation of an alliance among LNG buyers to lower the LNG selling price. However, Jiang Zhefeng (general manager of business development at CNOOC Gas & Power Limited) has said that no possibility of such an alliance exists since none of the participants would transparently cooperate with the other players.[242]

In April 2008 an official with the Dalian Municipal Government's energy department said that construction of PetroChina's LNG terminal in Dalian had begun, but the terminal would need to source LNG from the international spot market in its first year of operation until delivery of contracted long-term supplies from Australia commence in 2012.[243] Also in April 2008, Qatar gas signed 25 year LNG supply contracts with both CNPC and CNOOC. The 25 year arrangements saw Qatar gas sell 2 mt/y of LNG to CNOOC from the Qatar gas 2 project from 2009, and 3 mt/y of LNG to CNPC from the Qatar gas 4 project when it began operations in 2011. At that time, no price was officially revealed, but it was almost certain that the price would not exceed $10/mmbtu because CNOOC would not be able to bear the burden of a higher price. In 2007, the highest price CNOOC paid for spot cargo was $447/tonne, equivalent to $8.60/mmbtu. *Xinhua News Agency* reported that the price agreed with Qatar gas was likely to be no more than $8.60/mmbtu.[244] A CNOOC official said the firm may face difficulty in finding downstream buyers for the LNG the firm will receive from Qatar, thanks to rising LNG prices.[245] The difficulty of securing LNG supply sources drove China to be the front runner in the CBM-based LNG supply deal.

China's Adventurous CBM–LNG Deal

Peter Smith, writing in the *Financial Times* in March 2010, reported that:

> Royal Dutch Shell and PetroChina have moved to take a lead role in Australia's coalbed methane gas sector after making a joint offer for the Brisbane-based energy producer Arrow Energy,[246]valuing it at $3.4 bn. Arrow Energy is Australia's largest holder of coal-seam gas acreage with interests covering 65,000 square kilometres in Queensland and it has proved, probable and possible reserves of 11,042 PJ.[247]

Carola Hoyos and Ed Crooks commented that, considering that:

> the most important factor for any LNG project is securing buyers, Shell's partnership with PetroChina is an excellent choice, even though PetroChina had ended its agreement with Woodside Petroleum to buy about $41 bn worth of LNG over 20 years from its planned Browse LNG project in Western Australia … For PetroChina, the alliance with Shell will give the company a chance to learn the world's most extensive LNG experience, as well as acquiring knowledge of producing unconventional gas.[248]

Around the same time CNOOC was making plans to buy 3.6 million tons of LNG per year from the British gas producer BG Group's Curtis LNG facility in Queensland, Australia. It is the world's first purchase agreement for the supply of LNG from coal bed methane, and marks the first sale of LNG from coal seam gas to China. The deal is worth about $40 billion based on a crude oil price of $70 per barrel. CNOOC will acquire a 5 per cent equity interest in the reserves and resources of certain BG Group holdings in the Surat basin in Queensland. CNOOC will become a 10 per cent equity investor in the first of two liquefaction trains which will form the first phase of the Curtis project. BG Group and CNOOC will also build two LNG ships in China.[249] The above two LNG supply contracts based on coalbed methane reserves confirm that Chinese NOCs are becoming quite adventurous in their search to secure long-term LNG supply.

On 19 August 2009 CNPC entered a particularly expensive LNG deal. This was with ExxonMobil for the purchase of 2.25 mt/y of LNG from Australia's Gorgon LNG project over the next 20 years. The deal was signed between PetroChina International, an international trade subsidiary of CNPC, and Mobil Australia Resources.[250]

However, *China OGP*'s criticism of CNPC's entering into such an expensive contract was unprecedentedly harsh. *China OGP* pointed out the price of LNG in the deal was about $17.52/mmbtu, and this was much more expensive than the market price. *China OGP* argued that the LNG price was almost twice that which the Japanese utility firm CEP paid to another Gorgon shareholder, Chevron, in November 2005. At that time, CEP signed a deal for 1.5 mt/y of LNG import for 25 years at a cost of 10 billion Australian dollars. The price was around $7/mmbtu, considerably below CNPC's price of $17.52/mmbtu[251] (Table 5.30).

CNPC argued that the price was reasonable because it reflected inter-national natural gas prices and the long term nature of the negotiations. CNPC attributed the massive difference from India's Petronet deal to differences of delivery location and means of transportation. *China OGP*

Table 5.30: Comparison Between the Deals by CNPC and CEP

	CNPC	*CEP*
Total Value ($bn)	41	13.65
Service Period (years)	20	25
Annual expense ($bn)	2.05	0.546
LNG volume (mt/y)	2.25	1.50
FOB/tonne ($/tonne)	911	364
Exchange rate	6.8317	6.8317
FOB/tonne (yuan/tonne)	6224.44	2486.74
VAT added (10 per cent)	622.44	
DPV price (yuan/tonne)	6846.88	
MBTU/tonne	52	52
Calorific value (calorie/cubic metre)	9435	9435
Tonne/cubic metre	1388.87	1388.87
Price in yuan/cubic metres	4.48	1.79
Price ($/mmbtu)	17.52	7.0

Source: Qiu Jun and Yan Jinguang (2009), 2.

responded that according to the NDRC, the Chinese domestic pipeline natural gas price averaged 2.45 yuan/cubic metre in July 2009, less than the duty paid value (DPV) price of CNPC's contracted 4.48 yuan/cubic metre in the deal with ExxonMobil. The DPV price usually took no account of the cost of transportation, pipeline, and regasification; with these included the total price must be more than 5 yuan/cubic metre to be profitable. *China OGP* was concerned that the expensive imported price of LNG would put upward pressure on the domestic natural gas price, and was also likely to trigger powerful expectations of a future natural gas price rise. This would, in turn, be a signal to international natural gas sellers to raise prices in future negotiations. *China OGP* concluded that CNPC's reckless purchase deal posed a threat to the international LNG market, as the high price it was paying could become a benchmark for future LNG deals, distorting the market, and leading to unnecessary price fluctuations.[252]

Not surprisingly, by the end of December 2009 PetroChina decided to pull out of a $40 billion deal to buy natural gas from a project off Australia. Australia's Woodside Petroleum informed Australia's stock exchange that an early stage agreement for the Browse basin LNG project off Western Australia had not been settled by a 31 December 2009 deadline and had now lapsed.

Conclusion

A major problem in writing about the present state of natural gas in China is the lack of accurate recent data. For example, China's latest national resource survey was undertaken between 2003 and 2006. At the time of writing, therefore, the data is more than five years old. This may soon be remedied since the fourth survey is very likely to be undertaken in early 2010s.

Even if the absence of data makes some aspects of the situation hard to depict accurately, there is no doubt that the advance of the gas industry has been exceptionally rapid. The necessity of increasing natural gas use is being recognized increasingly by the Beijing authorities. In early 2010, Wu Yin, deputy head of NEA, said that China is set to increase exploration efforts, to build gas reserves, and to increase natural gas imports to meet the goal of raising the 4 per cent share of gas in China's energy balance to 8 per cent by 2015. This is an ambitious task but currently gas demand is outpacing domestic supply and the government is determined to stimulate domestic exploration and development to reduce reliance on oil-indexed imports.[253]

Natural gas shortage became a hot topic in the winter of 2009/10. In November 2009 the first shortage struck some central and eastern Chinese cities, and the necessity of the reform of the natural gas pricing regime was strongly highlighted. At the beginning of 2010, very cold weather in north China led to a severe gas supply shortage in Beijing. On 4 January 2010 gas consumption in Beijing reached a single day record figure of 53 mcm, some 8.86 mcm more than the daily consumption peak of the winter of 2008/9. It is also about 11 mcm more than the planned average daily supply in January 2010. Han Zhongchen, a deputy manager of CNPC's pipeline and natural gas unit, said that gas from Turkmenistan's Amu Darya gas project would arrive at Beijing in the second half of January 2010[254] (in fact, the inauguration ceremony for the gas supply from Turkmenistan was held at the end of 2009). Due to the supply shortage, Beijing Municipal Administration Commission (BMAC) decided to restrict gas use by industrial users – power plants and Beijing's central heating system – to avoid putting pressure on the maximum supply capacity of 52 mcm a day. Maximum use could endanger the pipelines linking the gas field and the towngas grid. However, Beijing's residential cooking gas was not on the restriction list. BMAC said some 289 gas boilers were on the restriction list.[255]

CNPC plans to construct ten natural gas storage tanks with a working capacity of 22.4 bcm from 2011 to 2015. During the years 2006–10, CNPC began and completed the construction of the Dagang gas storage tank group, which includes six tanks and storage, and working capacities of 3 bcm and 2 bcm respectively. CNPC also finalized the construction of the Jing 58 gas storage tank group which includes three tanks. China's current overall working capacity of gas storage stands at a mere 3 per cent of natural gas sales.[256]

China OGP pointed out that:

> the shortage was partly due to the cold weather but partly also to the rigid pricing regime. Considering that gas prices in China were half of international crude oil prices, it is necessary to undertake a gas price reform. Gas imports could ease the supply shortage, but the price burden is considerable.[257]

According to Liu Zhiguang, vice general engineer of CPPEI, the price of gas from Central Asia is pegged to international crude oil prices. When the crude price reaches $80/barrel, the after-tax natural gas price in Horgos Pass will be 2.2 yuan/cubic metre. Given the average pipeline transport fee of 1.08 yuan/cubic metre, the natural gas price will be 3.28 yuan/cubic metre when arriving at the towngas grids in the receiving cities. When Central Asian gas reaches Beijing, the price will be around 2.9–3.0 yuan/cubic metre. However, gas produced in the Tarim field sets the after-tax producer price at 0.522 yuan/cubic metre, while the gas price charged by Changqing oilfield is only 0.681 yuan/cubic metre. Changqing field is the major gas supplier to Beijing through the Shaanxi–Beijing pipelines. If the pipeline tariff is added, the price for end users in Beijing is about 2.05 yuan/cubic metre. PetroChina, operator of the pipeline and buyer of Central Asian natural gas, suffers huge losses because, under the present pricing system, the import price of the natural gas is higher than the price to the end users.[258]

There are a number of options for the pricing mechanism, including weighted average pricing (WAP), pricing by sources (PBS), and pricing pegged to international crude oil prices (PTC). Domestic natural gas suppliers such as PetroChina and SINOPEC, as well as mid-stream pipeline operators, prefer to adopt the weighted average pricing method. However, it is difficult to implement this method because foreign natural gas prices usually float in accordance with international crude oil prices. However, downstream towngas operators have their own proposal, and are demanding a pricing method covering both upstream and downstream prices. Under this system, if the upstream natural gas price is changed, downstream towngas operators could change the prices for the gas reaching end users. The impact of gas price reform on the fertilizer makers is also causing concern in the market. Due to the low natural gas price, the chemical fertilizer industry has experienced a boom over the last five years, and consequently the industry is witnessing serious overcapacity. The Chinese government has moved to end irresponsible investment in the industry. To remove the natural gas price subsidy for the industry is one of the possible ways of controlling this overcapacity. According to calculations by the CPCIA, if the natural gas price increases 0.2–0.4 yuan/cubic metres, the cost of producing 1 tonne of urea will rise 120–240 yuan.[259] In 2009, almost 80 per cent of methanol producers with gas as feedstock halted production, and the remaining 20 per cent are still producing but at a very low operational rate. If natural gas prices were raised by 20 per cent, about 90 per cent of methanol producers would have to shut down. Fertilizer producers are facing the same situation.[260] Any meaningful rise of gas prices would directly influence the production of agriculture, which is the last thing the Beijing authorities want to do.

Gas price reform is a very difficult issue to handle in China, but there is a consensus that a gradual increase of the gas price is inevitable. Another very sensitive issue is the security of gas supply. The ERI has projected that China would need 200 bcm of natural gas in 2015, double the 2008 level, and the figure would reach 300 bcm by

2020.[261] In 2011 NEA began to float the figures of 260 bcm of gas demand by 2015 and 400 bcm by 2020, which represent a compound annual growth rate of 19 per cent for 2011–15 and 9 per cent for 2016–20. PetroChina envisages China's domestic gas production could, by 2020, reach as much as 210 bcm, rather than the previously projected 120–150 bcm. Considering that China's domestic production in 2010 was 95 bcm, it will be a hugely ambitious task to produce over 200 bcm by 2020. The target domestic production figure indirectly confirms the planners' intention to maximize domestic production and so reduce the gap between demand and production which has to be supplied by import. The figure of 150 bcm production by 2020 looks more realistic, and it would be safe to project that the shortage of 100–150 bcm has to be covered by the import of LNG and pipeline gas, if China's gas demand reaches 250–300 bcm. A key question is whether China can afford to import increasing volumes of gas as LNG and by pipeline, at high oil-related prices. China's domestic gas prices are not keeping up with international prices and presumably cannot do so if international prices continue to be based on those of oil.

A stern reality is that there is no shortage of gas at high prices, that is, if China is prepared to buy LNG at $100 equivalent oil prices. The fact is that China cannot afford to import this gas unless it is subsidized, as there would not be enough consumers willing to pay the high gas prices. So is China right to build its massively expensive international import infrastructure whose functioning will require increasing subsidies?

China's gas industry can expand massively if the expansion is based on domestic production, conventional and unconventional, produced at costs of delivery which Chinese consumers can afford. If it has to depend on huge volumes of imports at prices equivalent to more than $100/barrel, these can be averaged with domestic costs, but only up to a point. Without gas price reform, the volume of gas imports will be significantly affected by the financial burden, and this is the main obstacle to the country's gas expansion.

Appendix 5.1: China's Four Main Gas Production Bases

Tarim Basin

Since the early 1950s, six packaged oil and gas fields and 20 industrial oil and gas-bearing structures have been found in the Tarim basin, with total reserves of 20.5 btoe.[262] The biggest discovery so far is still the Kela-2 gas field (Table A.5.1). In 2010, CNPC announced that production from Dabei block of the Kela gas field and from Tazhong 1 gas field, located in Bazhou Qiemo County, will add 1.5 bcm/y of gas.[263] In the same year, CNPC drilled a 7,764 metres deep oil well (Keshen-7) in Tarim basin, the deepest of this kind up to now.[264]

Table A.5.1: Gas Fields in the Tarim Basin

	Area Size (sq. km)	*Proven reserves bcm*	*Production capacity /Export capacity bcm*
Kela-2 Gas	47.1	250.61	10.0/10.0
Tazhong 4 Oil	35.7	11.93	
Yaha Oil/Gas	48.9	40.54	1.2/1.2
Kekeya Oil/Gas	27.5	31.36	
Hetianhe Gas	145.0	61.69	2.0/-
Lunnan Oil	36.6	4.03	
Yingmaili Oil/Gas	48.3	30.98	1.05/1.045
Yangtake Oil/Gas	18.3	27.43	1.0/0.99
Yudong-2 Gas	10.2	7.33	0.33/0.329
Donghetang Oil	16.5	1.37	
Hade 4 Oil	66.6	0.79	
Jilake Oil/Gas	52.5	13.68	0.486/0.485
Jiefangjudong Oil	14.0	3.44	
Sangtamu Oil	18.6	1.85	
Tazhong 16 Oil	24.2	0.13	
Tazhong 6 Gas	58.0	8.53	0.25/-
Dina-2 Gas	-	175.2	5.1/
Dabei III Gas	-	130.0	
Yakela-Dalaoba Gas	-	29.30	1.0
Tahe Gas	-		10.0

Sources: Kambara and Howe (2007), 87; PetroChina, West–East Natural gas Transportation Pipeline Project, 2002/3, 2; *China Daily*, 11 October 2007 ('Major Xinjiang gasfield found', www.chinadaily.com.cn/china/2007-10/11/content_6164935.htm); 'SINOPEC Northeast Company', http://english.sinopec.com/about_sinopec/subsidiaries/oilfields/20080326/3030.shtml.

Kela-2 field: The salient facts about this field have been outlined in an article by Jia Chengzao and others in the Chinese Science Bulletin. Kela-2 is located in the centre of Kelasu structural belt in Kuqa Depression. The main component of the natural gas is methane, whose content is higher than 97 per cent. It is dry gas whose source rock is Jurassic coal measure. The Kela-2 structural trap was formed during the Xiyu period and later became a reservoir. The late formation of the reservoir and the thick seal rock of Lower Tertiary gipsmantle are the reasons why the giant Kela-2 gas field has been well-kept. The abnormally high pressure of the Kela-2 gas field results from the

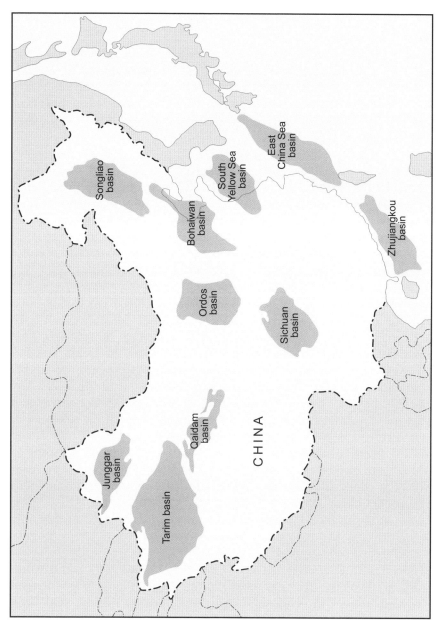

Map A.5.1: China's Gas Distribution by Basin

Source: revised from CNPC map.

strong structural compression in the northern part of the field during the Xiyu period.[265]

The State Reserves Commission conducted an evaluation of the gas reserve in the gas field in Korla (close to Kela) on 4 April 2000. It verified that it has a gas bearing area of 47.1 square kilometres and a proven reserve of 250.61 bcm (later the figure was changed to 284 bcm).[266] On 3 February 2004, PetroChina started to drill the first development well in the Kela-2 gas field. The first development well, code named Kela 2-3, was to reach 4,220 metres beneath the earth's surface. The other development wells, Kela 2-4, 2-7, and 2-8, were to follow. In December 2004, Kela-2 Gas Field was put into production, with a designed annual output of 10.7 bcm (a more recent figure than that appearing in Table A.5.1).[267] According to PetroChina, in the six years up to September 2010, Kela-2 field produced more than 50 bcm of gas. Seventeen gas wells had been put into operation, and the maximum daily gas output of a single well exceeded 5 mcm.[268]

Dina-2 field: China's largest gas condensate field, Dina-2 gas field, located at Kuqa in Tarim, is estimated to yield 5.1 bcm/y of gas, which is 40 per cent of the 12 bcm/y capacity of the first stage of the West–East gas pipeline (WEP). Dina also produces 0.3 mt/y of crude and 0.5 mt/y of other products such as LPG and lighter hydrocarbons. CNPC announced that the firm plans to supply Dina-2 gas to the WEP. The gas field has 175.2 bcm of cumulative proven geological reserves and 13.4 mt of proven condensate reserves, which makes it the second 100 bcm-level gas field in Tarim basin after Kela-2 gas field. By 2005, the WEP had signed take-or-pay contracts with 40 clients, 50 per cent of which were civilian users, 40 per cent industrial users, and 10 per cent from the chemical industry. CNPC planned to complete the build-up of production at Dina-2 by June 2009. In other words, Dina-2 was expected to produce 4.5 mt/y of oil equivalent, composed of 5 bcm/y of natural gas and 0.56 mt/y of condensate oil.[269]

Yingmaili field: According to PetroChina, on 27 April 2007 construction work at the Yingmaili gas field cluster, China's largest condensate gas field located in Xinjiang Uygur Autonomous Region, had been completed, and production had officially begun. The Yingmaili gas fields comprise three condensate gas fields, Yingmaili, Yangtake, and Yudong-2, with total proven original natural gas in place of 65.74 bcm and proven original condensate and crude oil in place of 26 mt. The construction, consisting of the investments required for internal gas gathering and transmission, and crude oil and gas treatment, along with an external gas pipeline, began in December 2005. The designed annual production capacity is 2.5 bcm of natural gas, 500 thousand tons of condensate oil, and 40 thousand tons of liquefied natural gas.[270]

Tahe, Yakela-Dalaoba fields: In the Tahe field, SINOPEC's second-largest oil field, the company has set a goal of adding 100 bcm of proven geological gas during the period 2011–15.[271] SINOPEC is producing 1.2 bcm/y in Tahe gas field and plans to increase this volume to 5 bcm in 2015 and further to 10 bcm in 2020.[272] In 2009, the field's production was 1.345 bcm. Following the addition of 135 mt of proven crude reserves in December 2008, the field has a recoverable reserve potential of 996 mtoe. For SINOPEC, the Tahe field is crucial in strengthening its upstream position so that it can diversify its operations away from the refining and petrochemicals sector. Indeed, Tahe is likely to boost SINOPEC's reserves by 5 per cent compared with the current 4 billion boe estimate, following the completion of further drilling work. On a national level the Tahe field is even more significant, given that it is one of China's few large fields that has yet to reach peak production. In 2008 it started contributing around 30 per cent of its output to CNPC's WEP. The Yakela-Dalaoba gas field is located in the Kuqa county of the Aksu region in the northern part of the Tarim basin. The fields are developed by **SINOPEC** Northwest China Company in this part of the county. Development efforts in the field started in August 2003 in order to promote the development of the county's agriculture by providing feedstock for the production of fertilizer. The discovery of the Yakela gas field in 1984 has given SINOPEC Northwest China cumulative reserves of proven natural gas condensate of 24.5 bcm in Yakela and 4.8 bcm in Dalaoba, as well as 8.86 mt of condensate oil. The field's annual handling capacity is 1 bcm.[273]

In Xinjiang province some gas discoveries have also been made in Junggar basin. CNPC has found a natural gas field with proven reserves of 100 bcm in northern Xinjiang Uygur Autonomous Region. The Kelameli field is the first with reserves of this size ever discovered around the Junggar basin, according to Chen Xinfa, general manager of Xinjiang Oilfield Company. The field is located 250 km from Karamay city. Junggar basin is said to have 2.5 tcm of gas in reserves, but less than 10 per cent have been proved. Xinjiang Oil planned to increase its annual gas production to 5 bcm by 2010. Natural gas output in Xinjiang is expected to reach 24 bcm in 2008. CNPC's natural gas production has grown by more than 20 per cent in three consecutive years.[274] In addition, a newly discovered Mahe gas field in Karamay has almost 30 bcm of natural gas, and its daily gas output is 1.47

mcm.[275] CNPC aims to invest 23 billion yuan on gas field development in Qinghai province during the period 2011–15 and plans to increase the annual gas supply to 15 bcm/y for the province.[276]

Ordos Basin

The accumulated proven natural gas geological reserves in the Ordos basin stand at 1.8625 tcm, of which 0.5883 tcm (31.6 per cent) have been developed.[277] At the beginning of 1999, CNPC Changqing Petroleum Exploration Bureau's Changqing field conducted an appraisal of Ordos basin's gas reserves. It concluded that the geological gas reserves of the field might reach 6–8 tcm, which would be 20 per cent of China's total. Up to the end of 1998, Changqing had reported cumulative proven gas reserves of 309.8 bcm and 121.1 bcm of controlled gas reserves. The field's initial goal was to find 1 tcm of proven reserves and to have a gas production capacity of 10 bcm/y by 2005.[278] Changqing field is the operator and gas supplier of the Shaanxi–Beijing natural gas pipelines, which deliver natural gas to Beijing and surrounding cities.[279] CNPC planned to build gas storage tanks with a total storage capacity of 12 bcm in its Changqing oilfield, which will become the largest gas storage unit of this type in China.[280]

In early March 2006 China further opened the doors to its large natural gas industry when PetroChina signed an exploration deal with Total. This deal, to tap the Sulige gas field in Ordos basin, was the second such agreement, following the 2005 pact between PetroChina and Royal Dutch Shell to explore for gas in Changbei natural gas field, which is in the same basin as Sulige. Both companies intended to spend 30 months and $20 million drilling in Sulige. Full production was expected to reach 400 million cubic feet a day, and the gas is intended to supply Beijing via pipeline for 20–30 years.[281] In 2004 Ordos basin produced 7.5 bcm of gas, 19 per cent of China's total gas production. Recoverable reserves from the basin amount to 290.4 bcm.[282]

Sulige-6 field: As described in *China OGP*:

> in late January 2001 PetroChina announced the discovery of a new gas field in Ih Ju League of the Inner Mongolia Autonomous Region, 700 km north of Beijing. This report from the State Oil and Gas Appraisal Office under the Ministry of Land and Mineral Resources indicated that the Sulige gas field has 220.4 bcm of proven gas reserves, of which 163.2 bcm are recoverable. In 2007 the field's proven reserves were 533.6 billion cubic meters. PetroChina's Changqing Oilfield Company believed that ultimate reserves in the field will exceed 700 bcm, making it China's single largest gas field. This confidence came after drilling Sulige-6 in the Sulige area,

which struck a high-yield gas flow at upper Palaeozoic. The peak daily gas output reached 1.2 mcm. Seven other wells sunk later all reported either high or medium level commercial gas flows. Unlike the Kela-2 gas field which is only 48.1 square kilometres, Sulige covers a gas bearing area of over 5,500 square kilometres and the area expects further expansion.[283]

In 2008, the western part of the field found another 580 bcm, making the gas field the first in China to have proven reserves exceeding one tcm.[284] By 2008 a total of 1,145 wells had been drilled and 21 gathering stations built. The target production in 2010 was 10 bcm/y. According to the development program of the Changqing Oilfield, Sulige will reach a production scale of 35 bcm/y in 2015, accounting for 70 per cent of the total gas output of Changqing.[285]

In May 2008 CNPC announced that it had invested 307 million yuan ($43.9) million in a new processing plant for natural gas in Sulige gas field. The plant, with 5 bcm of productive capacity, the biggest in East Asia at that time, went into operation on 27 June 2008. The first processing plant, which started operation in 2006, had a capacity of 3 bcm. In the summer of 2009, a new gas processing plant for Sulige gas field started full operation in Inner Mongolia, increasing the gas field's annual processing capacity to 13 bcm. The plant, the third of its kind built by CNPC in Sulige, was designed to process 5 bcm/y of gas and to become a main gas source for the Shaanxi–Beijing gas pipeline.[286] In late October 2010 CNPC announced that it was going to build the fifth natural gas processing plant in Sulige field with an annual processing capacity of 5 bcm, bringing the total annual processing capacity of Sulige field to 23 bcm.[287]

Changbei gas field: The field is located on the edge of the Maowusu Desert in the Ordos basin in Shaanxi Province and the Inner Mongolia Autonomous Region, and covers 1,588 square kilometres stretching from north Jingbian prefecture to Yulin prefecture in Shaanxi Province. On 18 May 2005 Shell signed drilling contracts and Letters of Intent for the award of Engineering, Procurement, and Construction (EPC) contracts for the development of the Changbei field. According to an article in *the Engineer*:

> Total development costs for the full lifecycle of the [Changbei] project will be about $600 million, covering the construction of the central processing facilities, inter-field pipelines and development drilling of about 50 horizontal and multilateral wells over 10 years. The contract for the drilling rigs and associated services covering the drilling of about 30 wells over the next six years was awarded to the No. 1 Drilling Company of Liaohe Petroleum Exploration Bureau. A four-year directional drilling contract was awarded

to Halliburton Energy Services (Tianjin) Ltd, while a three-year contract for drilling fluids and associated services was awarded to the Engineering Technology Institute of Changqing Petroleum Exploration Bureau.[288]

On 1 March 2007, Changbei went into commercial production. Petro-China and Shell are jointly developing the field under a production sharing contract, with Shell the field development operator.[289] Up to late September 2010, Changbei tight gas field had produced 3 bcm of gas.[290]

The estimated recoverable reserves of the field are 22 bcm of gas, with 'gas in place' of 73 bcm. The field reached its peak production capacity in 2009, producing 3.3 bcm of gas. The field is expected to last about 20 years, that is, up to 2026. It is expected to generate $5.3 bn in revenue (undiscounted) during its remaining life (calculating from 1 January 2010), to yield an IRR of around 30.4 per cent.[291]

Daniudi gas field: In early November 2005, SINOPEC completed the first phase of the exploitation its Daniudi gas field, located in the northern part of the Ordos basin with proven gas reserves of 261.5 bcm. During the first phase of the project it had a gas production capacity of 1 bcm; 248 wells are planned, of which 217 had been completed and 150 put into operation by June 2008.[292] The highest yield from a single well is 0.5 mcm. The gas transmission project includes the construction of a 3 bcm pipeline from Tabamiao in the Ordos basin to Yulin in Shaanxi province and related ground facilities.[293] In 2010, the field's production capacity was estimated at 2.3 bcm/y but is projected to reach 3 bcm/y by 2015 after adding 50 bcm of newly proven reserves.[294]

According to a SINOPEC official:

the company will have to transport gas produced from its Daniudi gas field through its own Yulin–Jinan pipeline, also known as the Shaanxi–Shandong gas pipeline ... SINOPEC has twice postponed the building of the Yulin–Jinan route because of uncertainty about the prospects for the Daniudi gas field. The company intended to build it in 2004, but dropped the plan after the central government persuaded it to build a much shorter line from Daniudi to Yulin, from where the gas is piped into PetroChina's trunk line to Beijing. The eastern section of Yulin–Jinan pipeline from Puyang in Henan to Jinan was due be in operation from November 2009. SINOPEC will source gas from its aging Zhongyuan oil and gas field to ease gas shortage in Shandong during the winter pending the planned availability of Daniudi Gas in October 2010.[295]

Sichuan Basin

An article by Sheng Li De summarizes the characteristics of Sichuan basin as follows:

[it] is the main gas area of China. The accumulated proven geological gas reserves stand at 1.5564 tcm, of which developed reserves account for 668.9 bcm (43 per cent of the total).[296]

In another article, Sheng Li De says:

It covers an area of 230,000 square kilometres. The evolution of this Meso-Cenozoic basin was influenced by both trans-Eurasia Tethys tectonism from the west and the circum-Pacific tectonism from the east. Of the 112 gas fields PetroChina and SINOPEC have explored in Sichuan–Chongqing area, 78 are in Dazhou, located in the eastern part of Sichuan [which] has natural gas reserves of 3.8 tcm.[297]

Longgang gas field: On 21 May 2006, CNPC officially opened its first exploration well in Yilong County. After drilling down over 6,500 meters it estimated daily gas production at about 1.2 mcm, and the sulphur content at only about 30 grams per cubic metre. CNPC completed the second and third exploration wells in November and December 2006, and declared afterwards that the exploration had revealed the existence of a large natural gas reserve in the area, but that its exact extent was still unknown. Encouraged by a huge reserves potential, PetroChina decided to accelerate the exploitation of the new Longgang gas field, situated on the boundary between three counties: Yilong, Yingshen, and Pingchang. According to the most conservative estimate the reserve will exceed 300 bcm but it may be as much as 500 billion, or even 1 tcm, which is equal to 1 billion tons of petroleum.

At a CNPC meeting, Xu Yongjie, general manager of Chuanqing Drilling Company, the CNPC subsidiary in charge of the Longgang field exploration, said:

the company would work hard to get exact numbers for the reserve to the public in October 2008. At present, China's largest gas reserves are at CNPC's Sulige field, with a proven 533.6 billion cubic meters, and the second largest is at Puguang gas field, discovered by Sinopec in the Sichuan Basin, with a proven 350 billion cubic meters as of February 2007.[298]

In 2007, Longgang gas field was described as the biggest natural gas discovery in China in the past five decades.[299] According to Han Xuegong, a senior consultant with CNPC, the new find, known as Longgang gas field, has two to three times the reserves of SINOPEC's Puguang field.[300]

It is safe to say that Longgang has proven reserves of between 700 and 750 bcm. The field's production could reach 4 bcm a year by 2010. According to Wood Mackenzie Consultants, 'this is easily the biggest gas discovery in South East Asia since 2007'. Wang Ying and Ying Lou said, in *Bloomberg online*:

PetroChina and Newfield Exploration Co., a US natural gas producer, signed a joint study agreement in late 2007 to develop the Weiyuan gas field in Sichuan. Ding Sheng, chief representative of Newfield's China unit, said on 17 January 2008 that the companies aim to complete the joint study by 2010.[301]

However, industry sources have suggested that PetroChina undertook the first well drilling, in 2011, without Newfield.

Puguang gas field: In early April 2006, SINOPEC announced that the company had discovered Puguang gas field, located in Dazhou city, north-west of Sichuan province. By the end of 2005, the field's proven recoverable reserves had reached 251.1 bcm, and in 2010 the field's accumulated reserves reached 405.1 bcm. SINOPEC planned to invest 40 billion yuan in the E&D of the Puguang gas field which is China's largest integrated marine gas field.[302] The company planned to produce 4 bcm/y of commercial gas in 2008 and the figure was projected to be 8 bcm/y in 2010. The purification plant in the field can process 12 bcm of natural gas annually.[303] Officials of Dazhou, in Sichuan Province, announced that a total of 3.8 tcm of natural gas deposits had been found in the western part of the Sichuan basin. The reserves at Dazhou include discovered proven exploitable reserves of 244 bcm.[304] It was reported in August 2008 that SINOPEC's important Puguang gas field had conducted a productivity test on its Pu 202-2 well, its first trial so far. In October 2008 the Energy Bureau of the NDRC said that the Puguang gas field had initiated a gas production test starting with the P302 2 well and including the nine development wells. *China OGP* reported that the whole production test would last for 13 days and yield 5.2 mcm gas, 0.78 mcm hydrogen sulphide, and 1,990 tonnes of sulphur dioxide.[305] In Xuanhan county, where Puguang field is located, the gas reserves are forecast to surpass 1.5 tcm, of which no less than 1 tcm of gas is recoverable. SINOPEC's initial plan indicated that the pipeline will travel through Sichuan, Chongqing, Hubei, and Henan terminating at Shandong. However, PetroChina was doubtful about the development of the gas field and the feasibility of the pipeline, due to uncertainty about both the reserves and the economic feasibility of the pipeline. SINOPEC changed its mind about supplying Puguang gas to Shandong province. Instead it aimed at supplying gas from Daniudi field in Inner Mongolia to Yulin city in Shaanxi province first, and then possibly extending to Shandong province.[306] After the Anping–Jinan gas pipeline came into operation in May 2006, SINOPEC's gas supply in Shandong province (which had been 0.325 bcm in 2005) was set to reach more than 0.6 bcm in 2006. In Shandong province, there are four other operating gas pipelines, namely, the Puguang (Henan)–Qingdao (Shandong) pipeline, the Dongping–Jinan (Shandong) pipeline, the Zibo–Laiwu (Shandong) pipeline, and the Jiaozhou–Laizhou (Shandong) pipeline. Together with the Anping–Jinan pipeline, this adds up to a total pipeline length in the Shandong province of 1,400 km.[307]

In fact the NDRC has vetoed SINOPEC's plan to build a gas pipeline from Puguang gas field to Jinan in Shandong. The NDRC directed SINOPEC to send the gas to Shanghai to ensure gas supply in the Yangtze River Delta area.[308] The Sichuan–Shanghai or Sichuan-to-East pipeline covers a distance of 1,702 km and runs from Puguang gas field in Dazhou (Sichuan Province), to the Qingpu District of Shanghai. An 842 km branch line will connect Yichang in Hubei to Puyang in Henan Province. The construction of the first 1,360 km section from Yichang in Hubei Province to Shanghai started on 22 May 2007 and it was completed by March 2010.[309] The investment was estimated to be 62.7 billion yuan ($9.44 billion).[310] Zhang Shaoping, director of the SINOPEC Jianghan Petroleum Administration Bureau, said SINOPEC would build a natural gas storage tank in Qianjiang county in Wuhan with a storage capacity of 300 mcm, and another tank (with unrevealed capacity) is to be built in eastern Jiangsu province.[311] In 2010 SINOPEC made a discovery of 875 bcm geological reserves in Yuanba block in Sichuan province, and a SINOPEC executive indicated that Yuanba gas field may be similar in size to the adjacent Puguang gas field. SINOPEC plans to build 6 bcm/y of gas production capacity in the field by the end of 2015.[312]

Chuandongbei gas field: In December 2007, CNPC signed a cooperative contract with the US firm Chevron for the joint exploitation of natural gas from Chuandongbei gas field, well-known for the high sulphur content of its natural gas. (In 2003 the hydrogen sulphide blowout of Luojiazhai gas field (see below) killed 243 local people.) The Chuandongbei field was scheduled to be put into operation in 2009 (later pushed back to 2011) and aimed at producing 2 bcm/y of gas by 2010 and 6 bcm/y by 2015.[313] In November 2009, Chevron and CNPC were given approval by the Chinese government to proceed with the development of the Chuandongbei (CDB) Natural Gas Project, which includes the Tieshanpo, Dukouhe, Qilibei, Gunziping, and Luojiazhai gas fields. The NDRC approved the first stage of the development plans for the $4.7 billion project.[314] Chevron has started building a gas field with a 3 mcm/d capacity in Xuanhan county, as part of the first development phase covering the block's Luojiazhai and Guanziping prospects. The $4.7 billion first phase also involves drilling 14 production wells in four well pads, two gas gathering stations, 11 valve boxes, and about 60 km of pipelines. The Chuandongbei block covers 1969 square kilometres and holds 176 bcm of proven gas reserves, including about 60 bcm in the Luojiazhai field.[315]

Luojiazhai gas field: According to *Chengdu Time*:

> on 14 August 2007 US Chevron has won a bid to cooperate with CNPC in developing natural gas in Sichuan basin. The cooperation would cover the Luojiazhai gas field that has high sulphur content, ranging from 7.13 per cent to 10.49 per cent. The field has a gas reserve of 58.11 bcm. CNPC expanded the area covered by the cooperation to four blocks in total in 2007. Chevron has won the bid over rivals such as France's Total SA, Royal Dutch Shell and Statoil ASA of Norway. Luojiazhai became CNPC's third natural gas development project with foreign partners.[316]

Besides Luojiazhai, the two companies are also developing the Tieshanpo and Duhekou gas fields, which have 73.3 bcm proven reserves with a sulphur content of 14.19 and 15.27 per cent respectively.[317]

According to *China OGP*, during the years 2006–10 PetroChina planned to build three natural gas purification plants in Dazhou, in the south-west of Sichuan. With a total investment of 16 billion yuan, the three plants were able to purify 24 mcm/d of gas with high sulphur content. Xuanhuan Luojiazhai purification plant, with a capacity of 9 mcm/d, was expected to be completed by the end of 2006. The Wangyuan Tieshanpo plant was due to be the second to come on stream in June 2008 with a capacity of 6 mcm/d. The third, the Xuanhuan Dukouhe plant with a capacity of 9 mcm/d, was due to complete its construction by the end of 2009.[318] In 2010, Chevron began construction of the project's first natural gas purification plant and started development of the Luojiazhai and Gunziping gas fields.

Offshore Gas

Yacheng 13-1 gas field: Yacheng 13-1 is the largest gas field in China's offshore region, with proven initial in-place gas and condensate reserves of around 98.2 bcm and 3.74 mcm, respectively. The field is located about 100 km south of Hainan Island in the South China Sea and is connected to the mainland by two major pipelines. The larger of these is a submarine line delivering 3 bcm/y of gas to Hong Kong (800 km away) where since 1994 it has been used for thermal power generation; the smaller delivers 0.5 bcm/y to Sanya City on Hainan Island for use as feedstock for chemical fertilizer plants.[319]

Chunxiao gas field: This field, located in the East China Sea 350 km from Ningbo, covers an area of 22,000 square kilometres and has a total of 70 bcm of proven natural gas reserves. On 7 August 2006, CNOOC put its Chunxiao gas field at the East China Sea into trial production. It can produce 2.5 bcm of gas. The field is composed of four gas wells and aims to supply the offshore gas to Shanghai and Zhejiang. To this end, offshore drilling platforms, underwater pipelines and receiving stations, gas-fuelled power plants, and towngas grids were built. China had spent 10 years in preparatory work on the gas field before 19 August 2003 when CNOOC, together with SINOPEC, signed three exploration and two development contracts with Shell and Unocal covering the Sag basin in the East China Sea.[320] The western parties later withdrew from the project due to the boundary dispute between China and Japan, and CNOOC decided to pursue the development alone. China and Japan have carried out several rounds of talks on resolving the disputes in the East China Sea since 2004. Although they reached a consensus to push for an early settlement of the issue while Chinese President Hu Jintao was on a state visit to Japan in May 2008, the dispute continues.[321]

Ledong 22-1 field: On 7 September 2009 CNOOC's website announced that:

> Ledong (LD) 22-1, an independent gas field of the Company, has success-fully commenced production. As of 2009, it is producing approximately 30 thousand cubic feet of natural gas per day via 5 wells. About 47 kilometres east of the producing field Yacheng 13-1 and 20 kilometres western of LD 15-1 gas field, LD 22-1 is located in the Yinggehai Basin of the Western South China Sea. The average water depth is about 93.5 meters. LD 22-1 is jointly developed with LD 15-1. After being further processed at the Dongfang 1-1 gas terminal, natural gas from LD 22-1/15-1 will be piped to customers in Hainan province including refinery plants, chemical plants and city gas. Peak production of LD 22-1/15-1 is expected to be around 150 MMcf per day. LD 22-1 together with LD 15-1 started its production in September 2009. When in full-scale production, LD 22-1/15-1 will become the second biggest independent gas field of the Company offshore China.[322]

Based on these gas fields, CNPC envisages that China's domestic gas production could reach 95 bcm in 2010, 210 bcm in 2020, and 300 bcm in 2030. CNPC argues that China would form a natural gas sup-ply pattern dominated by domestic natural gas.[323] In consideration of the limited proven gas reserves scattered around the country, however, developing a 200 bcm domestic production base in China by 2020 will be an extraordinary achievement.

Appendix 5.2.: Domestic LNG Facilities

China has been building small-scale LNG facilities since the 1990s. By the end of 2008 domestic liquefaction capacity reached about 4 mcm per day. The first installation for commercial use, the Zhongyuan LNG Plant, was completed in November 2001, but a year earlier the first peak-shaving demand LNG facility, the Shanghai Pudong LNG Plant, had been completed. The largest onshore liquefaction plant in China was installed in Xinjiang by Guanghui Group, with liquefaction capacity at 1.5 mcm per day. This LNG is usually transported to the south-eastern regions by railway or highway. Though the transportation cost is relatively high, the coastal regions need the LNG for peak-demand.[324] According to Guanghui Group, in China there are 22 mini LNG plants in operation, and 20 mini LNG projects built or under construction. Their total capacity in 2010 was 2.5 mt. The figure in 2015 and in 2020 will be 4.3–5.7 mt/y and 9.0 mt/y respectively.[325]

Table A.5.2: China's Domestic LNG Facilities

	Operation Time	Daily Supply (mcm)
Puyang LNG Plant of Zhongyuan oil field, Henan province	Nov 2001	0.15
Guanghui LNG Plant in Xinjiang province	Sep 2004	1.50
Fushan gas field LNG Plant in Hainan	April 2005	0.30
Weizhoudao LNG Plant of Xinao Group in Guangxi province	March 2006	0.15
Suzhou LNG Plant in Jiangsu province	Nov 2007	0.09
Shenran LNG Plant in Taian of Shandong province	March 2008	0.15
Guanghua LNG Plant in Jincheng of Shanxi province	Oct 2008	0.30
Ordos LNG Plant in Inner Mongolia	Dec 2008	1.00
CNOOC Hengqindao LNG Plant in Zhuhai, Hainan province	Dec 2008	0.60

Source: China Securities Journal (2010), 34.

In August 2010 it was reported that PetroChina planned to invest 300 million yuan to build an LNG supplying project in Lhasa. The scheme involves building gas-receiving facilities in the city's economic development zone, a 38 km urban gas grid, and gas filling stations. PetroChina reached agreement with the Tibetan authorities on building the LNG facilities in late 2009 as part of a makeshift plan to supply gas to Tibet before pipeline gas arrives in 2015. The LNG will be trucked to Tibet from Qinghai gas field in the Qaidam basin, via the Qinghai–Tibet highway or by a parallel railway. On 18 October 2011 it was reported that operations at the first phase of PetroChina's Qinghai liquefied natural gas (LNG) processing plant had begun. It was expected that the gas would be priced at 4 yuan ($0.60) per cubic metre.[326]

The newly established Huixin LNG plant in Dazhou, in Sichuan province, has already started selling its LNG. In July 2010, the plant produced 0.5 mcm/d of LNG (with feedstock from SINOPEC's Pu-guang gas field) which included 0.1 mcm/d for supplying SINOPEC and the remaining 0.4 mcm/d for other buyers in Yunnan province. When it reaches its full operational capacity, the plant will produce 1 mcm/d of LNG. The selling price of the LNG stands at 2.7–2.8 yuan/cubic metres, or about 4,000–4,150 yuan/metric tonne.[327] In August 2010, SINOPEC's first LNG service station was put into operation in Guiyang, in Guizhou province. The LNG service station has an annual gas filling capacity of 7.665 mcm, and it can provide an LNG filling service for 200 LNG city buses every day.[328]

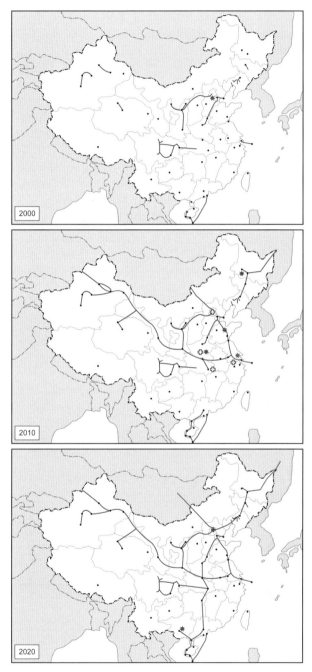

Map A.5.2: China's Gas Pipeline Network Development: 2000, 2010, and 2020
Source: CNPC.

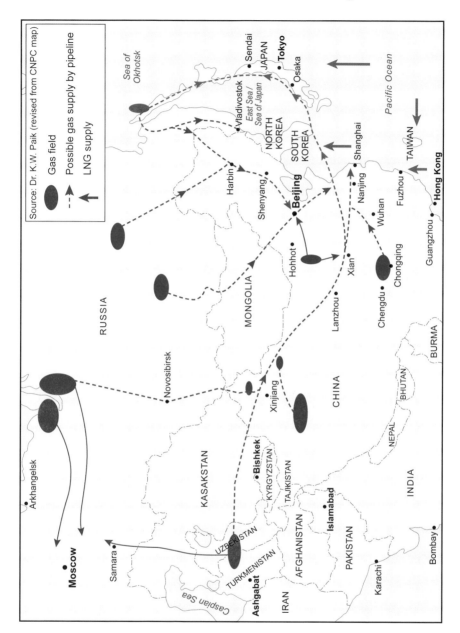

Map A.5.3: China's Gas Supply Envisaged during the Second Half of the 1990s

Source: Compiled by author

CHAPTER 6

CHINESE OIL AND GAS INVESTMENT IN CENTRAL ASIA

As China's own oil and gas industries became incapable of expanding their output to match the country's needs, important policy innovations, aimed at extending the sources of oil and gas, have demonstrated that China has more alternatives than were first believed. A question involving the relations of two powers (China and Russia) has been converted into a more complex one involving also the Central Asian Republics, which have become important energy suppliers to China. In addition, the export of Chinese capital has begun to play an important role in promoting energy supply – from Russia, Central Asia, and much further afield.

China's 'Going-Out' Policy

In response to the state's call to utilize 'two resources and two markets' (oil and gas: domestic and foreign), China's upstream monopoly, CNPC, has been engaged in international exploration and development since 1993, the year in which China became an oil importer. Fighting hard in recent years against the declining production of its own eastern oilfield, CNPC has adopted international E&D as a corporate strategy. Since 1991, CNPC and its subsidiaries have been examining petroleum E&D offerings in Canada, Peru, Venezuela, Indonesia, Thailand, Russia, Mongolia, USA, India, and Pakistan. In the spring of 1992, CNPC Canada paid 6.64 million Canadian Dollars to acquire a UTF research project with recoverable reserves of 22.84 mcm of asphalt. This was CNPC's first equity holding in a foreign oilfield. In October 1992, the first overseas barrel of crude was produced by CNPC Canada in the North Twinning Oil Field. Six CNPC subsidiaries were involved in this trial stage of foreign operations – CNODC, CNPC International, CNPC Canada, CNPC Latin America, MC & CNPC Oil (CNPC Central Asia), and CNPC Asia–Pacific.[1]

In 1997 CNPC entered into several important international deals. On the same day (4 June) that CNPC president Zhou Yongkang signed a production sharing agreement (PSA) with the Iraqi oil minister Rashid (to develop Al-Ahdab oilfield), CNPC vice president Wu Yaowen agreed to acquire a majority stake in Kazakhstan's third largest oil producer Aktyubinskmunai. This company held around 130 mt of oil, and in 1997 produced about 2.5 mt; in the expectation that output from this venture would reach 5 mt per year by 2000, CNPC agreed to invest $4.3 billion in Aktyubinskmunai over the next 20 years. Of this figure, $585 million was to be invested between 1998 and 2003, in return for a 60 per cent interest in the company. CNPC also planned to build a 3,000 km pipeline running from Tengiz oil field across Kazakhstan to the Chinese border, at a cost of $3.5 billion. The pipeline was to be constructed in three stages lasting six to eight years in total. The first leg, from the Kazakh city Chimkent to western China, would take about three years and the next two, linking Chimkent to Uzen by way of the Kazakh city of Aktyubinsk, a further four years.[2]

On 15 September 1997 CNPC's first shipload of crude oil from foreign exploration arrived at the north China port of Qinhuangdao. The 60,000 tonnes of oil, carried by the Chinese tanker Liu He, was purchased by CNPC in the south-east Asian market with the capital gained from selling its equity oil in Peru's Talara oilfield to the Peruvian state oil company PetroPeru. By 1997, CNPC was operating in nine countries – Peru, Sudan, Venezuela, Canada, Thailand, Kuwait, Iraq, Turkmenistan, and Kazakhstan. In the mid-1990s Huang Yan, vice president of CNPC, expected that by 2000 CNPC would have secured 12 mt of crude production capacity in its foreign operations, of which 5 mt was to be its equity oil.[3] It is hard to say if this expectation was fulfilled. It is better interpreted as setting the tone for the scale of the operation, rather than as a precise empirical prediction.

Between 1998 and 2000 there was, however, a sharp decline in foreign investment flows which can be explained by a combination of events:[4]

- the restructuring of the NOCs which the Chinese government carried out in 1998;
- the Asian financial crisis that spilled over to Russia and Latin America – in fact, when CNPC decided to pull out of a $400 million deal for developing the Uzen oil field in Kazakhstan, it blamed financial uncertainty in Asia;
- and overspending on the Kazakhstan and Venezuelan deals during the 1997–8 period, which forced Chinese NOCs to rethink their foreign investment strategies.

Since 2000, investments in Russia and Central Asia (RCA) and Africa have again been steadily increasing. In 2002, the target areas for China's 'Going-Out' strategy were Central Asia, Russia, the Middle East, and

Africa. Central Asia and Russia were receiving particular attention, because those areas could not only shorten supply lines and diversify China's current import configuration, but could also be of positive geopolitical significance.[5] To implement the Going-Out policy, the state oil companies were using the following vehicles: CNPC was using China National Oil & Gas Exploration & Development Corporation; PetroChina's international aliases were PetroChina International Ltd and PetroChina International Co.; SINOPEC was using the SINOPEC International Petroleum Exploration and Production Corporation; CNOOC was using CNOOC International Ltd; Sinochem incorporated Sinochem Petroleum Exploration & Production Co. Ltd in 2002 and started to focus on the upstream business. (Sinochem's overseas upstream activities began to show some results from 2007–8 onwards.)[6]

What was new in the new round of Going-Out was that China was warming to the idea that it could maximize its benefits by coordinating the activities of the government, oil companies, diplomatic channels, and the trading sector. As there was no Ministry of Energy, no Chinese government department was in a position to dominate the Going-Out strategy. The SDPC (now NDRC), SETC (now SASAC), and MLNR took responsibility for guiding and supervising the petroleum industry;[7] these were later joined by NEA and MOCOM and these top five institutions played the dominant role in supervising the petroleum industry.

Research carried out at Chatham House shows that during the period 1995–2006 investments were very concentrated in a limited number of countries. For example, Sudan, Nigeria, and Angola were the host nations for about 94 per cent of total NOC investments in Africa; and Russia and Kazakhstan accounted for 83 per cent of all investments in the RCA region.[8] Even though CNPC's total investments in the RCA region were worth over $9.1 billion, the majority of that was in the Central Asian Republics, Kazakhstan in particular, due to Russia's reluctance to open its upstream sector.

Chinese NOCs' M&A deals: 2008–10

During the second half of the 2000s Chinese NOCs' overseas expansion activities intensified. In 2008 and 2009 alone, Chinese cross border oil and gas acquisitions reached $28.0 bn. The country took advantage of the economic downturn and lower asset values to step up its global acquisitions and to finance more projects in the upstream, midstream, and downstream sectors. One of the current financing strategies is to use bilateral loan-for-oil deals with several countries as a means of securing long-term deals. According to industry sources, these loans add up to about $50 billion, or 70 per cent of the total value of investments by the three major NOCs since 2008. These loans were particularly effective at the time (the credit crunch of 2008–9) when several resource-rich countries were short of cash.

Table 6.1: Major Chinese Cross-border Oil and Gas Acquisitions, 2005–10

Date	Target Asset	Investor	Deal Value (US$bn)
June 2009	Addax Petroleum (Switzerland: 100%)	SINOPEC	7.24
Oct 2010	Repsol's Brazilian subsidiary (Spain: 40%)	SINOPEC	7.1
Nov 2010	Pan American Energy (Argentina: 60%)	CNOOC	7.06
April 2010	Syncrude (Canadian: 9.03%)	SINOPEC	4.7
Aug 2005	PetroKazakhstan (Kazakhstan)	CNPC	4.2
June 2006	Udmurtneft (Russia, 99.49%): 51% resold to Rosneft	SINOPEC	3.7
May 2010	Statoil's Peregrino (Brazil: 40%)	SINOCHEM	3.1
March 2010	Bridas Corp (Argentina: 50%)	CNOOC	3.1
March 2010	Arrow Energy (Australia: 100%), jointly with Shell	CNPC	3.1
April 2009	MangistauMunaiGaz (Kazakhstan: 100%): 50% resold to KMG	CNPC	2.6
July 2008	Awilco (Norway: 100%)	CNOOC	2.5
July 2008	Awilco Offshore (Norway)	CNOOC	1.0
Dec 2010	Occidental's Argentina unit (100%)	SINOPEC	2.45
May 2009	Singapore Petroleum (Singapore: 96%)	CNPC	2.4
Jan 2006	OML 130 Akpo (Nigeria), 45%	CNOOC	2.3
Nov 2010	Chesapeake (US: 33.3%)	CNOOC	2.2
Oct 2004	Shell's Angola Block 18 (50%)	CNPC	2.0
Oct 2006	Nations Petroleum (Kazakhstan: 100%)	CITIC	1.9
Sep 2009	Athabasca Oil Sands (Canada: 60%)	CNPC	1.9
Sep 2008	Tanganyika Oil (Syria: 100%)	SINOPEC	1.8
Sep 2006	EnCana's Ecuador asset 100%	CNPC & SINOPEC	1.4
July 2009	Angola Block 32 (Angola: 20%)	CNOOC & SINOPEC	1.3

Source: Crooks (2009d); Kong, B. (2010) 170–82; Jiang and Sinton (2011).

In 2008 and 2009, a number of the Chinese NOCs broadened the type of M&A deals which interested them: they no longer sought deals involving exclusively upstream assets, but began to show interest in the oil service and refining sectors as well. The first evidence of this was COSL's Awilco deal. In July 2008 COSL, a subsidiary of CNOOC, proposed a buyout of the Norwegian oil services company Awilco. The $2.5 billion deal would give COSL access to drilling technology and enable it to expand its international operations.[9] A second example of investment broadening was PetroChina's first overseas downstream asset buyout, summarized as follows by the *Financial Times*:

> PetroChina has agreed to pay $1 bn for 45.5 per cent of the Singapore Petroleum Company (SPC) in what will be the first major Chinese offshore purchase of a downstream energy company. This will be PetroChina's first cross-border acquisition of a public company, the first Chinese takeover of a publicly listed company in Asia and the largest public takeover in Singapore since 2001.[10]

Another characteristic of Chinese NOCs' Going-Out strategy in recent years is that the NOCs are beginning to join forces when bidding for overseas energy assets, a tactic aimed at cutting costs while maintaining China's strategy of stockpiling overseas assets. This tactic is seen in the acquisition of a 20 per cent stake in Angola's oil block 32, in which CNOOC International Ltd and SINOPEC International Petroleum Exploration & Production Corporation established a joint venture for the bid, worth $1.3 bn.[11] The increasing coordination of the big three reflects the country's haste to exchange its US dollar reserves for real overseas assets, a move designed to protect the aggregate value of its assets.[12]

The biggest prize of Chinese NOCs' Going-Out Strategy so far is CNPC's success in securing a strong position in both the super-giant Rumaila and Halfaya oil fields in the 2009 summer and 2009 winter bidding rounds. At the end of June 2009 CNPC and BP won a 20 year service contract to develop Rumaila, which is located in Kuwait and southern Iraq and is estimated to contain some 15 per cent of Iraq's oil reserves. The field, which has 17 billion barrels of reserves was, as of June 2010, producing 0.96 mb/d – making up 40 per cent of Iraq's oil production of 2.4 mb/d. The consortium, led by BP (38 per cent) with partners CNPC (37 per cent) and the Iraq government's representative SOMO (25 per cent), has undertaken to nearly triple the Rumaila field's output to 2.85 mb/d, which would make it the world's second largest producing oilfield.[13] In December 2009, the super-giant Halfaya oil field, with 4.1 bn barrels of proven reserves, was allocated to CNPC's consortium with Malaysia's Petronas and France's Total. After Rumaila, this was the second large field in which CNPC won development rights.[14] Clearly these two major oil deals were a major step forward for CNPC. SINOPEC, however, was barred from the Iraqi bidding round due to its involvement in the M&A deal for Addax Petroleum in June 2009.[15]

In 2009 alone, Chinese state-owned companies, including NOCs, spent a record $32 billion on energy and mining acquisitions to meet rising demand for resources in the world's fastest-growing major economy. Even

Sinochem Group stepped up its purchase of oil assets. In October 2009, Sinochem completed the 100 per cent acquisition of Emerald Energy[16] and in May 2010 the firm agreed to pay US$3.1 bn to Statoil ASA for 40 per cent of the Brazilian offshore Peregrino field.[17] All this activity is evidence that Chinese NOCs' foreign oil and gas expansion is very likely to continue. China's rapidly growing financing capacity puts the country in a very good position to handle this expansion. When it comes to M&A deals in Russia, however, only a few selected and preliminary coordinated deals are likely to be implemented in the near future (the details of the M&A deals in Russia will be discussed in the Chapter 7).

Table 6.2: Value of Chinese Overseas Petroleum Investment, 1992–2009 ($bn)

	Total	*Africa*	*MENA*	*RCA*	*Asia*	*SNA*
CNPC	25.257	7.599	5.555	9.182	0.810	2.111
SINOPEC	12.559	3.115	4.464	4.220	0.210	0.550
CNOOC	3.399	2.289	0.0	0.0	0.988	0.122
SINOCHEM	0.779	0.0	0.679	0.0	0.0	0.100
Zhenhua Oil	0.095	0.0	0.065	0.0	0.030	0.0
CITIC	1.997	0.0	0.0	0.0	1.997	0.0
Total	**44.086**	**13.003**	**10.763**	**13.402**	**4.035**	**2.883**

Note: MENA = Middle East and North Africa; RCA = Russian and Central Asia; and SNA = South and North America.

Source: Kong, B. (2010), 66.

The focus of Chinese NOCs' M&A moved to Latin America during the 2009–10 period. As shown in Table 6.2, the total investment in South and North America during the years 1992–2009 was a mere $2.9 billion. According to data compiled by Bloomberg, the $2.45 bn deal in December 2010 to purchase Occidental's Argentina unit took the annual Chinese bids for overseas energy assets to a record $38.8 billion in 2010. Prior to this deal, companies including SINOPEC and CNOOC had invested more than $13 billion in the South American oil industry in 2010. In November, CNOOC bought a stake in Argentina's Pan American Energy LLC for $7.06 billion. In October, SINOPEC agreed to pay $7.1 billion for a 40 per cent stake in Repsol YPF SA's Brazilian unit. That was China's largest overseas oil deal since SINOPEC bought Addax Petroleum for 8.3 bn Canadian Dollars (US$7.24 bn) in 2009.[18] The scale of Chinese NOCs' investment in Latin America reflects not only economic priorities but also political ones. But, despite the shift of priority from Africa to Latin America, what has not changed is Beijing's special focus on oil and gas supply from Central Asian Republics.

Central Asian Republics' Oil and Gas Supply Options

Crude Oil Supply Through the Sino-Kazakh Oil Pipeline

A close observer has summarized the advent of the China–Kazakhstan relationship as follows:

> At the end of 1991, China and Kazakhstan signed a five year inter-governmental economic and trade agreement that reduced tariffs on imports and exports and proposed 80 projects for greater cooperation. It was clear by 1994 that Sino-Kazakhstan trade has strong complementarities. In Xinjiang production exceeded demand. In 1991, gasoline production reached 32,800 b/d but consumption was only 18,300 b/d; diesel production was 31,000 b/d while consumption was 20,400 b/d. Beijing viewed Kazakhstan as a market for Xinjiang's petroleum products. These Xinjiang surpluses were needed in Kazakhstan, where demand for gasoline exceeded supply by 14,200 b/d, and by 27,000 b/d for diesel fuel. The deficits must be covered by imports which have traditionally come from Russia, an often unreliable source. Xinjiang's surplus petroleum products must be shipped by rail to the Lanzhou refinery, 1,000 miles away.[19]

In the spring of 1994, CNPC Vice President Zhang Yongyi, as part of Prime Minister Li Peng's business delegation to Kazakhstan, proposed cooperation in both the exploration and development of Kazakhstan's oil resources. In 1997, China made a major investment decision to purchase oil-producing assets in Kazakhstan. Dushanzi Refinery in Xinjiang became CNPC's first refinery to run foreign crude oil, with the arrival of the first batch of 1,700 tonnes of equity oil from Kazakhstan on 21 October 1997. The oil was CNPC's first equity oil production from the Kazakh Aktyubinsk Oilfield, in which CNPC has a 60 per cent stake. Another 150,000–200,000 tonnes of equity crude oil was expected to arrive from the field before the end of 2007. Under the contract signed in June 1997, CNPC promised to raise the Kazakh oilfield's annual production to 5 mt within five years (from 2.6 mt in 2007). In September 1997 China signed a contract with Kazakhstan to build two separate pipelines, to move crude oil to China and Iran. According to the September contract, CNPC also committed itself to develop Uzen field, which is located 50 km from the Caspian Sea and was estimated to be capable of producing more than 8 mt of oil annually between 1997 and 2002 period.[20]

In early June 1998, a mercaptan removing unit, installed by the Dushanzi Petrochemical Corporation in Xinjiang Uygur Autonomous Region, was projected to go into operation (Kazakh oil contains a high proportion of mercaptan). Capitalized at $210,000, the unit would enable Dushanzi, a subsidiary of CNPC, to blend 30 per cent of crude oil imported from Kazakhstan with 70 per cent of oil from the Tarim basin. Dushanzi started to process Kazakh crude oil from 1997, with an initial volume of 70,000 tonnes. The volume was expected to reach 500,000 tonnes in 1998. CNPC's equity oil from Kazakhstan was shipped to Dushanzi by rail. The oil products produced by Dushanzi were transported to other parts of Xinjiang, Shaanxi, and Sichuan provinces and to the Inner Mongolia Autonomous region.[21]

The Kazakhstan government announced that in September 1998 China and Kazakhstan would start to construct a transnational oil pipeline extending from Uzen oilfield to Xinjiang province. The 3,700 km pipeline was designed to transport 20 mt of crude per year from Uzen, the second largest oilfield in Kazakhstan. During the late 1990s Shell began to explore the possibility of parallel oil and gas pipelines from the Caspian Sea to Xinjiang province in a project named Lunar, which was never implemented. Like Shell, CNPC had no confidence in building massive, long-distance pipelines.

In the summer of 1999, the Chinese government decided to shelve the 3,277 km transnational oil pipeline between Kazakhstan and Xinjiang Uygur Autonomous Region, the three reasons being: low predicted economic returns, inadequate crude resources in Kazakhstan, and insufficient supporting infrastructure in Xinjiang province. In fact, CNPC's detailed feasibility study found that the crude supply from the Kazakh side would be no more than 7.6 mt per year (from Aktyubinsk and Uzen), far below the designed pipeline capacity of 25 mt. To make the pipeline economically viable, the crude supply to China would have to be priced at $5 per barrel – lower than crude purchased in the Middle East, something which was virtually impossible. Besides this, continued low oil prices in the late 1990s had a negative effect. Even though the investment for the pipeline was pegged at $2.4 billion, the Kazakh side would make no financial contribution whatsoever. In these circumstances, CNPC as the constructor and operator of the pipeline, told its Kazakh partner that the chances of the pipeline being constructed in the near future were very slim.[22] Significantly, in March 2001 when Kazakhstan's President Nursultan Nazarbayev promised the USA that Kazakhstan would reroute its oil to the BTC pipeline after its completion, the Chinese–Kazakh pipeline project was entirely put on hold, since the Kazakh government was unwilling to guarantee to the Chinese side the annual minimum volume of 20 mt of crude oil for the pipeline.[23]

The initiative for staged pipeline development: Kenkiyak–Atyrau section

Instead of constructing a major and long-distance crude oil pipeline between Kazakhstan and China at a stroke, China decided to pursue the Sino-Kazakh oil pipeline as a multi-phased development. In 2002, construction started on the Kenkiyak–Atyrau crude pipeline, the first Chinese pipeline project in Kazakhstan. The 448.8 km pipeline starts from Aktyubinsk oilfield and ends at the port city of Atyrau, from where the crude oil heads for Europe. CNPC's pipeline bureau sub-contracted one third of the construction from the Russian contractor of the pipeline. In the initial stage, the handling capacity of the pipeline was 6 mt/y, and with second phase completion, the pipeline capacity was designed to handle a 10 mt/y delivery. The Kazakh crude oil would be processed in Lanzhou refinery, after entering China via Alashan in Xinjiang.[24] In March 2003, the first section started operation. Thus a China–Kazakhstan pipeline, first proposed in 1997, but held up for six years until more oil reserves were found, was in operation at last.[25] A month later, SINOPEC's Changzhou refinery in Hubei province began refining Kazakh crude oil.[26]

Important M&A activities were carried out in Kazakhstan in 2003 by both CNPC and SINOPEC. CNPC expanded its asset base in Kazakhstan significantly. In May 2003 it won a tender for Kazakhstan's 25.12 per cent interest in the CNPC–AktobeMunaiGaz joint venture. CNPC undertook to invest US$4 billion in the project over 20 years, US$540 million of this in the first five years. By the end of 2002, CNPC had met its investment obligations. The total in-place reserves at the Zhanazhol and Kenkiyak fields were estimated at 570 mt, including about 140 mt of recoverable reserves. At the beginning of 2004, residual recoverable reserves at Zhanazhol stood at more than 78 mt of oil. At Kenkiyak, there were approximately 72 mt (above salt) and 28.3 mt (sub-salt) reserves of crude oil. In July 2004, CNPC announced the discovery of a rich oil source, which was producing 1,110 t/d of crude from the sub-salt field.[27]

Both CNPC's assets and SINOPEC's new license areas were located in the western part of Kazakhstan, close to the Caspian Sea. The transportation of the produced crude required a long-distance pipeline development. Therefore, Yakovleva explains:

> ... Chinese oil can enter the Atasu–Alashankou pipeline only after the Kenkiyak–Aralsk–Kumkol leg has been built and integrated into the Western Kazakhstan–China transnational trunk line. The Kumkol fields were the closest source of hydrocarbons in Kazakhstan to China. PetroKazakhstan began supplying oil to the Chinese market by rail in mid-2001. The

Canadians exported approximately 0.15 mt of oil to China in 2002, and the figure rose to 0.483 mt in 2003 ... Lukoil, which has a 50 per cent stake in the Turgai Petroleum joint venture, was using the Caspian Pipeline Consortium (CPC) pipeline to export oil produced at Kumkol. The company considered the Atasu–Alashankou project attractive and did not exclude the possibility that a portion of its oil would be redirected eastward when the pipeline is built. However, oil from Kumkol alone would not be enough to fill the pipeline. PetroKazakhstan and Turgai Petroleum's output in 2004 was projected to be 5.34 mt and 3.34 mt of oil respectively.[28]

To justify the 3,000 km crude oil pipeline development, CNPC had to secure at least 20 mt/y of crude oil supply, but it was no easy task to find production assets to fill the large-scale pipeline to China. Until Chinese NOCs managed to secure the 20 mt/y of crude supply within Kazakhstan, their attempts to persuade Russian oil companies to allocate or divert western Siberian crude to Kazakhstan, using the existing pipeline network between Kazakhstan and Russia, provided interim coverage of crude oil supply for the pipeline connecting west Kazakhstan and China's Xinjiang province. The Chinese, Yakovleva continues,

> hoped to spur Russian interest in the route with a swap offer. CNPC could supply oil from the Zhanazhol and Kenkiyak fields in western Kazakhstan to Atyrau and then to Samara, while the Russian companies could supply the same amount of oil via the Omsk–Pavlodar–Shymkent pipeline for shipment east to China. Beijing will need Russian oil before completing the connecting pipeline from Kenkiyak to Kumkol to bring its own oil from western Kazakhstan to western China. In 2003, CNPC's annual oil production in Kazakhstan was approximately 6 mt of which Aktyubinsk is the source of 5.5 mt and about 0.5 mt are from Buzachi through the Nelson Buzachi Petroleum B.V. joint venture.[29]

There was even some interest in a partnership between SINOPEC and CNOOC in the effort to secure a reliable crude supply from the western part of Kazakhstan.

> In March 2003 [Yakovleva writes] both SINOPEC International Petroleum and Production Corp, a subsidiary of SINOPEC group, and CNOOC reached an agreement with BG Group to buy the latter's 16.7 per cent stake in Agip Kazakhstan North Operating Company N.V.(Agip KCO), the operator for the project to develop the huge Kashagan field off the Caspian coast of Kazakhstan. The value of the deal was estimated at $1.23 billion. But the other members of Agip KCO, primarily the American and European companies, blocked the deal with the Chinese by exercising their pre-emptive right to purchase the stake.[30]

The exercise of first right of refusal by the western consortium members of the super-giant Kashagan project was not unexpected, as the stake was too high to lose for the western consortium members. The failure drove SINOPEC to enter Kazakhstan not by sea but by land.

> In late December 2003 [Yakovleva concludes] SINOPEC's subsidiary Shengli Oilfield acquired a 50 per cent stake in Big Sky Energy Kazakhstan (BSEK), which belongs to China Energy Ventures Corp, registered in Calgary.[31]

The case confirmed the difficulty of finding a quality oil producing asset in Kazakhstan.

CNPC's and SINOPEC's interest in purchasing oil producing assets in Kazakhstan reflected Beijing's yearning to maximize the benefit from constructing a long-distance crude oil pipeline between Kazakhstan and China. From a Beijing planner's point of view, to increase the portion of equity oil was the easiest way to maximize the benefit. In February 2004, CNPC and KazMunaiGaz decided to add the Alashankou–Dushanzi section, within the Chinese border, into the project. While the pipeline project proceeded, the two countries were studying the possibility of increasing the capacity of the trans-border Dostyk–Alashankou railway to 9 mt/y to enhance the business tie-ups between the two countries. At that time, the railway had a capacity of 7.5 mt/y and in 2003, 2.5–3.0 mt of Kazakh oil entered China by rail.[32]

In the early 2000s, Beijing had great expectations of the Angarsk–Daqing crude oil pipeline, but the collapse of Yukos in 2003 provided new momentum to reactivating the Kazakhstan oil supply option. Given the ups and downs in Sino-Russian energy cooperation, and China's determination to invigorate its undeveloped west, the China–Kazakhstan oil pipeline, whose strategic significance was of no less importance than the China–Russia pipeline, was speeded up. The Chinese side adopted a down-to-earth stance towards the China–Russia oil pipeline. A CNPC official said:

> China will receive around 15 mt/y of Russian oil by rail by 2006, and that is a volume not much smaller than the planned 20 mt/y initial throughput of the Sino-Russian oil pipeline. If the Sino-Russian line really evaporates, that won't cause much trouble to us in a purely economic sense.[33]

This remark confirmed CNPC's huge frustration about the absence of progress on the China–Russia oil pipeline, and CNPC had no choice but to accelerate the China–Kazakhstan oil pipeline development.

The Atasu–Alashankou section development

CNPC and KMG signed an agreement for the construction of the Atasu–Alashankou pipeline on 17 May 2004; construction started in late September.[34] In mid-July 2004 KazTransOil, (a state-owned oil transportation company and subsidiary of KazMunaiGaz[35]) and CN-ODC (China National Oil Development Corp,[36] a subsidiary of CNPC) announced that the two parties had established a parity joint venture to build the Atasu–Alashankou pipeline. The plan was to run the pipeline from the railroad oil loading rack at Atasu station in Karaganda Oblast to the Chinese border near the Druzhba/Alashankou railroad terminal. The pipeline route was to be Atasu–Agadyr–Akchatau–Aktogai–Ucharal–Alashankou, through Karaganda, Eastern Kazakhstan, and Almaty oblasts. Its total length was to be 988 km. The first phase was commissioned in 2006, and it is expected that the second phase will be completed by 2013. Annual capacity of the first phase of the 813 mm diameter pipeline is 10 mt and that of the second phase 20 mt.[37] China was pursuing two goals through this pipeline development: to supply the country with hydrocarbons and to diversify its oil supply sources. [38]

In August 2005, CNPC won a $4.18 billion bid for PetroKazakhstan, owner of Kumkol oil. CNPC's purchase of PetroKazakhstan marked a great leap for China into Kazakhstan's booming oil business; it was applauded in China as a major victory over the privately-owned Russian giant, Lukoil. The deal was the most expensive oil acquisition ever in Central Asia, providing CNPC with 150,000 barrels a day of oil production in central Kazakhstan, just 1,000 km from the Chinese border, plus access to promising exploration acreage. The price of $4.18 bn valued CNPC, with 390 million barrels of proven reserves, at around US$9 in enterprise value per barrel of oil equivalent. This compared with the $3.70 EV/boe paid by its listed subsidiary PetroChina to buy oil assets in August 2005, and with CNOOC's bid which was valued at around $8.10 EV/boe. PetroKazakhstan, which had won Kazakh oil reserves during privatization during the 1990s, bought a refinery at Chimkent.[39] Through the PetroKazakhstan acquisition, CNPC gained a 50 per cent stake in the Kazgermunai field, which produced 40,000 b/d of crude in 2005, in addition to the 150,000 b/d of crude oil produced by PetroKazakhstan in the Turgai basin.[40]

Two months later, in October 2005, CNPC completed the acquisition of 100 per cent of PetroKazakhstan (PK). China was entitled to do this, but the Kazakh government imposed a condition that 33 per cent of PK equity should be transferred or sold to KazMunaiGaz. That is the core of a preliminary agreement between CNPC and PK on 19 October. On 13 October the Kazakhstan parliament modified the country's energy law to permit government intervention in the PK sale. On 15 October CNPC reached an agreement with KazMunaiGaz to sell 33 per cent of PK's shares to KazMunaiGaz for $55 per share, a price equal to CNPC's initial offer. Meanwhile, CNPC had to share PK's Shymkent refinery with KazMunaiGaz on a 50–50 basis as the price of approval by the Kazakhstan government.[41] On 18 October the deal was finalized, but Lukoil argued it had a pre-emptive right to buy PetroKazakhstan's stakes in the joint venture Turgai Petroleum, which accounted for 20 per cent of PK's total reserves. Nonetheless, on 26 October the final green light was given to CNPC.[42]

In December 2005, the 246 km long Alashankou–Dushanzi crude oil pipeline was completed, ending the first phase of the China–Kazakhstan crude pipeline. The investment for the first phase of the pipeline totalled about 1,046 billion yuan. The pipeline was to be linked up with the 962.2 km long Atasu–Alashankou pipeline, which was put into operation on 16 December 2005. On 25 May 2006 the first flow of Kazakh crude oil arrived at Alataw Pass in Xinjiang, and the nine years' dream became a reality. The Alashankou–Dushanzi pipeline was designed to have an annual handling capacity of 20 mt when the second phase of the pipeline was completed in 2010. The maximum handling capacity is 50 mt. After the China–Kazakhstan crude pipeline is put into operation, Dushanzi Petrochemical will expand, with a refining capacity of 10 mt/y and an ethylene cracking capacity of 1.2 mt/y. The existing capacity was 6 mt/y and 0.22 mt/y respectively. The expansion work in Dushanzi, started in late August 2005, was scheduled for completion by 2008, with the total investment of 27 billion yuan. One merit of this crude pipeline development was that it admitted cooperation with Russia. In fact China, Kazakhstan, and Russia agreed on utilizing the China–Kazakhstan pipeline to transport Russian oil from west Siberian oilfields. In October 2005, Rosneft revealed its plan to export 1.2 mt of crude through the pipeline.[43]

The opening of the pipeline had a major economic impact, as *Global Insight* described:

> PetroChina aimed to make the Dushanzi refinery 90% dependent on crude oil imports from the Sino-Kazakhstan pipeline. The completion of the third and final Kumkol-to-Kenkiyak section of the China–Kazakhstan oil pipeline has paved the way for a significant increase in oil imports from Kazakhstan to Dushanzi. PetroChina is now planning to mothball the existing 120,000 b/d crude oil distillation unit at Dushanzi (which mainly treats crude oil from Xinjiang) and the Dushanzi plant is now being reconfigured to support greater imports of Kazakhstan crude oil, which is lighter than crude from Xinjiang but contains higher levels of sulphur. The Dushanzi expansion programme consists of construction of 20 processing units and 12 petrochemical processing units. This will ensure a more diverse range

of products to supply local markets in line with the government's strategy of promoting development in China's far-western regions.[44]

Zhou Dadi, then director of the Energy Research Institute under the NDRC, confirmed that China was concerned about the problem of filling the Atasu–Alashankou–Dushanzi pipeline from Kazakhstan to China. He said there were no guarantees that Russian companies would supply oil into the China–Kazakhstan oil pipeline, and China would regret it if the expensive pipeline was partially unused. However, he was optimistic on the long-term perspective, due to CNPC's production from Zhanazhol field together with the production from other companies.[45]

An unscheduled negotiation over the metering issue caused a delay in the oil flow. The two countries use different metering systems for the calculation of crude flow. In addition, oil filtering work was needed to remove some impurities introduced by the pipeline delivery.[46] But finally, on 29 July 2006, the China–Kazakhstan crude oil pipeline definitively started its commercial operation.[47]

Another sizable Kazakh oil asset acquisition was inevitable. In October 2006, China's state-controlled CITIC group agreed to pay 13.03 billion yuan, or $1.9 billion, for the Kazakh oil assets of Canada-based Nations Energy. Nations Energy (formerly Triton–Vuko) was founded in 1996 as a private Canadian company focusing on oil and gas projects in the Caspian region of the former Soviet Union. The company started in 1997 with hydrocarbon rights for the development of the Karazhanbas heavy oil field on the Buzachi peninsula (in the Mangistau Oblast, Western Kazakhstan) acquired through a share purchase agreement with the Kazak government. Nations Energy's Kazakh subsidiary, JSC Karazhanbasmunai, held the rights, until 2020, to develop the Karazhanbas Oil and Gas field in Mangistau Oblast. The field has proven reserves in excess of 340 million barrels of oil and production of over 50,000 b/d.[48]

Due to the shortage of Kazakh crude oil supply for the China–Kazakhstan crude pipeline, it was inevitable that Russian crude would be supplied to China through the pipeline. In 2006, Russia planned to export 1.2 mt of crude oil to China through the China–Kazakhstan oil pipeline. The Atasu–Alashankou pipeline can convey 5 mt of Russian crude oil to China. In 2004 Russia had exported six mt of crude to China by rail.[49] In November 2006, TNK–BP signed a contract to supply Russian oil to China within two months via Kazakhstan's Atasu–Alashankou pipeline. Rosneft also announced plans to supply oil to China along the route. The company said that it had agreed with the buyer to deliver 1.5 mt.[50]

Russia's Energy Minister Khristenko said, in the summer of 2007, that Russia had reached an agreement to export its crude to China through the China–Kazakhstan crude oil pipeline – but Russia had been saying this well before the construction of the pipeline had begun in September 2004. The pipeline had originally been designed to transport blended crude oil from Kazakhstan and Russia. If Russian crude oil reaches China by this route at least two advantages follow. The first is that crude oil supply for the pipeline, with an annual designed handling capacity of 10 mt, would be fully guaranteed. The second is that negotiations about the China–Russia crude oil pipelines would lose some of their urgency. China's stance was that if no progress were made on China–Russia oil pipeline negotiations, the development of both oil and gas pipelines from Kazakhstan would be encouraged.[51]

Kenkiyak–Kumkol section agreed

China and Kazakhstan announced the go-ahead for the construction of Phase II of the China–Kazakhstan crude oil pipeline and of the China–Kazakhstan natural gas pipeline project on 18 August 2007.[52] In the presence of the Chinese President Hu Jintao and his Kazakh counterpart Nursultan Nazarbayev, CNPC signed an agreement on the construction and operation of the Phase II project of the crude oil pipeline, doubling the handling capacity of the China–Kazakhstan crude oil pipeline from 10 mt/y to 20 mt/y. In the first half of 2007, Kazakhstan supplied 2.26 mt of crude oil to China through the pipeline and by August 2007 a total of 4.05 mt of crude oil had reached China since Phase I of the pipeline was put into commercial operation on 11 July 2006.[53]

In November 2007 Energy Minister Khristenko announced that Russia would start delivering up to 5 mt of oil to China via Kazakhstan as soon as 2008. There was simply no more available. Kazakhstan's development of new oil fields had fallen behind schedule, so its oil shortage was obvious. This explained why Kazakhstan had approached Russia for assistance. Minister Khristenko said that Russia and Kazakhstan had almost finished drafting an intergovernmental agreement on the Caspian oil pipeline. Russia's oil pipeline schedule did not include exports to China via Kazakhstan, but Russian oil supplied to Kazakhstan could be re-exported to China without export duty being charged.[54]

Despite the China–Kazakhstan crude pipeline's dependence on Russian crude supply, CNPC's verdict on its Kazakhstan oil business was positive:

> In April 2008, Zhou Jiping, then vice president of CNPC, said that with the joint efforts of employees from both countries, the eight projects invested in by CNPC in Kazakhstan (AktobeMunaiGaz, PetroKazakhstan,

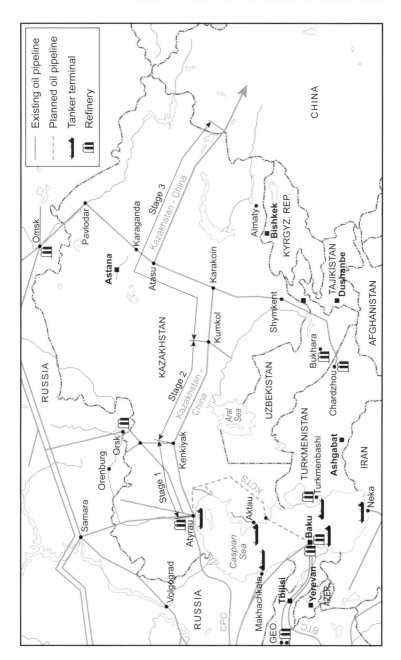

Map 6.1: Kazakh Oil Pipeline to China

Source: Slightly revised from IEA map

KAM Oilfield, North Buzachi Oilfield, ADM Oilfield, Kazakhstan–China Crude Pipeline, Northwest Crude Oil Pipeline and the Oil Products Sales Company) had achieved a great deal in exploration, development, refining, crude oil pipeline transportation and marketing. In 2007, CNPC produced 18.62 mt of crude oil and 4 bcm of natural gas from the eight projects.[55]

In the same month, Vyksa Steel Works (VMZ), part of the Unified Metallurgical Company (OMK) in the Nizhny Novgorod Region of Russia,[56] commenced deliveries of large-diameter pipes for the construction of the Kenkiyak–Kumkol oil pipeline, part of the second stage of the trunk oil pipeline from Kazakhstan to China. OMK announced that the company had delivered 34,000 tonnes of 813 mm diameter pipes for Kenkiyak–Kumkol during the months April–July 2008. The pipes, with an external anticorrosive polymer coating, have a thickness of 9.5 and 11.9 mm.

The Chinese Ambassador to Kazakhstan, Qing Gopin said in April 2009 that China was ready to build a refinery in Kazakhstan near the Chinese border. There are three refineries in Kazakhstan and their total capacity is between 13 and 15 mt of oil. Kazakh-Chinese companies operating in Kazakhstan produce about 20 mt and China's share in this accounts for 13 mt.[57] CNPC had expected that CNPC–AktobeMunaiGaz would produce more than 10 mt of oil and gas in 2010, but the project produced of only 3.11 mt of oil in the first half of 2011.[58] The Aktobe Oilfield Co. is Kazakhstan's fifth largest oil firm and has development licenses for two oil fields and three oil reserves.[59]

In late September 2009, Zhu Baogang, manager of the pipeline division of CNPC International Ltd (CNPCI) said that the company was scheduled to construct the second stage of the China–Kazakhstan Crude Oil Pipeline Phase II Project.[60] When its second phase was complete the China–Kazakhstan Crude Oil Pipeline would be 2,800 km in length; its oil transmission capacity would be upgraded to 20 mt per year. The 792 km length Kenkiyak–Kumkol section, with a 10 mt/y capacity was put into operation on 11 July 2009.[61] In 2009 China imported 7.73 mt of oil through the China–Kazakhstan oil pipeline. During more than three years of operation the amount of pipeline transportation of crude oil totalled 20.39 mt, 12 per cent of China's annual imports of crude oil.[62]

In short, Astana and Beijing have re-oriented the oil flows from Central Asia. During the visit by President Nazarbayev to Beijing in April 2009, the Chinese media reported that the Kazakhstan oil pipeline, which was already in operation, and the China–Kazakhstan natural gas pipeline, which was then under study,[63] were of historic significance, and the two countries had conducted genuine cooperation in both energy and non-energy sectors.[64] The development of crude oil pipelines between the two countries was a painful blow to Russia, as it has led to a collapse of its monopoly position as the sole transit country and buyer of central Asian energy resources. This was not what China had intended, but it was an important unintended consequence of Chinese determination to develop a stable crude supply source from the Central Asian Republics, in particular from Kazakhstan. China has every reason to place the highest priority on constructing a long-distance crude pipeline connecting oil assets on the shores of the Caspian with the Chinese refineries in Xinjiang province. Unlike the 1990s when there was no sure guarantee of filling the 20 mt/y of pipeline capacity, the development scheme of major oil fields in Kazakhstan, including Kashagan field, announced during the 2000s, provided an assurance to the Beijing planners as to the availability of crude supply for the 2,200 km long-distance pipeline.

It should be emphasized, however, that Chinese planners were not opportunistic in trying to guarantee crude supplies from the Central Asian Republics. China's approach was to identify the supply sources and construct the pipeline using their own investment. Chinese planners did not hesitate to commit a massive investment to the staged pipeline development connecting the Caspian Sea in Kazakhstan and the western part of China. This pipeline can be easily linked with China's domestic crude pipeline in Xinjiang province, and at the same time it will help the State Council's ambitious initiative for west China's economic development, which started in early 2000. The Beijing authorities will aim at maximizing crude supply from Kazakhstan in the coming decades, but China's attempts to increase the role of equity oil in Kazakhstan will meet with strong resistance from Kazakhstan's authorities. The possibility causes alarm in the Kazakhstan government. In May 2010, according to a BBC report:

> Kazakhstan energy minister Sauat Mynbayev said in the Kazakh parliament that China held a 50–100 per cent ownership stake in 15 companies in the energy sector. According to the Kazakh energy ministry, out of the 80 mt of crude oil which Kazakhstan is expected to produce in 2010 25.7 mt will go to China.[65]

If Moscow's stance towards the expansion of China's equity oil option in Russia had somewhat relaxed during the 2010s, the expected restriction on China's strengthened equity oil option in Kazakhstan will encourage Beijing to be more positive and aggressive about partnerships between Chinese NOCs and Rosneft and other major Russian oil companies.

Table 6.3: Conventional Oil and Natural Gas Resources in the Caspian by Country, end-2009 (bn bbl & tcm)

	Proven reserves		Ultimately recoverable reserves		Cumulative production		Remaining recoverable resources	
	Oil	NG	Oil	NG	Oil	NG	Oil	NG
Azerbaijan	7.0	1.4	29.9	4.4	11.7	0.3	18.2	4.1
Kazakhstan	39.8	2.0	78.2	6.1	9.2	0.4	68.9	5.8
Turkmenistan	0.6	7.9	19.5	14.2	3.6	2.3	13.9	11.9
Uzbekistan	0.6	1.7	5.5	5.2	1.1	1.5	4.3	3.7
Other Caspian*	-	0.2	1.4	0.3	0.2	0.0	1.3	0.3
Total	**48.0**	**13.2**	**134.4**	**30.3**	**25.8**	**4.5**	**108.6**	**25.8**
Share of World (%)	**3.5**	**7.2**	**3.9**	**6.5**	**2.3**	**5.0**	**4.7**	**6.9**

*   Armenia, Georgia, Kyrgyz Republic, and Tajikistan

Source: IEA (2010), 500, 524.

Natural Gas Supply Through the trans-Asia Gas Pipeline

The most painful blow to Russia's position in the Central Asian Republics was the decision to start the construction of the gas pipeline for deliveries of Turkmen gas to China. Beijing was fully aware of the danger of provoking Russia's anger by pursuing natural gas imports from Turkmenistan through a long-distance trans-national pipeline, and that is why China decided to develop a new gas field and build the pipeline itself. The equity gas option protected China from any unnecessary criticism by either Russia or the European Union. The gas from Turkmenistan has fundamentally changed the prospects for Sino-Russian gas cooperation.

Turkmen gas reserves
According to the Turkmen Government's 2030 Oil and Gas Industry Development Programme, the country's total reserves are 22.4 tcm, including 6.2 tcm in the Turkmen section of the Caspian Sea. Turkmenistan's gas reserves, like all those in Central Asia, have a high hydrogen sulphide content and other impurities, which makes them require extensive processing. In addition to existing production, an increase to 250 bcm/y, including 200 bcm/y for export, is envisaged by the 2030 Oil and Gas Industry Development Programme. Despite the massive proven gas reserves, achieving a large production target may not be easy. The ambitious target of 140 bcm production in 2015 and

240 bcm in 2020 have been widely quoted.[66] Turkmenistan's main gas producing area is in the south-east of the country. Fields there include Dauletabad field (the largest),[67] the Shatlyk field (the largest in Soviet times but in decline), and the newly-explored Yolotan–Osman field.[68]

Table 6.4: Turkmenistan's Gas Production and Export (bcm)

	Gas Production	*Gas Export*
2006	70	50
2010	120	100
2015	140	116
2020	240	140

Source: Ministry of Oil and Gas Industry and Mineral Resources of Turkmenistan, quoted in Denisova (2007), 46.

According to Turkmengeologiya, a Turkmenistan state company, recoverable gas resources in Turkmenistan in 2005 were 20.475 tcm. Exploratory work has exposed 147 gas and gas condensate fields in Turkmenistan, and the reserves have been estimated at 4.823 tcm. The most significant share of the reserves (4.449 tcm) is contained in 138 fields located in south-eastern and central Turkmenistan. The remaining 337.9 bcm are in western Turkmenistan, in nine offshore fields within the Turkmen sector of the Caspian Sea.[69] Turkmenistan was developing 51 fields with reserves of 2.513 tcm of gas. Twelve fields have already been prepared for development and their reserves are estimated at 279.5 bcm of gas. Seventy-three fields with gas reserves estimated at 1.895 tcm are being explored. Another 12 fields with gas reserves of approximately 135.1 bcm are temporarily closed.

Turkmengeologiya confirmed that the discovery of the Gunorta (South) Yolotan field, in the eastern part of the country, was one of the major events of 2004. In Gunorta, subsalt carbonate sediments yield high gas inflows of over 1 mcm/d. Very intensive exploration, together with 3D seismic surveys by CNPC, has been carried out to ensure the timely supply of produced gas to China, and China is set to receive the gas from the giant field from 2013. A promising new field was discovered in Kepe in 2004. Commercial gas inflow from the first well exceeded 0.2 mcm/day. Significant inflows of hydrocarbons (0.3 mcm and 30 tonnes of gas condensate per day) have also been found at Gutlyayak. At Bagadzha, the figure was 1 mcm/day. Prospecting wells are being drilled at the Yerburun, Narazym, Dortgul'depe, and South (or Yuzhny) Yolotan fields. China needs gas on a very large scale from Turkmenistan, in particular from Gunorta Yolotan field by 2013. To achieve the large-scale supply, on top of prospecting (exploration) wells, there should be supporting development wells for production. The first exploratory well was begun at the newly-discovered Burgutli site in the Transsunguz Karakum. Turkmenistan's gas reserves grew by 290 bcm between 2000 and 2004. From 2011 to 2020 Turkmenistan expects to add more than 2 tcm of gas to the total of known reserves.[70]

The discovery of the super-giant gas field Yuzhny-Yolotan

During the international conference Oil and Gas of Turkmenistan 2006 (held in September of that year) the Turkmen authorities revealed that Yuzhny-Yolotan field's reserves exceeded 1.5 tcm of gas and 17 mt of oil. According to Ashgabat's original plans, gas production at the first stage of field development should run to 15 bcm annually, and eventually the figure may rise to 43 bcm. A new exploratory well drilled at the field in October 2006 made a very large gas discovery and this changed its initial plans. Annual production may reach its engineering maximum at 200 bcm, based on the new reserves estimate of 7 tcm. In November 2006 the Turkmen President, Saparmurat Niyazov, announced the discovery of a giant gas field with reserves of 7 tcm in the Yuzhny-Yolotan structure. The initial estimate of reserves was 1.5 tcm. This giant field is located in immediate proximity to another vast gas field, Dauletabad, with commercial reserves of 1.3 tcm. The state concern Turkmengeologiya conducted a seismic survey and exploration drilling at the Yuzhny-Yolotan structure in 2004–5. The first exploratory well drilled, to a depth of 4,500 metres, in the autumn of 2005 produced natural gas flow at a daily rate of 1.4 mcm. In the region, there are promising structures near the Yashlar and Osman acreages. According to Ishanguly Nuryev, minister and chairman of Turkmengeologiya, this region will become the main resource base for boosting Turkmenistan's gas production in the period 2020–5.[71]

The Turkmen president, Kurbanguly Berdymukhamedov (who succeeded Niyazov on his death in 2006) told CNPC officials in Beijing on 8 August 2008 that preliminary data indicated natural gas reserves in Turkmen subsurface resources which were much higher than initially thought. Based on this revelation, Berdymukhamedov proposed that the Chinese authorities consider the possibility of purchasing an additional 10 bcm of natural gas from Turkmenistan each year in addition to the 30 bcm already envisaged in inter-governmental agreements. He said that by the end of 2009 a gas pipeline from Turkmenistan to China would be completed which would have the capacity to carry 40 bcm of gas. He added that an audit currently being carried out by Gaffney, Cline and Associates in England was another indicator of Ashgabat's reliability. CNPC secured a license to explore and extract hydrocarbons on land in Turkmenistan. Under Turkmen-Sino agreements, exploration would be carried out at the Samanpede, Yashyldepe, Metedzhan, and Gendzhibek fields, which are part of the contract territory of the company named Bagtyyarlyk. More than US$6 billion would be spent on the accelerated start of the field's development, as well as the modernization of existing infrastructure, to permit the production of 13 bcm of natural gas annually. The rest of the gas would come from the development of new gas fields. The construction of one of two gas treatment plants, to supply the transnational gas pipeline running from Turkmenistan to China, began in June 2008.[72]

Turkmenistan

The idea of exporting gas to China from Turkmenistan originated in in December 1992 when CNPC, together with the Mitsubishi, visited Turkmenistan to propose the so-called Energy Silk Route project, sometimes known as the Pan-Asian Gas project. In the same month, this pipeline project proposal was disclosed at a meeting of representatives from the Turkmengaz Association, Mitsubishi, and CNPC which was held in Ashkhabad. During the years 1993–5, CNPC, Mitsubishi, and the Turkmen government had studied the possibility of developing a 7,000 km gas pipeline connecting Turkmenistan with Japan via Uzbekistan, Kazakhstan, and China.[73] In 1993, CNPC established a company called Central Asia Corporation, which was directed by CNPC's vice president, Zhang Yongyi, to manage this project. The original plan, introduced in 1993, aimed at building an onshore pipeline from Turkmenistan to Tanggu, a port close to Tianjin, where the gas would be liquefied and then shipped to Japan. This project aroused CNPC's interest because the firm could play a significant role in piping out Xinjiang's gas to consumers in east and south China.[74]

In April 1994 when Prime Minister Li Peng made his tour through Central Asia, CNPC's vice president, Zhang Yongyi, signed letters of intent and generally boosted the idea of Sino-Central Asian oil cooperation with the Central Asian republics. In Turkmenistan, the plan for the Turkmenistan–China–Japan natural gas pipeline was signed and a letter of intent agreed between CNPC and the Turkmenistan Ministry of Oil and Gas, to set up a commission to study the matter.[75] On 22 August 1995, CNPC, Exxon, and Mitsubishi signed an agreement to launch a feasibility study for a 7,000 km pipeline from Turkmenistan via Kazakhstan and China (specifically Lianyungang, in Jiangsu province) to Japan. The estimated investment was $11.8 billion. Both Exxon and Mitsubishi were conducting oil and gas E&D in Xinjiang Uygur

Autonomous Region,[76] and at the end of 1996 their final report was prepared. The verdict was not favourable, recommending that no immediate action be taken on development.[77] After that, the option of supplying Turkmen gas to China was completely sidelined for some time.

It was during the 2003–5 period that Turkmenistan began once again to collaborate with China on upstream sector development, and contracted CNPC to provide oil and gas services at numerous deposits on the right bank of the Amu Darya river. CNPC wasted no time in confirming the presence of rich hydrocarbon reserves in the area. The initial proposal for the Central Asia–China gas pipeline was presented as the Kazakhstan–China gas pipeline, which was to follow the same route as the Kazakhstan–China oil pipeline. In June 2003, during the visit of the Chinese president, Hu Jintao, to Kazakhstan, agreements to expedite the appraisal of the project were signed.[78] In December 2005 the Turkmen President, Saparmurat Niyazov, even disclosed that China had agreed to buy the gas at a price of $80 per 1,000 cubic metres.[79]

Sino-Turkmen gas framework agreement signed
The first breakthrough was made in early April 2006 when Turkmenistan and China signed a framework agreement on the gas sector, providing for the construction of a gas pipeline to China with a designated capacity of 30 bcm/y, and the supply of gas for 30 years from 2009. The Amu Darya deposits, where the giant fields of Dauletabad (1.3 tcm) and Shatlyk (1.0 tcm) were discovered, were chosen as the resource base for the pipeline. China assumed the responsibility of reaching the necessary agreements with Uzbekistan and Kazakhstan on building a trans-regional gas pipeline.[80] After his six day visit to China, President Niyazov told his government staff that the natural gas pipeline would be a modern silk road which would make a considerable contribution to Turkmenistan's economy, and further strengthen the friendship between China and Turkmenistan.[81] According to the framework agreement between the two countries, the natural gas price would be set with reference to international prices, on the basis of reasonable and fair dealing, and the US dollar would be the preferred currency for payment. Since Turkmenistan exported natural gas to Russia at a price of $65 per 1,000 cubic metres (in 2006), it was assumed that a price lower than that for Russia, say $60, would be applied.[82]

In November 2006, Turkmenistan authorized CNPC participation in the development of the Gunorta Yolotan gas deposit. The Turkmen state concern Turkmengeologiya signed a three year service contract with the CNPC Changqing Petroleum Exploration Bureau, to drill 12 test wells to a depth of up to 5,000 metres at Gunorta Yolotan, at a cost of US$152 million. The sudden death of President Niyazov in December 2006[83] did not change Turkmenistan's commitment to China, and in July 2007 his successor, Berdymukhamedov, confirmed the previous gas sector agreements and signed a production sharing agreement with CNPC to develop the Bagtyyarlyk area on the right bank of the Amu Darya river.[84] The establishment of two working groups, one responsible for setting up Amu Darya Natural Gas Company and the other for the Central Asian National Gas Pipeline Company, contained some concrete additions to the framework agreement which Niyazov had signed, during his state visit to China in April 2006; in July 2007 two follow-up agreements were signed by President Berdymukhamedov on his first state visit to China. According to CNPC's CEO Jiang Jiemin, CNPC aimed to produce 13 bcm/y of gas from the right bank of the Amu Darya river in Turkmenistan; the remaining 17 bcm/y of gas would be supplied from other regions of Turkmenistan. In fact, some CNPC insiders were worried whether Turkmenistan had such large amounts of gas available for sale to China. As always, the other issue was the price. Even though CNPC did not disclose the price, it was generally assumed to be slightly lower than $180 per 1,000 cubic metres. A CNPC insider said that the price settled for the Turkmen gas was definitely lower than the price offered by Russia.[85]

To mark Berdymukhamedov's state visit to China in July 2007, *China OGP* published an interesting article on the further progress of the China–Turkmenistan gas pipeline. The point was that Russia was not willing to see Turkmenistan supply a huge amount of natural gas to China. Gazprom was worried that if more than a certain amount of Turkmenistan gas were exported to China, gas exports to Russia would no longer satisfy demand. In 2007, Russia imported 50 bcm/y of gas from Turkmenistan. However, Turkmenistan hoped to deepen energy cooperation with China as an important counter balance to its dealings with Russia. China could not, however, hope to gain too much. If Russia, Turkmenistan's largest gas buyer, raised the price to a reasonable level, then Russian influence over Turkmenistan would increase. In other words, the price would be an issue on which China would have to do some hard bargaining over the proposed China–Turkmenistan gas pipeline. Secondly, the USA also kept an eye on the energy resources of Central Asian countries and regarded Russia as its major rival in this area. So China also needed to take into account possible conflicts with the USA. In any case, however, China would not easily give up its efforts to source pipeline gas from Turkmenistan. *China OGP* concluded that the China–Turkmenistan gas pipeline would not be settled either as smoothly or as quickly as the China–Kazakhstan crude oil pipeline. Nor, however, would the China–Turkmenistan gas pipeline develop as slowly and as erratically as the China–Russia crude oil and gas pipelines had done.[86]

It was clear that Turkmenistan was seeking energy cooperation with other nations based on mutual economic benefits rather than on political alliances. The country was also eager to reduce reliance on Gazprom, and was willing to export its own rich natural gas to others at a lower price than the international one. But, since Turkmenistan was not willing to sell gas to Russia at a particularly low price, it made no sense for the country to sell gas to China at a low price either. Thus, China should prepare itself to be offered the gas at a price equal to the international one.[87]

China rewarded by Turkmenistan's stance against Russian monopoly

In August 2007, Turkmenistan, signalling a possible policy change, awarded China the rights to develop gas reserves in the east of the republic to feed an export pipeline that was being built to China.[88] An OIES study has described this breakthrough:

> …construction of the Turkmenistan–China gas pipeline was officially launched, while Turkmenistan granted CNPC a number of exploration and production licenses. CNPC Exploration and Development Company Ltd (CNPC E&D) is responsible for building the pipeline, which will run from Gedaim on the Turkmen-Uzbek border, 1,818 km through Uzbekistan and Kazakhstan, to Horgos in Xinjiang where it will connect to WEP II. The section of the Trans-Asian Pipeline on Turkmen Territory, which runs 188 km from the Malai gas field in the east of the country to the border with Uzbekistan, will be built by Stroytransgaz of Russia. In April 2008 CNPC also signed a deal with KazTransGaz on construction of the Kazakh section of the 1,300 km pipeline with annual throughput capacity of 30 bcm, linking the Uzbek-Kazak border point with Horgos, via Shymkent and Almaty.[89]

The Central Asia–China natural gas pipeline which was proposed as early as April 2005 made significant progress during 2007. On 17 July China's President Hu Jintao and Turkmenistan's President Berdymukhamedov were both in Beijing at the signing of a gas production sharing contract and a gas sales and purchase agreement between CNPC, the State Agency for the Management and Use of Hydrocarbon Resources of Turkmenistan, and Turkmengaz. This marked a further advance down the road opened by the framework agreement on gas cooperation signed between the two countries in April 2006.[90]

On 28 December 2007 PetroChina signed an agreement with

CNODC, a wholly-owned subsidiary of CNPC, under which the two companies would each inject eight billion yuan in cash into CNPC E&D, a 50 per cent-owned subsidiary of PetroChina. The funds would be provided for the construction of the planned Central Asia–China natural gas pipeline.[91] For this purpose, in November 2007 Trans-Asia Gas, a wholly-owned subsidiary of CNPC, was established, and the firm was responsible for the development and construction of the natural gas pipeline from Turkmenistan to north-west China via Uzbekistan and Kazakhstan.[92] In the summer of 2008, the construction of a gas treatment plant was started at one of the most promising gas fields – Samandepe, on the right bank of the river Amu Darya. The modern industrial complex with a capacity of 5 bcm of gas would be constructed by CNPC.[93]

During President Hu Jintao's visit to Ashgabat, Turkmenistan, after the Olympic Games, Turkmenistan and China signed five bilateral documents, including a framework agreement between Turkmengaz and CNPC. The Turkmen president said that the two countries had set fuel and energy as the top priorities for such cooperation. He added that:

> … thanks to our joint efforts, the Turkmenistan–China gas pipeline, originating on the right bank of the Amu Darya, where Turkmen geologists have discovered large oil and gas fields is starting to take shape. The future pipeline, along which China will receive not the 30 bcm originally planned but 40 bcm of gas, was rightly being described as the project of the century. This was not just because of its unprecedented length of approximately 7,000 kilometres, but by virtue of the role it would play in the global energy market and in the integration processes in the whole Asian geo-space.[94]

The two countries also signed a joint statement and agreed to form a Turkmenistan–China Cooperation Council, a decision backed by an inter-governmental agreement.

The gas price burden and the equity gas solution
The most frequently quoted estimates of the price which CNPC had agreed for gas imports from Turkmenistan were $180 and $195/1,000 cubic metres. A border price at this level would make the city gate price in China's coastal area definitely more than $10/mmbtu after the transportation tariff of $3.8/mmbtu was added. Despite this price burden, Beijing gave the green light for the development of the Central Asia–China Gas Pipeline. The key factor was the availability of the 'Equity Gas' option which would give CNPC enough of a cushion to balance the burden from this high border price. This was the fundamental difference between the Turkmen and Russian approaches to gas export. China had no hesitation in accepting Turkmenistan's Equity Gas

option, and this indirectly explained why China's gas price negotiations with Russia had still failed to break the stalemate.

Unlike its Sino-Russian counterpart, the Sino-Turkmen gas negotiations wasted no time, and focused on the development of the pipeline. It was driven by China's policy of impatience over what it called Russia's NATO (No Action Talking Only) stance. On 14 December 2009 at the inauguration ceremony of the Central Asia–China Gas Pipeline, held in the gas plant on the right bank of the Amu Darya River in Turkmenistan, the presidents of China, Turkmenistan, Kazakhstan, and Uzbekistan together turned on the flow of natural gas. The Central Asia–China Gas Pipeline starts at the Turkmen-Uzbek border city of Gedaim and runs through central Uzbekistan and southern Kazakhstan before reaching Horgos in China's Xinjiang Uygur Autonomous region. This gas pipeline has dual lines in parallel, each 1,833 km long. The diameter is 1,067 millimetres. Construction of the pipeline had commenced in July 2008 and line A became operational in December 2009 with line B following on 26 October 2010.[95] CNPC confirmed on 15 September 2010 that the annual transmission capacity of line B had reached 9 bcm and the company hoped to expand this to 15 bcm by the end of 2010, and then to 17.7 bcm by the first quarter of 2011.[96] A delivery capacity of 30 bcm per annum should be reached by the end of 2011.[97]

In the words of Vladimir Socor:

> … this breakthrough on the Central Asia–China gas pipeline, as part of Turkmenistan's policy of gas export diversification, undermined Russia's position not only in the European gas trade but also on two Asian fronts. These were the negotiations on future Russian imports of Turkmen gas (in scaled-down volumes) and on possible Russian gas exports from eastern Siberia to China.[98]

Chinese planners would not have taken the bold initiative of the equity gas option without the confirmation of the proven reserves of the South (or Yuzhny) Yolotan field. The sustainability of the gas supply was, for the Chinese planners, the most important factor justifying the pipeline.

In September 2010 CNPC announced that the Turkmen Amu project's gas production would hit 6.57 bcm before 1 October 2010. (Since the start of operations on 14 December 2009 the project handling capacity had reached 5.29 bcm.) The No. 1 natural gas plant had handled some 2.58 bcm of gas since the start of operations, including 2.32 bcm of gas supplied to the pipeline. The Amu project was due to reach a production capacity of 5 bcm in 2010 rising to 13 bcm by the end of 2011.[99] However, gas production from the South Yolotan field (first phase development) would probably be delayed by one to

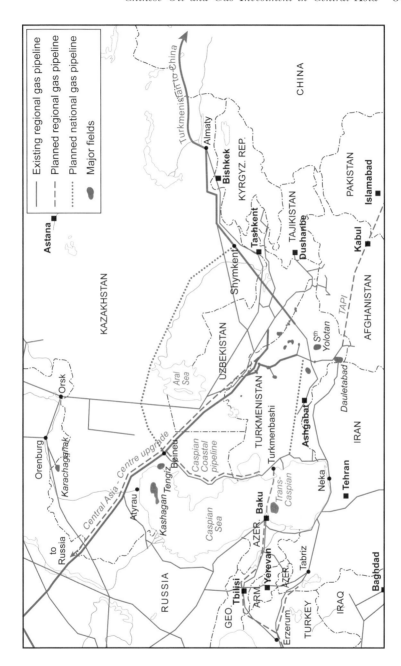

Map 6.2: Central Asia–China Gas Pipeline

Source: Slightly revised from IEA map

two years, to 2014–15. In 2009, Turkmengaz awarded construction and drilling service contracts worth close to $10 bn for Phase 1 development of South Yolotan to a consortium composed of CNPC, Petrofac, Gulf Oil & Gas, and LG International. In November 2010 Petrofac said there would be some delays for the EPC work for a gas plant to produce up to 0.97 bcf/d (10 bcm/y) of gas. China will import 30 bcm/y of full contractual volume from Turkmenistan by 2018–19.[100]

According to the NDRC's newsletter of 25 September 2010, CNPC has suffered huge losses on gas imports from Central Asia since it began importing gas from the region at the end of 2009. The NDRC did not give the exact figures, but confirmed that the losses were due to the price of imports being higher than their value on the domestic market.[101] This report further explained why the Chinese planners paid so much attention to the importance of the Equity Gas option from Turkmenistan.

On 1 March 2011 an NDRC website release (noted by *Reuters*), claimed that consensus had been reached between NDRC head Zhang Ping, top energy official Liu Tienan, and the visiting Turkmen deputy prime minister, Baymyrat Hojamuhammedov. Turkmenistan would supply 20 bcm/y of gas on top of the 40 bcm/y of the gas supply deal. The NDRC indicated that the inter-governmental framework agreement would be signed during the second half of 2011, although there were a number of factors, such as price and infrastructure issues, that could still derail the agreement. In parallel with the new deal with Uzbekistan in April 2011 (discussed below), this new gas deal with Turkmenistan to increase the export volume from 40 bcm/y to 60 bcm/y, if implemented as announced, will have huge implications for Sino-Russian gas cooperation.[102]

Uzbekistan

At the beginning of 2005, Uzbekistan's potential gas reserves were estimated by the UNDP at 5.9 tcm, 3.2 times more than the inter-nationally-recognized recoverable reserves. In total, 193 hydrocarbon fields had been discovered, of which 93 were under exploration and 67 had been prepared for exploration; geological prospecting work and infrastructure development were underway at the remaining 33. In the early 2000s, more than 75 per cent of production came from Shurtan, Zevardy, and Dengizkul-Khaouzak fields, and a further 20 per cent from the Alan, Kokdumalak, Pamouk, and Koultak fields. Uzbek gas is of relatively poor quality, containing high concentrations of sulphur, carbonic acid, and other acid and toxic gases. And Uzbekistan's largest gas deposits are characterized by high ground pressure of up to 60 mega pascals.[103] In the mid-2000s the country produced over 60 bcm/y of gas, of which around five bcm/y was exported to Kazakhstan, Kyrgyzstan, and Tajikistan. Around 8 bcm/y is delivered to Russia and 50 bcm/y is consumed in the country itself. No more was available for China unless the country increased its production.[104]

During President Hu Jintao's June 2004 visit to Uzbekistan, CNPC and Uzbekneftegaz reached an agreement on opportunities to develop a closer partnership between the two countries on surveying and drilling for oil and gas. A year later, during his May 2005 visit to China, Uzbekistan's president, Islam Karimov, announced the signing of a $600 million oil deal with China. This was part of a bid to promote bilateral energy cooperation as well as to improve the partnership between CNPC and Uzbekneftegaz. In the preliminary negotiation, CNPC was asked to guarantee $96 million for the Prophase E&D investment by establishing a joint venture, implementing a 50–50 shareholding arrangement focusing on the oilfields around Bukhara and Khiva, most of which are located in remote, hard-to-reach areas.[105]

Uzbekistan has attempted to broaden its international cooperation:

> In August 2006, Uzbekistan signed a production sharing agreement between Uzbekneftegaz, Lukoil, Petronas, CNPC and the two [South[Korean companies KNOC and POSCO. All companies have an equal share in the consortium, except for [South] Korea's 20 per cent stake which is divided between KNOC (10.2 per cent) and POSCO (9.8 per cent). CNPC and Petronas' participation in the consortium could secure gas supplies from the Aral field for the new pipeline, an idea backed by Uzbekistan. Aral Sea gas production is scheduled to start by 2012, eventually reaching a peak of about 25 bcm/y.[106]

Besides this, in June 2006 CNPC and Uzbekneftegaz signed an agreement on the creation of a joint venture named Mingbulakneft to develop the Mingbulak oil field in the Namangan region. Andizhanneft, affiliated with Uzbekneftegaz, and CNODC were the founders of the venture which they own on a parity basis. The Mingbulak oil field is located in the Fergana oil and gas region in the Namangan region of Uzbekistan and was opened in 1992. It has been developed by Uzneftegazdobycha, Uzbekistan's structural branch. This is CNPC's second Uzbek oil and gas project. CNODC created a subsidiary in December 2006 named CNPC Silk Road, composed entirely of Chinese capital. The subsidiary is to conduct exploration work in Uzbekistan and has a license to explore five investment blocks within the Ustyurt, Bukhara–Khiva, and Fergana oil and gas regions.[107]

The level of cooperation was upgraded on 30 April 2007 when Uzbekistan and China signed an inter-governmental agreement 'on

construction and operation of the Uzbek section of the Central Asia–
China Gas Pipeline'. The length of this pipeline will be 530 km and
the estimated capacity 30 bcm. In the summer of 2008 Uzbekneftegaz
confirmed that construction had begun near the village of Sayet in
Dzhondor district in the Bukhara region of Uzbekistan. The operation
of the gas pipeline, which has cost $2.0 billion, will be the responsibil-
ity of Asia Trans Gas, a joint venture set up by Uzbekneftegaz and
CNPC for design, construction, and operation of the pipeline. The
project involves the construction of two branches of the main gas
pipeline. According to the working schedule, the first branch and the
KS1 compressor station were completed before 31 December 2009 and
commissioned in early 2010. The second branch and two compressor
stations, KS2 and KS3, are scheduled for completion and commissioning before 31 December 2011.[108]

The Uzbek government announced on 27 July 2007 that the country
would give support to the construction of the gas pipeline. Jiang Jiemin,
CNPC president, said that the Chinese government had signed separate
agreements with Uzbekistan and Kazakhstan for the smooth construction of the pipeline from Turkmenistan.[109] On 28 January 2008 the
document establishing Asia Trans Gas Company was officially signed,
and the construction and operation of the Uzbek section of the Central
Asia–China Gas Pipeline could therefore proceed.[110] In June 2008
Uzbekneftegaz said CNPC and the Uzbek National Holding Company,
Uzbekneftegaz, had signed an agreement on the creation of a joint
venture (Mingbulakneft) to jointly explore and develop the Mingbulak
oil field. CNPC was due to invest $208.5 million in geological surveys
within five years.[111] In July 2008, CNPC started construction of a
natural gas pipeline from Uzbekistan (the Bukhara region) to Horgos
in Xinjiang.[112]

On 9 June 2010 CNPC signed a framework agreement to purchase
10 bcm/y of natural gas from Uzbekistan's Uzbekneftegaz State Holding Co. According to the agreement, the gas would be imported through
the Uzbekistan–China gas pipeline, though no time frame was given
for when imports would begin.[113] On 27 September during President
Dmitry Medvedev's visit to Beijing, the Lukoil CEO, Vagit Alekperov,
and the CNPC president, Jiang Jiemin, signed an agreement to expand
their strategic cooperation, and in October Alekperov indicated that
Lukoil may supply China with around 10 bcm per year of gas produced
in Uzbekistan as early as 2014.[114]

On top of this, during his official visit to Beijing in April 2011
President Islam Karimov signed an inter-governmental agreement on
construction and operation of the Uzbek segment of a third strand of

the Turkmenistan–China gas pipeline. The new pipeline will have a delivery capacity of 25 bcm/y, more than one third of Uzbekistan's production. The $2.2 bn construction cost will be financed with loans from China Development Bank.[115] It remains to be seen how the expansion of the Uzbek gas supply from 10 bcm/y to 25 bcm/y will affect the development timetable of China's WEP-III, IV and V pipelines and Sino-Russian gas cooperation.

Kazakhstan
Shamil Yenikeyeff provides the following description of the hydrocarbon situation in Kazakhstan:

> Kazakhstan's domestic hydrocarbon reserves are estimated at 3.3–3.7 tcm, of which 2.5 tcm are proven. Kazakhstan's gas potential could reach 6–8 tcm with further development of resources in the Caspian Sea. There are six major gas fields (Karachaganak, Tengiz, Kashagan, Zhanazhol, Imashevskoye, and Zhetybai) whose proven reserves range between 99 bcm in the smallest field and 1,370 bcm in the largest. Kazakhstan's Ministry of energy and mineral resources projected Kazakhstan would produce around 80 bcm by 2015 and 114 bcm by 2010.[116]
>
> …
>
> On 18 August 2007, Kazakhstan and China reached agreement on building and operating a new gas pipeline to China. On 8 November CNPC and KazMunaiGaz signed the basic principle agreement on the Kazakhstan–China Gas Pipeline construction and operation. Provisional agreements envisioned a Trans-Asian Gas Pipeline network consisting of two trunk lines. The first (running through Southern Kazakhstan) will be the Kazakh section of the Turkmenistan–China gas pipeline. The second will deliver gas from western Kazakhstan to western China.[117]
>
> …
>
> The Kazakh section of the Turkmenistan–China gas pipeline will run from the Uzbek–Kazakh border to the border between China and Kazakhstan via the Kazakh city of Chimkent, ending in Khorgos, in Xinjiang province. The Turkmenistan–China pipeline will consist of two parallel 1,067 mm diameter pipelines and five compressor stations capable of transporting 30 bcm/y. The project cost is estimated at more than $6.5 bn. CNPC will provide 100 per cent financing for the pipeline and hopes to start building the Chinese section by 2010.[118]
>
> The second planned gas pipeline to China, the Western Kazakhstan to Western China (Beyneu–Bozoy–Kyzylorda–Chimkent) initiated in 2005, will be 1,480 km long and have 1,016–1,067 mm diameter pipes and a projected annual capacity of 10 bcm/y. Kazakhstan and China initially planned to build the pipeline along a central route (Atyrau

(Makat)–Aktobe–Zhanazhol–Chelkar–Atasu–Dostyk–Alashankou) with a capacity of 30 bcm/y, projected to reach 40 bcm/y by 2015. Both parties have since cut capacity to 10 bcm/y due to economies. A preliminary cost estimate of $3.84 bn recouped over 30 years caused Chinese investors to pause in light of additional estimates that gas transported via this pipeline would cost almost three times as much as gas imported from Uzbekistan.[119]

In October 2010, the Kazakh Minister of Oil and Gas, Sauat Myn-bayev, said that the Beyneu–Bozoy pipeline project, regarded by Beijing as the second section of the Kazakhstan–China gas pipeline, may start in December 2010. He added that, once the report on the 'gas cap' on the Zhanazhol field was accepted by the State Reserves Commission, it should be possible to start construction work in December 2010 as planned, without the general approval of the Central Development Commission.[120]

Priority given to the Trans-Asia gas pipeline
According to Yenikeyeff:

> In February 2008, KazTransGaz and Trans-Asia Gas Pipeline Ltd (owned by CNODC) formed a joint venture to become sole operator of the Kazakh section of the Turkmenistan–China pipeline. Similar joint ventures with relevant Uzbek and Turkmen gas companies were established by Trans-Asia Gas Pipeline Ltd to operate their respective sections of the gas pipeline.[121]

Kazakhstan views the pipeline to China as an energy security project, ensuring that its domestic gas demand can be met without relying on imports. In early November 2008, CNPC and KazMunaiGaz signed a preliminary agreement regarding the construction of the Western Kazakhstan–Western China gas pipeline, which is planned to send 5 bcm/y of gas to China and roughly 5 bcm/y for use in southern Kazakhstan.[122]

A ground breaking ceremony to mark the start of construction of the Kazakhstan–China gas pipeline took place in Almaty region, near the 42 km post of the Almaty–Kapchagai highway on 9 July 2008. On the same day, Trans-Asia Gas Pipeline's general director, Beimbet Shyakhmetov, said that construction of the Kazakh–China gas pipe-line would cost $6–6.5 billion, even though there was no design and budget documentation relating to the compressor equipment, automatic controls, and auxiliary equipment. This costing referred to the phase of construction covering the Uzbek-Kazakh border to Horgos (1,300 km). He added that the overall cost for the gas transportation project between China and Turkmenistan was estimated at around $20 billion. Trans-Asia Gas, the operator of the gas pipeline, planned to raise at least $6.0 billion in loans by the summer of 2008, repayable in 15 years. A CNPC corporate guarantee would be in effect for the first five years. The project's recoupment period would come to 12 or 13 years, with funds being raised over a period of 15 years. There would be no state budget funds and all funds would be loans. The first segment of the pipeline would go from the Uzbek-Kazakh border to the Kazakh-Chinese border through Shymkent, the administrative centre of the South Kazakhstan region, and would reach Horgos. It was planned that this segment would be built during the 2008–9 period, with a capacity of 40 bcm/y and a length of 1,300 km. Construction works on the Kazakh section started on 9 July 2008 and the first phase was finished in July 2009. It was built by Asian Gas Pipeline Company, a joint venture of CNPC and KazMunaiGaz. The main contractors of this section were KazStroyService and China Petroleum Engineering and Construction Corporation. The second phase of construction started in December 2010. The second segment, covering Beyneu–Bozoi–Kzyr-Orda–Shymkent, was planned to have an annual capacity of 10 bcm, with a length of 1,480 km.[123]

The Kazakh deputy prime minister, Umirzak Shukeyev, confirmed that Prime Minister Karim Masimov had approved budget funding for the project in February 2008, and ordered economic and budget planning ministers to allocate the required funds in the country's 2009–11 three year budget. The first stage of the pipeline development was estimated to cost approximately $3.4 billion, while the construction and commissioning of the second phase was expected to cost an additional $389 million. The first phase would have an annual transmission capacity of 5 bcm and was expected to be completed between 2009 and 2011. The second phase would bring the pipeline capacity up to 10 bcm when completed – between 2011 and 2014.[124] On 10 October 2008 Masimov and Wang Dongjin, the vice president of CNPC, met and discussed the gas pipeline project and its potential early completion. The first of two parallel lines was completed in 2009 and the second line was completed by the end of 2010. The whole gas pipeline was designed to have a throughput capacity of 40 bcm per year. In September 2011, the two governments agreed to build the Kazakh section of Pipeline 'C' that will originate from Turkmenistan and cut through Uzbekistan. The capacity of Pipeline 'C' will rise to 25 bcm a year by December 2015, boosting the network's capacity to 55 bcm.[125]

In September 2011, the two governments agreed to build the Kazakh section of Pipeline 'C' which will originate in Turkmenistan and cut through Uzbekistan. The capacity of Pipeline 'C' will rise to 25 bcm/y by December 2015, boosting the network's capacity to 55 bcm.

Construction of the 1,305 kilometre Kazakh section is expected to begin in early 2012, and Pipeline 'C' will be operational by January 2014 with an initial capacity of 15 bcm a year.[126]

Yenikeyeff has argued that:

> ... construction of the Trans-Asia Gas Pipeline to China opened new opportunities for Kazakhstan in terms of linking all transit pipelines into a unified system. This might allow Kazakhstan greater freedom in terms of gas swaps with different producers and gas deliveries to southern Kazakhstan – provided that the West–South (Western Kazakhstan–Western China) branch of the Trans-Asia gas pipeline network goes ahead.[127]

The importance of development of the Trans-Asia gas pipeline to China lies in the fact that it is directly linked with China's domestic trunk gas pipeline. This is the big difference from Russia's still potential Altai supply project.

Conclusion

The introduction of long-distance pipeline gas from the Central Asian Republics is posing a real challenge to Gazprom's aim to monopolize pipeline gas supply to China. Sino-Russian gas cooperation will be severely affected in the coming decades by this Sino-Central Asian gas cooperation. The initiatives taken by China in its international oil and gas expansion were driven by the necessity of guaranteeing the sustainability of oil and gas supply to satisfy the country's soaring energy demand, and were supported by its ever-growing financial capacity, coming from its massive foreign exchange reserves.

The most remarkable progress made in the Central Asian Republics was through the oil and gas upstream asset purchase with the related pipeline development. Compared with Russian supply sources, the Central Asian Republics offered no special geographical advantage. What, then, caused the Chinese to give priority to the Central Asian Republics? China's stance towards Russia's oil and gas projects was palpably different from its stance towards the Central Asian Republics. What drove this difference?

First of all, the Chinese State Council's decision to promote western China's economic development in early 2000 opened the door for the development of a large-scale natural gas pipeline, the so-called WEP-I. Despite very limited proven gas reserves (well below the 1,000 bcm needed to justify the 4,000 km pipeline development), WEP-I was completed in 2004 and WEP-II was approved very shortly afterwards when China was authorized by Turkmenistan to take up the equity gas option for upstream gas sector development. In other words, China's domestic policy of constructing long-distance trunk pipelines helped to concentrate its focus on the development of oil and gas supply sources in the Central Asian Republics, in particular crude oil in Kazakhstan and natural gas in Turkmenistan. Even though there was no domestic crude oil pipeline to parallel WEP-I and II, the momentum was laid for the parallel trunk gas pipeline development (WEP-III and IV). Kairgeldy Kabyldin, the CEO of KazMunaiGaz, said at the Eighth Russian Oil and Gas Conference, that the Chinese government had proposed to increase the capacity of the pipeline running from Kazakhstan to China from 30–40 bcm/y to 60 bcm/y.[128]

What followed in spring 2011 were two announcements on the increase of gas supply: from 40 bcm/y to 60 bcm/y from Turkmenistan (in March 2011)[129] and from 10 bcm/y to 25 bcm/y from Uzbekistan. The increased volume alone requires another 35 bcm/y capacity WEP pipeline development. In April 2011, coincidently, China's NEA said that construction of the third, fourth and fifth West-to-East gas pipelines would begin in the next five years.[130] The most important message from the NEA's plan is the expansion of the WEP corridor from WEP-IV to WEP-V. The so-called 'WEP Corridor' development would allow China to maximize the scale of gas imports from both the Central Asian Republics and Russia.

What China has been looking for is a way of applying the Central Asian model to its natural gas imports from Russia; but Gazprom is defiantly refusing to give China opportunities to develop the value chain of businesses related to gas supply from Russia. In short, China's initiative in producing oil and purchasing gas assets in the Central Asian Republics was accepted by the Republics, while Russia continued to block oil and gas acquisitions by Chinese NOCs (even though in 2006 acquisition had been approved for the first time). A series of acquisitions of oil companies in Kazakhstan by Chinese NOCs and financial institutions, along with the equity gas option given to the Chinese NOCs, allowed Beijing planners to make a commitment to related pipeline development. The equity gas option was a huge incentive to the Chinese NOCs, as it allowed them to enter the value chain; upstream, midstream, and downstream activities could be established and this would be enough to cushion the financial burden imposed by the high price of the imported gas. The development of value chain businesses is the core of the Central Asian Model. However, this Central Asian Model was never viewed positively by the Russian authorities, and that was the reason why the loan for oil option (which became the core concept of the Russian Model) was developed by the Chinese planners. The following chapter will show what has been achieved through Sino-Russian oil and gas cooperation. Detailed research will reveal what were the limits of that cooperation.

CHAPTER 7

SINO-RUSSIAN OIL AND GAS COOPERATION SINCE 2000

During the first decade of the new century China became the world's largest user of primary energy. Given that its neighbour, Russia, has a large oil industry and massive gas reserves, the need for Sino-Russian energy cooperation seems compelling. If this has not yet taken the obvious form of large-scale sales of Russian gas to China, it has, not surprisingly, taken many other forms, traditional and innovative. Some of these have dynamic possibilities for the future, regardless of whether the major question of gas exports from Russia to China is resolved.

The Oil Sector

The Angarsk–Daqing Pipeline

The visit of China's Prime Minister Zhu Rongji to Russia at the end of February 1999 injected new enthusiasm into the Sino-Russian bilateral oil and gas pipeline programmes. Of the 11 agreements signed after this, the fourth regular meeting between Zhu and his Russian counterpart, Yevgeny Primakov, three concerned trans-boundary oil and gas pipelines (one oil and two gas). The oil-related accord was for a pre-feasibility study of the Angarsk–Daqing pipeline. According to a tentative proposal, this pipeline, with an expected diameter of 920 mm, would enter Chinese territory at Manzhouli in Inner Mongolia and be linked with the Daqing oil field, from where oil could be further shipped to inland provinces.

Even though up to that point all pipeline negotiations with Russia were exclusively handled by CNPC, in the late 1990s SINOPEC also began to explore the possibility of oil supply from Russia. Unlike CNPC's focus on natural gas supply from Irkutsk, SINOPEC was more interested in a crude oil pipeline. This had two possible routes. One was the east line which would pass through three north-east provinces along with the gas pipeline; the other was the west line which would pass across Mongolia and cross the Chinese border at Inner Mongolia. SINOPEC estimated that China would have a crude oil shortage of 6 mt in 2005 and 10 mt in 2010.[1] However, SINOPEC's initiative failed to receive the blessing of the Chinese leadership.

After the second session of the sub-committee on energy cooperation of the regular Sino-Russian Commission, held on 20 and 21 March 2000 in Beijing, a bilateral agreement was signed by SDPC's chairman Zeng Peiyan and the Russian Fuel and Energy Minister, Victor Kalyuzhnyi. Included in this agreement was a trans-national oil pipeline designed to move 30 mt/y of oil from West and East Siberia to China.[2]

Yukos–CNPC feasibility study for the oil pipeline
On 17 July 2001, an agreement on the design of a feasibility study for the construction of an oil pipeline from Russia to China was signed in Moscow (during President Jiang Zemin's visit). The agreement was signed for the Russian side by the Ministry of Energy, Transneft, and Yukos while the SDPC and CNPC signed for China. The pipeline length was to be 2,437 km, 1,642 km in Russia and 795 km in China: the cost of the project was $1.7 bn. Phase 1 (2005–10) would handle 20 mt/y and Phase 2 (2010–30) 30 mt/y.[3] The two sides agreed to surrender proportional profits for a compromise pricing formula. Once the deadlock on pricing was broken after Jiang's trip, CNPC as the state-designated negotiator for Sino-Russian oil and gas cooperation would conduct feasibility studies[4] with the Russian partners, Yukos and Transneft. Since pricing was viewed as the hardest part of the whole process to resolve, CNPC believed that doing the detailed technical work before the scheduled construction start-up in 2003 would help the whole project to proceed more smoothly. The pipeline would source crude supply from both West and East Siberia. Three major oil fields in West Siberia would contribute up to 20 mt and over 10 marginal oilfields would guarantee the rest. It would start to pump 20 mt of oil annually to north-east China to offset the output depletion at Daqing. PetroChina had already announced a plan to expand the refining capacity of Dalian Petrochemical from the existing 7.1 mt/y to 20 mt/y.[5]

China's top priority in 2001 was given to oil supply from Russia, rather than gas. CNPC had learned that the China–Kazakhstan crude oil pipeline development could not be implemented as planned due to limited crude oil availability. In parallel with CNPC's initiative regarding an Angarsk–Daqing crude oil pipeline, SINOPEC, and even Sinochem, had expressed their interest in exploration and development in Russia. Were it not for an order from China's State Development and Planning Commission (SDPC) giving CNPC exclusive authorization in any petroleum-related negotiations with Russia, SINOPEC might have talked the Russian side into redesigning the route of the transnational oil pipeline linking Angarsk to Manzhouli and then to Daqing. Having failed to bring Russian oil via pipelines to its refineries in north China, SINOPEC pinned its hopes on exploration and development, the Tomsk region in west Siberia being the company's target.

With rapidly depleting reserves and declining production, the authorities in Daqing began to cast their eyes towards the Verkhnechonskoye oil field in the Irkutsk region. Daqing signed an agreement with Yukos and Rosneft to inject both technology and funds into the Verkhnechonskoye oilfield, 1,000 km north of the Chinese border. In early September, in a parallel development, PetroChina inaugurated a wholly-owned subsidiary company called China Petroleum International (Exploration & Development) Company (CPIC). CPIC is wholly responsible for PetroChina's overseas E&D activities. While earlier cooperation with Russia had all been carried out by CNPC, now CPIC would manage Daqing's Russia adventure. Shou Xuancheng, the head of CPIC, replied very cautiously to questions about possible Daqing investment in Russia. Shou noted that 'cooperation is at a preliminary stage and no specific investment plan has been decided for the moment'. Separately CNPC was seeking to join the project to develop the Chayandinskoye gas field and Talakanskoye oil field, both located in the Sakha Republic. Resulting production would be sent via the putative Sino-Russian oil and gas pipeline to Chinese markets.[6]

Shi Xunzhi, the Chinese chief negotiator of the two bilateral projects, said the feasibility study could not be finished until 2003 – when the pipeline construction was planned to start. The key issue in their negotiation for the Chinese side was that Russia must establish reserves sufficient for 25 years of oil delivery. After a two-year negotiation, the two sides worked out a price formula based on the weighted prices of Brent, Minas, and Oman oil, designed to make Russian crude competitive with imported crude from other sources.[7]

As part of its negotiations with the management of Yukos in Moscow on 14 August 2002, CNPC expressed its readiness to provide credit for construction of the Russian section of the Angarsk–Daqing oil pipeline. Shortly thereafter, PetroChina announced its intent to invest $483 million in the construction of the pipeline and also expressed its wish to acquire a stake in the field from which oil was to be supplied. By placing its bet on Yukos, the clear favourite in oil and politics from 1996 to 2002, China appeared to have chosen a winning tactic for expansion in Russia. As soon as Yukos was displaced as a Russian heavyweight, the Talakanskoye field, the primary resource for the oil pipeline to Daqing, was turned over to Surgutneftegaz.[8] By that time CNPC had commenced the expansion of the Dalian refinery with a planned capacity of 20 mt/y to be reached by 2005.

Russian crude supply to north-east China and its implications for the related refinery revamping

PetroChina undertook to refurbish its north-eastern oil pipeline grid in preparation for the import of eastern Siberian oil. PetroChina had handed a feasibility study on the renovations to CNCP's in-house consulting company for advice. This study reviewed several options but the one certainty was that a 524 km, 10 mt/year pipeline would be built parallel to the two existing Daqing–Tieling lines, enhancing the total transportation capacity of the north-eastern oil pipeline grid to 55 mt/year. The new line would be used for Daqing oil transportation between 2005 and 2009 and then switched to Russian oil transportation from 2010 onwards. The State Development Planning Commission had agreed that all the Russian oil transported by the trans-national pipeline would be consumed in north-east China, to compensate for declining output of the Daqing and Liaohe oilfields.[9]

Table 7.1: China's North-Eastern Pipelines

	Diameter	Length	Capacity	Status
	mm	km	mt/y	
Daqing–Tieling 1	770	517	45	In operation
Daqing–Tieling 2	770	517	45	In operation
Tieling–Qinhuangdao	720	454	20	In operation
Tieling–Dalian	720	460	20	In operation
Tieling–Fushun	720	43.7	20	In operation
Fushun –Anshan	426	117	5	Abandoned
Anshan– Xiaosonglan	529	279	10	Never used

Source: Li Yuling (2003c), 1.

Due to their geographic proximity to the northern consuming markets, Dalian Petrochemical, WEPEC, Jinzhou Petrochemical, and Jinxi Petrochemical had been assigned the task of refining the Russian oil, which would be 20 mt/y before 2010 and 30 mt/y thereafter. In 2003 it was projected that prior to 2010 the Russian oil would mainly be processed in the two refineries in Dalian – 15 mt/y by Dalian Petrochemical, with refining capacity stands of 7.1 mt/y, and 5 mt/y by WEPEC. At that time Dalian Petrochemical was overhauling its refining units and enhancing its capacity to 20 mt/y to accommodate the Russian oil that was expected to arrive in 2005. After 2010, Jinzhou Petrochemical and Jinxi Petrochemical, with a combined capacity of 11 mt/y, would use the additional 10 mt/y of Russian oil.

PetroChina was thinking of assigning part of the Russian oil to Jilin Petrochemical (which could process 5.6 mt/y of crude), partly in

Table 7.2: China's North-Eastern Refineries

	Refinery Type	Key Products	Capacity mt/y	Sulphur content of crude processed
Daqing Petrochemical	Fuels - lube oil- petrochemical	Lube oil, wax oil	6	Low
Daqing Refinery	Fuels – lube oil	Lube oil, wax oil, additive	8	Low
Harbin Petrochemical	Fuels	-	3	Low
Jilin Petrochemical	Fuels - petrochemical	Basic organic materials	5.6	Low
Qianguo Petrochemical	Fuels	-	9.2	Low
Fushun Petrochemical	Fuels – lube oil - petrochemical	Wax oil, Lube oil, Surfactant	5.5	Low
Liaoyang Petrochemical	Fuels - petrochemical	Fibres	5.5	Low
Liaohe Petrochemical	Fuels	Bitumen, lube	4.1	Low
Jinzhou Petrochemical	Fuels	Needle Coke	5.5	Low
Jinxi Petrochemical	Fuels	-	5.5	Low
Dalian Petrochemical	Fuels – lube oil	Lube oil, wax oil	7.1	Low
WEPEC	Fuels	-	5	High

Source: Li Yuling (2003c), 3.

order for it to upgrade its refining capacity, and partly to lift its loss-making petrochemical companies out of the red. Since WEPEC was the sole refinery with the ability to process high sulphur content oil, a huge amount of capital would be needed to revamp the refining units of Dalian Petrochemical, Jilin Petrochemical, Jinzhou Petrochemical, and Jinxi Petrochemical. The refurbishment of Dalian Petrochemical required 9.8 bn yuan, and Jinzhou Petrochemical and Jinxi Petrochemical together required about 5 bn yuan.

PetroChina was considering building a product pipeline to north China, in order to shift the possible surplus of oil products resulting from Russian oil imports in north-east China, where the market was relatively satiated. While around 15 mt/y of products would be shipped from Dalian to the south, about 6.5 mt of products were expected to flow into the north China market in 2005. The products were aimed at SINOPEC, which dominated the north China market. According to PetroChina, the freight cost of transporting products from north-east

China to north China by railway then stood at 130 yuan/tonne. The cost would fall to around 50 yuan/tonne after the pipeline was built.[10]

To prepare for the increase in imports, PetroChina started an expansion of its Liaoyang refinery in north-eastern China, its main centre for processing Russian oil.[11] The projected increase in oil imports from Russia led PetroChina to undertake a significant increase in oil storage capacity. It boosted oil storage capacity at the Daqing field in order to prepare for greater imports of Russian crude via the ESPO, which will terminate in Daqing. PetroChina has installed two tanks for offloading Russian oil, and eight more were planned for completion by the end of 2010, boosting Daqing's crude oil storage capacity by nearly 10 million barrels.[12]

All this is evidence that CNPC was totally convinced the Russian crude oil would be supplied to north-east China as planned. CNPC had never expected any delay to, let alone suspension of, the Angarsk–Daqing pipeline plan, but this was the beginning of a pipeline saga which started with the initiative taken by Transneft in 2002.

Transneft's Low Profile but Bold Initiative

On 2 April 2002 Transneft gave a first public presentation in Vladivostok about its project to build an export pipeline from Angarsk to the Russian Far East coast. This envisaged that the Angarsk–Nakhodka crude pipeline would have a total length of 3,765 km, a diameter of 1020–1200 mm, and 50 mt/y of delivery capacity. The pipeline's loading piers would be capable of handling tankers of up to 300,000 tonnes deadweight, with 4 mcm tank capacity. There would be between 22 and 29 pumping stations (depending on the route) and it would cost $5.0–5.2 billion. Transneft's initiative was very appealing to the Primorskii Territory administration and the two parties signed a declaration of intent to build the pipeline. In the words of Yuri Likhoyda, then deputy governor of Primorskii Territory:

> ... revenues from the pipeline alone can keep the territory going. Its operation will generate 20 billion rubles in taxes a year ($645 million), while at present the revenues flowing into the budget stand at around 12 billion rubles ($387 million). For this reason, we are prepared to take on the Khabarovsk Territory, which is lobbying for the option of ending the pipeline in their port of Vanino, rather than in our port of Nakhodka.[13]

According to Anatoly Bezverkhov, head of the Transneft Construction and Strategic Development Department, who headed its delegation to Vladivostok, the initiative had been endorsed by President Putin and the

Russian government. A week later Putin received Transneft's president Semyon Vainshtok in the Kremlin, when the main topic of discussion was the Baltic Pipeline System (BPS) and Putin also approved Transneft's route; but according to Sergei Grigoryev (the Transneft VP and spokesman) the company was inclined to end the pipeline in the port of Nakhodka on Dong Hae (East Sea in Korean), or the Sea of Japan. Administration officials made no public statement to this effect in order to avoid possible embarrassment, because of the fact that Russia had earlier signed documents with China providing for construction of the Angarsk–Daqing pipeline. The Transneft feasibility study was carried out by its subsidiary Giprotruboprovod, a Moscow based Institute for Trunk Pipeline Design. According to Bezverkhov, the project would be financed by Transneft's own funds and through loans.[14]

What drove Transneft to put forward the Angarsk–Nakhodka crude pipeline? On 3 April 2002 Transneft's vice president, Sergei Grigoryev, said that:

> ... global considerations rather than the cost of the project is what one should have in mind. The Chinese pipe will be cheaper to build, but it would provide access to just one market. And what if all of a sudden they no longer needed our oil? The Pacific pipe ensures access to markets in the USA, Japan, South Korea, south-east Asia, Australia, and China too.

This was the core idea behind Transneft's bold initiative.[15]

In 2002, Nakhodka's capacity stood at 4.2 mt of oil and oil products per year. If adequately modernized, the port would have the capacity to handle 150,000–500,000 tonne super tankers, which would require either three new floating tanker facilities connected to the shore by a new 3.5 km subsurface pipeline, or five existing oil transfer piers which would need to be modernized. The pipeline and port oil terminals would cost $4–5 billion; the pipeline alone would cost $2.3–2.56 bn (depending on the route). The project would span 23 years, including three years to complete the construction of the pipeline. Analysts believed that during the initial 10 years of the loan payback period, a tonne of crude oil would cost a minimum of $25–29 to be pumped from Angarsk to Nakhodka. At a later stage, tariffs were expected to go down to $14–15 per tonne.[16]

Based on the preliminary feasibility study by Transneft, Russia welcomed Japan's initiative taken in early 2003 and outlined below. Transneft's 2002 initiative indirectly confirms that Moscow was already considering the alternative to the Angarsk–Daqing crude oil pipeline project being pursued by Yukos and CNPC, well before the collapse of Yukos.

Japan's Initiative and China's Response

In early January 2003, it was reported that President Putin and Prime Minister Koizumi were expected to announce an agreement to cooperate on a 4,000 km pipeline to export oil from Siberia to East Asia. This $5 billion venture would provide the first significant outlet for Russian energy production to East Asia and eventually to the US west coast, reducing those regions' reliance on Middle East producers.[17] The plan, supported by Transneft, forced the Russian government to reconsider a scheme from the Russian oil group Yukos to supply eastern Siberian oil to China. The Transneft plan was believed to be preferred by Moscow. Another assumption was that the scheme was likely to be favoured by the USA as it would open Russian oil supplies to a wider market than China.

On 13 January 2003 a Japanese government official said that preliminary studies had indicated that the project was attractive, and that several state financial institutions would be involved in backing it. This is the strongest indication yet of Japanese determination to push ahead with the pipeline in the face of opposition from Yukos, which was in discussions to build the alternative $1.8 bn, 2,400 km pipeline from Angarsk to Daqing.[18]

In the wake of Transneft's proposal, CNPC's president, Ma Fucai, told a Russian journalist that the Angarsk–Daqing line would be a win–win project for both countries. On the Russian side, the FOB price of the crude exported to China through the line would be $0.5/barrel higher than that for sales to the European market, in sharp contrast to early rumours that China's toughness on the oil price had stalled the pipeline. He also revealed that preparations for the import of gas from Russia were going smoothly. At a conference held in Seoul in January 2003, representatives from Russia, China, and South Korea agreed to wrap up the feasibility study of the gas line by June.[19]

In early March 2003 Iwao Okamoto, the head of Japan's Energy Agency, met Igor Yusufov, Russia's energy minister, to argue for a link from Angarsk to Nakhodka. The talks came amid signs of a compromise ahead of a Russian government cabinet meeting on 13 March, which was scheduled to decide between the Pacific link and the competing project running from Angarsk to Daqing. Russian officials indicated that they were interested in combining a single 50 mt pipeline system which would have two branches, the main one to Nakhodka (Kozmino) along with a smaller spur from Skovorodino for CNPC. On 4 March, *China Daily* said that the authorities might agree to the extension of the pipeline to Japan as long as Russia met its commitments to supply adequate quantities of oil to China. Japan was said to be considering ways to support the pipeline it favoured, via direct government support in the form of soft loans, or by providing finance from the Japan Bank for International Cooperation (JBIC).[20]

On 13 March, the Russian government notified CNPC that it would build both pipelines (the Angarsk–Daqing line and the Angarsk–Nakhodka line) although the Daqing line would be given priority. A PetroChina official remarked that 'most of the dust related to the contest (between the two lines) has settled'.[21] That was when *Xinhua* reported that PetroChina was preparing to revamp its north-eastern oil pipeline grid, to prepare for the arrival of eastern Siberian oil.

On 27 April 2003 President Vladimir Putin and his counterpart the Chinese president, Hu Jintao, signed a joint declaration which recognized that 'energy cooperation is of overwhelming importance to both countries'. In the statement, the heads of the two countries agreed that carrying out oil and gas projects on the scale of the Angarsk–Daqing pipeline should be the cornerstone of bilateral energy cooperation. This was the first time that Putin had aired his stance on the Russia–China pipeline since the appearance of the so-called Daqing or Nakhodka debate which started in April 2002. Putin's stance signalled the Russian government's eventual determination to push ahead with the line, as well as revealing something of the state of Sino-Russian relations. As a statesman internationally famed for his pragmatism, Putin's final agreement on the line was consistent with Russia's national interest.[22] The Russian government formally needed the tug-of-war – even though Prime Minister Kasyanov had announced on 29 April that Moscow would prioritize the construction of the China bound line, it was still looking for an excuse not to make a definitive choice between the two proposed pipelines (Angarsk–Daqing and Angarsk–Nakhodka).[23]

Russia and China took another step on 28 May 2003 when Yukos and CNPC clinched a general agreement on the main principles and understandings for a long-term contract to supply oil to China via the Angarsk–Daqing pipeline. Under the agreement, Russia would supply 20 mt/y of oil to China for five years after the pipeline became operational in 2005 and the volume would rise to 30 mt/y in the following 20 years. The 25-year deal would involve a total volume of 700 mt, then estimated to be worth $150 billion. On the same day Yukos and CNPC also signed a $1.1 bn contract to export 6 mt of Russian oil by rail between 1 June 2003 and 1 June 2006, as a stopgap before the pipeline started operating.[24]

Japan offered a renewed financial package worth $7 bn in a further attempt to persuade Russia to build an oil pipeline across Siberia to the Pacific, instead of the rival route into north-east China. On top of the $5 bn to support construction of a 4,000 km pipeline to Nakhodka, Japan proposed a further $2 bn for the development of Siberian oilfields, with some aid in the form of low-interest loans.[25] On 2 September Russia's Ministry of Natural Resources announced that it would reject both suggested routes of the Angarsk–Daqing pipeline, citing environmental reasons. The northern route of the pipeline, which would run along the Baikal–Amur railway, would pass through Russia's Tunkinsky Natural Park, and the southern route, proposed by Yukos, would run along the southern shore of Lake Baikal. The Russian side even refused to attend the routine meeting of the sub-committee on energy cooperation of the Sino-Russian Commission on 25 August.[26] Prime Minister Mikhail Kasyanov stated that Russia would live up to its promises about the Russia–China pipeline, but three to four months would be needed to make the pipeline 'technically and environmentally' viable and that meanwhile Russia would increase its oil supply to China by rail from the current 3 mt/year, to 4.5–5 mt/year by 2005 as compensation for the pipeline delay.[27]

The 2003 pipeline saga drove the Beijing planners to rethink their strategy. Around this time China decided to reactivate its Central Asian oil supply option. It reached agreement with Kazakhstan on starting construction of the Atasu–Alashankou pipeline, the second phase of the Kazakhstan–China oil pipeline. A PetroChina executive in charge of the pipeline said that 'the Kazakhstan–China pipeline also tops our agenda' and the Central Asian option was 'no longer a backup to the Russia–China pipeline'.[28] This was the confirmation that the Chinese authorities' patience was running out and that they had decided to prioritize pipeline development with Central Asia. Unaware of the change in China's stance towards the pipeline from Russia, Prime Minister Kasyanov said on 16 December that if the Angarsk–Daqing line and Angarsk–Nakhodka line could not be combined into one, two separate pipelines would be built. One is tactical and the other is strategic, he said; the two lines are complementary.[29]

The Rise and Collapse of Yukos

Yukos was the basis of Mikhail Khodorkovsky's rise to fortune. It was a by-product of the Loans for Shares Scheme. Khodorkovsky was one of the 12 founding members of the so-called Menatep (which means Intersectoral Centre of Scientific Technical Progress), which agreed to take the government's stake in Yukos (a petroleum company that had been spun out of Rosneft in November 1992) as collateral for its loan.[30]

Khodorkovsky took over 88 per cent of Yukos stock with a payment of $350 million, even though the market value of Yukos was as much as $3–5 bn. Because of its rapid expansion, Yukos made a number of enemies, including Transneft.

It is not an exaggeration to say that the seed of Yukos's tragedy was sown when Khodorkovsky not only decided to take on Transneft by threatening to end its monopoly in the European part of Russia, but also began a campaign to construct a pipeline between East Siberia and China. Sakwa pointed out that 'The Angarsk–Daqing pipeline would have broken Transneft's effective monopoly on oil transport, and thus removed one of the last instruments whereby the state could control the company'.[31] As mentioned earlier, Yukos had signed a 20 year oil supply contract with China on 28 May 2003. Khodorkovsky and Yukos acted as if they were sovereign powers, whereas, from Putin's point of view, what Khodorkovsky had done (in effect, make foreign policy towards China) was the state's and his (Putin's), not an oligarch's, prerogative. Khodorkovsky pushed his luck too far. With regard to the Northern oil buyout,[32] he attacked CEO of Rosneft, Sergei Bogdanchikov's overpayment for the asset. However, Putin protected Bogdanchikov's stance and decided to punish Khodorkovsky. On 25 October 2003 Khodorkovsky was arrested at Novosibirsk airport and in December 2004, Yukos's most valuable asset, Yuganskneftegaz, was sold at yet another rigged auction for $9.35 bn to an obscure and hitherto unknown entity named the Baikal Finance Group, which turned out to be a shell company owned by Rosneft.[33] As discussed later, the collapse of Yukos and the sales of Yuganskneftegaz offered a unique opportunity for CNPC to lend $6 bn to Rosneft, which had massive reserves but minimal production capacity. Ironically, CNPC's mistake of taking Yukos too seriously subsequently led to its large-scale lending to the State oil company Rosneft.

Moscow Wisely Stays in the Middle

In the spring of 2004, China adopted a tough stance towards the ESPO (East Siberia–Pacific Ocean)[34] oil pipeline route. CNPC said China would not participate in the building of the Taishet–Nakhodka oil pipeline without a branch to Daqing. Taishet–Nakhodka without the Daqing branch line would be an economic disaster in which China would not invest. Japan expected to win a high-stakes struggle with China over the route of a $10 bn oil pipeline from the Russian Far East to north Asia, but had no other leverage which might persuade Russia to fully commit itself to the Taishet–Nakhodka line. Sergei Grigoryev, vice president of Transneft, highlighted again that the whole idea of the ESPO oil pipeline was to avoid having too many of their eggs in one basket. This was not so much a pipeline to Japan as a pipeline to all the countries in the Asia–Pacific region.[35] Moscow did not give any sign as to whether Japan's wishes would be fully reflected in its final decision.

In the summer of 2004, Russia was seemingly mapping a new export route, namely the Taishet–Nakhodka line. A feasibility study on this new route was underway, and related work was due to conclude in July. Based on the study, the Russian government would make a final decision on the direction of the pipeline before Prime Minister Wen Jiabao's visit to Moscow in September.[36] However, China's position was very clear. PetroChina repeated that it would not participate in building the Taishet–Nakhodka oil pipeline without a branch to Daqing.[37] During the sixth session of the sub-committee on energy cooperation of the regular Sino-Russian Commission in Beijing on 25 August 2004, Russia's Minister of Industry and Energy, Viktor Khristenko, said there was no alternative to the Taishet–Nakhodka line. The Transneft pipeline feasibility study had been presented to relevant state agencies in late July 2004 for technical and ecological evaluation.[38]

In September 2004, following Chinese Prime Minister Wen Jiao's visit to Russia, the key problems remained unanswered, despite media reports that Sino-Russian energy cooperation had made substantial progress. Prime Minister Wen signed a joint communiqué with his Russian counterpart, Mikhail Fradkov, pledging full-scale cooperation in the energy sector. The joint communiqué included the following points: i) Russia would continue to strengthen oil and gas cooperation with China; ii) Russia would fully evaluate the route of the Far Eastern oil pipeline and consider extending the pipeline to China irrespective of the route; iii) both sides agreed to increase oil trade by rail, with exports expected to reach 10 mt in 2005 and 15 mt in 2006; iv) it was agreed to produce a natural gas development plan as soon as possible, but no reference was made to any specific project.[39]

On 25 September 2004 German Gref, Minister of Economic Development and Trade Development, disclosed that the Russian government had decided to adopt the Taishet–Nakhodka pipeline with a branch line to China. But China wanted the branch line to have priority. *Xinhua News Agency* reported that the stalling of the Angarsk–Daqing pipeline had taught the Chinese government a profound lesson: never put all your eggs in one basket.[40] China felt the necessity to find an alternative in advance. When Prime Minister Wen Jiabao met Kazakhstan's Prime Minister Daniyal Akhmetov (at the third meeting of the Shanghai Cooperation Organization (SCO) in Bishkek in September 2004), he said that Sino-Kazakh energy cooperation was a win–win situation,[41] indirectly confirming China's frustration about Sino-Russian oil pipeline development.

Putin Elaborates Russia's Stance

On 14 October 2004, Vladimir Putin began a state visit to China, accompanied by a large delegation which included several ministers, military officials, and five regional governors from the Russian Far East. Before leaving Moscow, Putin made it clear that he would not be rushed into deciding between two projects: an $18 billion pipeline from the Kovykta gas fields in Siberia and another pipeline, which was also being sought by Japan. 'I hope you will understand me,' he said 'when I say sincerely and frankly: first of all, we need to satisfy our own national interests – we should develop Russia's Far Eastern territories'.[42]

On the second day of this summit meeting a total of 13 documents were signed following talks between Presidents Putin and Hu Jintao in Beijing. They included CNPC and Gazprom's strategic cooperation agreement,[43] a joint presidential declaration, an additional agreement on the eastern sector of the Russia–Chinese border, an intergovernmental protocol on travel by Russian and Chinese ships in the area near Tarabarov and the Bolshoi Ussuriisky Islands, and a protocol on the end of talks on opening the two countries' markets to each other's goods and services.[44] However, President Putin left many key questions about energy supplies to China unanswered. In particular, the oil pipeline issue was not discussed at all during the visit. Arkady Dvorkovich, head of Putin's team of economic experts, confirmed that building a pipeline to China was not a pressing task on Russia's agenda.[45]

President Putin elaborated:

> The development of pipelines in Eastern Russia will be done in the interests of the Russian regions concerned and the country's partners. One issue is routes for movement by our pipeline transport. Most of all, we have to proceed from our national interests, we have to develop the Russian Federation's Far Eastern territory, so we have to plan and implement major infrastructure projects in those areas. So, any final decision will be made proceeding from these considerations, but we are taking the interests of our partners into account … We think that the Chinese People's Republic, as a consumer of this energy source, is not only interested in transportation infrastructure (including pipeline transportation) developing in the Russian Federation's east region, but also in the development of this transportation infrastructure as a stimulus to additional geological prospecting work. In this way Russian energy resources are fully evaluated, so that Chinese

consumers, considering plans for developing China's economy, can know how on what scale and, in what timeframes from Russia they can import resources from Russia … We have an absolutely open dialog with our Chinese partners. We are aware of China's interest in the stable receipt of energy resources from Russia. And Russia is interested in China being a reliable and stable partner with growing needs for energy resources … [Putin insisted that] there are neither political, ideological nor economic problems here that could hamper our development of relations in the sphere of energy … We want to and will be working with China, keeping in mind the prospects for the development of its northern, north-western and north-eastern territories, and we will be working also on the matter of shipping raw energy materials to China.[46]

Russia's decision was not to show any favouritism towards to either Japan or China, but to prioritize its own interest.

Russia's Strategy of Delay

The impact of the decision for a $6bn loan for oil to Rosneft in early 2005 was considerable. Russia's stance not to favour either China or Japan caused the latter to rethink its strategy. In spring 2005, Japan's trade minister Shoichi Nakagawa threatened that, if Moscow pursued a plan to build a spur to China first, Tokyo would withdraw its offer to help to finance a $11.5 bn oil pipeline from Siberia to Russia's Pacific coast. Russia's energy minister Viktor Khristenko signed an order calling for the first $6.5 bn phase of the pipeline to be laid from Taishet (in Irkutsk) to Skovorodino (close to the Chinese border in the Amursk region) by late 2008. A vice president of Transneft said that the company did not need Japanese loans because, according to Cyrus Ardalan, the vice chairman of Barclays Capital, Transneft would have no problem raising finance for the ESPO oil pipeline. Transneft's size, state ownership, and its strategic role in the Russian economy combined to make it an attractive counterpart.[47]

During his visit to Russia from 30 June to 3 July 2005[48] President Hu Jintao and President Putin worked out strategic cooperation agreements on oil through the medium of their state-owned oil companies, CNPC with Rosneft, and SINOPEC with Rosneft.[49] CNPC and Rosneft agreed to study opportunities for increasing oil deliveries to China under mutually profitable conditions, while Rosneft and SINOPEC expected to establish a joint venture to carry out joint geological exploration and study in Sakhalin Block 3. One week later President Putin confirmed that Russia would prioritize China over Japan as the recipient of oil supplies from the Taishet–Nakhodka oil pipeline.[50] However, the Russian authorities still did not confirm whether priority would be given to the development of the pipeline between Skovorodino and Daqing oilfield.

In November 2005, the Russian Prime Minister's visit to Beijing ended with a joint pledge of possible cooperation in space exploration, nuclear power, and natural gas, but no agreement on a crude oil pipeline. At the closing press conference, Russia's Prime Minister Fradkov and his Chinese counterpart, Wen Jiabao, gave no indication of a deal having been reached on a planned pipeline to export the still undeveloped East Siberian oil to the Pacific coast via China. Fradkov said that the construction of thermal and nuclear power plants was the main priority for Russian investments in China. There was agreement that exports of Russian oil to China by rail should reach at least 105 million barrels (15 mt) in 2006.[51]

Green Light for the East Siberia–Pacific Ocean (ESPO) Pipeline

The Russian government and President Putin were determined to go ahead with the ESPO oil project, partly as a means of avoiding dependency on one single buyer of oil, namely China, partly as a way to boost Russian influence in the Pacific region, and partly because construction of the ESPO pipeline would also speed up socio-economic development in the eastern part of the country. The development of the eastern regions of Russia is particularly significant because of the worsening demographic situation and the creeping Chinese immigration which creates fears of losing control over the region.[52]

Minister Khristenko described the ESPO oil pipeline as a window to Asia:

> The 'eastern project' provides an outlet to the Pacific coast. Herein lies its essence, it is diversification of the supply routes. The European market will remain the key market place for Russia and at the same time the share of Asia–Pacific countries in Russian oil exports will grow from the current 3 per cent to 15–18 per cent in 2015.[53]

This confirms the importance that the Russian leaders gave to developing the ESPO pipeline.

Finally, on the last day of 2004 the Russia's Prime Minister Mikhail Fradkov signed Directive no. 1737-r which gave the go-ahead for the design and construction of the ESPO oil pipeline system. [54] The proposal had come from Minpromenergo and Transneft. Their joint proposal was coordinated with key federal agencies and the justification for investment in the project had received favourable reviews from the State Environmental Expert Review Committee.

Fradkov gave Minpromenergo the task of coordinating, monitoring, and supervising pipeline design and construction, and of reporting on progress every six months. The prime minister also charged MNR with drawing up a programme for geological study, and assigning East Siberian and Far East hydrocarbon deposits to subsoil users. MNR would coordinate the programme with Minpromenergo and the Ministry of Economic Development and Trade (MEDT) and the MNR would grant final approval. Minpromenergo, MEDT, and MNR, in collaboration with Transneft, were asked to make proposals about the construction stages of the ESPO pipeline system by 1 May 2005.[55]

In summary, Prime Minister Fradkov gave the following instructions:

• Minpromenergo was to coordinate, monitor, and supervise pipeline design and construction and report on progress every six months;
• MNR was to draw up plans for a geological study and to assign fields to subsoil users;
• MNR was to coordinate the programme with Minpromenergo and MEDT, with MNR granting the programme its final approval;
• Minpromenergo, MEDT, and MNR, in collaboration with Transneft were to determine the construction stage of the East Siberia–Pacific crude oil pipeline system by 1 May 2005;
• Ministry of Transport (Mintrans) and Ministry of Defense to develop arrangements for entry and exit of vessels to and from Perevoznaya's harbour to ensure navigational safety.

Fradkov assigned to Mintrans, with the help of the open-joint stock company Russian Railways, the task of developing and implementing measures for delivery of construction materials for ESPO. He also directed the Federal Tariff Services (FTS) to set oil transport tariffs to include the costs of pipelines from western Siberia to Taishet. He suggested that FTS include in the tariffs the financing costs for loans for the design and construction of the ESPO.

ESPO Construction

On 26 April 2005, Minister Khristenko, signed a decree (no. 91) establishing the stages of construction for the ESPO oil pipeline system. This was the start of a new chapter of East Siberia's development.

> I order [Khristenko said] the approval of a Transneft proposal on the stages of construction of the ESPO oil pipeline with a total capacity of up to 80 mt per year. This puts into effect Russian government Directive No. 1737 signed on 31 December 2004 and resolves issues connected with

the construction of [an] ESPO oil pipeline, taking into consideration the opinion of the Ministry of Economic Development and Trade.[56]

The first phase would be the construction of an oil pipeline along the route Taishet (Irkutsk Region)–Ust-Kut Area (Irkutsk)–Kazachinskoye (Irkutsk)–Tynda (Amur Region)–Skovorodino (Amur Region).[57]

On 10 November 2005, Minpromenergo submitted an integrated operational schedule to design and build the ESPO oil pipeline system. The government appointed Transneft as controller of ESPO pipeline design and construction. The project feasibility study had been shown to Rostekhnadzor (Federal Service for Environmental, Technological and Nuclear Supervision), the primary state expert review board, and to other supervisory bodies for environmental impact assessment.[58] The result of the review and the approval of the feasibility study were scheduled for completion on 30 December 2005. Rostekhnadzor was to deliver its conclusion by 28 December. Rosprirodnadzor and Rosnedra, two departments of the MNR, were to deliver theirs by 11 November 2005.

Transneft's feasibility study aimed at completing the construction and installation of the linear part of the pipeline before the end of August 2008; oil pumping stations were to be put in place before 10 September and the terminal in Perevoznaya (rather than Kozmino) was to be completed before 8 November. Commissioning of the first launching complex was scheduled for 1 November 2008.[59] The MNR, however, had serious objections to Transneft's feasibility study. First, the MNR objected to a segment of the route that would run 800 metres from the edge of Lake Baikal. Secondly, it opposed the choice of Perevoznaya Bay as the terminal point for the route. The Perevoznaya area has great biological diversity and is the habitat of the endangered Far Eastern leopard. Even though Prime Minister Fradkov accepted the project with Perevoznaya Bay as the final destination on 31 December 2004, the MNR continued to oppose Perevoznaya Bay as the terminal. In October 2005 President Putin chided the government for baseless foot-dragging regarding the project, and this intervention from the Kremlin caused the MNR to lay aside its objections.[60]

On 6 January 2006 President Putin announced that first phase construction of the ESPO oil pipeline project would start in the summer of 2006. 'The decision has been taken' he said and 'I believe that by April all the agreements will have been concluded and we can anticipate beginning this summer'.[61] And so, in March 2006, Transneft and Pet-roChina signed a memorandum of understanding for the preparation of a feasibility study on the construction of a crude oil pipeline from Russia's Skovorodino to the Russia–China border.[62]

Challenges to the Pipeline Route

Before the actual development, a last hurdle had to be removed. On 26 April 2006 President Putin ordered the pipeline's route to be changed to take it farther from Lake Baikal, the world's largest freshwater lake. The original route would have taken it to within 875 yards of Baikal, raising fears that the lake's unique ecology could be at risk in the event of a rupture. Dmitry Ogulchansky of Transneft said that the decision to reroute the pipeline at least 42 km north of Lake Baikal could extend its length by 1,250 km.[63]

Greenpeace explained the Environmental Concerns:

> Russia's environmental activists were seriously alarmed by the fact that Transneft's route would pass Lake Baikal (a UNESCO World Heritage site) at no more than 800 meters and the terminal will be built on the Amur Bay in Primorye, home to the remaining population of 30 Amur leopards. It is hard to design a route that would do more damage. The pipeline would cross about 50 large and small rivers, tens of motor and railways, as well as seismically active areas with seismic activity of up to 10–11 magnitude. This, along with passages through high mountain ranges, extreme climate and ecological geographical conditions poses a serious threat to construction, use of the pipeline and the provision of its safety. The pipe route will cross the largest river of the Baikal basin – Verkhnyaya Angara River. A rupture of the pipeline and pollution of this river will eventually result in the pollution of the Baikal Lake itself. The route is also going to cross the Amur River. The risk of accidents resulting in oil spills is 17 times higher in the Amur Bay than if an alternative terminal site is selected near Nakhodka or elsewhere in south Primorskii Krai. No other region in Russia has such a high density of protected areas.[64]

Greenpeace's argument against the route was correct. In late February 2004, the government of the Sakha Republic had in fact proposed a new route to Nakhodka for the crude oil supply.[65] The Sakha government also proposed to lay the pipeline in one corridor with the Chayanda–Nakhodka gas line and switch it in future to a pipe linking the Kovyktinskoye, Dulisminskoye, and Yaraktinskoye fields in the Irkutsk Region. The Sakha government claimed that the route proposed by Transneft was ecologically hazardous. More than 1,100 km of the route would pass through the axial part of the Baikal reef zone, of which 100 km would cross a zone prone to earth tremors with a magnitude of over 9.0, according to Yegor Borisov, chairman of the Sakha government. The Sakha Republic's President Vyacheslav Shtyrov argued that the route proposed by the Republic was backed by the MNR, Gazprom, Surgutneftegaz, scientific centres, and the majority of ministries and governmental agencies. So, recognizing the Republic's interest, President Putin (at a conference on development of transportation infrastructure in the Far East held in Khabarovsk on 26 February 2004) had invited Vyacheslav Shtyrov, the president of the Sakha Republic, to submit development studies for a new pipeline.[66]

In fact, MNR had been prepared not to approve construction of the pipeline in such close proximity to Lake Baikal. It had repeatedly asked Transneft to submit a comprehensive plan for ensuring the safety of a pipeline system that would traverse territory listed by UNESCO as a World Heritage site. The MNR was also opposed to building an oil-loading terminal in Perevoznaya Bay, where the uniqueness of the climate and biology provided grounds to reject the plan.[67] The disputes over the pipeline landed on the desk of President Putin in April 2006, and he asked Konstantin Pulikovsky, new head of Russia's Technology Supervision Department, for an opinion. Pulikovsky was given the right to decide the final route of the pipeline, or to modify the current construction scheme.[68] Putin gave the final verdict on the pipeline route, and consequently the initial route through Buryatia in East Siberia was replaced with a more northern one going via the Sakha Republic.[69]

At last all the hurdles were crossed and Transneft did not look back. By April 2007 it had built a third (over 900 km of the 2,700 km) of its planned pipeline to China and was on track to extend the route to the Pacific coast. On 20 July 2007 the Russian government ordered that the FTS, MIE, and MEDT must set the initial price of oil transportation by 15 December 2007.[70]

In July 2007, MIE's deputy Minister Andrey Dementyev told a Cabinet meeting that plans to build an oil pipeline to the Pacific coast would most probably remain on hold until at least 2015. He added that the government expected oil companies to be able to produce 40 mt of crude in the area by 2015, whereas the pipeline's second part had the capacity for 50 mt. Russia could supply six per cent of the Asia–Pacific market from the pipeline.[71] This cautious remark signalled a delay in the second leg of the ESPO oil pipeline development.

Transneft's Management Change

On 12 September 2007, Semyon Vainshtok, CEO of Transneft, stepped down after President Putin appointed him head of the preparation committee for the 2014 Sochi Winter Olympics. On 12 October 2007, Nikolay Tokarev (a director of Zarubezhneft), became the new CEO of Transneft and soon declared that problems with the construction of ESPO would delay the commissioning of the first stage until at least the fourth quarter of 2009. Tokarev pointed out that neither Krasnodarstroytransgaz nor Stroisystema had performed well, and that consequently the construction schedule for the first stage was in ruins: the linear section was 40 per cent completed (the plan called for 67 per cent), while the oil-pumping stations were only 30 per cent ready (the plan was 68 per cent). The situation at Stroisystema was somewhat better, but only 40–50 per cent of what had been expected had been achieved.[72]

In mid-April 2008, Andrey Dementyev, former deputy minister of the MIE, said that the draft plan of the ESPO pipeline branch to China had been virtually prepared, but the construction of the branch line would not begin until Rosneft and CNPC had reached an agreement on the volume and cost of the oil. At the end of May, President Dmitry Medvedev said Russia and China had reached agreement in principle on the construction of a branch of the ESPO pipeline to China.[73]

The Interim Alternative: Oil to China by Rail

Until the spur line from Russia's Skovorodino to Daqing field was completed, crude oil supply from Russia to China by rail would continue to be important. Before its total collapse, Yukos was desperate to save this lucrative oil trading business by rail. In February 2004, Yukos had signed contracts to supply oil to China in 2004–5. In 2004, it was set to supply 3.86 mt to CNPC and 2.55 mt to SINOPEC, and the figures in 2005 would be 5.5 mt for CNPC and 3–3.5 mt for SINOPEC. In late March, representatives of CNPC and SINOPEC conducted negotiations with the Yukos management over a seven year contract to supply 15 mt of oil to China per year. The contract envisaged annual deliveries of 10 mt of oil for CNPC and 5 mt for SINOPEC.

On 27 March, 2004 the Russian Railways Co (RZD)[74] president, Gennady Fadeyev, and the Yukos board chairman, Simon Kukes, signed an agreement on cooperation between the two companies spanning the period to 2011 and including the organization of oil shipments to China by rail, which would enable Yukos to increase oil export volumes dramatically. According to Fadeyev, Yukos oil shipments by Russian railways were set to reach 6.4 mt in 2004, 8.6 mt in 2005, and almost 15 mt in 2006. Considering that the volume of Russia's contribution (by rail) to FSU crude exports is set to decline, as shown in Table 7.3, the scale of Russian crude oil exports to China by rail is very likely to decline significantly once the ESPO spur line to Daqing field is completed.

Table 7.3: Russia's Contribution to FSU Crude Exports (mt)

| | Transneft System | | Russian Rail | Kazakhstan Rail | | Atasu–Alashankou |
	Total	of which Russian		Total	of which China	
2000	128.477	116.353	14.069	-	-	
2001	142.134	125.817	15.644	-	-	
2002	147.240	128.887	25.759	-	-	
2003	159.305	139.548	38.223	0.820	0.817	
2004	188.738	167.253	36.382	1.165	0.810	
2005	204.922	181.061	20.791	0.795	0.795	
2006	206.435	182.335	15.640	1.570	1.570	1.845
2007	205.695	182.440	14.580	0.875	0.875	4.825
2008	196.906	173.267	14.250	0.840	0.840	6.014
2009	197.188	171.825	14.090	0.840	0.840	7.736

Source: RC (2010), 110.

Yukos supplied China's SINOPEC with 0.28 mt of crude oil per month (a total of 2.5 mt) in 2004 through the Naushki border crossing in Buryatia and across Mongolia to the Chinese city of Erenhot. This route is more profitable than that for supplies to CNPC via Zabaikalsk-Manchuria (on the Russia–China border). Alexandr Sapronov of Yukos-RM explained that the distance from Angarsk, where the Transneft pipeline ends, to Naushki is 700 km and the transportation tariff is just over $30 per tonne. In comparison, a tariff of almost $60 per tonne is charged on the Zabaikalsk route which, starting from Angarsk is 1,500 km long. Beginning in 2005 the number of trains passing daily through the border at Zabaikalsk-Manchuria increased from 14 to 20 and at Grodekovo-Suifenhe from 10 to 12.[75]

ESPO's First Stage Operation and the Spur Line Development

On 4 October 2008, Transneft commissioned the first 1,100 km segment of the ESPO oil pipeline. The ceremony was held at pump station No. 10 at the Talakan oil field. Deputy Prime Minister Igor Sechin, MEI Minister Sergei Shmatko, and Sakha president Vyacheslav Shtyrov attended. 'Today', Sechin said, 'is the first day in the life of the ESPO system. The 1,100 km segment from Talakan to Taishet is ready for operation'. He added that the resource base needed to fill the ESPO oil pipeline would be created and the fields in eastern Siberia alone would be able to supply 40–45 mt/y of crude oil in the future.[76]

According to Alexei Sapsai, director general of the ESPO PMC:

> … all of the land along the pipeline's first leg had been deforested, cleared and prepared and that 2,450 km of pipe had been laid; a stretch of 1,776 km had been hydro-tested, and a 1,094 km pipeline section had been filled with oil. Seven oil pumping stations (including Taishet, Rechushka in the Irkutsk region and near Aldan in Yakutia) will be built along the 2,700 km Phase 1 Taishet–Skovorodino pipeline. In addition to the station in Taishet, two other stations will have tank farms.[77]

ESPO required a massive investment, and the unprecedented scale of the ongoing global financial crisis caused a serious funding problem for Russia's state-owned companies. Russia had no choice but to negotiate a massive loan from China for both Transneft and Rosneft. A Russian source said:

> … the Chinese side offered [no-]nonsense lending terms. They did not set a fixed rate of interest but wanted to provide the funds at LIBOR+5%, a floating rate; and they proposed to give money as a banking loan. They also demanded multiple guarantees for these loans, guarantees about revenue and oil supply.

In short, the talks collapsed due to a disagreement over the interest rate for the loans, China insisting on a floating rate while Russia wanted a fixed one.[78] A compromise was reached in February 2009 when representatives of Rosneft and Transneft, and of CNPC and the China Development Bank, signed an agreement in which the Russian companies received $25 billion in return for Russian crude oil supplies to China and an assurance on the building of the spur pipeline to China. According to Sergey Sanakoyev, an inter-governmental commission expert, Chinese banks normally finance large projects under LIBOR (London Interbank Offered Rate) plus 300 to 500 basis points (3–5 percentage points). The level in March 2009 was LIBOR + 4.24–6.24 per cent. The China Development Bank agreed on a fixed interest rate of 5.5–6.0 per cent which even Igor Sechin acknowledged was low, observing that 'we do not have such rates right now'.[79]

In April 2009, Rosneft said it would finish building a pipeline from its giant Vankor field in the Krasnoyarsk territory to the Transneft pipeline network by mid-April. Rosneft CEO Sergei Bogdanchikov boasted that 'we'll complete the last weld by 15 April and we've literally 6 km left'. He also said that Rosneft and Transneft would sign an oil-trading contract by 10 April, under which Rosneft planned to sell the crude to Transneft, which would then repay China's loan.[80]

The ESPO Spur Line to China

On 21 April 2009, in Beijing, Russia's deputy prime minister, Igor Sechin, and his Chinese counterpart, Wang Qishan, signed an inter-governmental agreement on oil cooperation, whereby the two countries would build a spur from the ESPO pipeline to China.[81] The first joint in the planned 63.8 km pipeline spur from the Russian town of Skovorodino to the border with China was welded on 27 April in the presence of Igor Sechin and Wang Dongjin, vice president of CNPC. The second phase, which would increase the line's capacity to 80 mt a year, 'should be completed as scheduled in 2014', Sechin said.[82] On 18 May 2009 the official start of construction on Chinese territory took place in Mohe, in Heilongjiang province, at which Chinese Vice-Prime Minister Wang Qishan said that the oil pipeline was an important strategic project in oil and gas cooperation between Russia and China, and also a bridge of friendship and cooperation between the two countries.[83] In the early stages of ESPO, the volume of production from East Siberia would not be sufficient to fill the pipeline. Therefore, to fill the ESPO (pipeline capacity), in addition to the existing Omsk–Irkutsk pipeline, oil from East Siberia, the Tomsk region, and Khanty-Mansi in west Siberia would need to be provided. To facilitate west Siberian supply, the Omsk–Irkutsk pipeline had to be connected with ESPO at Taishet where ESPO will start.[84]

In June 2009, Rosneft's annual general meeting approved the $15 billion loan from the Chinese Development Bank. The ESPO crude price will be based on market quotes for Russian crude at the port of Kozmino – or Primorsk if volumes via Kozmino are too small to warrant a market quote. In the event that less than 12.5 mt are shipped to the Kozmino port, the difference in the volume will be calculated using quotes for Urals oil at the ports of Primorsk and Novorossiisk. At the end of December 2007, Transneft registered the enterprise Spetsmornefteport Kozmino in Nakhodka, as part of the programme for creating a bulk oil port on the Dong Hae (East Sea in Korean) or Sea of Japan coast.[85] The oil supplies started from January 2010. The total value of the 9 mt of oil supplied as part of the Rosneft–CNPC agreement would be $65 billion based on an average oil price of $50 per barrel for 20 years. The 6 mt supplied via Transneft would be worth $27 billion. Rosneft expected to pay an average rate of 5.69 per cent annual interest on the $15 billion over 20 years.[86]

In November 2009, the Russian authorities announced that all the onshore and offshore facilities for oil exports at the Port of Kozmino had been completed. The facilities approved included a 430-m oil pier, two harbour fleet piers with mooring posts for tankers, 17 km of process pipelines, and 250 km of cable utility lines. Also completed at the site were a booster pumping station, oil quantity and quality metering stations, a specimen oil storage site, and a drainage water accumulating tank. At the same time, the first trains carrying more than 8,000 tonnes of oil left Skovorodino, endpoint of the first stage of the ESPO, for delivery to Kozmino. According to a representative of the Vostokneftetrans Company, oil is supplied to Skovorodino from Yakutia.[87]

On 23 November 2009, Rosneft sold the first ESPO cargo through a tender auction, with the buyer emerging as the Finnish company International Petroleum Products OY at a 50 cents premium to average *Platts* Dubai published prices for December. Rosneft awarded the tender to IPP OY for 100,000 metric tonnes of ESPO crude for 27–29 December loading FOB Kozmino basis. A total of 15 companies participated in the tender, with two companies bidding at positive differentials.[88]

ESPO First Stage Completed

Finally, on 2 December 2009, the opening of a new oil terminal at Kozmino, near Nakhodka, brought Russia closer to its strategic goal of diversifying its energy exports.[89] The new $1.7 billion Pacific oil terminal provided access to the major energy importers, Japan, China, and South Korea. More than 3 mt were exported from the new terminal during the first quarter of 2010,[90] with volumes expected to triple over coming years when the pipeline is completed. Prime Minister Putin, who pressed the button to get Siberian oil flowing into the first tanker for delivery to an Asian customer in Hong Kong, stressed the strategic importance of greater access to Asian markets as follows:

> This is definitely a serious event. It's a strategic project, because it gives us access to completely new markets – the Asian and Pacific markets, which are growing and have a huge potential. Today Russia is present on these markets, but on a very small scale. Today's opening gives us completely new possibilities.[91]

This ceremony marked the end of four years of work spent in constructing the pipeline and the port of Kozmino, worth a combined 420 billion rubles ($14 billion), and so reducing the industry's reliance on the European market.[92] Another ceremony was held only eight months later. On 29 August 2010 Russia began filling the spur of the ESPO pipeline to China at an opening ceremony attended by Prime Minister Putin, who this time said:

Of course, our cooperation with China is not limited to just hydrocarbons ... but speaking of energy, Russia is China's main partner in the field of peaceful use of nuclear energy, and equipment supplies here amount to billions of dollars.

The opening of the spur is seen as a key to the diversification of Russia's exports and to using the Asia–Pacific market as a counterbalance to the traditional European destinations.[93] On 27 September 2010, Presidents Medvedev and Hu Jintao attended a ceremony to launch the completed Skovorodino–Daqing line.[94]

When the second stage of the ESPO oil pipeline development is operating, more Asian buyers will take the high quality crude from Kozmino port, near Nakhodka. In early March 2010, GS Caltex, which has the second largest refinery in South Korea, announced that the company had imported ESPO oil from Vladivostok. The estimated import volume is at least one cargo of over 730,000 barrels and the negotiated price is rumoured to be around $2 less than Dubai crude price.[95] ESPO's technical specifications are better than Middle East crude in terms of sulphur and API degree, and the shipping distance is much shorter (Kozmino is only 1,000 km from South Korea, while the shipping distance to the Persian Gulf is 11,000 km), so there is a good chance that the ESPO deliveries could expand significantly. The main obstacles are the availability of the supply and sustainability of quality. If these two factors can be guaranteed, Japanese and South Korean refineries will have no hesitation in reducing their heavy dependence on Middle East crude supply. New Nippon Oil has projected that ESPO crude could be a benchmark crude in Asia (and if Sakhalin offshore's Sokol crude is added, the volume could reach 1.4–1.5 mb/d), and could have a significant effect on Asia's crude supply.[96]

ESPO Crude Challenges the Asian Premium

North-east Asia's consumers have traditionally had to pay the so-called 'Asian premium' despite the fact that a buyer was purchasing the same grade of oil as a European consumer.[97] Over the last 20 years, on average the premium was $1.2 per barrel but in 2008 it reached $8.10 per barrel. The arrival of Russian supplies has reduced the premium and TNK–BP has said that:

... the expected increase in Russia's transportation capacity in the coming years would allow oil producers to re-direct volumes between western and eastern destinations, for the first time providing flexibility for producers' sales strategies. Russia is expected to bring on line 78 mt/y of new transportation capacity by 2014, with crude production estimated to grow by nearly 40

Table 7.4: ESPO Specs Comparison

	API	*Sulphur %*
ESPO	34.8	0.62
Brent	37.5	0.46
Forties	40.6	0.59
Dubai	30.4	2.13
Oman	32.95	1.14
Urals	31.55	1.30
Sokol	39.7	0.17
Vityaz	34.4	0.22

Source: Platts (2010a).

mt/y by that time. Russian producers will be able to make decisions on crude supply directions depending on where the price is better.[98]

Russia's Ministry of Energy has suggested creating an oil quality bank. Transneft is studying the possibility of allocating five grades of oil within this framework. It would partly be graded according to its source, which affects its sulphur content and density. While maintaining the current method of mixing grades, the Ministry is offering, by 2010, to distinguish crude transported by the Druzhba pipeline via Ust-Luga (with a sulphur content of 1.5 per cent), from that delivered through the ports of Primorsk (1.4 per cent), Novorossiisk (1.3 per cent), and Tuapse (0.58 per cent), as well as via ESPO (0.75 per cent).[99] This percentage, from the Ministry of Energy is higher than the 0.62 per cent reported by *Platts*.

Taxation and SPR Storage

In late November 2009, the Russian government announced an immediate exemption from export duty for East Siberian crude. The exemption is designed to stimulate development in the remote oil province. A total of 13 separate fields in East Siberia are eligible for the zero rate; in mid-December, a Customs Union commission decided to expand the list of oil fields to 22 and amend the quality classification of oil subject to this rate, to take effect on 19 January 2010. A government resolution of 25 December setting export duties for January did not prescribe a mechanism for how the list was to be expanded. A resolution setting export duties for February applied the zero rate to 22 oil fields.[100]

The long-awaited zero rate will be applied to crude with a density of between 694.7 and 872.4 kg/cubic metres at 20 °Celsius (equivalent approximately to API gravity of 30–70 at 15.5 °Celsius) and with a sulphur content of between 0.1 and 1 per cent. The Energy Minister strongly hinted in late October 2009 that the zero rate was likely to last for a minimum of five to seven years, but the government has yet to make a final decision on this.

Platts reported that ESPO barrels are currently priced at a differential to a commonly used benchmark (*Platts* Dubai), but due to its location, ample production levels, and wide equity ownership, it has elements that could, over time, help it become a major price indicator of spot oil volumes in Asia. Currently the terminal at Kozmino has a tank farm with a capacity of 350,000 cubic metres (close to 2.6 million barrels) and a loading capacity of 300,000 b/d. Vostoknefteprovod, a company affiliated with Transneft, was created to operate the new ESPO. The Skovorodino delivery station has a storage capacity of 80,000 tonnes and a loading installation for 82 rail tanks, with a current loading capacity of 35,000 t/d, which is expected to increase to 43,000 t/d (close to 323,000 b/d) in the near future.[101]

In addition to facilitating ESPO crude exports to north-east Asian countries such as South Korea and Japan, the 2.6 million barrels of tank farm development at Kozmino could also play a role in providing a large scale SPR storage provider. As discussed in Chapter 5, Beijing is committed to increase the SPR storage tank capacity, the ultimate goal being to have a storage capacity of about 40 days' consumption (with an aim of building up oil reserves to the equivalent of 100 days of consumption by 2020, after building the second and third phases of SPR bases). If the current expansion of the motor industry continues during the coming decade, the chance of a serious shortage of SPR storage capacity is very acute. Without a bold contingency plan, it will be not be easy for Beijing to cope with the SPR storage development in coming years. In this context, an SPR storage base near Kozmino, built jointly by the importing countries (South Korea, Japan, and China) based on the UNDP GTI (Greater Tumen Initiative), could serve as a regional SPR storage. If a consensus on SPR development is reached among the member countries, this would provide an opportunity for regional multilateral cooperation for SPR storage development within the GTI area.[102] Considering that both China's and Russia's conventional policy is based on bilateral rather than multilateral cooperation, this type of initiative could be a difficult option for both Beijing and Moscow. Nonetheless, political blessings from the highest level might open the door for a pilot SPR project which could fundamentally change the degree of Sino-Russian oil cooperation. Not only would the level of cooperation be upgraded but also a new agenda would be added. If that occurred, then the forthcoming decade could witness a very different chapter in Sino-Russian oil cooperation.

Second stage ESPO development will have a major impact on Sino-Russian oil cooperation; but it is too early to predict the volume which will be taken by Chinese NOCs or oil traders when the second stage of the ESPO is implemented. As the potential for further discoveries of oil fields in East Siberia is quite high, it will not be difficult to achieve filling the 1.0 mb/d, or 50 mt/y, capacity pipeline to Kozmino port. How quickly this can rise to 1.6 mb/d, or 80 mt/y, depends on the results of comprehensive exploration. More crude oil will be supplied by ship from Kozmino to China's coastal areas, and Sino-Russian oil cooperation will be strengthened by this increased oil trading and related cooperation in the downstream sector.

Factors Related to ESPO

The tariff issue
The economic viability of ESPO is heavily dependent on the tariff scale fixed by Russia. In 2005, Transneft had made a preliminary estimate of per tonne tariffs to transport crude oil via Taishet–Perevoznaya, generously endowed with government concessions including the railroad tariff for the Skovorodino–Perevoznaya section; the estimate was $47/ tonne in the initial phase though after the repayment of project loans, that tariff might drop by half. Over the operational life of the pipeline, average tariffs were projected to be $38/tonne. Transneft's calculations were based on an assumption that oil companies can sustain world prices of $25 a barrel. Transneft receives the tariff of $47/tonne from Rosneft, which supplies western Siberian crude oil by rail to its refinery in Komsomolsk-na-Amur. As of 2005, Russia's investment company Aton projected that the cost of delivering crude oil by rail to Black or Baltic seaports is comparable – between $40 and 45 per tonne ($5.5–6.0 per barrel).[103]

On 3 November 2005, Saenko, head of the Department of Energy Policy at Minpromenergo announced that the tariff for westward oil transport would be $22–24 per tonne, while oil shipped eastward via a combination of oil pipeline and railroads would be roughly $100 per tonne. He added that, even when the eastern pipeline was operating from Taishet to Perevoznaya Bay, the tariff would be $38.80. To deliver oil to Taishet from Western Siberia would cost another $11–12. Transneft confirmed these figures, saying that the tariff on oil transportation via the ESPO oil pipeline would ultimately be $49.90.

Table 7.5 shows the scale of the rise in tariffs from their low point in 1998. Since then unit tariffs, measured in fees per 100t-km have climbed 6.8 times in dollars (as of 2009). Measured per barrel of

Table 7.5: Transneft's Tariffs, 1998–2010

	Transportation Revenue $bn	Revenue per barrel $	Revenue per barrel-mile $cents	Revenue Per 100t-km $
1998	0.907	0.42	0.03	0.13
1999	0.947	0.43	0.03	0.14
2000	1.298	0.57	0.04	0.19
2001	2.020	0.81	0.05	0.25
2002	2.551	0.93	0.07	0.31
2003	3.469	1.15	0.08	0.35
2004	4.405	1.35	0.10	0.43
2005	5.488	1.68	0.11	0.52
2006	6.771	1.98	0.13	0.59
2007	7.760	2.29	0.15	0.68
2008	9.805	2.94	0.20	0.92
2009	9.580	2.87	0.20	0.89
2010*	12.214	3.61	0.25	1.12

*   estimate

Source: RC (2010), 177.

crude shipped, they have also risen 6.8 fold. This simply reflects the fact that the distances over which crude for export has been hauled have risen faster than production. Despite these rises, Transneft's tariffs remain low: for instance the company's unit tariff fees of 0.17 cents per barrel-mile in 2007 still compare very favourably with the roughly 0.23 cents per barrel-mile in US inter-state pipelines, and 0.54 cents per barrel-mile charged by the Caspian Pipeline Consortium.

At the end of February 2008 the FTS (Federal Tariff Service) sent the Russian government a tariff estimate for oil throughput via the ESPO oil pipeline. According to FTS calculations, the tariff for the first and second stages amounts to $38.80 per tonne. This was the same as the tariff prepared in 2005. At that time the cost of the project was estimated at $11.5 billion. After three years, the cost estimate of the first stage from Taishet to Skovorodino alone doubled from $6.6 billion to $12.3 billion. Also included in the first stage of the project is the Kozmino terminal with a construction cost of at least 46 billion rubles (roughly $2.0 billion).[104] According to the ESPO Project Management Centre (PMC), construction of the second stage will cost about $13–14 billion. As for the tariff, Denis Borisov, an analyst with the investment company Solid, has calculated that to indemnify Transneft expenses for the entire pipeline, the tariff should be $55–57 per tonne. If the tariff remained at $38.80 and the costs were at $11.5 billion as originally estimated, the project would have a payback period of 17 years. At the higher tariff, recovery improves to between five and seven years. Experts, however, concluded that at the higher tariff nobody would use ESPO. Until completion of the second stage, oil companies would rather pay rail fees for oil transport from Skovorodino to Kozmino. The rail rate has not yet been calculated, but the provisional figure is $60 per tonne via all routes. The overall charges for oil delivery to Kozmino could exceed $100 per tonne.[105]

Transneft failed to finalize the tariff for ESPO pipeline by the end of 2008 and was still struggling to find the optimal formula during the first half of 2009. Denis Volkov, head of oil and gas sector regulation at FTS said that 'at the moment there is a certain computation that the network tariff for transportation on ESPO will be in the region of $30–32/tonne'. That price was calculated in February when the exchange rate was 36 rubles to $1, but Volkov said that the financial crisis and ruble devaluation of 2008 made a fundamental difference to the ESPO tariff.[106]

In the autumn of 2009 Transneft was proposing that ESPO should be divided into eastern, western, and central export tariff zones, with $34 per tonne charged for oil transported via the eastern zone, $48 per tonne via the western zone and about $42 per tonne via the central zone. The eastern zone will pump oil from the Talakan field, the western zone from the Vankor field (including oil pumping via the current system of oil pipelines) and the central zone from fields in southern Krasnoyarsk. This proposal was filed with the FTS. Transneft meanwhile has drawn up an estimate of the ESPO's network tariff for 2010, based on a transportation volume of 15 mt rising to 30 mt starting in 2011, including 15 mt to be pumped to China.[107]

In late December 2009, the FTS proposed setting a separate tariff for transhipping oil at the port of Kozmino, the terminus of the ESPO, at 155 rubles per tonne. According to an FTS spokesperson, Anna Martynova, this is a separate tariff that is needed to charge for tanker shipments made before the network tariff has been set. FTS has approved a through-transportation fee for crude deliveries via ESPO of 1,598 rubles/tonne ($52.68/tonne or $7.21/barrel). The ESPO tariff includes the services for crude deliveries via the pipeline, by railroad, and for crude re-loading at terminals, including re-loading at the Kozmino sea port for onward export.[108] The tariff rose to 1815 rubles/tonne ($61/tonne) from 1 December 2010, but this was still less than half Transneft's estimated economic cost of $130/tonne.[109]

In 2010, FTS announced that the tariff for transporting oil through the ESPO pipeline system to China would be 1,651 rubles per tonne from 1 August.[110] The proposed crude oil price was set to the bench-mark Brent rate with a discount of $3 per barrel. In November 2007, Rosneft increased the proposed price by $0.675 per barrel (consequently the discount in relation to Brent was reduced to $2.325). It was not possible to reduce it any further. According to the new contract (under which Rosneft has to deliver to China 300 mt over the 20 years) the price will be subject to monthly variations based on quotations from the Argus and *Platts* agencies for the port of Kozmino.[111]

Vostok refinery

On 17 December 2007 the Rosneft CEO, Sergei Bogdanchikov, said that the firm was aiming to build an oil refinery at the coastal point of ESPO on its own and did not need to attract partners for the project. Rosneft also selected a construction platform at Cape Yelizarov, several kilometres from Kozmino, and has prepared a preliminary feasibility report.[112] While the terminus had a capacity of 15mt/y, the pipeline could carry 30 mt. Original plans called for delivering the 15mt dif-ference to China on a 67 km branch pipeline. If the branch line issue was not resolved by the time ESPO was ready for start-up in late 2009, Transneft believed that the extra oil could be delivered to refineries in Komsomolsk and Khabarovsk. Rosneft, the owner of the Komsomolsk Oil Refinery which has a capacity of 7.3 mt (due to rise to 8 mt in 2010), and West Siberian Resources, which owns the Khabarovsk Oil Refinery with 3.5 mt capacity, could take as much oil from the ESPO as they are capable of refining.[113]

In 2008, Rosneft estimated that Vostok or East Refinery (the provi-sional name) would cost $5–7 billion. Rosneft considered constructing the refinery in two stages, each having an annual capacity of 10 mt. Exports will account for 90 per cent of the refinery's production. Plans call for the first stage of the refinery to begin operation by 2012. A 17 km pipeline would deliver oil products from the refinery to the port, and an additional tank farm with a total volume of 800,000 cubic metres will allow for an increase in capacity. A new ocean terminal at Vostok Bay, which is located only two kilometres from the refinery, can accommodate tankers of up to 150,000 tonnes.[114]

The target location was the Cape of Yelizarov near Nakhodka. The first stage construction costs come to $4.9 bn, with expected comple-tion in 2013. The estimated capacity is 20 mt/y, twice that initially suggested. The second stage is expected to cost $9bn. At the end of March 2010, however, the Federal Service for Supervision of the Environment, Technology and Nuclear Management (Rostekhnadzor) published a negative conclusion from a commission of experts, following the state ecological examination of the design documentation on the refinery. Primorsk environmentalists have opposed the construction of the refinery, claiming that its proposed location would cause irreparable damage to the environment. From the point of view of both ecological safety and economic feasibility, it would be better to locate the refinery east of the port of Vostok – towards Prudikha and its oil terminal in Krakovka Bay – or in the eastern part of Nakhodka. Prudikha is located in immediate proximity to already operating ports and a railway.[115] At the end of 2010, Rosneft board approved a general concept for construction of a petrochemical plant near Nakhodka. Planned throughput capacity is 3.4 mt/y, consisting of naphtha and liquefied hydrocarbons from the Komsomolsk and Achinsk refineries and also from the Angarsk Petrochemical company. However, it is not clear what Rosneft's real intention is.

Loans for Oil

The first $6 billion loan to Rosneft

At the beginning of 2005, Russia's Vedomosti reported that CNPC was willing to lend Rosneft up to $6 billion to finance its purchase of Yuganskneftegaz. The two sides discussed a loan that would be secured against deliveries to China of some 48.4 mt (or 354 million barrels) of oil over the next six years.[116] According to the Federal Energy Agency's chief, Sergei Oganesyan, Rosneft received a $6 billion credit from Russian banks to purchase YNG, and Vnesheconombank in turn received the funds for this credit from an array of Chinese banks in exchange for oil supplies to China.[117]

Xinhua News Agency also reported that Rosneft had signed a supply contract with CNPC and a relevant transportation agreement with the Russian Railway Company to supply 50 mt of crude by rail to China by 2010. *Xinhua* added that CNPC had tried to bargain for a $25–30/barrel crude import price, and an increase in its holding of Yugansk's shares from 20 per cent to 25 per cent plus one share. The exports were to start on 1 February 2005 and would amount to an estimated total of 4 mt in 2005.[118] It was also reported that analysts had said that allowing CNPC to take a stake in Yugansk would help to provide a veil of legitimacy to a transaction which otherwise amounted to nationalization. A 20 per cent share of Yugansk equity might cost the Chinese as much as $4 billion.[119]

Clearly China's decision to provide a $6 billion loan both provided a lifeline to Rosneft and laid the ground for a successful Rosneft IPO a year later. It is no coincidence that a number of agreements between

Rosneft and the Chinese state oil companies were signed in 2005. Through these, China learned the true influence of lending billions of dollars to the desperate Rosneft. In May, the Russian Ambassador to China, about to leave Beijing after 13 years' service, indirectly confirmed the impact of the $6 billion loan with the prediction that 'first of all, oil will go to China, and then it will go east [that is, to Japan and South Korea] in the second stage'.[120]

On 19 July 2006, CNPC announced that it had bought 66.2242 million shares in Rosneft for $500 million ($7.55 per share). This is equivalent to about 0.5 per cent of Rosneft (deduced from the Russian company's statement that 1.38 billion shares sold in the IPO represented 13 per cent of its capital). CNPC had put in an order for $3 bn worth of shares but received only a sixth of this;[121] without the $6 bn loan even this minor equity purchase would not have been conceivable.

The second loan for oil

It is no coincidence that from early 2009 China began to change the world's pattern of oil trade because of its vast foreign exchange reserves. In February 2009 China and Russia signed a series of deals on oil pipelines, and long-term crude oil trade, collectively known as the 'loan-for-oil'. China has agreed to provide $25 billion in loans ($15 billion to Rosneft and $10 billion to Transneft) in exchange for shipments of oil; Russia committed itself to exporting 300 mt of crude oil to China between 2011 and 2030. China's state media regarded the loan-for-oil deal as a win–win arrangement. Russia will be taking a major step forward in diversifying its oil exports while China would be ensuring a stable oil supply from Russia's enormous and conveniently located oil fields, thus diversifying its oil imports. The 300 mt of crude oil will satisfy about 4 per cent of China's current demand, and it is worth about $160 billion.[122]

Russian crude supply to China, starting from 2011, is likely to be priced in accordance with spot market prices, according to *China Business Post*. The news cited a CNPC official as saying that China and Russia will set oil prices in line with spot prices rather than at a fixed $20/barrel delivered to a port, as market rumours had suggested. The source added that the price would be linked to the crude quality. Jiang Yi, an expert on Russian issues from the Chinese Academy of Social Sciences, said that 'China–Russia oil negotiations have not made significant progress in recent years due to differences over oil prices. To China, this loan-for-oil is mainly about the pipeline. Once the pipeline is completed, the Chinese side will be at an advantage'.[123]

China's new venture with Kazakhstan, however, deviated from the 'loan-for-oil' formula. The $5 billion loan from CNPC will give Chinese oil firms a 50 per cent stake in MangistauMunaiGaz (MMG), Kazakhstan's biggest private oil and gas company. This deal is more like a 'loan-for-oil assets' transaction than one of 'loan-for-promised oil supply' as in the case of the previous three contracts. CNCP will receive half of the oil that will be produced by the jointly-owned MMG. This model is more in line with the Chinese government's preference for financing acquisitions, since it gives Chinese NOCs direct ownership of resources – in contrast to the other three deals under which Chinese NOCs could only extend loans to foreign NOCs for guaranteed oil supplies or possible special access to future exploration projects.[124]

The Gas Sector

Kovykta Gas Export Initiative

During the years 2000–2, two major decisions were made that would fundamentally affect the expansion of China's natural gas sector. The first was the decision on the first West–East Pipeline (WEP-I) development in early 2000 and the second, in August 2002, was the decision to give the green light to both Guangdong and Fujian LNG terminals (See Chapter 5). In addition, Kogas joined the Sino-Russian Irkutsk gas supply project in November 2000, and that laid the ground for a three year feasibility study on the Kovykta gas pipeline project. The three-party study was CNPC's highest priority, but many obstacles such as the question of the route and the price were still to be resolved.

Why did China choose the eastern route? One major reason was that it preferred to minimize political risk and to save paying transit fees by avoiding a transit country. Another reason was that the economic benefits brought by the pipeline could make Mongolia suddenly richer than Inner Mongolia, a potentially negative influence on stability in the autonomous region. The third reason was that China would be able to channel the economic benefits of the pipelines to the north-east, a region desperately in need of the economic and social benefits that Russia–China pipelines can bring. In short, China chose to inject vitality into its north-eastern economy rather than to prioritize its relationship with Mongolia.[125]

Given the choice of the eastern route for the oil pipeline, the viability of a parallel gas pipeline was greatly strengthened. CNPC understood that there had been internal disagreements among the shareholders of the Kovykta gas project which had paralysed progress for a period. The Chinese side waited patiently, believing that it was in a strong position as the major customer for the gas. In April 2002, the Russian government convened a meeting on the Kovykta project which reached two decisions: Russia would resume negotiations with China, and Rusia Petroleum would be the coordinator and lead the negotiations with CNPC. Rusia Petroleum's president visited Beijing together with a Kogas consortium representative. During the visit, it was decided that Rusia Petroleum and CNPC would start the next phase of the work, consisting of planning the mode of cooperation, the overall project design, the pipeline route, the gas price, and the scale and timetable of supply. Negotiation on price would be the hardest problem.[126] It seemed impossible that this feasibility study could be completed before the autumn of 2002.

In April 2002, *Xinhua* reported that further uncertainties had arisen in relation to the Russia–China–South Korea gas pipeline project. The Russian media indicated that the Russian government was considering revoking BP's license and adopting Gazprom's proposal that all Kovykta gas should be consumed domestically. Nonetheless, the three-party feasibility study continued. A couple of months later, *Xinhua* raised the questions of gas prices, pipeline routing, BP's export license, and the uncertainty about the future of the Russia–China oil pipeline, which was bound to affect the fate of the gas pipeline.[127]

Events during the year of 2002 greatly allayed the Chinese energy planners' concerns about the scale of Russia's gas reserves. As discussed in Chapter 3, in 2002 Moscow's Central Commission for Reserves of the Ministry of Natural Resources approved the revised figure of Chayanda gas proven gas reserves as 1,240 bcm. In addition to this, in October 2002, Gazprom and the Sakha Republic Government launched a joint venture bid for the development license for the Chayandinskoye field and other fields in the Sakha Republic. Gazprom's initiative on Chayanda gas was not very well advertised, but it is safe to say that the confirmation of 1,240 bcm proven reserves heightened the confidence of Gazprom and the Moscow decision makers about a UGSS development in the Asian part of Russia.

At the end of 2003 the result of the three year feasibility study on Kovykta gas supply to China and South Korea via the Yellow Sea confirmed that the project was technically and economically viable, and it was submitted to the respective governments for approval. *Xinhua's* opinion was that Gazprom's initiative was a political and economic gimmick designed to blackmail BP. In early 2003, Gazprom began to show strong interest in Kovykta, and the Ministry of Natural Resources vowed to revoke BP's license. BP offered to sell an 11 per cent equity stake in Kovykta, but that was not good enough for Gazprom. This turnaround was no surprise to CNPC. Compared with the oil pipeline, the gas line was less urgent. In fact, the Chinese side has long held the belief that if the oil line was not built, then the gas line might also be abandoned.[128]

It is worth noting that one of the key issues in the feasibility study was the pipeline route. The media reported that a Russia–China–South Korea gas pipeline would certainly bypass North Korea. Senior sources in PetroChina, however, let it be known that the routing of the line and the price issue had not been decided.[129] In Korea during the first half of 2003 very serious consideration was given to gas supply from Sakhalin to the Korean Peninsula, as an alternative to Kovykta gas supply to South Korea through the Yellow Sea.[130] In addition to the enormous political implications of this approach (because it would have had to be associated with some solution to the problem of North Korea), it also affected the Russian authorities' stance on exports to north-east Asian countries.

Until then, only a few specialists understood what was at the core of Russia's gas policy towards Asia. The first hint of it came in the Russian statement 'Energy Strategy for the period until 2020',[131] but an even more obvious clue came from the Gazprom CEO, Alexey Miller, in his keynote speech at the Twenty-second World Gas Conference in early June 2003.[132] His presentation showed two LNG plants, one at Vladivostok and another at Vanino, in addition to Korsakov on Sakhalin Island; but there was no mention of the Kovykta project as the export source. This speech, however, did not attract much attention, and no one at the time saw it as a revelation of the real blueprint for Russia's gas exports to Asia.

A hidden factor that might have affected the fate of Sino-Russian gas pipeline cooperation came to light in a November 2003 *Xinhua* report. A common question asked in industrial circles was: if the Chinese are shown to be able to manage the West–East gas pipeline single-handedly, why are foreign companies still needed? At a start-up ceremony in Jingbian, Shaanxi on 1 October, a CNPC official asked: 'Why should we share a ripening fruit (the West–East pipeline) with someone that has neither laboured nor sweated for it?'[133] This remark strongly implied that chances for the Shell-led international consortium were getting slimmer. Consequently Gazprom's attempt to penetrate China's gas market through WEP-I development was set to be frustrated.

Further negative news came in early 2004 in the form of the collapse of the deadlocked negotiations between PetroChina and the Shell-led international consortium on the WEP-I,[134] the workshop set up by the foreign consortium having already been disbanded. This signalled that foreign companies' involvement in the WEP-I pipeline might quite possibly have ended. Gazprom's strategy to enter the Chinese gas market also collapsed. On 8 March, 2004 Zhang Guobao, deputy director of the NDRC, had already indicated that disagreement on the terms of the joint venture and the related division of resources had resulted in the negotiations stalling. Even though Zhang depicted the project as proceeding quite smoothly, PetroChina announced the termination of its negotiations with western partners in early August.[135] Without delay, Gazprom informed Kogas that it would not support a pipeline crossing the Yellow Sea. This could be considered Gazprom's Korean card and was a good example of how the interplay of pipeline politics affected Russian pipeline gas exports to north-east Asian countries.

Gazprom's Asian UGSS Initiative and the Sakhalin-1 Option

During the second half of 2004, the Strategic Development Department of Gazprom began to float the so-called 'Discussion Package'; this appeared under the self-explanatory title of the 'Interagency Working Group to Develop a Programme for Creating a Unified Gas Production, Transportation and Supply System in East Siberia and the Far East with Potential Exports to markets in China and other countries in Asia and the Pacific'. It took three years to finalize this UGSS scheme, and finally, on 3 September 2007, the Ministry of Industry and Energy approved the 'Eastern Programme' worth 2.4 trillion rubles. The Russian newspaper *Kommersant* reported that 'the dates of the field's development and exports will depend on the results of Gazprom's negotiations with buyers in China and South Korea. China's rejection of Russia's terms would kill the export component of the programme'.[136] Gazprom's focus was centred on the evaluation of the three main options – the west, central, and east lines.

Even though CNPC entered into a strategic cooperation agreement with Gazprom in October 2004, ExxonMobil confirmed on 2 November that it was in talks with CNPC about potential sales of gas from the Sakhalin-1 project. *Xinhua* reported on the same day that Gazprom had said it was considering cooperation with ExxonMobil in exporting gas to Asia. ExxonMobil's chairman, Lee Raymond, said that the Sakhalin-1 consortium was considering several options including both LNG and pipelines.[137] These reports indirectly confirmed that neither ExxonMobil nor CNPC were ready to accept Gazprom's exclusive position as a sole negotiator for Russia's gas export to Asia. The process of getting national approval for the three-country study on Kovykta gas exports to China and South Korea, was completely overshadowed, especially in Russia, by Gazprom's ambitious Eastern Programme. On 15 April 2005 Russia's Industry and Energy Ministry announced that it regarded the 'east route' scheme as the best choice for transporting Russia's East Siberian and Far East gas to China and South Korea. In other words, China's hope for a Kovykta gas pipeline began to look like an illusion, which was a big disappointment for Beijing. [138]

In late autumn 2006, CNPC signed a preliminary agreement with ExxonMobil to purchase natural gas from the Sakhalin-1 project.[139] For the Kremlin and Gazprom, this was an indirect challenge to Gazprom's monopoly gas export negotiating position with Asia–Pacific countries. It is hard to tell whether CNPC was naïve and had simply underestimated the obstructive role that Gazprom was playing.[140] In any case, Moscow predictably opposed this initiative. A year later, Moscow's final approval of the East Gas Programme confirmed that Gazprom has been given formal priority over all gas supplies in the Asian part of Russia.

Gazprom's 2005 Beijing Interview

The absence of any special announcement on natural gas pipeline cooperation during the July 2005 summit was compensated for by the press conference given by Gazprom's deputy CEO, Alexander Medvedev, which was held in Beijing on 21 September 2005. He said that the talks with CNPC were aimed at identifying which of two routes should take priority, and fixing the timing of the start of any deliveries. He added that the question was at the initial stages of negotiation. 'We don't like it when pipelines are lying empty', he said, 'so we should identify which is the priority project. The Russian and Chinese sides would each be responsible for building the sections of the pipelines in their respective countries'. He also mentioned that Gazprom would also be interested in investing in infrastructure in China, if the government allowed it. He highlighted the importance of the price issue by saying that any price of the resource sold in the Chinese market would have to take into account the 'realities' of the international market for gas and LNG. Interestingly, he mentioned that Gazprom would in the future consider supplying LNG to China from Sakhalin and was also interested in investing in power plants in China.[141]

The Chinese media highlighted Medvedev's assurance (given to allay Chinese fears about Russia's supply ability) that Gazprom was able to supply more gas than any other possible supplier to China. There is a consensus that Russian gas will be supplied to China first through the eastern route from Sakhalin to Harbin, capital of Heilongjiang province. However, Medvedev said that Gazprom and CNPC were still evaluating the two alternative options, the eastern and western routes.[142]

According to Gazprom's blueprint, both routes would be designed with a capacity of 20–30 bcm/y. The company hoped to start supplying gas to China in 2010. If the two routes were built, the Russian gas supply to China would reach about 60 bcm/y. Gazprom suggested that the gas pipeline be financed by Russia on the Russian side and by China on the Chinese side. Aside from the pipeline gas supply option, Gazprom did not rule out the possibility of directly providing LNG to China. Gazprom also revealed its interest in participating in China's power generation sector. Medvedev repeated that the company was eager to join hands with the country's power companies, especially in the gas-for-power generation sector.[143]

CNPC's Lukewarm Response

CNPC was fully aware of the importance of diversifying gas import sources. Negotiations over the project of building a gas pipeline to carry Central Asian Republic gas to China had gone smoothly and quickly, which in fact had imposed some pressure on Gazprom. As a matter of fact, the rapid construction of the China–Kazakhstan crude oil pipeline should have warned Russia that in the gas sector (as in oil) China would not 'put all its eggs in one basket'. CNPC said that the China–Kazakhstan gas pipeline and the east China–Russia gas pipeline, together with the WEP, could possibly become three parallel lines, together fulfilling the same task of carrying natural gas to gas consumption centres such as east and central China, with the possibility that the three could be connected.[144] While Gazprom was merely talking, China was moving forward with gas exploration and development in Uzbekistan. What the Chinese leaders wanted to show was that they could convert words into actions. *Xinhua News* cynically suggested that the case of a deal to supply natural gas to China was similar to that of the China–Russia oil pipeline: China was doomed to a marathon round of negotiations. In December 2005, sessions of the Joint Gazprom/CNPC Working Group and Joint Coordinating Committee led to the signing of a contract to engineer an underground gas storage facility in the oil dome Zhen-11.[145] This result, however, was not enough to change the Chinese gas planners' negative perception towards Gazprom's 'too much talk without real action'.

Putin's over-Advertised trip to China

In March 2006, on the eve of a trip to China, President Putin said that cooperation in the energy sector was one of the most important elements of Sino-Russian trade and economic relations, and that it was growing successfully and had good long term potential.[146] President Hu also referred to the nature of the cooperation. In a televised speech he said that China and Russia should convert the current trade-centred pattern of cooperation into one giving more emphasis to production and processing.[147] The most important agreements reached in the energy sector during the March 2006 visit were:[148]

- an agreement on 'The Main Principles of Establishing Joint ventures on the Territory of the Russian Federation and the People's Republic of China to deepen Cooperation in the Oil Sphere between Rosneft and CNPC';
- a protocol between Transneft and CNPC on 'Building a branch from the East Siberia–Pacific Ocean Oil Pipeline to China';
- an agreement between ROA UES of Russia and the State Grid Corporation of China to draft a feasibility study for a project to supply electricity from the Russian Federation to the PRC;
- a memorandum between Gazprom and CNPC on natural gas supplies and the construction of two gas pipelines.

This last agreement mentioned above (Protocol on the Supply of Natural Gas from Russia to the People's Republic of China) to construct two gas pipelines to China, with the capacity to move 60–80 bcm of gas annually, formed the centrepiece of Putin's state visit to China. Gas would flow to China beginning in 2011, and Gazprom's CEO, Alexey Miller, said that the western pipeline with an annual transmission capacity of 30 bcm would be considered first.[149]

On 12 May 2006 Zhou Jiping, vice president of CNPC, visited Gazprom and met its deputy CEO, Alexander Medvedev, and both sides agreed to arrange for future commercial talks on the proposed gas pipelines. In order that the first deliveries of west Siberian gas should reach to China by 2011, they aimed to complete commercial negotiations by the end of 2006, but again no breakthrough was made.[150] The fourth meeting of the joint coordinating committee (JCC) between Gazprom and CNPC was held in St. Petersburg at the end of July. This time, Zhou's counterpart was the JCC co-chairmen, Alexander Ananenkov, deputy chairman of the Gazprom management committee. On the Russian side, therefore, the person responsible for gas exports (Medvedev) was not the same as the person responsible for strategic international relations (Ananenkov).[151]

But the Chinese planners were disappointed and continued to think that the commitments they were hearing from Russia on gas supply were not deliverable. Gazprom continued talking. In November 2007, Medvedev said:

> Two factors, the adoption of a programme for the development of the gas industry in East Siberia and the Far East and the fact that Gazprom expects to complete the deal on joining the Kovykta project in the near future, naturally influence gas sales to countries of the East. As the deal has not yet been completed, we have only preliminary calculations on how Kovykta may affect our plans … The dates of completion of the construction of an eastern section of a gas pipeline from Russia to China will be determined in commercial negotiations. But in any case, the Kovykta field development plans have been presented to the Natural Resource Ministry.[152]

The Chinese planners were getting frustrated.

No Breakthrough by 2009

China Central TV reported that Putin's first visit to Beijing since becoming Russian prime minister in May 2008 had been a productive one. The visit coincided with the sixtieth anniversary of the establishment of Sino-Russian ties, and Russian Language Year in China. Putin's visit was seen as representing a closer and more dynamic relationship. *China Daily* also reported that Russian firms planned to sign deals valued at more than $5.5 bn during Prime Minister Putin's October 2009 visit to Beijing, ranging from a $0.5 billion loan agreement between China's Development Bank and its Russian equivalent, VEB, to finance joint projects in transport, infrastructure, construction, and mineral extraction.[153] The coverage was typically diplomatic, but the price negotiations made no progress towards a compromise.

The *Financial Times* reported that Prime Minister Putin signed $3.5 bn of deals during the talks with Prime Minister Wen, but the two countries failed to make substantial progress on a key deal for Russian gas supplies to China, signing only a preliminary agreement that left unanswered questions about prices and the source of supplies. The *Financial Times* added that the gas supply agreement had been the centrepiece of the talks, which Prime Minister Putin had hailed as part of a growing strategic partnership that had withstood the financial crisis which had left the Russian economy floundering compared to that of its eastern neighbour. The Gazprom CEO, Alexey Miller, said that the agreement could open the way for the supply of 70 bcm of gas per year to China from fields in Siberia and Sakhalin, but the CNPC president, Jiang Jiemin, said there had been no deal on the volume of gas to be supplied or on the price. The Russian deputy prime minister, Igor Sechin, said that agreement over prices could be reached in 2010 and deliveries could begin in 2014 or 2015.[154] Igor Sechin added that in 2009 Russia's shipments of coal to China would reach a value of $1.0 bn.[155]

The only positive development relating to gas sector cooperation was that in October 2009 the China–Russia Investment Co Ltd (CRICL), a Sino-Russian joint venture, acquired a 51 per cent stake in the Russian oil and gas company Suntarneftegaz. This joint venture will be granted exploration and development rights to the Berezovsky and South Cheredeisky gas fields, which have total reserves of more than 60 bcm. The CRICL plans to invest $300 million in developing three gas fields, which are located near the proposed west Siberia–Pacific gas pipeline.[156]

The Altai Pipeline Project

In September 2006 Bogdan Budzulyak, Gazprom's director of transportation and storage, suggested that the projected costs of the western route (Altai gas pipeline) had grown from the planned $4 billion to $5 billion.[157] The authorities of the Altai Republic approved a project – designed by the St. Petersburg-based institute Giprospetsgaz, a Gazprom subsidiary – to build a trunk gas pipeline from Western Siberia to China. The pipeline route would be from Altai Territory to Shebalinsky, then to the Chuisky tract, the Ongudaisky and Ulagansky areas, and then stretching along the Kosh-Agachsky area to the Chinese border. The total length of the gas pipeline on the territory of the Gorno-Altai region would be 500 km.[158]

However, some confusion was created by a report that Russia's Fuel and Energy Ministry had said that the Altai project had essentially been removed from the draft gas development strategy up to 2030. The report, authored by Gazprom and released on 7 October 2008, stated that the possibility of implementing the Altai project still existed, but only if economic conditions were satisfactory, if a trade agreement with China could be agreed, and if the environmental constraints were resolved. The report added that:

> … compared with deliveries of LNG, pipeline gas deliveries are substantially less competitive given that the distance to the destination market in China is in excess of 6,000 kilometres. Moreover, the project would compete against pipeline gas delivered from Turkmenistan to China via Uzbekistan and Kazakhstan. China has already attained a certain level

of understanding with Turkmenistan on the issue of organizing natural gas deliveries. Another challenge for Altai is achieving netback parity with exports to Europe from the fields in western Siberia that supply the pipeline. China does not currently have the economic foundation for market-based pricing that would ensure the efficiency and competitiveness of Russian gas deliveries.[159]

Nevertheless, the Altai supply option was not abandoned. In November

Map 7.1: Altai Gas to China

Source: Gazprom.

2008 a Gazprom delegation led by the Deputy CEO, Alexander Medvedev, held separate commercial talks with CNPC in Shanghai on Russian natural gas shipments to China. The delegations discussed prices, volumes, routes, and the timeframe. Gas was to be delivered via two routes with 30 bcm to be channelled through the Western and 38 bcm via the Eastern route. The Western option could be implemented relatively quickly, given the existence of available and prepared reserves, a developed infrastructure, and gas-processing capacities.[160]

In December 2009 another Gazprom press release also confirmed that the Altai project was still alive: the negotiating group led by Medvedev held another round of commercial talks with CNPC's delegation about organizing natural gas supplies to China. The parties discussed the actions which were needed in order to supply gas from Russia to China via the 'western' corridor. The talks included the pricing issue and technical aspects of supply.[161] The priority given to Altai gas export to China is a clear reflection of Moscow's determination to construct a pipeline that would allow Gazprom to divert its surplus European volume to China. However, China continued to be disappointed by the suspicious and mistrustful Russian strategy.[162]

So after all there was no breakthrough on Sino-Russian gas cooperation by 2009, while by contrast Sino-Russian oil cooperation had achieved another breakthrough due to the $25 bn loan for oil. Sino-Russian gas cooperation during the years 2003–9 was a virtually empty glass, while Sino-Russian oil cooperation was at least a half-full glass. It is not an exaggeration to say that Rosneft's pragmatic approach towards China saved the endangered relationship through the 2005 and 2009 loans, while Gazprom's rigid stance towards China, based on its monopoly status, caused the delay in reaching a gas price compromise and gave China a strong incentive to prioritize the option of gas supply from the central Asian republics, in particular from Turkmenistan.

To the Chinese planners, Gazprom's push for the Altai project rather than East Siberian gas (or Vostok gas) export to China raises serious questions about Gazprom's hidden intentions. Why Altai first? Starting the Altai project would allow Gazprom to reallocate more gas to China when demand from the European market shrinks. Both European observers and the Chinese share this view but Russians do not. The gas supply from the Altai project could give Gazprom swing supplier status, similar to the position of oil exports to Asia via the ESPO pipeline. Given that current gas reserves in west Siberia are set to decline sharply in the coming years, the development of the massive gas reserves in the Yamal Peninsula in the Arctic Circle will be accelerated.[163] Stern points out that, given the facts that first Yamal production would start in the third quarter of 2012, that independent gas production was rising rapidly, and that at the same time Gazprom's markets (domestic, CIS, and European), were recovering only uncertainly, a surplus of gas was more likely than a deficit. What is very clear is that the cost burden is set to grow. As mentioned in Appendix 7.1, Gazprom used to highlight Bolshekhetskaya region of West Siberia as the supply source for China. However, Bolshekhetskaya development, whose potential production is projected to be up to 30–40 bcm/y, is no longer led by Gazprom but by Lukoil.[164] Even though Gazprom will not comment on any alternative supply source for the Altai exports, it looks certain that another source will eventually be needed for the western route. Due to the perception that the Altai project fits very effectively with Gazprom's ambition to tame both European and Chinese consumers, China continues to veto the Russian monopoly's ambitious plan to prioritize Altai gas exports. Strictly speaking, China has more attractive gas supplies from central Asia, and wishes to prioritize the eastern pipeline. China knows only too well that Altai exports can only go ahead if WEP-III is allocated for gas supply.

No Compromise in 2010

Even though no breakthrough over the gas price negotiation was made during the first half of 2010, Gazprom floated the idea of two spurs (one from the Amur area, the other from Primorsk) to China from its ambitious Eastern Gas Programme. Gazprom has no intention of offering a discounted gas price to north-east Asian gas consumers because it wants to receive similar prices to those it charges to European customers, possibly even more. For Khabarovsk, in 2010 Gazprom offered a price of 10,000–12,000 rubles ($330–400) per 1,000 cubic metres. This would be higher than that charged to Europe. Stern has pointed out to the author that at an oil price of $100–120/bbl, the 2011 price of Russian gas to European countries would be $350–450 per 1,000 cubic metres, and with such fuel costs, much Far Eastern industry would not be competitive. Mikhail Korchemkin, director of East European Gas Analysis, has estimated that an economically justified price for gas deliveries on the SKV pipeline (which will cost Gazprom 75 billion rubles to build) would be $350–500 per 1,000 cubic metres.[165]

In late January 2010 Medvedev, the deputy head of Gazprom, stated that the company would sign a contract with China in 2011 to supply it with natural gas starting from 2015.[166] A delay in reaching a price compromise was confirmed in early February by a *China Daily* report that China and Russia were still in protracted price negotiations.

'Natural gas supply is still under discussion', said Sergey Tsyplakov, Russia's trade representative in China; 'the two sides are still negotiating prices before any major projects can be launched.' But it was more than three years since the two sides had signed a preliminary gas deal, yet the 2010 agreement had failed to resolve the disagreements over pricing, and conditions that had so far blocked concrete progress.[167] At Gazprom's 2010 AGM, Medvedev said Gazprom now aimed to reach a gas price accord by the middle of 2011, and Deputy Prime Minister Igor Sechin indicated that the formula for pricing Russian gas deliveries to China might be agreed by September 2010.[168]

On 26 September 2010, Gazprom announced that it had agreed the major terms of a natural gas delivery contract with CNPC. According to the press release, the legally-binding agreement sets out the key commercial parameters of future Russian gas deliveries to China on the 'western route'. These parameters include volumes and dates, take-or-pay levels, the period for increasing deliveries, and the level of guaranteed payments. The actual export contract was due to be signed in mid-2011, and deliveries were scheduled to begin in late 2015. The legally-binding contract would run for 30 years and involve the delivery of 30 bcm per year.[169] The companies had reached agreement on everything except the price of gas. But, after this, a contract was expected to be signed before July 2011 (although this has not happened) and Medvedev stated that Gazprom planned to begin construction of the Altai pipeline at the end of 2011.[170]

The Price Lesson

It is well known that Russia was very unhappy about China's strategy of using the coal-based bench mark for the pipeline gas price negotiation (See Chapter 3). Immediately after the three year feasibility study on gas from Kovykta was completed, it was rumoured that Gazprom was very much opposed to its early development because the feasibility study had revealed the details of production costs at Kovykta. At that time, both CNPC and Kogas had offered to pay around $20–30/1000 cubic metres while Rusia Petroleum was expecting $100/1000 cubic metres.[171] Gazprom had no interest in supplying cheap Kovykta gas to China, and made it very clear that the firm had no intention of accepting CNPC's coal-based benchmark price and would apply the international market price to its gas exports to China.

It was reported in January 2008 that, starting in 2010, PetroChina would pay Turkmenistan a wellhead price of US$195/1000 cubic metres, based on a crude price of $45 per barrel. This price would set a benchmark for Gazprom gas exports to China and test their affordability in southern and eastern China.[172] This compared with Gazprom's gas sales price to Europe of more than $250/1,000 cubic metres in the same year.[173] CNPC officers complained that Gazprom was asking the Chinese buyer to accept the European border price, which would make the city gate price too high for Chinese consumers. Some of the statements made about the suggested price were contradictory. In June 2007 the CNPC delegation, during a meeting with Gazprom in St Petersburg, strongly suggested that China could not afford to pay more than $100/1,000 cubic metres,[174] even though in November 2006 a PetroChina executive said it could afford to pay up to $180/1,000 cubic metres.[175] According to a deputy manager of PetroChina Natural Gas and Pipeline Co., the gas price at the Chinese border town of Khorgos, including the pipeline transportation fee, would be US$245/1,000 cubic metres, averaging 2.02 yuan per cubic metre ($8.2/mmbtu) after tax. Including the transportation fee for China's domestic section of the pipeline, Turkmen gas at the city gate would be priced at 3.1 yuan per cubic metre ($12.6/mmbtu) (based on the crude price of $100/barrel level). [176]

Another related part of the story was the collapse of the LNG supply deal between PetroChina and Woodside Petroleum at the beginning of 2010.[177] The price burden was too high. In September 2007, PetroChina had signed two long-term LNG purchase contracts with Shell and Woodside. The price in PetroChina's Woodside contract has been revealed as US $10/mmbtu.[178] Gazprom's Medvedev, commented that 'China paying market prices is good for our discussion'. CNPC and PetroChina officials, however, argued that the figure was beyond what China could afford to pay and insisted that the acceptable level was around $7.0/ mmbtu.[179] In the spring of 2008, two major deals were signed by PetroChina and CNOOC for 5 mt/y of LNG from Qatar, but no clue as to the price was made. This confirmed that the Beijing planners were beginning to recognize the necessity of paying global market prices for LNG supplies, despite the price burden.[180] However, *China OGP* argued forcefully that 'given that there is no possibility for CNOOC to pay more than $10.0/mmbtu, the actual price is anticipated to be $6–8/mmbtu in an environment where LNG supply tends to be tighter'.[181] This report confirmed that China had problems in accepting the LNG price being paid by Japan and South Korea.

Price continued to be the key factor in assessing the real progress of Sino-Russian gas cooperation. In November 2007 Medvedev said, after a three-hour meeting with the CNPC vice president, Zhou Jiping, in Beijing, that 'the price has not yet been agreed upon. The negotiations are proceeding at a high professional level'. He added that 'an important subject at the talks was the prices of pipes and services, which neither Gazprom nor CNPC were satisfied with. This refers to both the global market and the internal markets of the two countries.' Medvedev said that the talks had made significant progress in the development of a price formula, but raised this point: 'China buys oil and oil products at world prices, so why should it buy gas cheaper?'[182]

Significant changes, however, were taking place in CNPC's stance towards the gas price negotiations. First of all, it completely abandoned the strategy of using a coal benchmark for gas prices. Secondly, it began to use an LNG benchmark, despite the problem that LNG prices differ by coastal province. CNPC's price negotiation strategy became much more realistic: using the price at which Gazprom sells natural gas to Europe as the starting point in determining the price formula for Altai gas. The CNPC representatives believed that the formula suggested by Gazprom would make Altai gas more expensive than European prices, something which CNPC certainly could not accept. The Chinese were surprised by a report alleging that they had agreed to buy gas at $195 per 1,000 cubic meters from Turkmenistan. Chinese negotiators said in a private conversation that this was a tactic by Turkmen experts, who wanted to push the gas price up for Russia.

By the summer of 2009, Russia had suggested that its plans to build gas pipelines to China were seriously delayed because of the failure to reach a pricing deal with Beijing:

> Gazprom deputy chief executive Alexander Ananenkov observed that 'no one is talking about gas supplies to China in 2011 anymore'. This statement came as Russia's President Medvedev and Prime Minister Putin met China's President Hu Jintao during his state visit to Russia in June 2009. Pricing had been the main stumbling block since Moscow and Beijing [had] agreed to build two pipelines to ship up to 80 bcm of West and East Siberian gas to China in 2006. China had since then agreed to buy gas from the former Soviet republics of Kazakhstan and Turkmenistan, with supplies due to begin in the next few years. Ananenkov suggested that a decline in demand in Russia and abroad, because of the global economic crisis, may prompt it to postpone the launch of the giant East Siberian Kovykta gas field until after 2017. There was speculation that this situation might strengthen Beijing's hand at talks with Gazprom and hinder Gazprom's desire to diversify away from the European markets where it already supplies a quarter of demand.[183]

But in late 2009 came news of a possible breakthrough. On 16 October the Russian Ambassador to China, Sergey Razov, said in an interview with *China Business Post* that China and Russia had reached an agreement on the price of natural gas exported from Russia to China. It would, he said, be pegged to a package of oil prices in the Asian market, and he also emphasized that the economic returns from natural gas exports to China should not be less than those of exports to Europe.[184] The euphoria, however, was short-lived. In early 2010, as mentioned earlier, Gazprom reverted to the position that the price would not be fixed until the summer of 2011.

In late September 2010, Gazprom and CNPC signed the extended major terms of natural gas supply from Russia to China. The document sets the key commercial parameters of the forthcoming natural gas delivery to the PRC market via the 'western' route: the volumes and the timeframe for export start-up, the take-or-pay level, the supplies build-up period, and the guaranteed payment level. The agreement is legally binding.[185]

More details of the negotiations followed:

NDRC deputy chairman Zhang Guobao confirmed that disagreement persists between China and Russia over the construction of pipelines for Russian gas shipments, with China prioritizing the construction of the eastern route and Russia the western route. Zhang said that 'China receives gas from western pipelines already, including the pipeline from the Xinjiang Uyghur Autonomous Region in west China which takes gas to eastern China, and also pipelines from Turkmenistan and Kazakhstan. Therefore an increase in gas deliveries to Xinjiang is not so important for China'. He added that 'concerning the eastern route, it is intended to deliver gas to north-eastern China which has a population of over 100 million, and which is experiencing serious gas shortages. Gas shipments via an eastern gas pipeline may solve the problem of gas shortage, including for a large number of local industries. Gas is not supplied there via a single pipeline. Russia has assumed a rigorous position and does not want to discuss this issue … Russia insists on building the western stretch of the gas pipeline first because gas is to be delivered from western Siberia to Altai in Xinjiang, just 90 km from the Russian border'. Zhang also mentioned that the gas price has not yet been negotiated. Of course, we are interested in the western pipeline as well. But Russia has proposed a price of over $300 per 1,000 cubic metres. We buy Central Asian gas at a price far below Russia's proposed price – $200–210 per 1,000 cubic metres.[186]

China OGP reported that a CNPC insider had disclosed that the acceptable price for China was $150/1,000 cubic metres with the conditions that China makes the investment in the construction of the natural gas pipeline, and that the price of natural gas imported from Turkmenistan is taken into account.[187] The remarks by Zhang and the gas price from CNPC clearly confirm that the gap between Russia and China was still very wide and, without a settlement on the route and the price, no breakthrough for Sino-Russian gas cooperation in 2011 was likely.

Gazprom responded by re-playing the Korean card to set up a benchmark gas price for its gas exports. As no gas price settlement between Gazprom and CNPC had been reached, Gazprom was keen to introduce this Korean card as a source of indirect pressure on the Chinese negotiators. In early November 2010 the Gazprom CEO, Miller, announced that his company would supply no less than 10 bcm/year of gas to South Korea starting in 2017. He added that the study of additional gas supplies agreed in 2009 had been completed. Crucial agreements on the switch to the next stage of the deal (covering commercial talks on the terms of gas sales and purchase agreements) had been reached at a 10 November meeting with the Kogas CEO. The commercial talks were to start in December. Miller also said that it had been agreed that all prices would be pegged to a Japanese basket of petrochemicals. He went on to confirm that:

> … in the course of pre-project research, we have analysed options for gas supplies to the Republic of Korea. In particular, we considered pipeline, pressurized and liquefied gas supplies. The companies analysed these options and presented them to each other for consideration. The final supply method will be determined through further negotiations … among other options, pipeline gas supplies to the Republic of Korea are possible.[188]

As to financing the pipeline construction, Miller said that:

> … there is an agreement that everything that concerns the construction of facilities on Russian territory is to be financed by Russia, and what concerns the building of facilities in the Republic of Korea should be financed by the Republic of Korea.[189]

The decision on the gas pipeline will be made soon enough to allow physical supplies of Russian gas to the Republic of Korea to commence in 2017.[190] Gazprom's Korean card, however, can be easily sidestepped if Beijing agrees to take 30–38 bcm/y of gas for their north-eastern provinces, without the 10 bcm/y supply for Korea. The stumbling block, however, is the price, not the volume for China.

Unusually, in mid-November 2010 Gu Jun, the deputy director-general of the National Energy Administration's international department, commented on the issue at a news briefing on Prime Minister Wen Jiabao's visit to Russia and Tajikistan between 22 and 25 November 2010. Gu said that, although companies from both sides had made many efforts to solve the problem, a certain difference still existed over the pricing of natural gas imported from Russia, ($100 per 1,000 cubic metres was mentioned). Gu called for the two sides in the price talks to demonstrate additional sincerity.[191] It is very rare for a briefing on a prime minister's trip to Moscow to focus specifically in this way on a single concrete issue such as the gas price; and a clear indication that China was very concerned about the delay of the gas price deal.

In January 2011, at the Third Annual China LNG Terminal Summit 2011 held in Shanghai, more information emerged. Xia Yishan, chairman of the Center for Energy Strategy Studies, said during this event that the price difference had dropped from $300/1,000 cubic metres to $260/1,000 cubic metres and that the gas talks would be finalized by October 2011.192 In fact, big expectations centred on the summit between President Medvedev and President Hu Jintao in June 2011. But again they were disappointed and the summit failed to release good news on the gas price deal, since the price difference was hardly reduced at all (Gazprom insisting on $350 and CNPC talking about $235–250/1,000 cubic metres).[193]

As a way of bridging the price gap, an OIES study suggested that if Gazprom and CNPC accepted that Russian gas imports should be initiated via the eastern rather than the western route, it would be possible to bridge the $100/1,000 cubic metres price difference. The study suggested that if the city-gate price in Shanghai implied by the Turkmen import price is used as a benchmark, then CNPC could afford to pay Russia just over $250/1,000 cubic metres for gas via the western route, but could increase this price to $315 for gas from East Siberia, thanks to the lower transport distance and cost.[194] If this ingenious suggestion were combined with a prepayment of $40 bn by CNPC,[195] the price gap could be resolved.

Sino-Russian M&A Deals and Strategic Partnership: Failures and Successes

CNPC's Attempts to Take Over Slavneft and Russia's Brutal Response

Russia was not an easy country for CNPC to enter with the hope of developing upstream assets. *Xinhua News Agency* reported that CNPC was invited by the Kremlin in late November 2002 to participate in the bidding to purchase a controlling stake in Slavneft, Russia's eighth biggest oil company, and that before the application deadline of 15 December 2002 it was the sole foreign contender. Things turned against CNPC over the following few days, however, when the Russian legislature held a special session to question CNPC's eligibility. By 255 votes to 63, the Duma passed a bill preventing any foreign company which was more than 25 per cent state-owned to participate in privatizing Russian enterprises.[196]

In late 2002, according to people familiar with the incident, a Chinese official was kidnapped in Moscow when CNPC was preparing to bid for the Russian government's interest in Slavneft. CNPC withdrew from the auction soon after the event. The official was a member of a CNPC delegation visiting Moscow for the auction of Slavneft. The official was kidnapped at the airport, bundled into a car and driven away, apparently as part of a Russian effort to convince CNPC to pull out of the bidding. The official was freed when CNPC dropped out of the auction. Two Russian oil companies, Sibneft and Tyumen Oil, went on to buy Slavneft for $1.86 billion – far short of the $3.0 billion that some analysts had said CNPC was prepared to pay for the Russian government's 75 per cent equity stake.[197] No other press media covered this story. It was a very costly experience for CNPC and a very strong signal that Chinese entry to Russia's upstream sector would never be easy.

CNPC's Ill-Fated Attempts to Purchase Stimul

Soon afterwards, CNPC made a second attempt to purchase an upstream asset in the Russian Federation. On 12 December 2003 it announced that it had acquired 61.8 per cent of the Orenburg oil and gas company Stimul for $200 million. Prior to the deal, the offshore company Victory Oil (Netherlands) had held a controlling stake. Orenburg Gazprom, a Gazprom subsidiary, continued to hold the remaining 38.2 per cent of Stimul shares. The Stimul opportunity had appeared in the summer of 2003 when Victory Oil decided to put its Stimul shares on the market. Besides CNPC, Bazovy Element (Russia), Total, Maersk (Denmark), and TNK–BP expressed their interest in purchasing the Stimul shares. However, Gazprom enjoyed a pre-emptive right to buy the controlling interest in Stimul.[198]

The project was relatively small and CNPC did not expect further problems from this acquisition deal to arise. In fact, it believed it was close to a deal to buy a controlling stake in Stimul. According to *China Business Post*, CNPC submitted an application to purchase the Stimul stake to the Russian government for approval, but heard nothing for more than two months (normally such a procedure takes one month). The best guess is that Gazprom might have used its pre-emptive rights, as well as its political influence on the Russian government, to forestall CNPC's purchase.[199] The double failure to purchase Russia's oil assets convinced CNPC that the only way to have any access to Russia's upstream assets was to work with the state companies Rosneft (for oil) and Gazprom (for gas).

Rosneft's Strategic Partnerships with CNPC and SINOPEC

Despite the failures of the M&A deals, in February 2005 Beijing agreed to lend $6 bn to Rosneft and it did not take long to see the positive influence of this loan. On 6 June 2005 Rosneft and CNPC concluded a long-term cooperation agreement. The CEOs of the two companies signed the document which expressed their intention to participate in full, long-term, and mutually beneficial cooperation and in joint development. The companies resolved to identify new ways of expanding oil exports to China on mutually profitable commercial terms, including the possibility of using the Atasu–Alashankou and Taishet–Skovorodino pipelines.

Rosneft's press release revealed more of the specifics of its cooperation with China:

> Under the agreement, the parties expressed their interest in increasing the volumes of Sakhalin natural gas supplied to China. CNPC is to apply its best efforts toward reaching agreement on deliveries of gas from the Sakhalin-1 project (in which Rosneft is a participant) on mutually profitable commercial terms. It also intends to sign documents on the project's implementation with project participants in the fall of 2005.
>
> ...
>
> [dated 7 June 2005] Rosneft and the China Petrochemical Corporation (Sinopec Group) signed an agreement on the establishment of a joint venture for exploration and surveying of the Veninsky sector (Sakhalin-3 project) on 1 July. The agreement was signed by Rosneft CEO Sergei Bogdanchikov and SINOPEC CEO Chen Tonghai.[200]

The shares in the proposed joint venture with SINOPEC were Rosneft 49.8 per cent, SINOPEC 25.1 per cent, and Sakhalin Oil, 25.1 per cent. In 2003, Rosneft won a five year license to pursue geological exploration of the Veninsky block of the Sakhalin-3 project, and in February 2006, the Ministry of Natural Resources transferred the license from Rosneft to Venineft to make sure the latter would have the right to explore the block.[201]

On 30 August 2005 Rosneft signed an interim financing agreement with SINOPEC for Sakhalin offshore exploration. According to the document, SINOPEC was to finance 75 per cent of expenditure during the stage of geological exploration at the Veninsky block (which is estimated to contain 169.4 mt of oil and 258.1 bcm of gas). Rosneft was to be responsible for financing 25 per cent of the expenditure from its own resources.[202]

On 29 March 2007 Rosneft and SINOPEC established a Corporate and Shareholder Agreement related to their joint work in exploring and developing the Veninsky block in the Sakhalin-3 project. According to the document signed on 26 March 2007 in Moscow, Rosneft International Limited and SINOPEC Overseas Oil and Gas Limited would become the owners of Venin Holding Limited which was established in October 2006. Venin Holding Limited would in turn be the sole shareholder of Venineft, the license owner and operator of the Sakhalin-3 project. Rosneft would have a 74.9 per cent stake in the project, with the remaining 25.1 per cent going to SINOPEC. In other words, Rosneft excluded Sakhalin Oil Co. from the project.[203]

Drilling of North Veninskaya well No.2 at the Veninsky area in 2009 enabled a more accurate estimate of reserves at the North Veninskoye field – which was discovered in 2008 as a result of the drilling of a first well. Russian C1 and C2 reserves at the North Veninskoye gas condensate field are estimated at 34 bcm of gas and 2.8 mt of condensate. Drilling of Veninskaya well No.3 discovered the modestly sized Novoveninskoye oil and gas condensate field.[204]

Zhang Guobao's Rare Display of China's Frustration

From Beijing's viewpoint, signing a strategic partnership was one thing, but real progress has been another. By early 2006, the lack of progress had made Beijing planners very frustrated. Zhang Guobao, vice director of the National Reform and Development Commission (NDRC), said in an interview with *Interfax* that, despite being China's fourth largest oil supplier, Russia did not understand China's energy requirements, lacked competitiveness, and had been unwilling to cooperate fully with China in the energy sector. Zhang supplied a series of examples:

> [He] recounted his experience of a trip to Russia. Zhang said Ma Fucai, former head of CNPC had been offered a deal in which China would pay for the reactor technology for the Lianyungang nuclear power station in Jiangsu province in exchange for a commitment to build the oil pipeline from Siberia to China. Zhang was authorized by the Chinese government to pay $1.4 bn to purchase the facilities, but despite going from one government department to another, he was unable to find someone who could negotiate the deal. He said it was a wasted trip.
>
> …
>
> It has been difficult to sign an agreement over proposed oil pipeline between

the two countries, with Russian opinion changing like the weather forecaster, one day saying they have reached an agreement, the next saying there is no agreement at all.

...

Russia has also been unwilling to allow China to purchase oilfields in western Siberia. 'China is willing to invest in Russia. But the question is will Russia let us?', Zhang asked.

...

[He] also mentioned that China is willing to allow Russian oil companies to construct oil refineries in China, but Russian companies have been slow to take an interest in refinery construction in China.[205]

Thus Zhang made clear that the amount of practical cooperation in his view was still insignificant and the whole process was unsatisfactory.[206] This rare display of open concern was provoked by China's frustration with the slow progress of Sino-Russian oil and gas cooperation.

SINOPEC's Udmurtneft Breakthrough

The strengthened upstream sector relationship between Rosneft and SINOPEC, however, did move into the realm of acquisitions. In March 2006, in Beijing, Rosneft and SINOPEC signed a memorandum of cooperation and agreed to jointly acquire some of the assets of Udmurtneft, which had been established in 1973 on the basis of an oil and gas producing department of the same name, and had been operating as an open joint stock company (OJSC) since 1994. Udmurtneft was the main oil producing enterprise in the Udmurt Republic, producing over 60 per cent of the oil in the Volga Ural region. In 2005, the company produced 5.98 mt of oil (43.6 million barrels), with daily production standing at 16,400 tons (115,000 barrels). On 31 December 2005, Udmurtneft's proved reserves stood at 78.4 mt (551 million barrels), proven and probable reserves (or at more than 131 mt (922 million barrels) of oil equivalent, as estimated by DeGolyer and MacNaughton).[207]

In April 2006, Rosneft and SINOPEC concluded an option agreement under which Rosneft would acquire Udmurtneft shares from SINOPEC in the event that SINOPEC won the tender to purchase Udmurtneft assets from TNK–BP. In fact, the condition of the Udmurtneft sale was that SINOPEC would give 51 per cent of Udmurtneft to Rosneft free of charge. SINOPEC had outmanoeuvred Gazprom, which had announced on 19 June that it was prepared to join the Hungarian firm MOL in a joint bid, as long as MOL agreed to cover the lion's share of the cost.[208]

On 20 June 2006 SINOPEC received the green light. This deal became the first major acquisition of an oil producing company in Russia by a Chinese oil firm. In August SINOPEC Overseas Oil and Gas Limited, a subsidiary of SINOPEC, finalized the deal to purchase 99.49 per cent of the ordinary shares of Udmurtneft.[209] *Xinhua News Agency* saw this deal as a milestone, being the first entry into the Russian upstream market by a Chinese company.[210] The deal was an indication that foreign companies could only enter Russia's oil and gas sector through joint ventures with selected Russian partners, and then only with a minority role.[211]

On 11 November 2006, Rosneft and SINOPEC signed a shareholder agreement setting out the principles of joint management, and finalized an agreement to transfer the Udmurtneft shares to the joint venture. At their extraordinary meeting on 12 December Udmurtneft shareholders elected a new nine person board of directors containing five Rosneft and four SINOPEC representatives.[212] According to Rosneft, SINOPEC alone would pay for the acquisition, and would reimburse itself out of cash flows. This move marked China's strategic entry into the Russian oil sector.

Vostok Energy: CNPC's Entry to Russia's Upstream Sector

On 16 October 2006 in Moscow, Rosneft, and CNPC signed a protocol to create Vostok Energy (VE). This joint venture's charter capital was set at 10 million rubles (approximately $377,400); 51 per cent held by Rosneft and 49 per cent by CNPC, with a five person board of directors, three from Rosneft and two from CNPC. Its main objective was to ensure geological survey and exploration operations on Russian territory, search for mineral fields, and obtain licenses for various types of subsoil use. It was intended to facilitate growth in foreign investment in the Russian Federation and create new employment opportunities for Russian citizens. The Rosneft CEO, Bogdanchikov, said in Beijing that VE should produce no less than 10 mt/y of oil in the following three to five years, adding that VE should operate as close as possible to the route of the ESPO oil pipeline then under construction.[213]

CNPC showed an interest in the Vankor project, but Rosneft's vice president, Mikhail Stavsky, indicated that Rosneft's venture with CNPC would be interested in the Yurubcheno–Tokhomskaya oil region.[214] CNPC's proposal for joint development of Vankor with Rosneft was repeated in 2007 but there was no positive feedback.[215] This confirmed that it was still too soon for CNPC's choice of upstream target to be fully accepted by Rosneft.

In July 2007, Vostok Energy won auctions to develop two hydrocarbon fields in the Irkutsk region. Vostok Energy bid 399.5 million rubles ($15.61 million) for the Zapadno-Chonsky, with an initial payment of 85 million rubles ($3.32 million), and 780 million rubles ($30.48 million) for the Verkhneichersky, with an initial payment of 100 million rubles ($3.91 million). The Zapadno-Chonsky field, in the Katanga district of Irkutsk region, is located 120 kilometres west of the East Siberia–Pacific Ocean Pipeline (ESPO). The field's prospective C3 resources, as entered on the state register of mineral resources, amounted to 5 mt on 1 January 2005. Preliminary D1 prospective resources included 30 mt of oil and 15 bcm of gas. The Verkhneichersky field is in the Katanga district, 250 kilometres north-east of Ust-Kut. ESPO is 90 kilometres away to the east. The field's D1 predicted reserves amount to 140 mt of fuel equivalent, including 50 mt of oil and 90 bcm of gas. The Preobrazhensky field reserves are estimated at 72 mt of oil and 70 bcm of gas.[216] Even though the potential reserves are relatively big, the proven reserves of VE's two fields are small.

Nikolai Syutkin, the general director of Vostok Energy said that these were the first licenses to have been won by the company. He added that Vostok Energy would concentrate on East Siberia, but consider other regions, too. He also reported that his company had submitted bids for the Ignyalinsky and Vakunaisky fields to be auctioned in September. On 15 August 2007 the Irkutsk subsurface resources department, Irkutsknedra, confirmed that Rosneft and Vostok Energy had submitted bids to acquire the Ignyalinsky and Vakunaisky hydrocarbon sections to be auctioned on 12 September 2007.[217] This was the same day on which Irkutsknedra would also auction the Ust-Iglinsky section claimed by Tekhenergo, Metall-Geologia, and Vostok-Mineral, and the South Kytymsky section claimed by Antei, Metall-Geologia, and Vostok–Mineral. The deadline for submitting bids expired on 10 August and the deadline for paying deposits was 7 September. The winner would gain the right to combined use (geological prospecting and extracting) of hydrocarbons in the licensed section for the following 25 years. The initial one-time payments on the Vakunaisky and Ignyalinsky sections were 665 million rubles and 330 million rubles, respectively.[218] Resulting from the 12 September auctions, Gazprom Neft, the oil-producing arm of energy giant Gazprom, reported that its subsidiary, Kholmogorneftegaz, had won a tender to prospect and develop two oil fields in East Siberia.

According to Rosneft:

… in 2009 LLC Vostok Energy carried out 593 linear km of 2D seismic exploration, including 345 linear km at the West Chonsky license area and

248 linear km at the Verkhneichersky area. Drilling of a prospecting well at the West Chonsky area is scheduled in 2010.[219]

In September 2010 CNPC disclosed that the company would bid for tenders to develop Russian oil fields only on a joint basis with Rosneft.[220] In late November 2010, it was reported that CNPC and Rosneft were discussing the possibility of developing operations in the Sea of Okhotsk near the port Magadan. According to Igor Sechin, the deputy prime minister, the two companies agreed that Vostok Energy would acquire several small- and medium-sized fields in East Siberia. Included in the talks were CNPC's possible participation in exploration activities in the Magadan shield.[221] The timing of this report was interesting because it could be regarded as a token of Moscow's appreciation for the approval of the Tianjin refinery project in August 2010. It will take time to see the scale of the production from the fields being explored by VE.

Chinese-Russian Eastern Petrochemical Company: Rosneft's Entry to China's Downstream Sector

On 10 November 2006 Rosneft's CEO, Bogdanchikov, announced in Beijing that the registration of the second joint venture between Rosneft and CNPC for oil refining and oil products marketing in China, Chinese-Russian Eastern Petrochemical Company (CREPC), was close to being concluded. According to Sergei Goncharov of the Rosneft Representative office in China, CNPC, via its PetroChina subsidiary, would hold 51 per cent in this joint venture and Rosneft the remaining 49 per cent. Part of the venture would be an oil refinery in China with an annual capacity of 10 mt. Apart from the oil refinery, Rosneft would wish to have no fewer than 300 gasoline stations in China.

By the end of 2007, the joint venture was expected to acquire its first several dozen stations. It was estimated that the oil refinery would require approximately $3.0 billion to build. The construction or purchase of 300 gasoline stations would cost a further $150–200 million. Denis Borisov, an investment analyst, assumed that the gasoline network would enable the joint venture to get extra margins from retail sales. The difference in wholesale and retail prices in China was $1–2 per barrel. Nevertheless, there was no profit guarantee as Beijing regulated prices for refined products. Gasoline was 15 cents per barrel cheaper in China than in Russia. Until the Chinese government cancelled state regulation of gasoline prices at the end of 2007, Rosneft had no plan to enter the retail market.[222] The green light was given in 2010.

In late August 2010, Zhang Guobao, head of China's National Energy Administration (NEA), said this joint venture refinery would receive

70 per cent of its oil from Russian companies, while the remaining 30 per cent would be procured from the international market. According to the Russian energy minister, Shmatko, the refinery would receive its oil via the ESPO spur pipeline from 1 January 2011. Sergei Andronov, head of Rosneft's oil and oil product export department, said that the Russian oil would be purchased at Kozmino and supplied through the ESPO pipeline at ESPO brand oil spot prices. This variant required a feasibility study, since the refinery was configured in such a way that it required blends of different oils.[223] In September 2010, Igor Sechin said that both CNPC and Rosneft had agreed to start the study on a $5 bn oil refinery project in Tianjin Municipality. The CREPC joint venture aimed at building a refinery with potential annual crude throughput of 13 mt. After the feasibility study, design, and construction, the second stage would involve building a network of 500 filling stations in northern China. Chinese deputy Prime Minister Wang Qishan said that the refinery would be built by 2015 and would receive 70 per cent of its oil from Russia and the rest from Arab countries.[224] In May 2011, construction work finally started.[225]

SINOPETRO's Venture with Yuzhuralneft

On 21 November 2006 SINOPETRO, a Russian subsidiary of CNPC, and Yuzhuralneft, a company registered in the Orenburg Region, announced yet another deal to form a joint venture. Yuzhuralneft, which was founded in 1992, specializes in geological exploration and production of oil at small fields with hard-to-recover reserves. The company's assets included 20 wells yielding some 2,000–3,000 tonnes of crude oil per month. Yuzhuralneft planned to ramp up production and build an oil-loading rail terminal. Yuzhuralneft assets would form the backbone of the Sino-Russian company, and CNPC would act as investor. The equity shares and size of the supposed investment were not disclosed, but analysts estimated that the Chinese inputs might total no more than $7–7.5 million.[226] In early 2008, it was reported that TNK–BP bought out Yuzhuralneft.[227] Even though the scale of venture was small, it was one of the reported failures of CNPC's Russian business development.

RusEnergy's Equity Stake in Suntarneftegaz

In April 2009, the Sino-Russian RusEnergy Investment Group acquired 51 per cent of the Sakha Republic-based Suntarneftegaz through its Sino-Russian Investment Energy subsidiary. The investment agreement signed by Suntarneftegaz provided for the participation of Chinese investors in the share capital of the company, on condition that it made long-term investments for the development of deposits controlled by Suntarneftegaz. During 2010 and 2011, RusEnergy was to invest almost $300 million in the development of deposits. According to Russian experts, the 51 per cent of Suntarneftegaz probably cost about $200–300 million, and it is unlikely that this transaction would have taken place without Kremlin approval. Suntarneftegaz was established in 2006, and owns licenses for Yuzhno-Berezovsky and Cheredeisky blocks in eastern Siberia with total gas reserves of 60 bcm. Annual aggregate production should amount to almost 3 bcm.[228]

CIC's Equity Stake in Nobel Oil

On 16 October 2009 CIC (China Investment Corp)[229] announced the purchase a 45 per cent stake in the Nobel Oil Group, established in 1991 with three operating oil fields in Russia possessing estimated reserves of 150 million barrels (20.5 mt) of oil. Nobel Holdings Investments owns all of the assets of Nobel Oil in Russia. The $300 million investment was to be completed in two phases. In the first phase, completed by the end of September 2009, CIC spent $100 million to purchase its stake in the Russian oil company and $50 million on the operating expenses of the oil fields. Based on the purchase agreement, CIC would hold 45 per cent of the company, the Russian company 50 per cent, and the remaining 5 per cent would go to a Hong Kong investment group named Oriental Patron, owned and managed by former Chinese government officials. This acquisition appears to be a back door route to obtaining listing of the Russian company in Hong Kong. The HK-based financial firm Oriental Patron Financial Group is a shareholder in both Nobel Holdings and the Hong Kong-listed Kaisun Energy Group.[230]

All of the above transactions tell us that, apart from SINOPEC's Udmurtneft purchase, the only successful Sino-Russian deals have been relatively small. Given that the Russian authorities have seen no pressing need to invite foreign partners to participate in Chayanda, Kovykta, and Sakhalin-3 gas developments,[231] Chinese upstream partnership with Gazprom seem very unlikely.

Conclusion

The fortunes of Sino-Russian oil and gas cooperation during the period 1993–2010 have been mixed. There is a solid basis for such cooperation because Russia has large oil and gas resources while China's market for these commodities is very large and expanding. The results so far have been fairly successful in the oil sector, but very unsuccessful in the gas sector. Oil sector cooperation is on the right track and there are plenty of opportunities for further expansion: cooperation in the natural gas sector is still a hostage to the repeated failures to reach a settlement over prices, which may be the most visible sign of deeper problems in the relationship.

Successful Sino-Russian oil cooperation has been created by China's imperative need to increase oil imports. This helps to explain the systematic approaches which Chinese planners have made to secure crude oil supplies from both Russia and the Central Asian Republics. China started to purchase oil-producing assets in Kazakhstan from 1997, but its acquisitions never reached the 20 mt/y production capacity level. As shown in Table A.7.2, Beijing pursued oil and gas supply from both Russia and the Central Asian Republics simultaneously. During the 1999–2002 period, Beijing's highest priority was to build the Angarsk–Daqing crude pipeline. Nonetheless, in 2002, well before the collapse of Yukos in 2003, Beijing authorized CNPC to begin the development of the Kenkiyak–Atyrau oil pipeline. When it became very clear during 2004 that the Angarsk–Daqing pipeline development was to be replaced by the ESPO initiative, China lost no time in starting the Atasu–Alashankou oil pipeline section, completing it in 2005. In the same year, CNPC purchased the Petro-Kazakhstan oil asset, and made a $6 bn loan to Rosneft. In other words, during the first half of the 2000s, virtually equal emphasis was given to the Angarsk–Daqing crude pipeline and to the China–Kazakhstan crude pipeline, but the collapse of Yukos diverted Beijing's attention towards pipeline development with Kazakhstan.

During the second half of the 2000s, China stepped up its attention to oil and gas supply from the Central Asian Republics. In fact, Beijing was greatly encouraged by Turkmenistan's approval of CNPC's upstream participation in the giant South Yolotan field in 2006, and in the following year the construction of the Kenkiyak–Kumkol pipeline was agreed.[232] By the end of 2009 almost 3,000 km of pipeline connecting west Kazakhstan and west China was completed. After 12 years of negotiations and preparation work, Kazakh crude supply to western China through the China–Kazakhstan crude pipeline also became a reality at the end of 2009. In the case of Sino-Russian oil pipeline development, the Skovorodino–Mohe–Daqing pipeline came on stream at the end of September 2010.[233]

In May 2010, the BBC reported that:

Kazakhstan energy Minister Sauat Mynbayev told Kazakh parliamentary deputies that China held a 50–100 per cent stake in 15 companies working in Kazakhstan's energy sector. According to the Kazakh energy ministry, out of 80m tonnes of crude oil which Kazakhstan was expected to produce in 2010, 25.7m would go to China.[234]

A year later the *Financial Times* reported that 'Chinese companies control 22 per cent of Kazakh oil output, less than US oil majors which have a 24 per cent share'.[235] Thus, after 12 years of intensive searching and massive investment to secure oil producing assets, China managed to secure well over 20 mt/y of production capacity, enough to justify the development of a 3,000 km pipeline from the Caspian side of Kazakhstan to its eastern border with China's Xinjiang province. The aim of the Chinese planners was a value chain business, maximizing equity oil participation and constructing the related crude pipeline infrastructure linked to pipeline networks in China. This could be called a 'Central Asian Model' which justifies China's massive investment in oil and gas development in the Central Asian Republics, in particular Kazakh oil and Turkmen gas.

However, the Central Asian Model did not work in Russia because the Moscow authorities initially rejected any attempt by China to purchase equity in Russia's oil producing assets, and allowed only very limited approval for Chinese upstream asset purchase. The Slavneft auction and the Stimul case showed the Russian government's determination not to allow Chinese NOCs any equity participation; but the Udmurtneft takeover by SINOPEC confirmed that a deal coordinated in advance would not necessarily be vetoed by Moscow. This is how the loan for oil option was used by the Chinese planners, who had not previously realized the importance of financing in Russian oil deals. 'Loans for Oil' or 'Financing the target supply source to make sure of long term supply security' became the core formula of the 'Russian Model'. The $6 bn loan in 2005 and $25 bn loan in 2009 were created to solidify the basis for Sino-Russian oil cooperation.

As for the gas sector, Sino-Russian cooperation during the 2000s took the form of 'No Action, Talking Only (NATO)' while Sino-Turkmen gas cooperation, on the other hand was moving more in the direction of Sino-Uzbek and Sino-Kazakh gas cooperation. The major breakthrough was made by the Turkmenistan government's decision in 2006 to allow CNPC to take an upstream position in Turkmenistan's gas exploration and production, together with the related gas pipeline development. Beijing never tried to apply a 'loan for gas' option to settle its long-delayed gas price negotiations with Russia. In the spring of 2006, CNPC's expectations were, however, raised by the announcement that about 60–80 bcm/y of gas supply from Russia to China would be available from 2011; but there was still no sign of any compromise on the border price issue. Russia insisted that China pay the European border price, which meant that by the time the gas reached many cities in China, the cost would be even higher than the European price.

The price CNPC had accepted for Turkmen gas was not cheap, but at least the Turkmenistan authorities had allowed CNPC to exercise the equity gas option, thus alleviating the financial burden. The equity gas option allowed the CNPC planners to cushion the financial burden of the high import price. But Gazprom completely ruled out equity gas for CNPC and offered no alternative to offset the financial burden of a very high price.

The only chance of breaking this stalemate lies in a change in the Chinese authorities' response since Gazprom is unlikely to change its hard line. Strictly speaking, CNPC has no authority to make any compromise on price negotiations. The one compromise it made during the 2000s was to abandon its coal-based benchmark price and adopt an LNG-based benchmark. There are a number of different LNG import prices for the receiving terminals at Guangdong and Fujian provinces and at Shanghai and Dalian (Liaoning province), but it will not be easy to find a mutually acceptable LNG benchmark price. The ultimate decision on the final price must come from the price department under the NDRC. Even though the NDRC will certainly pursue a step-by-step gas price reform, this will not increase prices sufficiently to narrow the gap between the domestic gas price and the prices of imported LNG. It will take some time to adjust China's domestic gas price to transnational pipeline gas prices based on any kind of European price level.

As a way of alleviating the price burden of Russian gas imports, CNPC explored the possibility of applying the Russian Model for a breakthrough in Sino-Russian gas cooperation. However, Gazprom's deputy CEO Alexander Medvedev bluntly dismissed talk of a Chinese loan as part of the gas export supply deal, and insisted that Gazprom would have no problems financing pipeline construction.[236] These remarks seemed to rule out a transfer of the Russian Model from oil to gas. As mentioned earlier, however, a month after Medvedev spoke, a report that an upfront payment of $40 bn for gas pipeline development seemed as if it might signal a change in Gazprom's stance.

With Gazprom prioritizing Altai (West Siberian) over East Siberian and Far East gas for the Chinese market, however, CNPC is likely to try to use the leverage of the West–East Pipeline (WEP) III against Gazprom. The choice between Altai gas and Central Asian gas for WEP-III will fundamentally determine the future of Sino-Russian gas cooperation in the 2010s. The recently completed WEP-II was constructed for Turkmen gas supply to China, but nothing has yet been decided about the supply source for WEP-III. If gas supplies from the Central Asian Republics are increased from 40–60 bcm/y, however, this is likely to exclude Altai gas from WEP-III. The allocation of another 30 bcm/y supply from the Central Asian Republics to WEP-III, straight after the completion of WEP-II, will guarantee a significant delay for Altai gas. These alternatives for WEP-III were not properly appreciated until the possibility of increasing supply from the Central Asian Republics was revealed during 2010.

Broadly, the verdict on Sino-Russian oil cooperation during the 1993–2010 period is positive, even though there is great potential for increasing its scope once the upstream and downstream joint ventures start to operate. The scale of crude oil supply from Russia to China is very likely to expand once the second stage of the ESPO pipeline is completed. Apart from the 24 mt/y crude supply, 15 mt/y by pipeline from Skovorodino to Daqing, and 9 mt/y for Tianjin refinery, Chinese buyers will aim to take extra crude from Kozmino. Invisible competition for Russian oil between Japan, South Korea, and China looks inevitable, but at the same time there is a real chance of promoting multilateral cooperation in the region. One possibility lies in the development of SPR (Strategic Petroleum Reserves) storage linked with ESPO crude supply to Kozmino. If the current rate of expansion of the car industry in China continues, the chance of a serious shortage of SPR capacity is very high. Beijing needs a contingency plan for SPR storage development and a facility near Kozmino, based on the UNDP GTI (Greater Tumen Initiative) regime, could serve as a regional scheme. Political agreement at the highest level would open the door for a pilot project of SPR in a GTI area, and this could fundamentally change the level of Sino-Russian oil cooperation.

By contrast, Sino-Russian gas cooperation during 1993–2010 was a big disappointment, and many western observers are quite sceptical about whether there will be any large-scale gas supply from Russia to China during the second half of the 2010s. In mid-2011, the continued stalemate in gas price negotiations suggested that a breakthrough in Sino-Russian gas cooperation will not be made without political agreement at presidential or prime ministerial level. Any further delay in Russian gas supply to China will restrict the expansion of China's pipeline gas market. The importance of Sino-Russian gas cooperation cannot be overemphasized since gas expansion in China will help to slow its dependence on coal. In 2009 alone China imported as much as 130 mt of coal, despite its own enormous production. Even though

China has launched a strong drive in favour of renewable energy, its heavy dependence on coal will continue in the coming decades.[237] The point, however, is whether anything gas can do, in terms of domestic development and imports, will be able to stop the increase in coal demand in absolute terms, and as a percentage of primary energy in China.

Table 7.6: The Mis-match of Chinese and Russian Approaches: the Story of the 2000s

Oil and Gas

Russia	*China*
Was willing to engage in endless negotiations, not in a hurry to conclude a deal	Wanted serious negotiations followed by a quick agreement and construction
Insisted on being in charge of all developments on its own territory, reluctant to allow foreigners to take significant equity but needed major external investment capital	Insisted on being in charge of the full value chain – major equity participation – to dictate timing of development; unwilling to contribute substantial investment on any other terms, but was forced to compromise in the oil sector due to desperate need

Gas Only

Russia	*China*
Adjusted negotiating position relative to Europe (prices) and South Korea (supplies)	Adjusted negotiating position relative to availability of alternative supplies of LNG, Central Asia and Myanmar, and in long-term perhaps also domestic (shale) gas
Wanted first project to be Altai to take advantage of surplus production in Western Siberia	Wanted first project to be eastern project (Kovykta)

Source: Author

It is no exaggeration to say that the different outcomes of Sino-Russian oil and gas cooperation during the 2000s were caused by the mis-matches of Chinese and Russian approaches (Table 7.6). Sino-Russian oil cooperation was driven by desperate necessity, but Sino-Russian gas cooperation was blocked by very different expectations on both sides about cooperation. Russia's priority was supply from Western Siberia (via the Altai pipeline) and maximizing the export price. China's priority from the west was Central Asian supply with equity participation; from the east, the priority would be supply from East Siberia to its north-eastern provinces. During the 2000s, Russia was very slow in grasping the key message from China, which was that, if the terms for Russian gas were not favourable, alternative supplies of LNG (from a variety of sources) and pipeline gas from Central Asia and Myanmar could be arranged within a relatively short time frame. In other words, given its other options, China was unwilling to import Russian gas on Gazprom's terms.[238] As a result, by 2015 China could be importing at least 70 bcm/y of gas, none of which is likely to come from Russia.

Coal imports from Russia provided a similar example. China recently began to import coal on a large scale, and the figure during 2009–11 was recorded at 130 mt, 165 mt, and 182 mt respectively, of which coal imported from Russia in 2009 and 2010 was 12 and 13 mt respectively. A study by Stanford University pointed out that 'China's coal buying behavior follows the logic of a "cost minimizer" and China's coal imports will fluctuate according to the arbitrage differentials between domestic and international coal prices'.[239] This clearly reflects China's very pragmatic stance towards the business of energy trading, and has significant implications for Sino-Russian gas price negotiations.

To Chinese planners, gas imports by pipeline from Russia remain extremely important. Despite the failure to reach agreement on prices by mid-2011, both sides still have expectations that their differences can be resolved. As argued earlier, if the upfront payment offer is combined with the option of gas export from East Siberia (rather than West Siberia) to China, there is a very good chance to bridge the price gap. The outcome of Sino-Russian gas negotiations in 2011 is a major question since it will fundamentally affect the level of Sino-Russian energy cooperation over the coming decades, either positively or negatively or perhaps both.

Appendix to Chapter 7

Table A.7.1: Review of Gazprom's Asian Policy

1997.02	Gazprom CEO, Rem Vyakhirev, announced the company's intention to formulate a comprehensive policy for penetrating the lucrative Asian gas market. He said that he saw a prime market for Gazprom's growth in Asia where the gas market was absolutely empty or devoid of competition. Some of the production from the $40 billion Yamal project could be diverted to Asia.[240] This was the first announcement of Gazprom's Asia Policy.
1997.06	Vyakhirev revealed a detailed blueprint for Gazprom's 'new' Asian initiative in a speech delivered to the World Gas Conference. Vyakhirev stated that Gazprom was supporting a number of proposals to develop East Siberian gas for domestic consumption and for exports to East Asia. Vyakhirev's view was that short-term (immediate) Asian demand for Russian hydrocarbons could be met by LNG imports. However, after the 2005–7 period, Asia would require additional sources of natural gas reserves. To this end, a new gas production facility would be constructed in the East Irkutsk region at some point after the year 2000. The production centre would eventually be linked by trunk pipelines to China, North and South Korea, and Japan.
1997.08	Gazprom and CNPC signed an agreement on 'cooperation' in the gas sector.
1997.10	Valery Remizov, deputy chairman of Gazprom, confirmed the fact that Gazprom had yet to decide the exact means by which west Siberian resources would be exported to China. The options under consideration included: i) the construction of a 6,000 km pipeline to Shanghai; ii) the construction of a 'new' terminal in southern China to facilitate increased exports of LNG.
1997.11	Russia's President Boris Yeltsin, attended the 'Fifth Russian-Chinese Summit' in China. During the visit, the First Deputy Premier Boris Nemtsov, and his Chinese counterpart, Li Lanqing, signed a Memorandum of mutual understanding in the principal areas of economic, scientific, and technological cooperation. The memorandum gave top priority to 'large-scale' energy projects as a means of creating a solid and substantial basis for a long-term relationship between Russia and China. The 'large-scale' energy projects were: i) Irkutsk–China gas pipeline project; ii) West Siberia–China gas pipeline project; iii) electricity exports from Irkutsk to China.
1997.12	Gazprom and CNPC ratified the 'Memorandum on Negotiations between Gazprom and CNPC' on the implementation of the project for the delivery of Russian natural gas to the eastern areas of China.
1998.02	The Second Session of the Russia–China Commission on the preparation of Governmental Summit meetings on a Regular Basis was held from 16–17 February 1998. The Commission, co-chaired

by B. F. Nemtsov, and Li Lanqing, entrusted the parties' relevant departments with the task of assisting in the implementation of the project for the construction of the gas pipeline from West Siberia to the eastern areas of China.

1998.07   Gazprom and the administration of the Tomsk region signed a five-year cooperation agreement to enhance industry infrastructure. According to the agreement, Gazprom would provide assistance to the Tomsk region in the following areas: i) the construction of a pipeline system linking homes in the rural and urban districts of Tomsk to the main gas distribution network; ii) the construction of a number of gas pipelines in West Siberia; iii) the promotion of gas engine fuel for transportation.

1998.08   Gazprom announced that the results of a preliminary feasibility study on West Siberian–China exports were promising. More specifically, the Bolshekhetskaya cavity region of West Siberia contained approximately 3 tcm of natural gas reserves, including 0.75 tcm of Category C1 reserves, 0.6 tcm of C2 reserves, and 1.2 tcm of C3 reserves. This quantity of natural gas reserves would be sufficient to sustain natural gas exports to China of approximately 30 bcm/y, thereby justifying the construction of a 6,714 km pipeline to the potentially lucrative Chinese market.[241]

1998.11   Vyakhirev disclosed the details of two promising export options to China at the Summit Conference in Kuala Lumpur: i) The Altai project: envisaging the export of West Siberian gas to the Shanghai region of China via Xinjiang province; ii) The Baikal project: envisaging the export of West Siberian gas to the Shanghai region via a 6,467 km pipeline passing through Krasnoyarsk, Irkutsk, Mongolia, and Beijing.

1999.02   Both governments agreed to undertake FS work on Western Siberian gas exports to Shanghai areas, China.

1999.10   Vostokgazprom, based in the Tomsk region, was formed by Gazprom (49%) and the Hungarian-based General Banking & Trust Co Ltd (51%) which was partly owned by Gazprom through its Gazprom bank.

2001.03   PetroChina announced that 19 western companies passed the pre-launch qualification evaluation; Gazprom was included in this evaluation list.

2001.05   PetroChina submitted the appraisal report on the seven investment proposals for the West–East gas pipeline project (out of 19 companies) to the SDPC, and in June 2001 the SDPC approved the selection.

2001 summer Gazprom renewed work on the East Programme, the full name of which became 'The programme of Creation in Eastern Siberia and in the Far East a Uniform System of Extraction, Transportation of Gas and Gas Supply with Possible Gas Export to the Markets of China and Other Countries of the Asian Pacific Region'.

2002.06   The institutional basis for gas development in Eastern Siberia and the Far East changed fundamentally in June 2002 when the Russian government issued Decree No. 975-r, which instructed the Ministry of Energy and Gazprom to draw up a programme for a unified system of production, transmission, and distribution of gas in Eastern Siberia and the Far East.

2002.07   On 4 July 2002 PetroChina Co finally signed the Memorandum of Understanding on JVs with three international partners for the West–East pipeline project to share project risks and to take advantage of their international experience, financial strength, and technical and operational experience. The three partners were: i) Royal Dutch Shell Group + HK & China Gas Co; ii) ExxonMobil Corp + CLP Holdings; iii) Rao Gazprom + Stroytransgaz

A government order of 16 July 2002 authorized Gazprom and the Energy Ministry to prepare this document jointly and the same order gave Gazprom the task of co-ordinating the implementation of the programme. The government agreed with the main features of the programme on 13 May 2003.

2003.03   The first draft of the programme to create the eastern branch of UGSS was prepared jointly by Russia's Ministry of Industry and Energy and Gazprom, and it was approved.

2003.06   Gazprom CEO Alexey Miller's key note speech at the 22nd World Gas Conference suggested gas production from East Siberia and Sakhalin Islands would be 26 bcm in 2010, and the figure could reach 110 bcm in 2020. He added that Gazprom had been authorized by the Government to co-ordinate the establishment of a united system for gas production and transportation. His presentation also indicated two LNG plants at Vladivostok and Vanino, in addition to Korsakov in Sakhalin Islands, and it confirmed where Gazprom's real intention lay.

2004.03   Gazprom signed an agreement with the Chongqing Municipal Gas Group to build a 200 mcm underground natural gas depot. The proposed location was the Jiangbei district of the Municipality. The estimated investment was more than 200 million yuan.

2004.05   China Gas Holdings Limited signed a LOI with Gazprom for joint development of gas projects in China, and for Gazprom to become a strategic investor in the Chinese piped gas provider. China Gas handled a total of 25 midstream and downstream joint venture projects in China, and its market share in China was around 20% in 2010.

2004.06   Gazprom officially announced that the firm would remain Russia's only gas exporter, and its 100% subsidiary Gazexport would negotiate with potential customers on volumes, schedules, and gas price formulae.

2004.07   PetroChina/CNPC announced that the firm had decided to terminate joint venture negotiations with foreign groups – Shell, ExxonMobil, and Gazprom – with regard to their involvement in a $18 billion West–East pipeline development. It was a major setback to Gazprom's plan to connect West Siberian gas with the West–East pipeline when China's domestic gas resources in Xinjiang province became exhausted.

2004.08	In response PetroChina's decision to terminate the joint venture negotiations with foreign groups Gazprom stated, during the firm's meeting with Kogas in Moscow in early August, that Kovykta gas could not be supplied to China and South Korea. Gazprom made it very clear that the firm would pursue the gas pipeline heading to Nakhodka and then to South Korea via the offshore route.
2004.09	Gazprom's Strategic Development Department made a presentation on 'Eastern Siberia & Far East Natural Gas Production and Transportation options Economic Feasibility Study' in Irkutsk. This then became the so-called 'Discussion Package for the Interagency Working Group to develop a Programme for creating a Unified Gas Production, Transportation and Supply System in Eastern Siberia and the Far East with potential exports to markets in China and other countries in Asia and the Pacific'.
2004.10	On 14 October Gazprom and CNPC agreed to sign a strategic partnership agreement in Beijing.
2004.11	Gazprom and CNPC agreed to launch continual consultation under joint working groups created for key avenues of cooperation, and the first meeting of the Gazprom and CNPC joint co-ordinating committee was held in Sanya City. The meeting aimed at the implementation of points agreed in strategic cooperation deals between the parties.
2005.09	Alexander Medvedev, deputy chairman of Gazprom's management committee, confirmed that the company was negotiating with CNPC to build two pipelines to transport up to 60 bcm across the border annually.
2005.10	Gazprom won a contract to design an underground gas depot in China, and the contract was signed during a visit to China by a Gazprom delegation led by the company's deputy chief executive, Alexander Ananenkov. (The agreement was signed in Beijing on 14 October 2004, during a visit by Russia's President Vladimir Putin. The agreement made provision for a variety of forms of cooperation, including the organization of Gazprom gas exports from Russia to China. The joint coordinating committee, which had the task of coordinating the implementation of the agreement, set up a general joint working group and separate working groups for the principal forms of cooperation.) Gazprom said in a press release that the design project was to be entrusted to VNIIGAZ, Gazprom's chief research and development centre. The visit included regular meetings of a joint working group and a joint coordinating committee of Gazprom and China National Petroleum Corporation (CNPC). At the meetings, the two companies discussed progress made in the implementation of a strategic cooperation agreement in 2005 and approved a plan for 2006.
2006.03	On 21 March 2006 a memorandum between Gazprom and CNPC on natural gas supplies and construction of two gas pipelines was signed. The agreement (Protocol on the Supply of Natural Gas from Russia to the People's Republic of China) to construct two gas pipelines to China, with capacity to move 60–80 bcm of gas annually

and costing $10 billion, formed the centre piece of Putin's state visit to China.

2006.07–08   On 31 July and 1 August, within the framework of the agreement for strategic cooperation between Gazprom and CNPC, the 4th meeting of the joint co-ordinating committee (JCC) between Gazprom and CNPC was held in St. Petersburg.

2007.06   On 15 June Russian government's Commission on the Fuel and Energy Sector (FES) instructed MIE to approve the East Programme.

2007.09   Russian minister of industry and energy, Viktor Khristenko, signed the approval order (no. 340) creating the Eastern Gas Programme (EGP).

2008.04   In accordance with the Russian Government Directive of 16 April 2008 the license for the Chayanda oil and gas condensate field was granted to Gazprom as the owner of the Unified Gas Supply System.

On 25 April, Gazprom invest Vostok[242] (the wholly-owned subsidiary of Gazprom which was established in July 2007 to implement Gazprom's eastern Russian programmes) was mandated to implement the East Gas Programme for Gazprom. The company was later charged also with the promotion of Gazprom's projects as part of the Development Programme for an integrated gas production, transportation, and supply system in Eastern Siberia and the Far East. In this role it also aimed at the large-scale development of gas exports to China and other Asia–Pacific countries.

2008.09   On 1 September 2008 the Russian government issued orders for Sakhalin-3's three blocks – Kirinsky, East Odoptinsky, and Ayashsky – be handed to Gazprom.

2009.02   On 18 February the first Russian liquefied natural gas (LNG) plant as part of the Sakhalin-2 project was launched.

2009.03   On 10 March the Russian government adopted the investment certificate for the comprehensive investment project permitting the elaboration of the project documentation on South Yakutia development.

2009.07   On 2 July Gazprom commenced exploration drilling in Kirinskoye field, and on 31 July Khabarovsk hosted the celebrations dedicated to welding the first joint of the Sakhalin–Khabarovsk–Vladivostok gas transmission system.

2009.10   During 12–14 October CNPC and Gazprom signed a preliminary agreement that left unanswered questions on price and the source of supplies. Gazprom CEO Alexey Miller said the agreement could open the way for the supply of 70 bcm of gas per year to China from fields in Siberia and Sakhalin, but CNPC president Jiang Jiemin said there had been no deal on the volume of gas to be supplied, or on the price.

2009.11   During 1–2 November in Shanghai, Gazprom's delegation had a round of commercial negotiations with CNPC. Gazprom deputy CEO, Alexander Medvedev, and CNPC vice president, Wang Dongjin, agreed upon formation of the working groups of experts from Gazprom and CNPC to address joint participation in the

	projects for setting up gas chemical complexes in the Far East.
2009.12	Gazprom announced that on 22–7 December a Gazprom delegation led by Alexander Medvedev, would hold the year's final round of commercial talks relevant to arranging Russia gas supplies to China via the western and eastern routes with CNPC in China.
2010.09	Gazprom announced in a press release that it had signed a deal extending the terms of a natural gas delivery contract with CNPC. According to the press release, the legally-binding agreement sets out the key commercial parameters of future Russian gas deliveries to China on the 'western route'. These parameters include volumes and dates, take-or-pay levels, the period for increasing deliveries, and the level of guaranteed payments. The actual export contract was due to be signed in mid-2011, and deliveries were scheduled to begin in late 2015. Under the current terms, the contract would be valid for 30 years and involve the delivery of 30 bcm per year. That is, the companies had reached agreements on everything but the price of gas.

Source: Paik, K-W. (2002b); JBIC (2005); The section on the Sino-Russian energy cooperation chronology during 1993–2009: from various issues of *Russian Petroleum Investor*, *Interfax China Energy Weekly*, *Interfax Russia & CIS Oil and Gas Weekly*, and *China OGP*.

Table A.7.2: Sino-Russian vs Sino-Central Asian Republics' Oil and Gas Cooperation

	Sino-Russia		Sino-Central Asian Republics	
	Oil	*Gas*	*Oil*	*Gas*
1992	Irkutsk exploration acreage			FS on Turkmenistan gas
1995				CNPC, Exxon, & Mitsubishi started a joint FS for Turkmen gas to China
1997		Preliminary FS on Kovykta gas	CNPC bought Aktyubinsk & Uzen oil fields, Kazakhstan	
1999	FS on Angarsk–Daqing line	Preliminary FS on both Kovykta gas and Altai gas	China decided to shelve 3,000 km oil pipeline project	
2002	China chose Guangdong & Fujian LNG supply sources China ruled out pipeline passing through Mongolia			
	CNPC's forced withdrawal from Slavneft's bidding process		CNPC started to construct Kenkiyak–Atyrau oil pipeline	
2003	Japan's oil pipeline route initiative	Russia's Energy 2030 strategy	Kenkiyak–Atyrau oil pipeline completed	CNPC began to explore the three options of gas supply from Kazakhstan
	CNPC's failed attempt for Stimul		CNPC bought North Buzachi field - SINOPEC and CNOOC attempted to buy BG's 16.7% equity of Kashagan	
2004	ESPO towards Nakhodka direction decided	CNPC–Gazprom strategic cooperation agreement	CNPC decided to build Atasu–Alashankou pipeline	
	PetroChina terminated its negotiations with western IOCs on WEP.			

Table A.7.2: continued

	Sino-Russia		Sino-Central Asian Republics	
	Oil	*Gas*	*Oil*	*Gas*
2005	China's $6 bn loan to Rosneft & Rosneft's Veninsky deal with SINOPEC		- CNPC bought Petro-Kazakhstan oil asset - Atasu–Alashankou section completed	CNPC began to focus on gas exploration in Uzbekistan
2006		Gazprom announced 60–80 bcm/y of gas export to China		- Turkmenistan and China signed a framework agreement on the gas sector - Turkmenistan authorized CNPC participation in development of the Gunorta Yolotan gas deposit
	- SINOPEC bought Udmurtneft - CNPC bought Rosneft's equity - Upstream JV Vostok Energy & Downstream JV Eastern Petrochemical Co established			
2007			Construction of Kenkiyak–Kumkol agreed	
2009	China's $25 bn loan for oil to Rosneft and Transneft	No gas price agreement made yet	Kenkiyak–Kumkol section completed	Ceremony of the Central Asia–China Gas Pipeline held
2010	Skovorodino–Daqing line connected	No breakthrough on gas price talks		The gas import volume from 40 bcm/y to 60 bcm/y decided

Source: devised and compiled by the author.

CHAPTER 8

CONCLUSION

Nearly a quarter of a century after it first emerged, the question of the price and other aspects of large-scale Russian gas sales to China remains tantalisingly unresolved, even though on many occasions the endgame of negotiations seems to have been reached. The delays are a measure of the relative size of the gains and losses involved in what would be a gigantic international contract. Neither party has yet been prepared to make the decisive compromise which would bring about an enormous change, not only in the relations between these two countries but also in the world energy market.

At the beginning of the 1990s the author had great expectations that the changed environment which followed the collapse of the cold war would usher in a new pattern of energy cooperation – specifically in oil and gas – consisting of a combination of bilateral and multilateral cooperation, in the north-east Asian region. This had been inconceivable during the cold war, but the changed environment in the 1990s should have been very favourable to bilateral Sino-Russian oil and gas cooperation, and its extension into multilateral cooperation in northeast Asia for the first time in the region's history. Progress, however, has been very slow and frustrating. We have seen much talk and many studies, but few real examples of cooperation. Two decades later, this book is an attempt to analyse the achievements of Sino-Russian oil and gas cooperation during the 1990s and 2000s, to explain why these achievements were so different from what had been anticipated, and to foresee how the developments of the last two decades might evolve in the 2010s. This book has been structured around these fundamental questions and this chapter outlines its conclusions.

What was achieved through Sino-Russian oil and gas cooperation during the 1990s? Through the two MOUs, signed in 1994 and 1997, both Russia and China affirmed their deep interest in constructing long-distance oil and gas pipelines from the oil and gas fields of Russia's East Siberia to China's densely populated northern provinces, targeting, in particular, the region of the Beijing–Tianjin Municipalities. The most tangible and important result was the three feasibility studies for an Angarsk–Daqing crude oil pipeline and for two natural gas pipelines (one to carry Kovykta gas to Beijing and the other to carry Altai gas to western China). The agreements to carry out these studies were signed in early 1999 by both Russian and Chinese prime ministers. The three studies were important because they reflected Chinese commitment to pursuing pipeline development, even though the negotiators were not convinced about the scale of proven reserves in the designated supply sources. In principle, China was interested in Russia's oil and gas because geographical proximity would allow the option of relatively cheap and safe pipeline transportation; but it was not sufficiently desperate to commit itself to projects before feasibility studies had been carried out.

Oil

The preparation process continued during the first half of the 2000s but Sino-Russian oil and gas cooperation during that period produced very few concrete results. There were a number of meaningful deals in the middle years of the decade and finally, during the second half of the 2000s, cooperation began to produce some tangible achievements – even though these were still not of the magnitude hoped for during the 1990s, and they came at a significant price. During the first half of the 2000s, the widely advertised Angarsk–Daqing pipeline project became a casualty of the Yukos affair, which involved the arrest of Khodorkovsky and the consequent collapse of Yukos. This imposed a heavy cost in the form of the complete suspension of the Angarsk–Daqing project, and it took a further two years for Russia to give its final approval to the first section of the East Siberia–Pacific Ocean (ESPO) oil pipeline.

The most tangible achievement was the completion of the first section of the ESPO line at the end of 2009, along with the spur pipeline to China at the end of August 2010. The first stage of the line is capable of delivering 0.6 mb/d from Taishet to Skovorodino and it is being projected that it should reach 1.0 mb/d by 2016 and 1.6 mb/d by 2025. With the completion of the Mohe–Daqing section in October 2010, 0.3 mb/d of crude oil started to flow to China at the beginning of 2011. Another major and tangible step forward is the Tianjin refinery joint venture between CNPC and Rosneft, the construction of which started in May 2011.[1] Russia has to supply 70 per cent of the oil to the joint venture refinery, whose capacity is 13 mt/y, meaning that another 9 mt/y of Russian crude has to be supplied to China from 2015, when the construction work is completed. By the mid-2010s China alone could receive a total of 24 mt/y of oil from

Russia. Due to the decline of production from the oil fields in three north-eastern Chinese provinces (Heilongjiang, Jilin, and Liaoning), in particular the decline of the Daqing field, crude supply from Russia to Heilongjiang was China's highest priority. Even though the rate of decline of Daqing has slowed somewhat, Chinese planners were anxious to maximize the scale of Russian imports in order to diversify Daqing's sources of oil supply. In view of China's heavy dependence on oil imported by sea, pipeline supplies became a matter of urgency for the Chinese planners as part of their strategy of diversification.

Is there any possibility of further increasing oil supply from Russia to China? For the time being, the chances look slim. In 2008 the Siberian Scientific Research Institute (SSRI) of Geology, Geophysics, and Mineral Resources gave a strong warning that a level of 80 mt a year in East Siberia could not be reached until 2025, even though the Ministry of Industry and Energy wanted to commission the second stage of the ESPO pipeline in 2015–17, by which time East Siberian deposits should be providing nearly 56 mt to ESPO. SSRI pointed out that increasing production to this level and maintaining it for 30 years implied the production of 1.5 bt, while in 2008, the volume of explored reserves amounted to only 0.52 bt. SSRI suggested that the state should transfer almost 200 blocks to subsoil users for development, in order to attain the 1.5 bt of oil reserves; but by the beginning of 2008 only 70 deposits had been distributed. It is not clear how long it will take before the proven reserves at the main production bases in East Siberia and Krasnoyarsk region can reach these levels. Around the mid-2010s, there will be no difficulty in achieving a production level of 50 mt/y, considering that Vankor's peak production will reach 25 mt by 2014, while Verkhnechonskoye will reach 9 mt by 2015–17, and Talakanskoye will reach 6 mt by 2016. The remaining 10 mt/y could be covered by a number of scattered fields in East Siberia and the Krasnoyarsk region, including Yurubcheno-Tokhomskoye field where peak production could reach 20 mt. There is, however, no guarantee that China would get a higher allocation of Russian oil exports from Kozmino, because other north-east Asian consumers, such as Japan and South Korea, are anxious to secure bigger volumes of Russian supply.

Nevertheless, the prospects of Sino-Russian oil cooperation in the next two decades are much more favourable than those of Sino-Russian gas cooperation. Two factors have caused oil sector cooperation to receive the highest priority in Sino-Russian energy relations. First of all, China had no choice but to enter into negotiations on crude oil imports from Russia, due to a sharp decline in production at Daqing oil field. Even though the decline in the production rate was not as severe as the early projections had suggested, Beijing planners had to find an alternative supply source, and Russian crude supply by pipeline was an ideal option. Since some of the refineries in the north-eastern provinces had already been refurbished to receive the Russian crude oil, China had to secure at least a minimum volume of crude oil from Russia. This was the reason why China offered a $6 bn loan for oil in early 2005, a loan which saved Rosneft's IPO scheme. The second loan for oil, in early 2009, involved a total of $25 bn for both Rosneft and Transneft, and it covered 300 mt of oil supply between 2010 and 2030. It was a win–win deal for both Russia and China: Russia secured a massive loan in exchange for the sizeable supply of crude to China, and in return China secured its crude supply from Russia.

Secondly, the price negotiation on the crude deal between China and Russia did not pose any major obstacle (even though, in the period 2009–11, there was a renegotiation of the original price agreement in the wake of the crude oil price increase). Since international oil pricing had already been accepted in China, there was no obstacle to finding a price formula that was mutually acceptable to both countries. From the Chinese leadership's viewpoint, the reliability of crude supply was the top priority, and they were ready to take any step necessary to increase the volume of imports from Russia. Despite a sequence of failures of acquisition deals in Russia – the Slavneft deal in 2002 and the Stimul deal in 2003 – and despite the absence of any possibility of equity ownership in Russian oil fields, Beijing struggled to secure a minimum of crude supply, and in early 2005 introduced the loan for oil, a method which was applied again when China needed to secure a larger volume of oil supply in early 2009.

In short Sino-Russian oil cooperation was driven by China's need to secure its crude supply from Russia. This need will remain very strong in the coming decades, and that is why Beijing dragged its feet with regard to the approval of the Tianjin joint refinery venture between CNPC and Rosneft; it was to make sure that 70 per cent of the crude for the refinery would be covered by Russian crude supply.

When the second stage of the ESPO pipeline project is completed by the mid-2010s, the combination of ESPO and Sakhalin-1 and 2's crude supply to China and north-east Asian countries will be at least 1.3 mb/d. ESPO's entry into the north-east Asian market has challenged the 'Asian premium' on crude from the Middle East. If the comprehensive exploration work during the 2010s justifies the expansion of ESPO's second stage, bringing its supply capacity to 1.6 mb/d, ESPO's role in diversifying the supply sources of north-east Asian consumers will be even more significant. So it is regrettable that the momentum of the comprehensive exploration which took place between 2005 and 2008 flagged in the wake of the 2008 global financial crisis. It remains to be seen how and when this momentum in East Siberia will be revived.

Sino-Russian oil cooperation in 2000s can be summarized as follows:

- China, very much disappointed by the failure of the Angarsk–Daqing pipeline during the first half of 2000s, did not fully understand the internal and external political dynamics which had led Russia to take this decision;

- A desperate oil supply need, and lack of really large-scale alternatives (in Central Asia) forced China to agree to major investments, as well as financing supply and infrastructure, without being permitted to take any equity positions in Russia's upstream projects;
- China's proactive stance towards Sino-Russian oil cooperation did not reflect its genuine trust in Russia, but expressed the urgency of the Chinese oil supply situation;
- In the end the Russians got most of what they wanted, that is, major infrastructure development in East Siberia plus more diversity in its oil exports to Asia;
- China did not get the massive and secure quantities of oil which it wanted.

Natural Gas

In the ten years after 2000, Sino-Russian cooperation in the natural gas sector showed very few tangible advances; some announcements during the second part of the decade turned out to be overly optimistic. Following the political problems arising from the TNK–BP ownership of the Kovykta gas field, Moscow decided to give priority to the development of Chayanda gas and the surrounding four major gas fields in the Sakha Republic which were allocated to Gazprom.[2] Even though the 1,791 bcm of reserves in these fields is not as big as Kovykta's 2,000 bcm, Chayanda gas alone is sufficiently plentiful to allow Gazprom to pursue a 30 bcm/y long-distance pipeline development. The confirmation of Sakha Republic reserves makes the situation very different from the 1990s, when Beijing was still seriously concerned about whether the proven reserves were sufficient for a 3,000–4,000 km pipeline development. Once Gazprom had acquired the Kovykta field in 2011, the company had ample resources to supply 30 bcm/y of gas to China and South Korea, as long as the price issue could be resolved. China's gas demand during the 2000s increased significantly and the development, in particular, of the West–East Pipeline (WEP-I) during the first half of the decade laid solid ground for China's natural gas expansion. The decision to construct WEP-II, to facilitate a large volume of gas imports from Turkmenistan during the second half of 2000s, was instrumental in developing the WEP corridor (which includes WEP-III, IV, and V) during the Twelfth Five Year Plan.

Gazprom prioritized Altai (West Siberian) gas export to west China. The development of exports from Altai was not regarded very positively by the Beijing authorities since they gave much higher priority to the supply of East Siberian gas to north-eastern China. Even though it was not Beijing's favoured supply option, the first Altai initiative had been convincing and had it been sufficiently attractive, Beijing would not have hesitated to allocate Altai gas to the WEP-III pipeline. However, since China has decided to prioritize Central Asian (in particular Turkmen) gas as an equity supply source, Altai gas no longer looks like a 'must-have' option for China. Altai gas must overcome this problem if it is to access the Chinese gas market by the mid-2010s. Even if the price issue were to be settled in 2011, supplying Altai gas before 2016 may not be easy, and 2017–18 would probably be a more realistic date.

The verdict on natural gas sector cooperation is therefore not favourable, mainly because Beijing's need for pipeline gas from Russia is not as desperate as its need for Russian oil. Natural gas is still regarded as the most expensive fuel source for power generation, and power accounted for only around 15 per cent of China's total gas consumption in 2008. Natural gas tends to be treated, as a peak load, not a base load, energy source. Even though Beijing argues that in the future more natural gas will be used in the 'gas for power' sector, the financial burden of using gas for power is too great without reform of the distorted electricity, gas, and coal pricing system. It is the city gas sector, rather the 'gas for power' sector, which is receiving pipeline gas from the Central Asian Republics.

This is the reason why CNPC cannot accept the oil-related border price which Gazprom is demanding. Chinese planners find Gazprom's demand excessive, because CNPC cannot increase domestic gas prices, which are strictly controlled by the NDRC's price department. As soon as it became clear that this price stalemate would continue, Beijing made the final decision to construct the WEP-II pipeline in order to bring gas from Central Asia. The equity gas option offered by the Turkmen authorities was enough to compensate for the burden of the high border price for imports.

Gazprom's current stance is that Russia will move on to export its oil and LNG to 'Asia' taking the view that if China wants to buy, that would be fine, but if not, other countries will be happy to do so. At the same time, Russia (unlike Central Asia and many other countries in the world) refuses to allow China to own any part of the field and pipeline development. This rigid stance has been a significant part of the reason why Beijing has not accepted Gazprom's commercial terms.

The Beijing planners are fully aware of the risks involved in Gazprom's strategy of prioritizing Altai rather than East Siberian gas exports, and they are very uncomfortable about Gazprom's 'swing supplier' strategy. After the global financial crisis of 2008, the EU's appetite for Russian gas contracted, and this drove Gazprom to a more aggressive Asian gas export policy. China had not bargained for a gas supply which, in accordance with Gazprom tactics, was shared with the European gas market. But, for as long as East Siberia remained without a developed pipeline structure, Altai gas exports fit neatly into Gazprom's strategy of switching its European gas exports to China. The concept is similar to the ESPO, which allowed Russia to export its crude oil to the Asian market directly, and not just depending on European buyers.

Western media and energy security specialists argue that Putin is using gas exports as a blackmail weapon against European buyers. Whenever gas supplies to Ukraine or Belarus have been suspended, Putin's hostile stance has been mentioned. When gas suspension occurs, Russia wants to re-direct the gas exports to China. So far this has been a verbal threat to European buyers, but it will become much more real once the necessary Altai route pipeline infrastructure is completed. However, if gas exports from Russia to China via the Altai route begin, the western media will point out how the Altai export option enhances Russia's negotiating position (vis-à-vis European buyers). The Chinese planners, however, have no wish to be blamed for 'robbing' the Europeans of their gas when in fact they would prefer to buy Russian gas not from Altai but from East Siberia. The key point is that the Chinese do not need Altai for the WEP system to work because they can obtain Central Asian gas. They need East Siberian gas because regional gas capacity in the three north-eastern provinces of China is relatively small and, without access to East Siberia or Sakhalin, the alternative is large-scale LNG imports.

In summary, Sino-Russian gas cooperation in the first decade of the century was so limited because Russia tried to replicate its experience with oil exports, but found China unwilling to agree. This unwillingness was due to four main factors: first, Russia refused to allow equity in fields or pipeline projects, and therefore refused China any control in the value chain, which is what the Chinese wanted; second, Russia demanded unattractively high prices; third, China had alternative import options (the Central Asian Republics, Myanmar, and LNG imports) as well as the potential to expand domestic production; and, fourthly, there was a lack of trust on both sides. Russia wanted to avoid depending completely on China as a market, and China wanted to avoid over-dependence on Russia as a source of supply. The failure of the price negotiations is a reflection of all of these problems.

The most recent attempt to strike a Sino-Russian gas price deal was made in June 2011, but no positive result was reached despite negotiations and preparations between deputy prime ministers Igor Sechin and Wang Qishan, in parallel with the negotiations between Gazprom and CNPC. The price gap was still too wide although, at the time of writing, both parties are set to continue the negotiations. Compromise, however, will require changes of stance on both sides, and if these do not occur by the end of 2011 it could mean a dead-end for Sino-Russian gas cooperation. The Russian side may have under-estimated the level of confidence of their Chinese opposite numbers, who believe that their supply alternatives give them substantial leverage. As long as this assessment remains, China is likely to offer the minimum, rather than the maximum, opportunities for pipeline gas supply from Russia over the next two decades.

Likely Evolution of Oil and Gas Trade up to 2030

What, then, will emerge from Sino-Russian oil and gas cooperation in the coming two decades? It will not be easy to predict to what extent cooperation will grow, but the boundaries can be drawn by looking at three possibilities – the 'business as usual' scenario, the optimistic scenario, and the pessimistic scenario). The three scenarios are calculated using a projection of oil and gas trading between the two parties, and together they depict the limits of Sino-Russian oil and gas cooperation in the coming two decades.

Under the 'business as usual' scenario, the scale of oil trading between Russia and China will be around 25–30 mt/y in 2020. Of this 25–30 mt/y, 15 mt/y will be supplied through the pipeline extending from Skovorodino to Daqing, and 9 mt/y will be supplied by ESPO to the Chinese-Russian Eastern Petrochemical Company's Tianjin-based refinery. The remaining 5 mt/y will be imported by rail or through the China–Kazakhstan oil pipeline. The volume is unlikely to change by 2030 as long as the ESPO delivery capacity remains at 50 mt/y.

In the case of natural gas exports from Russia to China, Russia

Table 8.1: Sino-Russian Oil Trading Projection, 2020 and 2030 (mt/y)

	2020	*2030*
Business as usual scenario	25–30	25–30
Optimistic Scenario	30–35	40–45
Pessimistic Scenario	15	15

Source: Author's projections

Table 8.2: Sino-Russian Gas Trading Projection, 2020 and 2030 (bcm/y)

	2020	*2030*
Business as usual scenario	30	40
Optimistic Scenario	68	68
Pessimistic Scenario	0	20

Source: Author's projections

will fail to reach 60–70 bcm/y in 2020, even if China's gas demand is as high as 300 bcm. The scale of gas exports will be, at most, 30 bcm/y by 2020 thorough a single pipeline. The potential market for Russian pipeline gas to China will be significantly squeezed, as China will give priority to 40–60 bcm/y of gas supply from the Central Asian Republics, and more imports of LNG. China's LNG imports will reach well over 60 bcm/y by 2030, but supply from Russia is very unlikely to be more that 10 bcm/y. The volume of gas supply from Russia to China by 2030 is likely to be around 40 bcm/y (30 bcm pipeline gas and 10 bcm LNG).

Under the optimistic scenario, the scale of oil trading between the two countries could reach 30–35 mt/y, of which 15 mt/y would be transported to Daqing and 9 mt/y to the Tianjin refinery. The remaining 6–11 mt/y of oil could be supplied from Kozmino to China's coastal areas by ship, as supply from East Siberia increases. In the case of natural gas supply to China, if a breakthrough on the export price is made in 2011[3] and China agrees to import Russian gas from two pipelines, the total volume could be as much as 68 bcm/y, of which 30 bcm/y of Altai gas would be delivered by 2016–18, reflecting Russia's prioritization of the Altai pipeline, and the remaining 38 bcm/y of East Siberian gas could be delivered by 2018–20 due to slower pipeline development from East Siberia.

Under the pessimistic scenario, the scale of oil trading between the two countries would stay at 15 mt/y with the plan to supply 9 mt/y to the Tianjin refinery being shelved due to the repeated failure of the price negotiations. There would be no extra crude supply from Kozmino to China's coastal areas by ship, due to the failure of the ESPO oil pipeline's capacity to reach 80 mt. In the case of natural gas supply, the continued failure to reach a price compromise would minimize the quantity of gas exports to China. The volume of pipeline gas imports could be considerably reduced, perhaps from the targeted 30 bcm/y to zero by 2020 and 20 bcm/y by 2030. In this case China would import well over 50 bcm/y of LNG and accelerate the development of domestic coalbed methane and shale gas.

National, Regional, and Global Consequences

National Consequences

Sino-Russian oil and gas cooperation has the potential to be a hugely significant initiative. However, the achievements of the 2000s did not fulfil the hopes of the 1990s and now, although the prospects of Sino-Russian oil cooperation during the 2010s and 2020s still look promising, gas cooperation still hangs on the results of the gas price negotiations.

What will be the national, regional, and global consequences of the success or failure of Sino-Russian oil and gas cooperation in the coming two decades? As far as Sino-Russian oil cooperation is concerned, no significant change in the current approach is likely during the 2010s. As discussed earlier, however, if the Arctic route trading option becomes a reality during the 2020s, oil trading between Russia and China through the Arctic route cannot be ruled out. Nearly half of China's GDP is thought to be dependent on shipping. Since the journey from Shanghai to Hamburg via the Northern Sea Route (along the north coast of Russia from the Bering Strait in the east to Novaya Zemlya in the west) is 6,400 km shorter than the route via the Malacca Strait and the Suez Canal, oil transportation via the Arctic route, if its commercial viability is proved, would be a very attractive option for China. Such a shift would have huge implication for the sea lanes through the Malacca Strait.

China's efforts to increase domestic gas production, including CBM and shale gas, will intensify, regardless of the fate of Sino-Russian gas cooperation. To the extent that cooperation fails to make progress in the early 2010s, China's dependence on gas supply from the Central Asian Republics and on LNG supply will intensify. Such a delay would have a negative impact on Beijing's plan to use East Siberian gas to revitalize the economic development of its north-eastern provinces in the 2010s. It is worth noting that the option of LNG supply from Novatek's Yamal field via the Arctic Sea to China during the 2020s will definitely be explored, in parallel with crude oil trading via the Arctic route. If the result is positive, the option of LNG supply from the super-giant Shtokman field in the Barents Sea via the Arctic route to China could be also explored.

Regional Consequences

What will be the regional consequences of Sino-Russian oil and gas cooperation? Jorge Montepeque, global director of markets reporting, *Platts*, has argued that:

> … [n]ot only is the ESPO crude stream steadily becoming a more important regional stream, it is gaining attention from many regions, including the Middle East … ESPO has become a key indicator of price in north Asia, in large part because the crude has quality and volume attributes that could, over time, lead it to become a major flat price indicator of spot oil volumes in Asia.[4]

Up until now, the east Asian region has been dependent on oil imports from the Middle East, with pricing based on Dubai crude, which in turn is closely linked to the Brent price. The oil being shipped to the region via Russia's ESPO will fundamentally change that.[5] This was not conceivable a decade ago.

The biggest beneficiaries of Gazprom's hard-line stance on gas prices are the Central Asian Republics. The importance of the Trans-Asia Gas Pipeline development lies in the fact that the development of 40 bcm/y of new pipeline supply has ended Russia's near monopoly position as a purchaser of gas from the central Asian republics. Unlike central Asian oil, in relation to which Russia is merely a transit country, gas from the central Asian region is not simply transported across the territory of Russia, but is intended for re-sale by Russian companies to Ukraine and Europe at considerably higher prices than those at which it is purchased. The profitability of this operation comes from more than the simple transport of oil. Jonathan Stern has pointed out that this was the case prior to 2009, but in that year Russia massively cut its imports of Turkmen gas, principally because of reduced market demand due to the recession. Also in that year, purchase prices increased to something like 'European netback levels'. As a result, it is highly unlikely that Gazprom will want to import any central Asian gas which is not part of PSA/JV arrangements.

But in 2006 when it signed the contract, Beijing was well aware of the danger of provoking Russia's anger by seeking the diversion of gas which was already being produced, from Turkmenistan to China, Nonetheless, China decided to develop a new gas field and build the pipeline itself. This so-called 'equity gas initiative' allowed China to be totally free from any pressure from Russia and, as mentioned above, to be compensated for the financial burden of the relatively high price for gas imports.

Table 8.3: The Characteristics of China's Central Asian Model vs the Russian Model

	Central Asian Model	*Russian Model*
Oil Sector	- Oil asset buy-out or - Oil company buyout mainly in Kazakhstan	- Loan for Oil (2005 & 2009) - Oil company buyout (2006, Udmurtneft) - Upstream Oil JV (Vostok Energy) allowed
Natural Gas Sector	- Equity gas in Turkmenistan and Uzbekistan - Pipeline construction - Value chain business development possible	- No equity gas in upstream and midstream allowed. - But loan for gas between Gazprom and CNPC is being reported (as of July 2011)[6]

Source: devised by author.

Equity participation became a cornerstone of the Chinese NOCs' Central Asian upstream and mid-stream natural gas import business model. As discussed earlier, the Russian model is dictated by Gazprom's insistence that Chinese investors should not be allowed to use the equity gas option in Russian Federation territory, and that no exception can be made to the rule that prices must be oil-related as in Europe. The difference between the Central Asian Model and the Russian Model was the availability of the equity gas option – in other words the opportunity of participation in the value chain in the up-stream and mid-stream sectors.

Vladimir Putin has professed his awareness of Chinese gas needs and insists that not only does Russia maintain close contact on energy with China. 'We also offer', he has said, 'to expand cooperation with them. We do not think that the prospective gas pipeline from Turkmenistan to China will damage our plans'.[7] However, the large scale (40 bcm/y) supply of Turkmen gas to China was, in effect, confirmation that Gazprom's ambitious plan to monopolize pipeline gas supply to China had failed, and that the Beijing planners were determined to diversify China's sources of gas imports, in particular by importing from the Central Asian Republics. In 2010, Russia learned that the Chinese government was proposing to increase the Central Asian import capacity from 30–40 bcm/y to 60 bcm/y.[8] The Moscow authorities and Gazprom were always fully aware of the danger that delay in Altai gas exports to China would lead to the allocation of WEP-III to Turkmen or Uzbek gas; but they suffered from a tendency to underestimate the scale of Turkmenistan's proven gas reserves. The news that UK auditor Gaffney, Cline & Associates (GCA) had confirmed Turkmenistan's claims that its South Yolotan field is the world's second-largest gas field, with reserves of up to 21.2 tcm, must have come as a shock to Moscow.[9]

As discussed in Chapter 6, it was announced during March and April 2011 that gas supplies from Turkmenistan and Uzbekistan to China were to be increased. On top of this, China's NEA confirmed in April that year that Beijing would pursue the construction of WEP-III, IV and V during the 2011–15 period. The message to Gazprom and Russia in this was not necessarily a positive one, since there was no guarantee from Beijing that the WEP-III will be allocated for Altai gas exports. The failure to reach a price deal during the St Petersburg June summit between Presidents Medvedev and Hu Jintao drove Russia to 'play the Korean card', that is, to set up a benchmark price for Russia's gas exports to the Asian market. In late August, President Medvedev's summit meeting with DPRK leader Kim Jong-il at Ulan-Ude announced progress towards a long-stalled deal to build a pipeline from Siberia across North Korea to South Korea.[10] If, however, this was a Russian message to China that it should end the delay in signing a gas deal with Russia, it quickly backfired when China announced that it would by 2015 build 7,000 km of new pipelines from central Asia, which would double China's gas imports from the Central Asian Republics,[11] from 30 to 60 bcm/y. In other words, WEP-III will indeed be allocated to Central Asian gas supply rather than to Altai gas, which must face further delay. Given that the volume of central Asian gas exports to China could rise to as much as 95 bcm (60 from Turkmenistan, 25 from Uzbekistan, and 10 from Kazakhstan) the total volume China could eventually take from West Siberia and the Central Asian Republics could reach 125 bcm/y. As WEP-II and III are already committed to transmit 60 bcm/y from the Central Asian Republics, China needs two more trunk pipelines to import all the remaining 65 bcm/y of gas. This may be the reason why, in the spring of 2011, Beijing for the first time included WEP-V in the list of projects to be undertaken during the coming five years, even though it is not clear whether China can afford to import as much as 125 bcm of pipeline gas by 2020, since the projected demand before 2020 is unlikely to reach the 100 bcm/y level. It is more realistic to conceive of WEP-V as supplying needs after 2020 rather than before. The decision to accelerate the construction of WEP-IV to increase pipeline gas imports from 60 bcm to 90 bcm will be affected by the availability of LNG for China's coastal provinces. If the price of LNG import is too high, the desirability of another 30 bcm/y increase in pipeline gas before 2020 cannot be ruled out. Even if the green light for WEP-IV is given, however, the implicit competition between Altai gas and central Asian gas will be inevitable. The worst scenario for Altai gas is that China decides to allocate WEP-IV to the remaining 35 bcm/y from the Central Asian Republics, rejecting Altai gas altogether.

Moscow was perfectly aware of the urgency of reaching a price agreement. In April 2011 President Medvedev said that the gas deal was 'the single most important economic issue between the two countries, and should be resolved by the middle of 2011',[12] but the St. Petersburg summit still failed to strike the deal. If the failure continues beyond the end of 2011 this will be a huge set-back for Sino-Russian gas cooperation.

The Central Asian Republics have become the biggest beneficiaries of the many ups and downs of Sino-Russian oil and gas cooperation. The north-east Asian region failed to derive any benefit from Sino-Russian gas cooperation during the 2000s, even though the development of ESPO's first stage was a real gain both for Sino-Russian oil cooperation and for countries in the region – such as Japan and South Korea. Will Sino-Russian oil and gas cooperation contribute to multilateral cooperation in the north-east Asian region in the coming two decades? Despite the changed political environment, the north-east Asian region has failed to witness the long-awaited improvement in multilateral cooperation in the region's energy infrastructure development, in particular the Russian Far East region's long-distance natural gas pipeline development. Both Russia and China have had a preference for their conventional bilateral stance on cooperation, rather than for a serious exploration of untested multilateral cooperation options. And Russia's resource nationalism has kept the door closed to multilateral cooperation during the last two decades. The possibility of trilateral or multilateral cooperation on a Chayanda–Khabarovsk–Vladivostok gas pipeline has not yet been explored, even though trilateral cooperation over Kovykta gas exports was at least explored during the 2000–3 period, before it failed.

Gazprom's stance towards regional cooperation is a very important factor. In June 2011, Gazprom's deputy CEO Ananenkov said that the company saw no pressing need to attract foreign partners to operate and develop Sakhalin-3, Kovykta, and Chayanda; but the company was interested in foreign capital for petrochemical projects.[13] A month after the March 2011 earthquake in Japan, Gazprom and the Japan Far East Gas Co. (which includes Itochu, Japex, Marubeni, Inpex, and Cieco) agreed to study the possibilities of cooperating to build an LNG facility and related petrochemical plant. Gazprom later announced firm plans to construct an LNG plant with a capacity of 10 mt/y in Vladivostok by 2017.[14] However, this instance of bilateral cooperation is unlikely to be expanded into a regional multilateral scheme for the development of a joint petrochemical plant.

Another potential area of multilateral cooperation in the north-east Asian region lies in the strategic petroleum reserves (SPR) storage development in the Greater Tumen River Delta area. In September 2009, the first meeting of the Energy Board of the UNDP GTI was held in Mongolia with the aim of studying options for cooperation among the member countries (Russia, China, Mongolia, and North and South Korea). As discussed in Chapter 2, China aims to expand its SPR capacity rapidly. So far its efforts have been confined to SPR development within Chinese territory. China, however, cannot construct the storage unilaterally, since there are a number of constraints to be taken into account. If China's oil demand rises more rapidly due to an explosion of private car ownership, it will be impossible to increase SPR storage at the same time as satisfying a huge increase in the demand for oil; this could be a real possibility during the 2030s. As Beijing is paying special attention to the development of the 'Chang Ji Tu'[15] region of China (Changchun City, part of Jilin City, and the Tumen River), there is a possibility of strategic storage development in the Greater Tumen Initiative area through joint action by the GTI member countries. Even though a storage project based on multilateral cooperation could easily be rejected, as each country pursued its national interest, Russia and China would be the biggest beneficiaries of a sizable SPR facility in the UNDP GTI area. The UNDP GTI regime is an internationally recognized institution and both Russia and China are full members, even though Japan is not yet a member. Unlike the case of a natural gas pipeline project, Japan could also be a beneficiary of this SPR project. If it is promoted as a flagship GTI energy project, it could form the basis of an unprecedented multilateral cooperation project in the north-east Asian region. But, without China, no trilateral or multilateral oil or gas cooperation in north-east Asia is either conceivable or realistic.

Past Problems, Future Questions, Global Consequences

Among the questions which are provoked by the relatively unsuccessful Sino-Russian oil and gas cooperation of the 2000s are: i) was the problem that 'neither side understood what the other side wanted'? or ii) was it that 'each side does understand what the other side wants but will not compromise'? and, iii), if it is the latter, then is the problem fated to continue into the 2010s? It seems fairly clear in this case that both sides do understand what the other side wants, but the problem was the unreadiness to compromise, particularly in the case of gas.

Another set of major questions concerns the future: i) will the Chinese be sufficiently desperate for Russian gas supplies that they will make concessions? ii) will the Russians be so desperate to create a major Asian pipeline market that they will make concessions to get it? or, iii), will both sides in the end decide to rely on Russian Far East LNG supplies?

However these questions are answered, it is clear that there is a major difference between oil and gas in the way they affect China, Russia, Sino-Russian relations, and the rest of the world. First, Russian oil supplies from East Siberia and the Russian Far East are significant, but will not fundamentally change either Chinese dependence on Middle East oil supplies or global oil supply trends, unless new oil discoveries are so abundant that they justify a second ESPO pipeline. Unless the relationship between the two countries improves, much Russian oil from East Siberia and its Far East is very likely to go to the major Asian oil importers such as Japan and South Korea; but again this will not fundamentally change the global oil situation. Second, by contrast, Russian gas reserves in East Siberia and the Far East are so huge and stranded (that is, without nearby markets) that they could transform the gas industry in China. Russia could export 150–200 bcm/y to China by pipeline from fields which otherwise will continue to be stranded for many decades. Large-scale Chinese pipeline imports could expand with little delay, because fields and pipeline routes have been extensively studied and China has the investment capital to finance such projects. This will not happen either on the same scale or at the same speed if both countries decide to rely on LNG.

The current outlook is that much of the oil potential will be fulfilled, but this will not make a huge difference to China or to the global oil market, and that the gas potential will largely not be fulfilled and so the Chinese gas market will be much smaller than it otherwise could have been and large Russian gas reserves will remain stranded for many decades. But if the current outlook changes and the potential is more completely realized, it could make a huge difference to the global gas market. Failure to achieve large-scale gas pipeline imports from Russia would force China to significantly expand LNG imports. This would increase the competition for LNG supplies not just between LNG importers in north-east Asia (Japan, South Korea, and Taiwan) but also for other buyers of LNG in regions as far away as Europe. A failure of the Sino-Russian gas relationship will therefore deprive both countries of a potential win–win solution to their energy and development problems, and increase future global rivalry in the market for LNG.

BIBLIOGRAPHY AND OTHER SOURCES

Bibliography

This is a bibliography of works cited and related published materials. It consists of books, chapters in books, academic journals, special reports, monographs, conference and workshop presentations, professional and governmental publications, and authored newspaper and newswire articles.

They are arranged in alphabetical order primarily by author's name and secondarily by date. Books, chapters in books, and articles in academic journals are listed by name of author, year of publication, followed by full bibliographical details and sometimes relevant web addresses; articles in other periodicals are listed by name of author followed by bibliographical details which include the month or more detailed date of publication. Where an author has more than one article in the bibliography they are shown in order of date of publication.

Aalto, P. (2012). Aalto, P., ed., Russia's Energy Policy: National, Interregional and Global Dimensions (Cheltenham: Edward Elgar, 2012 forthcoming).

Aden, N., Fridley, D., and Zheng, N. (2009). 'China's Coal: Demand, Constraints and Externalities', Ernest Orlando Lawrence Berkeley National Laboratory, LBNL-2334E. (http://china.lbl.gov/sites/china.lbl.gov/files/LBNL-2334E.pdf)

Ahn, S-H. and Jones, M. T. (2008). 'Northeast Asia's Kovykta Conundrum: A Decade of Promise and Peril', National Bureau of Asian Research (NBR)'s *Asia Policy*, no. 5, 105–40.

Aleksashenko, S. (2010). 'Russia's Budget Dilemma', *Carnegie Endowment for International Peace, International Economic Bulletin*, 19 May. (http://carnegieendowment.org/publications/index.cfm?fa=view&id=40817)

Anderlini, J. (2009).'Chinese oil major in $1bn offshore deal', *Financial Times*, 25 May 2009.

Andrews-Speed, P. and Dannreuther, R. (2011). *China, Oil and Global Politics*, London: Routledge.

Aron, L. (2003). 'The Yukos Affair', *American Enterprise Institute for Public Policy Research* (website). (www.aei.org/outlook/19368)

Aslund, A. (2007). *Russia's Capitalist Revolution: Why Market Reform Succeeded and Democracy failed*, Washington, D.C.: Peterson Institute for International Economics.

Bai, J, and Chen, A. (2009)'PetroChina adds refining facilities for Russian oil', *Reuters*, 16 July 2009. (www.reuters.com/article/idUSPEK21059320090716)

Bai, J. and Miles, T. (2011).'China demand for fuels to plateau in rest of 2011 – NEA', *Reuters*, 22 April 2011. (http://uk.reuters.com/article/2011/04/22/china-energy-idUKL3E7FM05Z20110422)

Baidashin, V. (2002). 'Dueling Pipelines', *RPI*, May 2002, 17–24.

Baidashin, V. (2004a). 'Budgeted spending on Sakhalin-1 Project for 2004 endorsed at $1.37 billion: ExxonMobil Milestone', *RPI*, May 2004, 27–34.

Baidashin, V. (2004b). 'Yukos to Quadruple Rail-Based Oil Exports to China to 15 Million Tons by 2006: Railway Bridge to China', *RPI*, June/July 2004, 12–21.

Baidashin, V. (2005a). 'Railway Road to China', *RPI*, February 2005, 35–41.

Baidashin, V. (2005b). 'Sakhalin 2–Made in Russia', *RPI*, March 2005, 31–37.

Baidashin, V. (2005c). 'New Siberian Mega-Basin', *RPI*, August 2005, 23–28.

Baidashin, V. (2005d). 'Sakhalin-3 Attracts Majors', *RPI*, August 2005, 50–54.

Baidashin, V. (2006a). 'Eastern Pipeline – the Rubicon Is Crossed', *RPI*, February 2006, 21–7.

Baidashin, V. (2006b). 'Environmental Approval: Positive', *RPI*, May 2006, 24–9.

Baidashin, V. (2006c). 'Planned Pipeline Sparks Interest in East Siberian Fields', *RPI*, June/July 2006, 44–9.

Baidashin, V. (2006d). 'Surgutneftegaz Plans for East Siberia', *RPI*, September 2006, 31–7.

Baidashin, V. (2006e). 'Sakhalin – Far East Offshore Centre?', *RPI*, October 2006, 42–9.

Baidashin, V. (2007a). 'Rosneft Moves Slowly on Chinese Agreements', *RPI*, February 2007, 10–16.

Baidashin, V. (2007b). 'Rosneft Declares ONGC a Strategic Partner', *RPI*, March 2007, 5–10.

Baidashin, V. (2007c). 'Will There be Sufficient Crude Oil for the ESPO pipeline?', *RPI*, June/July 2007, 12–17.

Baidashin, V. (2007d). 'Gazprom Neft Offers Statoil Sakhalin Position', *RPI*, September 2007, 5–10

Baidashin, V. (2007e). 'Eastern Gas Programme Approved', *RPI*, November/December 2007, 10–15.

Baidashin, V. (2008a). 'ESPO Faces Delay', *RPI*, February 2008, 28–32.

Baidashin, V. (2008b). 'Russia Commits to Strong Exploration Programme', *RPI*, May 2008, 11–15.

Baizhen, Chua. (2011). 'China, Kazakhstan Sign Accord to Expand Gas Pipeline Network', Bloomberg Newswire, 8 September 2011. (www.bloomberg.com/news/2011-09-08/china-kazakhstan-sign-accord-to-expand-gas-pipeline-network-1-.html)

Bal, M. (2010). 'Turkmen tactics', Industrial Fuels and Power, 7 April 2010. (www.ifandp.com/article/003226.html)

Balzer, H. (2005). 'The Putin Thesis and Russian Energy Policy,' *Post-Soviet Affairs*, July–September, 210–25.

Barges, I. (2004a). 'Chinese State Corporation CNPC Makes a Successful Acquisition in Russia: Chinese Ventures', *RPI*, April 2004, 38–45.

Barges, I. (2004b).'Politics Key Hurdle in Developing Energy Relations Between

Russia and China: Great Russian–Chinese Wall', *RPI*, October 2004, 38–44.

Barnett, A. D. (1993). *China's Far West: Four Decades of Change*, Boulder: Westview Press.

Bellacqua, J. (2010). Bellacqua, J, ed., *The Future of China Russia Relations*, Kentucky: The University Press of Kentucky, 2010.

Belton, C, and Wagstyl, S. (2009). 'Battle erupts over budget crisis', *Financial Times*, 6 February 2009.

Belton, C. (2007). 'Kremlin power grows in TNK–BP Siberian gas deal', *Financial Times*, 23 & 24 June 2007.

Belton, C. (2008). 'The Putin Defence', *Financial Times*, 29 December 2008.

Belton, C. (2010). 'TNK–BP gas field saga shows difficulties of doing business in Russia', *Financial Times*, 7 June 2010. (http://blogs.ft.com/beyond-brics/2010/06/07/tnk-bp-gas-field-saga-shows-difficulties-of-doing-business-in-russia/)

Belton, C. and Dyer, G. (2009). 'Russian gas supply deal stalls', *Financial Times*, 14 October 2009.

Bergsten, C. F., Freeman, C., Lardy, N. R., and Mitchell, D. J. (2009). *China's Rise: Challenges and Opportunities*, Washington, D.C.: Peterson Institute for International Economics and Centre for Strategic and International Studies.

Berrah, N., Feng, F., Priddle, R., and Wang, L. (2007). *Sustainable Energy in China: The Closing Window of Opportunity*, Washington DC: The World Bank.

Blagov, S. (2004). 'Russia stirs up Sakhalin projects', *Asia Times*, 4 February 2004 (www.atimes.com/atimes/Central_Asia/FB04Ag01.html).

Blagov, S. (2005). 'China knocking on Russia's door', *Asia Times*, 6 July 2005.

Blagov, S. (2006). 'Russian Oil to flow to China even before pipeline completed', *Eurasia Daily Monitor*, vol. 3, no. 84, 1 May 2006. (www.jamestown.org/single/?no_cache=1&tx_ttnews%5Btt_news%5D=31637)

Blair, B., Chen Y., and Hagt, E. (2006). 'The Oil Weapon: Myth of China's Vulnerability', *China Security*, summer issue, 32–63.

BP Statistical Review of World Energy (various years).

Bradshaw, M. (2010). 'A New Energy Age in Pacific Russia: Lessons from the Sakhalin Oil and Gas Projects', *Eurasian Geography and Economics*, vol. 51, no. 3, 330–59.

Bradsher, K. (2009). 'China Outpaces US in cleaner Coal-Fired Plants', *NYT*, 10 May 2009. (www.nytimes.com/2009/05/11/world/asia/11coal.html)

Bradsher, K. (2010). 'China's Energy Use Threatens Goals on Warming', *NYT*, 6 May 2010.

Bryanski, G. (2009a). 'Russia launches Far East pipeline, eyes Exxon gas, *Reuters*, 31 July 2009. www.reuters.com/article/idUSLV7139820090731(www.reuters.com/article/idUSLV7139820090731)

Bryanski, G. (2009b). 'Russia's Putin launches new Pacific oil terminal', *Reuters*, 28 Dec 2009. (http://uk.reuters.com/article/idUKLDE5BR00F20091228)

Calder, K. E. (2004). 'The Geopolitics of Energy in Northeast Asia', presented at Korea Energy Economics Institute, Seoul, 16–17 March.

Campaner, N. and Yenikeyeff, S. (2008). 'The Kashagan Field: A Test Case for Kazakhstan's Governance of Its Oil and Gas Sector', IFRI, October

2008. (www.ifri.org/?page=detail-contribution&id=182&id_provenance=97)

Chan, Yvonne, (2009). 'China's first clean coal plant underway', Businessgreen, 29 June 2009.

Chazan, G. (2009). 'Sinopec is in talks to buy Addax: Surging Oil Price is fueling pursuit of U.K.-Listed exploration firm with fields in Northern Iraq', *Wall Street Journal*, 15 June 2009. (http://online.wsj.com/article/SB124499901463513231.html)

Chen Aizhu, and Graham-Harrison, E. (2008). 'UPDATE 2-China reshuffles energy sector, little change seen', *Reuters*, 11 March. (http://uk.reuters.com/article/2008/03/11/china-energy-commission-idUKPEK25296020080311)

Chen Aizhu. (2011). 'China set to unearth Shale Power', *Reuters Special Report*, 20April. (http://graphics.thomsonreuters.com/AS/pdf/chinashaledk.pdf).

Chen Dongyi. (2006). 'CNOOC broadens LNG viewpoint', *China OGP*, 15 December 2006, 19–20.

Chen Dongyi. (2007a). 'Sino-Turkmenistan gas pipeline expects further progress', *China OGP*, 15 July 2007, 17–19.

Chen Dongyi. (2007b). 'Sino-Turkmenistan gas cooperation strikes ahead among uncertainties', *China OGP*, 1 August 2007, 26–8.

Chen Dongyi. (2007c). 'CNPC settles first LNG source', *China OGP*, 15 September 2007, 1–3.

Chen Dongyi. (2007d). 'Sino-Kazakhstan energy cooperation is heated', *China OGP*, 1 September 2007, 4–5.

Chen Dongyi. (2008a). 'CNPC injects capital for advancing Central Asia–China gas pipeline', *China OGP*, 15 January 2008, 23–4.

Chen Dongyi. (2008b). 'Sinopec outlines four strategies for developing into a globalised oil company', *China OGP*, 1 March 2008, 3–7.

Chen Dongyi. (2008c). 'CNPC possibly secures overseas gas for Dalian LNG project', *China OGP*, 1 March 2008, 24–6.

Chen Dongyi. (2008d). 'Sinopec's planned Shandong LNG project moves ahead', *China OGP*, 15 June 2008, 9–10

Chen Mingshuang. (2006). 'Future Oil & Gas Resources of China', presented at 7th Sino-US Oil and Gas Industry Forum, 11–12 September 2006, Hangzhou.

Chen Minshuang. (2004). Future Oil & Gas Resources of China, presented at 7th Sino-US Oil and Gas Industry Forum, 11–12 September, Hangzhou.

Chen Wenxian. (2004a). 'Silk Road revisited', *China OGP*, 15 September 2004, 1–2.

Chen Wenxian. (2004b). 'Chinese Premier's energy visit to Russia', *China OGP*, 1 October 2004, 1–2.

Chen Wenxian. (2004c). 'Sino-Kazakhstan pipeline starts construction', *China OGP*, 1 October 2004, 2–3.

Chen Wenxian. (2005a). 'China and Russia's gas ties tightened', *China OGP*, 1 October 2005, 1–3.

Chen Wenxian. (2005b). 'Chinese and Canadian firms to jointly tap CBM in Guizhou', *China OGP*, 1 October 2005, 18–19.

Chen Wenxian. (2005c). 'China stresses energy development in next five years',

China OGP, 1 November 2005, 5–7

Chen Wenxian. (2006a). 'China and Canada cooperate on Xinjiang's CBM E&D', *China OGP*, 1 January 2006, 12.

Chen Wenxian. (2006b). 'Chinese and Canadian firms to jointly develop CBM resources in Anhui', *China OGP*, 15 March 2006, 23–4.

Chen Wenxian. (2006c). 'Sino-Kazakhstan crude pipeline under operation', *China OGP*, 1 June 2006, 24.

Chen Wenxian. (2006d). 'Necessary for Puguang gasfield to serve Shandong market?', *China OGP*, 15 June 2006, 7–9 & 12.

Chen Wenxian. (2006e). 'Gas price awaits a further adjustment', *China OGP*, 15 November 2006, 1–3.

Chen Wenxian. (2007a). 'CNPC's big oil/gas discoveries floating in the air', *China OGP*, 1 June 2007, 5–8.

Chen Wenxian. (2007b). 'Hard to separate gas pipelines from E&D and terminal marketing', *China OGP*, 1 June 2007, 8–9.

Chen Wenxian. (2007c). 'Towngas market expansion for China Gas and CNPC', *China OGP*, 1 August 2007, 8–9.

Chen Wenxian. (2007d). 'A V-shape W–E gas pipeline to carry Central Asian gas to east and south China', *China OGP*, 1 September 2007, 1–4.

Chen Wenxian. (2007e). 'Natural gas to promote as really expensive gas', *China OGP*, 15 September 2007, 5.

Chen Wenxian. (2008). 'A new era for China's CBM industry', *China OGP*, 15 June 2008, 1–5.

Chen, E, and Miles, T. (2010). 'China still stuck with Russia over gas price', 10 February 2010.

Cheng Guangjin. (2010). 'Gas still under discussion', *China Daily*, 10 February 2010. (www.chinadaily.com.cn/world/2010-02/10/content_9454139.htm)

Chernyshov, S. (2002). 'Key to Yakutia', *RPI*, September 2002, 29–36.

Chernyshov, S. (2003a). 'Battle Lines', *RPI*, April 2003, 31–36.

Chernyshov, S. (2003b). 'Gazprom Prevails Again', *RPI*, May 2003, 13–19.

Chernyshov, S. (2003c). 'Rancor over Vankor', *RPI*, August 2003, 31–7.

Chernyshov, S. (2004a). 'Despite Completion of Feasibility Study, Kovykta Project's Prospects Remain Uncertain: Cloudy Outlook', *RPI*, January 2004, 33–7.

Chernyshov, S. (2004b). 'Predictions of Rosneft and BP for Sakhalin Shelf Potential Fail to bear Out: First Setbacks', *RPI*, February 2004, 18–22.

Chernyshov, S. (2004c). 'Sakhalin-3 Ruling Illustrates Government's Cooling Attitude toward Foreign Investors', *RPI*, April 2004, 6–12.

Chernyshov, S. (2004d). 'Proposed Eastern Oil Transportation Project Changes Name and Route: Long-Suffering Projects', *RPI*, April 2004, 15–22.

Chernyshov, S. (2004e). 'Russian Ministry of Natural Resources Decides to Sell Yakutia's Subsoil Blocks: Long-Awaited Seed-up', *RPI*, August 2004, 5–12.

Chernyshov, S. (2004f). 'TNK–BP Makes New Concessions to Gazprom in the Kovykta Project: Collapse of TNK–BP Strategy', *RPI*, September 2004, 44–9.

Chernyshov, S. (2005a). 'Gas to the East', *RPI*, February 2005, 19–22.

Chernyshov, S. (2005b). 'Eastern Pipelines Progress', *RPI*, March 2005, 44–8.

Chernyshov, S. (2005c). 'TNK–BP Shifts Strategy for Kovykta', *RPI*, May 2005, 31–4

Chernyshov, S. (2005d). 'Surgutneftegaz: the Recluse Awakens', *RPI*, August 2005, 29–34.

Chernyshov, S. (2007). 'East Siberian Projects Face State Pressure and Lack of Oilfield Services', *RPI*, September 2007, 11–17.

Chernyshov, S. (2008). 'Projections Improve for Verkhnechonskoye', *RPI*, February 2008, 23–27.

China Energy Statistical Yearbook, Beijing: China Statistics Press, 1998 and 2001.

China OGP, *Xinhua News Agency*, China Natural Gas Report : A 2002 Update (Beijing : *Xinhua News Agency*, 2002).

China Securities Journal. (2010). *China Natural Gas Report: 2010*, Beijing: *Xinhua News Agency*.

China Statistical Yearbook, Beijing: China Statistics Press, Annual.

Chow, E. et al. (2010). Pipeline Politics in Asia: The Interaction of Demand, Energy Market, and Supply Routes, *NBR Special Report*, no. 23.

Christoffersen, G. (1998). 'China's Intentions for Russian and Central Asian Oil and Gas', *NBR Analysis*, vol. 9, no. 2.

Christoffersen, G. (2008). 'East Asian Energy Cooperation: China's Expanding Role,' *China and Eurasia Forum Quarterly*, volume 6, no. 3, 141–68

Chun Chun-Ni. (2007). 'China's Natural Gas Industry and Gas to Power Generation', Institute of Energy Economics, Japan, July 2007. (http://eneken.ieej.or.jp/en/data/pdf/397.pdf).

Chung Chien-Peng (2004). 'The Shanghai Co-operation Organisation: China's Changing Influence in Central Asia', *The China Quarterly*, December.

Chung, J. and Tucker, S. (2006). 'China National Petroleum eyes £3 bn take in Rosneft', *Financial Times*, 5 July 2006.

Clover, C and Buckley, N. (2011). 'Shades of difference', *Financial Times*, 13 May 2011.

Clover, C. (2008). 'Onward to 1998', *Financial Times*, 27 October 2008.

Clover, C. (2009). 'Russia to back floored rouble', *Financial Times*, 3 February 2009.

Clover, C. and Belton, C. (2008). 'Retreat from Moscow: Investors take flight as global fears stoke a Russian crisis', *Financial Times*, 18 December 2008.

CNPC RIE & T. (2008). 'China's Natural Gas Market: Today and Tomorrow', CNPC Research Institute of Economics & Technology, presented at a Tokyo conference, 5 December 2008.

Cognato, M. H. (2008). 'China Investment Corporation: Threat or Opportunity?', in 'Understanding China's New Sovereign Wealth Fund', *NBR Analysis*, vol. 19, no. 1, 9–36.

Cohen, B. J. (2009). 'Sovereign Wealth Funds & National Security', *International Affairs*, July Issue, 713–31.

Considine, J. I., and Kerr, W. A. (2002). *The Russian Oil Economy*, Cheltenham: Edward Elgar.

CPCC (2003). *China Petroleum and Petrochemical Industry Economics Research Annual Report 2003*, China Petroleum Consulting Company.

Credit Suisse Equity Research Report (Asia Pacific/China, Gas Utilities). 2006. 'Asia Coalbed Methane Sector: Massive gas supply potential', 18 October.

Crooks, E. (2006a). 'BP and Rosneft sign US$ 700m Sakhalin deal', *Financial Times*, 23 November 2006.

Crooks, E. (2006b). 'The 'elephant project' still on thin ice', *Financial Times*, 12 December 2006.

Crooks, E. (2007). 'End of dispute means BP can continue to do business in Russia', *Financial Times*, 23 & 24 June 2007.

Crooks, E. (2008). 'China's move signals offshore ambitions', *Financial Times*, 8 July 2008.

Crooks, E. (2009a). 'Gazprom battles to restore its reputation', *Financial Times*, 8 January 2009.

Crooks, E. (2009b). 'Shell savors bittersweet first LNG delivery from Sakhalin-2', *Financial Times*, 1 April 2009.

Crooks, E. (2009c). 'CNOOC signs 20-year BG deal', *Financial Times*, 14 May 2009.

Crooks, E. (2009d) 'China's oil ambitions take it to new frontiers', *Financial Times*, 3 July 2009.

Crooks, E. and Kwong, R. (2007). 'PetroChina has the right to look Exxon in the eye', *Financial Times*, 7 November 2007.

CSCAP (2010). 'Going East: Russia's Asia-Pacific Strategy', the Russian National Committee of the Council for Security Cooperation in Asia Pacific (CSCAP), *Russian in Global Affairs*, Oct/Dec, no. 4, 2010.

Daly, T. (2011). 'China tests Russian Resolve with gas prepayment offer', Nefte Compass, 14 July 2011. (http://www.energyintel.com/pages/Eig_Article.aspx?DocId-727329-)

Demongeot, Maryelle (2006). 'Russia's Sakhalin crude oil to reroute Asian trades', *Reuters News*, 3 July 2006.

Demongeot, M. (2009). 'The Asian oil premium? Almost gone, no coming back', *Reuters*, 23 April 2009. (www.reuters.com/article/2009/04/23/us-asia-oil-premium-analysis-idUSTRE53M1Y020090423).

Demytrie, R. (2010). 'Struggle for Central Asian energy riches', *BBC Online*, 3 June 2010. (http://news.bbc.co.uk/1/hi/world/asia_pacific/10175847.stm).

Denisova, Irina (2007). 'Turkmen Sensation', *RPI*, January 2007, 44–8.

Department of Communications and Energy (1997). State Planning Commission of P.R.China, *1997 Energy Report of China*, Beijing: China Prices Publishing House.

Dittmer, L. (1990). *Sino-Soviet Normalization and Its International implications, 1945–1990*, Seattle: University of Washington Press.

Dow Jones Deutschland (2010). 'TNK–BP: Kovykta Field Operator Files For Bankruptcy', 3 June 2010. (www.dowjones.de/site/2010/06/tnkbp-kovykta-field-operator-files-for-bankruptcy.html).

Downs, E. S. (2007). 'China's Energy Bureaucracy: The Challenge of Getting the Institutions Right,' in Meidan (2007, 64–89).

Downs, E.S. (2004). 'The Chinese Energy Security Debate', *The China Quarterly*, no. 177, 21–41.

DRC (2004). *Research on National Energy Strategy and Policy in China* (Zhonguo Nengyuan Fazhan Zhanliu Yu Zhengce Yanjiu), Development Research Centre (DRC), China State Council, 2004, Economic Science Press, Beijing.

Duan Zhaofang (2010). 'China's Natural Gas Market Outlook', presented at the 4th CNPC/IEEJ Press Conference of Oil Market Research (10 December).

Duce, J. and Wang, Y. (2010). 'PetroChina Plans $60 Billion of Overseas Expansion', *Bloomberg News*, 29 March 2010. (www.bloomberg.com/apps/news?pid=20601087&sid=autWB1u6AAR8&pos=5)

Dyer, G and Hoyos, C. (2010). 'BP and Sinopec join forces in shale gas talks', *Financial Times*, 18 January 2010.

Dyer, G, and Mitchell, T. (2008). 'COSL in $2.5 bn deal for Awilco', *Financial Times*, 8 July 2008.

Ebel, R. E. (2005). *China's Energy Future: The Middle Kingdom Seeks Its Place in the Sun*, Washington, D.C: The CSIS Press.

Ebina, M. (2006). 'Pipeline Project a Necessity for Japan and Russia', *RPI*, October 2006, 50–3.

Ebinger, C. K. and Zambekakis, E. (2009). 'The Geopolitics of Arctic Melt', *International Affairs*, November, 1215–32.Economy, E. C. (2011). 'China's Energy Future: An Introductory Comment', *Eurasian Geography and Economics*, vol. 52, no. 4, 461–3.

Economy, Elizabeth C. (2004). *The River Runs Black. The Environmental Challenge to China's Future*, Ithaca, N.Y.: Cornell University Press.

ECS (2008). 'Fostering LNG Trade: Role of the Energy Charter', Energy Charter Secretariat, 2008.

ECS (2009). Fostering LNG Trade: Development in LNG Trade and Pricing, Energy Charter Secretariat.

Egyed, P. (1983). 'Western Participation in the Development of Siberian Energy Resources: Case Studies', *Carleton University East–West Commercial Relations Series Research Report*, no. 22.

Ellman, M. (2006). *Russia's Oil and Natural Gas: Bonanza or Curse?*, London: Anthem Press.

Erikson, A. S., and Collins, G. B. (2010). 'China's Oil Security Pipe Dream: The Reality, and Strategic Consequences of Seaborne Imports', *Naval War College Review*, vol. 63, no. 2, 1–24.

Erochkine, V. and Erochkine, P. (2006). *Russia's Oil Industry: Current Problems and Future Trends*, London: The Centre for Global Studies.

Fan Wenxin (2000). 'Materialising the WGTE Programme', *China OGP*, 1 April 2000, 3–5.

Faucon, B. and Smith, G. (2007). 'BP joint ventures sells stake in gas field to Gazprom', *Wall Street Journal*, 25 June 2007.

Feng Yujun (no date). 'Russia's Oil Pipeline Saga', (www.bjreview.cn/EN/200430/World-200430(A).htm).

Fishelson, J. (2007). 'From the Silk Road to Chevron: The Geopolitics of Oil Pipelines in Central Asia', The School of Russian and Asian Studies, 12

December 2007). (www.sras.org/geopolitics_of_oil_pipelines_in_central_asia

Fjaetoft, D. B. (2009). 'Russian Gas – Has the 2009 economic crisis changed Russian gas fundamentals?', Institute for Economies in Transition, Bank of Finland, *BOFIT Online*.

French, P., and Chambers, C. (2010). *Oil on Water: Tankers, Pirates and the Rise of China*, London: Zed Books.

Fridley, D. (2002). 'Natural Gas in China' in *Natural Gas in Asia: The Challenges of Growth in China, India, Japan and Korea*, Wybrew-Bond, I. and Stern, J. eds., Oxford: Oxford University Press, 2002. 5–65.

Fridley, D. (2008). 'Natural Gas in China', in Stern, J. ed. 2008, 7–65.

Fu, C. (2007). 'Changbei gas field starts', *China Daily*, 2 March 2007. (www.chinadaily.com.cn/bizchina/2007-03/02/content_818128.htm)

Gaiduk, I. (2005). 'The Ministry of Natural Resources Reorients', *RPI*, January 2005, 23–9.

Gaiduk, I. (2006). 'Asia as Alternative to Europe', *RPI*, May 2006, 6–11.

Gaiduk, I. (2007a). 'Gazprom and TNK–BP Settle Kovykta', *RPI*, August 2007, 33–9.

Gaiduk, I. (2007b). 'Gazprom Moving Ahead in the Russian East', *RPI*, September 2007, 38–43.

Gaiduk, I. (2007c). 'Russian State Companies Continue to Strengthen Position', *RPI*, November/December 2007, 16–20.

Gaiduk, I. (2008a). 'New Transneft Leadership Revising Approach to Projects', *RPI*, April 2008, 40–45.

Gaiduk, I. (2008b). 'Gazprom Set to Control Strategic Gas Reserves', *RPI*, June/July 2008, 11–17

Gaiduk, I. (2008c). 'Repsol wants to join Rosneft on Sakhalin', *RPI*, September 2008, 17–22.

Gaiduk, I. (2009a). 'Russia and China Reach Oil Export', *RPI*, March 2009, 41–6.

Gaiduk, I. (2009b). 'Russia Turns Oil Flow Eastward', *RPI*, May 2009, 33–9.

Gaiduk, I. (2009c). 'Russia Adopts an Energy Strategy through 2030', *RPI*, October 2009, 5–11.

Gaiduk, I., and Kirillova, E. (2006). 'Potential for the Asian Leap', *RPI*, May 2006, 35–41.

Gaiduk, I., and Kirillova, E. (2007). 'New Russian Companies Vie for Major Status', *RPI*, February 2007, 17–19.

Garnett, S. W. (2000). *Rapprochement Or Rivalry?: Russia–China Relations in a Changing Asia*, Garnett, S. W. ed., Washington, D.C.: Carnegie Endowment for International Peace.

Gazprom (2004). 'Discussion Package for the Interagency Working Group to develop a Programme for creating a Unified Gas Production, Transportation and Supply System in East Siberia and the Far East with potential exports to markets in China and other countries in Asia and the Pacific', Gazprom, Strategic Development Department, September 2004. (Unpublished.)

Glazkov, S. (2003). 'Great Sale', *RPI*, February 2003, 14–19.

Glazkov, S. (2004a). 'Russia Targets Fast-Track Licensing of Petroleum Projects for Eastern Pipelines', *RPI*, January 2004, 27–32.

Glazkov, S. (2004b). 'Gazprom's Delay in Joining Kovykta Project Boosts Prospects for Rail-based Exports: Alternative Approach', *RPI*, February 2004, 31–7.

Glazkov, S. (2004c). 'Rosneft's Sakhalin Activities Show How Foreigners Can Gain Access to Russian Reserves: Model for Investors', *RPI*, May 2004, 35–8.

Glazkov, S. (2004d). 'ExxonMobil to Begin Supplying Gas to Russia's Khabarovsk Territory under PSA: Gas for Russia', *RPI*, August 2004, 34–9.

Glazkov, S. (2005). 'Which Direction for Vankor Oil?', *RPI*, October 2005, 43–6.

Glazkov, S. (2006a). 'Rosneft Joins Verkhnechonskoye', *RPI*, April 2006, 43–7.

Glazkov, S. (2006b). 'Eastern Pipeline Will Provide New Options', *RPI*, June/July 2006, 20–25.

Glazkov, S. (2006c). 'Gazprom pushes into Sakhalin Projects', *RPI*, August 2006, 28–34.

Glazkov, Sergei. (2007a). 'Far Eastern Intrigues', *RPI*, January 2007, 16–20.

Glazkov, S. (2007b). 'Gazprom to Liquefy Sakhalin gas?', *RPI*, February 2007, 24–30.

Glazkov, S. (2009). 'Developments in the Main Russian Oil and Gas Regions', RPI, March 2009, 13–18.

Glazkov, S. (2010). 'Gazprom Eastern Gas Program Taking Shape', *RPI*, April 2010, 11–17.

Goldman, Marshall (2008). *Oilopoly: Putin, Power and the Rise of the New Russia*, Oxford: Oneworld.

Gorst, I. (2005). 'China takes a great leap forward into its neighbour's oil business', *Financial Times*, 23 August 2005.

Gorst, I. (2011a). 'China–Uzbekistan: Gas Diplomacy', *Financial Times*, 22 April 2011. (http://blogs.ft.com/beyond-brics/2011/04/22/china-uzbekistan-gas-diplomacy/)

Gorst, I. (2011b). 'Oil: Kazakhs fear China at the gate', *Financial Times*, 11 May 2011. (http://blogs.ft.com/beyond-brics/2011/05/11/oil-kazakhs-fear-china-at-the-gate/#axzz1VnMyNYeJ)

Gorst, I. (2011c). 'Russia, China, two Koreas: gas games', *Financial Times*, 26 August 2011. (http://blogs.ft.com/beyond-brics/2011/08/26/china-calls-medvedevs-gas-bluff/#axzz1WAg8nMS4).

Gorst, I. and Dyer, G. (2009). 'Pipeline brings Asian gas to China', *Financial Times*, 14 December 2009. (www.ft.com/cms/s/0/38fc5d14-e8d1-11de-a756-00144feab49a.html).

Gorst, I., Dombey, D., and Morris, H. (2007). 'Turkmenistan opens gas and oil fields to west', *Financial Times*, 27 September 2007.

Grace, J. D. (2005). *Russian Oil Supply: Performance and Prospects*, Oxford: Oxford University Press.

Graham-Harrison, E. (2008). 'China–Turkmen gas price seen setting new benchmark', *Reuters*, 22 January 2008.

Grigorenko, Y. (2004). 'Offshore Oil Provides a Powerful Pull for Shelf Development: Locomotive for the Shelf', *RPI*, October 2004, 20–25.

GSGIR (2011). 'China: Energy: Gas', Goldman Sachs Global Investment

Research, 25 May 2011.

Gurt, M. (2011). 'Turkmen S.Iolotan gas field is world's No.2 – auditor', *Reuters*, 11 October 2011. (www.reuters.com/article/2011/10/11/gas-turkmenistan-idUSL5E7LB06V20111011).

Gvarstein, J. P. (2010). 'Shale Gas – An Emerging Factor in the Chinese Energy Mix', Statoil, 2010. (http://xynteo.com/uploads/Jens-PetterKvarstein.pdf)

Handke, S. (2006). 'Securing and Fuelling China's Ascent to Power: The Geopolitics of the Chinese–Kazakh Oil Pipeline', *Clingendael International Energy Programme*.

Hanson, P. (2009). 'The Resistible Rise of State Control in the Russian Oil Industry', *Eurasian Geography & Economics*, vol. 50, no. 1, 14–27.

Harrison, S.S. (1977). *China, Oil, and Asia: Conflict Ahead?*, New York: Columbia University Press.

Hart, T. G. (1987). *Sino-Soviet Relations: Re-examining the prospects for Normalization*, Aldershot: Gower.

Haukala, H. and Jakobson, L. (2009). 'The Myth of a Sino-Russian Challenge to the West', *The International Spectator*, vol. 44, no. 3, 59–76.

Hausmann, R. and Rigobon, R. (2002). 'An Alternative interpretation of the "resource curse": theory and policy implications', presented at the Conference on Fiscal Policy Formulation and Implementation in Oil Producing Countries by the IMF on 5–6 June.

Helmer, J. (2005). 'China to get first crack at Russian oil: Putin', *Asia Times*, 16 July 2005.

Henderson, J. (2010). Non-Gazprom Gas Producers in Russia, *Oxford Institute for Energy Studies (OIES) NG* 45.

Henderson, J. (2011a). 'The Strategic Implications of Russia's Eastern Oil Resources', *Oxford Institute for Energy Studies, WPM 41*.

Henderson, J. (2011b). Domestic Gas Prices in Russia – Towards Export Netback?, Oxford Institute for Energy Studies (OIES), NG 57.

Henderson, J. (2011c). The Pricing Debate over Russian Gas Export to China, *Oxford Institute for Energy Studies (OIES)* NG 56.

Higashi, N. (2009). Natural Gas in China: Market Revolution and Strategy, *IEA Working Papers Series*.

Hill, F. and Gaddy, C. (2003). *The Siberian Curse: How Communist Planners left Russia out in the Cold*, Washington D.C.: Brookings Institution Press.

Hook, L. (2010a). 'Doubts over Chinese coal-bed methane', *Financial Times*, 28 August 2010. (www.ft.com/cms/s/0/cfd6258a-b38d-11df-81aa-00144fe-abdc0.html#axzz1NBfizQM3)

Hook, L. (2010b). 'Home supplies to cut imports', *Financial Times*, 13 September 2010.

Hook, L. (2011). 'Latin moves in China's rush for oil', *Financial Times*, 5 April 2011. (www.ft.com/cms/s/2/05230786-5ec8-11e0-8e7d-00144feab49a.html#axzz1cs1Fwegd)

Houser, T. (2008). 'The roots of Chinese oil investment abroad', *Asia Policy* 5, 141–66.

Hoyos, C, and McGregor, Richard. (2008). 'China Sings two big LNG deals with Qatar', *Financial Times*, 11 April 2008.

Hoyos, C. (2006). 'PetroChina signs exploration deal with Total', *Financial Times*, 3 March 2006.

Hoyos, C. (2009). 'Shell wins 'gold rush' Iraqi oilfield auction', *Financial Times*, 12/13 December 2009.

Hoyos, C. and Crooks, E. (2010a). 'World leader has chosen a well-placed partner', *Financial Times*, 9 March 2010.

Hoyos, C. and Crooks, E. (2010b). 'A foot on the gas', *Financial Times*, 12 March 2010.

Hoyos, Carla. (2010). 'Europe the new frontier in shale gas rush', *Financial Times*, 8 March 2010.

IEA (2000). 'China's Worldwide Quest for Energy Security', International Energy Agency, Paris: IEA.

IEA (2002). 'Developing China's Natural Gas Market: The Energy Policy Challenges', International Energy Agency. Paris: IEA.

IEA (2007). *World Energy Outlook 2007*, International Energy Agency. Paris: IEA.

IEA (2009). *World Energy Outlook 2009*, International Energy Agency, Paris: IEA

IEA (2010). *World Energy Outlook 2010*, International Energy Agency, Paris: IEA.

IEA (2011a). Special Report: 'Are we entering a golden gas usage?'

IEA (2011b). *World Energy Outlook 2011*, International Energy Agency. Paris: IEA.

Illarionov, A. (2004). 'Russia's Latest Auction Farce Eerily Familiar', *Moscow Times*, 21 December 2004.

Inozemtsev, V. (2009). The 'Resource Curse' and Russia's Economic Crisis, *Chatham House REP Roundtable Summary*, 10 March. (

Jack, A. and Rahman, B. (2003). 'Japan to push case for Siberia–Pacific pipeline', *Financial Times*, 5 March 2003.

Jack, Andrew. (2003). 'Japan offers pipeline funding', *Financial Times*, 14 January 2003.

Jakobson, L. (2010). 'China prepares for an ice-free Arctic', SIPRI Insights on Peace and Security, no. 2010/2, March.

JBIC (2005). *The Future of the Natural Gas Market in East Asia, Vol 1* (Chapter 2: 'The Implications of China's Gas Expansion towards the Natural Gas Market in Asia'), Japan Bank for International Cooperation (JBIC), January 2005.

Jensen, J. T. (2011a). 'Natural Gas Pricing: Current Pattern and Future Trends', a Presentation to the Beijing Energy Club, Shanghai, 18 February.

Jensen, J. T. (2011b). 'Emerging LNG Market Demand: China', a presentation to the LNG Value Chain Conference, Rotterdam, 15 June.

Jensen, R. G., Shabad, T., and Wright, A. W. (1983). *Soviet Natural Resources in the World Economy*, Chicago: The University of Chicago Press.

Jentleson, B. W. (1986). *Pipeline Politics: The Complex Political Economy of East–West Energy Trade*, Ithaca: Cornell University Press.

Jia Chengzao et al. (2002). 'Petroleum geological characteristics of Kela-2 gas field', *Chinese Science Bulletin*, vol. 47, supplement 1, December issue, 94–9. (www.springerlink.com/content/d86543163125231w/fulltext.pdf).

Jia Chengzao, and Li Qiming. (2008). 'Petroleum geology of Kela-2, the most productive gas field in China', *Marine and Petroleum Geology*, April–May, 335–43.

Jiang Lurong (2006). 'China starts new round of price hike of gas', *China OGP*, 15 September 2006, 1–3.

Jiang Wenran (2009). 'China Makes Strides in Energy "Go-out" Strategy', *China Brief* (The Jamestown Foundation), vol. 9, no. 15.

Jiang, J., and Sinton, J. (2011). 'Overseas Investments Chinese national Oil Companies: Assessing the drivers and impacts', *IEA Information Paper*.

Jiao, W. (2011). 'China, Russia vow to boost relations', *China Daily*, 14 April 2011. (www.chinadaily.com.cn/china/brics2011/2011-04/14/content_12322482. htm)

Jie Mingxun (2010). 'Status & Outlook of Unconventional Gas Development in China', presented at China Oil and Gas Industry Summit (under Thirteenth China Beijing International High-Tec EXPO, 28 May.

Kalashnikov, V. D. (2004). 'Russian Far East Energy Sector Development and Cooperation Strategies towards Northeast Asia', presented at 2004 SRC (Slavic Research Centre)'s Summer International Symposium on Siberia and the Russian Far East in the 21st Century: Partners in the Community of Asia, 14–16 July, Sapporo, Hokkaido.

Kalici, J. H. and Goldwyn, D. L. (2005). *Energy and Security: Toward a New Foreign Policy Strategy*, Kalici, J. H. and Goldwyn, D. L. ed. Baltimore: The Johns Hopkins University Press, 2005.

Kambara, T. and Howe, C. (2007). *China and the Global Energy Crisis: Development and Prospects for China's Oil and Natural Gas*, Cheltenham: Edward Elgar.

Karaganov, S. (2011). 'Russia's Asian Strategy', *Russia in Global Affairs*, 2 July. (http://eng.globalaffairs.ru/pubcol/Russias-Asian-Strategy-15254).

Kedrov, I., and Gaiduk, I. (2009). 'China Becomes Russia's Main Energy Partner', *RPI*, November/December 2009, 35–40.

Kempton, D.R. (1996). 'The Republic of Sakha (Yakutia): The evolution of centre–periphery relations in the Russian Federation', *Europe–Asia Studies* 1996, vol. 48, no. 4, 587–613.

Kirillova, E. (2006). 'Gazprom Squeezes TNK–BP Out of Kovykta', *RPI*, November/December 2006, 51–6.

Kirillova, E. (2008a). 'Krasnoyarsk Emerging as an Oil and Gas Region', *RPI*, August 2008, 24–30.

Kirillova, E. (2008b). 'JOGMEC Eyes East Siberian Resources', *RPI*, October 2008, 22–7.

Kirillova, E. and Gaiduk, I. (2009). 'Russia Offers Japan Joint Energy Initiatives', *RPI*, June/July 2009, 5–10.

Kleveman, L. (2003). *The New Great Game: Blood and Oil and Central Asia*, New York: Atlantic Monthly Press.

Kokhanovskaya, Y. (2006). 'Official: Helium will delay Kovykta work', *Moscow Times*, 10 April 2006.

Kong, Bo (2010). *China's International Petroleum Policy*, Santa Barbara: ABC-CLIO, LLC.

Konovalov, S. (2008). 'YANR Becoming Strategic Centre for Russian Gas Extraction', *RPI*, September 2008, 11–16.

Kontorovich, A. E. and Eder, L. V. (2009). 'Oil and Gas from Russia will be delivered to the Asian-Pacific Market', in the proceedings of the 11th International Conference on 'Northeast Asian Natural Gas and Pipelines', organized by NAGPF & APRSJ, 27–8 October 2009, Tokyo.

Kramer, A. E. (2006). 'For China, a long wait for Russian gas supply', *New York Times*, 21 March 2006.

Kramer, A. E. (2009a). 'Putin's Grasp of Energy Drives Russian Agenda', *New York Times*, 28 January 2009. (www.nytimes.com/2009/01/29/world/europe/29putin.html).

Kramer, A. E. (2009b). 'Falling Gas Prices Deny Russia a Lever of Power', *New York Times*, 15 May 2009. (www.nytimes.com/2009/05/16/world/europe/16gazprom.html).

Kravets, V. (2006). 'Evenkia Awaits Development', *RPI*, January 2006, 17–23.

Kroutikhin, M. (2004). 'Russian Bear in China Shop: Gazprom leaves no room for foreign players in eastern Siberia', *The Russian Energy*, vol. 3, no. 118, June 2004.

Kryukov, V. and Moe, A. (1996). 'Gazprom: Internal Structure, Management Principles and Financial Flows', London: The Royal Institute of International Affairs.

Kynge, J. (2006). *China Shakes The World: The Rise of a Hungary Nation*, London: Phoenix.

Larionov, V. (1995). 'Deposits of Natural Gas in the Sakha Republic (Yakutia), Current Situation and Prospects of their Use in the Context of the Republic's Energy Policy, paper presented at the International Conference on NANGP', convened by NPRSJ, 3 March, Tokyo.

Li Xiaohui (2009). 'Guangdong to step up distribution of natural gas pipeline network', *China OGP*, 15 February 2009, 11–12.

Li Xiaohui (2010a). 'Gas pipeline interconnection and construction, a way to ease China's gas shortage', *China OGP*, 1 February 2010, 13–14.

Li Xiaohui (2010b). 'CNPC to step up implementation of CBM strategy', *China OGP*, 1 February 2010, 16–18.

Li Xiaohui (2010c). 'Second-batch SPR bases will allow China to feed 100-day demand', *China OGP*, 1 April 2010, 3–5.

Li Xiaohui (2010d). 'China's blueprint for refining industry', *China OGP*, 15 May 2010, 6–8.

Li Xiaohui (2010e). 'Competition in CBM industry aggravated by natural gas price adjustment', *China OGP*, 15 June 2010, 1–3.

Li Xiaohui (2010f). 'China, Russia in new era of energy cooperation', *China OGP*, 1 October 2010, 4.

Li Xiaohui (2011). 'China CBM industry to see booming development in 2011–15', *China OGP*, 1 June 2011, 7–9.

Li Xiaoming (2002a). 'West–East pipeline framework agreement signed', *China OGP*, 15 July 2002, 4–8.

Li Xiaoming (2002b). 'Review of Sino-Russian oil and gas cooperation', *China OGP*, 15 August 2002, 6–8.

Li Xiaoming (2004). 'CNOOC–Sinopec trading venture becomes the fifth Chinese state oil importer', *China OGP*, 1 July 2004, 1–2.

Li Yuling (2002a). 'Sinchem wins first overseas upstream project', *China OGP*, 1 February 2002, 7.

Li Yuling (2002b). 'Chinese tycoon launches private W–E gas project', *China OGP*, 15 June 2002, 13–15.

Li Yuling (2002c). 'China builds pipeline in Kazakhstan', *China OGP*, 15 June 2002, 15.

Li Yuling (2002d). 'BP's gas ambition beyond W–E pipeline', *China OGP*, 15 July 2002, 9 and 14.

Li Yuling (2003a). 'CNPC retreats from Slavneft auction', *China OGP*, 15 January 2003, 1–3.

Li Yuling (2003b). 'A U-turn of the Russia–China oil pipeline', *China OGP*, 1 March 2003, 4–5.

Li Yuling (2003c). 'Preparing for the arrival of Russian oil', *China OGP*, 1 April 2003, 1–3.

Li Yuling (2003d). 'FS of Russia–China–South Korea gas pipeline goes on amid upstream uncertainties', *China OGP*, 15 April 2003, 4–6.

Li Yuling (2003e). 'Moscow makes up its mind on Angarsk–Daqing oil pipeline', *China OGP*, 15 May 2003, 2–3.

Li Yuling (2003f). 'Russia, China step closer on clinching Angarsk–Daqing deal', *China OGP*, 1 June 2003, 1–3.

Li Yuling (2003g). 'Russia–China–South Korea gas pipeline: fraught with question marks', *China OGP*, 15 July 2003, 1–3.

Li Yuling (2003h). 'CNPC: China awaits Russia's official decision on the Angarsk–Daqing pipeline', *China OGP*, 15 September 2003, 3–4.

Li Yuling (2003i). 'PetroChina: confident in the economics of the Angarsk–Daqing pipeline', *China OGP*, 1 October 2003, 1–2.

Li Yuling (2003j). 'Shell's bargaining power in W–E gas pipeline dwindles', *China OGP*, 1 November 2003, 4–5.

Li Yuling (2003k). 'Route of Russia–China–South Korea gas pipeline still undecided', *China OGP*, 15 November 2003, 5–6.

Li Yuling (2004a). 'China, Kazakhstan to sign transborder pipeline contract in May', *China OGP*, 1 March 2004, 1–2.

Li Yuling (2004b). 'China to import LNG from Iran', *China OGP*, 15 March 2004, 1–3.

Li Yuling (2004c). 'Sino-foreign negotiations on W–E gas pipeline on the verge of falling through', *China OGP*, 15 March 2004, 5–6.

Li Yuling (2004d). 'CNPC's Stimul project in Russia at stake', *China OGP*, 15 March 2004, 12–13.

Li Yuling (2004e). 'CNPC eyes ownership of Atasu–Alashankou–Dushanzi', *China OGP*, 15 April 2004, 1–3.

Li Yuling (2004f). 'Wrangles within Russia cloud Russia–China–South Korea gas pipeline', *China OGP*, 15 April 2004, 3–5.

Li Yuling (2004g). 'New route, old question: where will the Russian oil go?', *China OGP*, 1 July 2004, 2–4.

Li Yuling (2004h). 'PetroChina: Co-investment with Japan in a No–Daqing Taishet–Nakhodka pipeline unlikely', *China OGP*, 1 August 2004, 1–2.

Lieberthal, K., and Oksenberg, M. (1988). *Policy Making in China: Leaders, Structures, and Processes*, Princeton, New Jersey: Princeton University Press.

Lieberthal, K.G., and Herberg, M. (2006). 'China's Search for Energy Security: Implications for US Policy', *NBR Analysis*, vol. 17. no. 1, 5–42.

Lin Fanjing (2005). 'Certain for a higher natural gas price in China', *China OGP*, 1 December 2005, 1–4.

Lin Fanjing (2006a). 'Xinjiang–Guangzhou gas pipeline, possible?', *China OGP*, 1 January 2006, 11.

Lin Fanjing (2006b). 'PetroChina to build two oil product pipelines for central China market', *China OGP*, 1 February 2006, 7–9.

Lin Fanjing (2006c). 'How long from imagination to reality?', *China OGP*, 15 March 2006, 7.

Lin Fanjing (2007a). 'Heading toward Central Asia with two-handed preparations', *China OGP*, 1 June 2007, 31–3.

Lin Fanjing (2007b). 'China's LNG industry embarrassed by high-cost import', *China OGP*, 1 October 2007, 1–3.

Lin Fanjing (2008a). 'China's pipelines construction to enter heyday', *China OGP*, 15 January 2008, 3.

Lin Fanjing (2008b). 'Updates of CNPC-launched pipelines', *China OGP*, 1 March 2008, 13.

Lin Fanjing (2008c). 'Thorough Reshuffle of China's energy sector ahead', *China OGP*, 15 March 2008, 1–3.

Lin Fanjing (2008d). 'Uncertainties in front of China's newly founded State Bureau of Energy', *China OGP*, 1 April 2008, 4–5.

Lin Fanjing (2008e). 'CNPC, CNOOC ink worthwhile deals with Qatargas', *China OGP*, 15 April 2008, 1–3.

Lin Fanjing (2008f). 'China's CBM industry asks for more favorable policies and investment', *China OGP*, 15 April 2008, 18–21.

Lin Fanjing (2008g). 'China pays record-high price for LNG spot imports in May', *China OGP*, 1 July 2008, 1–3.

Lin Fanjing (2009a). 'China to launch natural gas pricing reform this year, insider', *China OGP*, 15 March 2009, 1–3.

Lin Fanjing (2009b). 'Update of China's pipeline construction', *China OGP*, 15 April 2009, 3–7.

Lin Fanjing (2009c). 'China's towngas market reshuffling with newcomers participating', *China OGP*, 15 May 2009, 7–8.

Lin Fanjing (2009d). 'Chinese oil firms team up for overseas acquisitions', *China OGP*, 1 August 2009, 7–9.

Lin Fanjing (2009e). 'Sichuan-to-East pipeline pricing structure signals start of China's gas price reforms', *China OGP*, 1 September 2009, 1–3.

Lin Fanjing (2009f). 'China launches building of 2nd phase SPR bases', *China OGP*, 1 October 2009, 13–14.

Lin Fanjing (2009g). 'CNPC steps up integrating towngas in a bid to aggressively expand business', *China OGP*, 15 December 2009, 9.

Lin Fanjing (2010a). 'PetroChina targets domestic gas sources to hedge overseas supplies', *China OGP*, 15 January 2010, 11–12.

Lin Fanjing (2010b). 'China's natural gas pricing reform may further delay under inflationary pressure', *China OGP*, 1 March 2010, 2–4.

Lin Fanjing (2010c). 'Sinopec paves the way for growth of natural gas', *China OGP*, 15 July 2010, 5.

Lin Fanjing (2010d). 'Oversupply of refining capacity, a long-term issue for China', *China OGP*, 1 August 2010, 1–5.

Lin Fanjing and Lin Wei (2008). 'A race for towngas distribution', *China OGP*, 15 August 2008, 4–7.

Lin Fanjing and Mo Lin (2006). 'A higher price, a healthier market', *China OGP*, 1 January 2006, 4.

Liu Haiying (2002). 'China, Russia resume talks on Irkutsk gas development', *China OGP*, 15 May 2002, 1–2.

Liu Haiying (2004). 'Russia's solution to pipeline impasse', *China OGP*, 1 January 2004, 1–3.

Liu Haiying and Li Xiaoming (2002). 'Going-out rouses reflections on national oil strategy', *China OGP*, 15 April 2002, 1–3.

Liu Honglin, et.al. (2009). 'Shale gas in China: new important role of Energy in 21st Century', presented at 2009 International Coalbed & Shale Gas Symposium. (www.petromin.ca/sites/petromin/files/media/reports/shale%20gas/Shale_Gas_in_China_Tuscaloosa_2009.pdf).

Liu Xiaoli (2011). 'China's Oil Outlook and Oil Security Issue', presented at an International Conference on Russian–Asian Oil Summit, organized by Vostok Capital, 18–20 May 2011, Singapore.

Liu Yanan (2008a). 'China Petroleum Reserve Center launched to secure energy safety', *China OGP*, 1 January 2008, 8–9 and 15.

Liu Yanan (2008b). 'CNOOC reshaped as an integrated energy giant', *China OGP*, 1 March 2008, 8–9.

Liu Yanan (2008c). 'China's 4th largest oil company strives for more living space', *China OGP*, 1 May 2008, 5–7.

Liu Yanan (2008d). 'Competitive mechanism expected in China's CBM industry', *China OGP*, 15 June 2008, 1–2.

Liu Yanan (2009). 'China to edge out 50 mln tons of 'teapot' refining capacity by 2011', *China OGP*, 15 May 2009, 5–6.

Liu Yiyu and Zhou Siyu (2011). 'China pushes to develop green economy', *China Daily*, 23 November 2011. (www.chinadaily.com.cn/business/2010-11/23/content_11594441.htm).

Lo, B. (2008). *Axis of Convenience: Moscow, Beijing and the New Geopolitics*, London & Washington: Chatham House & Brookings Institution Press.

Lukin, A. (2003). *The Bear watches the Dragon: Russia's Perceptions of China and the Evolution of Russian–Chinese Relations Since the Eighteenth Century*, Armonk, New York: M. E. Sharpe.

Lukin, O. (2004). 'East Siberian Incentives Needed', *RPI*, June/July 2004, 11–18.

Lukin, O. (2005a). 'Trekking East with Gas', *RPI*, March 2005, 56–60.

Lukin, O. (2005b). 'East Siberian Incentives Needed', *RPI*, June/July 2005, 11–18.

Lukin, O. (2005c). 'MNR Forcing TNK–BP to Abandon Licenses for Kovyktinskoye and Verkhnechonskoye Fields: Deferred Offensive', *RPI*, September 2005, 13–20.

Ma Xin. (2008). *National Oil Company Reform from the Perspective of its Relationship with Governments: The Case of China*, a PhD thesis submitted to the University of Dundee (Centre for Energy, Petroleum & Mineral Law, and Policy: CEPMLP).

Mai, T. (2005). 'Power Panel', *China Daily* Business Weekly, 6–12 June 2005.

Mainwaring, J. (2011). 'Urals Energy temporarily abandons Well no. 51 at Petrosakh', *Proactiveinvestors*, 23 September 2011. (www.proactiveinvestors. co.uk/companies/news/33505/urals-energy-temporarily-abandons-well-51-at-petrosakh-33505.html).

Manning, R. A. (2000). *The Asian Energy Factor*, New York: Palgrave.

Mao Yushi, Sheng Hong, and Yang Fuqiang (2008). 'The True Cost of Coal', Greenpeace. (www.greenpeace.org/eastasia/PageFiles/301168/the-true-cost-of-coal.pdf).

Marcel, V. (2006). *Oil Titans: National Oil Companies in the Middle East*, London: Chatham House, 2006.

Marcel, V. and Xu, Y. (2008). 'Key Issues for Rising National Oil Companies', prepared for KPMG International.

McDonald, J. (2006). 'Chinese state oil giant buys $ 500 million stake in Russia's Rosneft', *Associated Press*, 19 July 2006.

McGregor, R, and Hoyos, C. (2004). 'Pipeline pullout embarrasses PetroChina', *Financial Times*, 4 August 2004.

McGregor, R. (2004). 'Putin resists Chinese pressure to approve big oil projects', *Financial Times*, 15 October 2004.

McGregor, R. (2005). 'Gazprom in talks over new China pipelines', *Financial Times*, 22 September 2005

McGregor, R., Pilling, D., and Ostrovsky, A. (2004). 'Japan likely to win on pipeline route', *Financial Times*, 24 March 2004.

Medetsky, A. (2007a). 'Pacific Pipeline Delayed Until 2015', *Moscow Times*, 20 July 2007.

Medetsky, A. (2007b). 'Turkmens Tack On 2nd Gas Price Hike', *Moscow Times*, 28 November 2007.

Medetsky, A. (2009). 'Putin Launches Pacific Oil Terminal', *Moscow Times*, 29 December 2009.

Meidan, M. (2007). *Shaping China's Energy Security: The Inside Perspective*, Michal Meidan, ed., Paris: Asia Centre – Centre études Asie.

Meidan, M., Andrew-Speed, P., and Ma Xin (2007). 'Shaping China's Energy Security: Actors and Policies', in Michal Meidan, ed., *Shaping China's Energy Security: The Inside Perspective*, (Asia Centre – Centre études Asie, 2007), 33–63.

Meyer, H. (2006). 'Russian Oil Pipeline construction begins', Associated Press, 28 April 2006.

Miles, T. and Beck, L. (2006). 'Russia opens energy tap, China wants more', *Reuters*, 22 March 2006.

Miller, A. B. (2003). 'Eurasian Direction of the Russia's Gas Strategy', presented at 22nd World Gas Conference Tokyo 2003, Key Note Address (KA 1-5),

4 June 2003.

Milyaeva, S. (2008a). 'Future Oil Production: Officials Are Optimistic: Oil Companies Are Not', *RPI*, May 2008, 19–24.

Milyaeva, S. (2008b). 'Investors Eye East Siberia', *RPI*, June/July 2008, 40–47.

Milyaeva, S. (2008c). 'Russian Deposit Licensing Unresolved', *RPI*, October 2008, 28–33.

Milyaeva, S. (2010). 'Gazprom Begins Push into the Asian-Pacific Region', *RPI*, May 2010, 17–21.

Minakir, P. A. (2007). *Economic Cooperation between the Russian Far East and Asia-Pacific Countries*, P. A. Minakir, ed., Khabarovsk: RIOTIP.

Ministry of Energy, Russian Federation (2010). *Energy Strategy of Russia: for the period up to 2030* (approved by Decree N 1715r of the Government of the Russian Federation, dated 13 November 2009). Published by Institute of Energy Strategy, Moscow, 2010.

Mironova, I. (2010). 'Russia gas in China: Complex Issues in Cross-Border Pipeline Negotiations', *Energy Charter Secretariat*.

Mitchell, J. (1996). *The New Geopolitics of Energy*, London: The Royal Institute of International Affairs.

Mitchell, J., and Lahn, G. (2007). 'Oil for Asia', *Chatham House Briefing Paper*, London, Chatham House.

Mitrova, T. (2011). 'The Domestic Context: Russian Gas Production', presented at an International conference on 'Russian Oil and Gas: New Trends and Implications', organized by Chatham House, London, 28–9 March.

Miyamoto, A. (1997). *Natural Gas in Central Asia: Industries, Markets and Export Options of Kazakhstan, Turkmenistan and Uzbekist*an, London: The Royal Institute of International Affairs.

Mo Lin (2004). 'Russia's changing mind', *China OGP*, 14 November 2004, 6–7.

Morse, R.K. and He Gang (2010). 'The World's Greatest Coal Arbitrage : China's Coal Import Behavior and Implications for the Global Coal Market', Programme on Energy and Sustainable Development, Freeman Spogli Institute for International Studies, Stanford University, Working Paper No. 94, August 2010.

Motomura, M. (2008). 'Japan's Energy Relations with Russia and Central Asia', presented at Clingendael Conference on 'The Geopolitics of Energy in Eurasia: Russia as an Energy Lynch Pin', The Hague, 23 January 2008.

Motomura, M. (2010). 'Evaluation of ESPO Crude in Japan and Japan's Strategy for East Siberia development', presented At Oil Terminal 2010 conference organized by Vostok Capital at St. Petersburg, 25–6 November.

Murphy, D. (2003). 'CNPC Official Kidnapped During Slavneft Auction, Sources Say', *China Energy Report, Dow Jones Newsletters*, 14 March 2003.

NAGPF (2004). 'Proceedings of 4th Forum', Northeast Asian Gas and Pipeline Forum conference held in Ulan Baartar, 16–18 August 1998.

NAGPF (2005, 2007). 'A Long-Term Vision of Natural Gas Infrastructure in Northeast Asia: 2005 version and 2007 Version', September 2005 and September 2007.

NAGPF (2007). 'Proceedings of 10th Forum', Northeast Asian Gas and Pipeline

Forum, Novosibirsk, 18–19 September 2007.

NAGPF (2009). 'Proceedings of 11th Forum', Northeast Asian Gas and Pipeline Forum, Tokyo, 27–8 October 2009.

NAGPF (2011). 'Proceedings of 12th forum', Northeast Asian Gas and Pipeline Forum, Ulaanbaatar, 29–30 August 2011.

NEAGPF, APRSJ (2009). 'Proceedings of the 11th International Conference on Northeast Asian Natural Gas and Pipeline: 'Multilateral Cooperation in Natural Gas and Pipeline in Northeast Asia', Northeast Asian Gas and Pipeline Forum and Asian Pipeline Research Society of Japan (organizers). 27–8 October 2009, Tokyo.

Nemtsov, B. and Milov, V. (2008). 'Putin and Gazprom: An Independent expert report', Moscow, translated from the Russian by Dave Essel, 1–28. (www.europeanenergyreview.eu/data/docs/Viewpoints/Putin%20and%20 Gazprom_Nemtsov%20en%20Milov.pdf).

Nezhina, V. (2009a). 'First Russian LNG Plant Headlines Shelf Developments', *RPI*, March 2009, 19–23.

Nezhina, V. (2009b). 'Gazprom Changes Priorities', *RPI*, August 2009, 44–50.

Nezhina, V. (2009c). 'Vankor begins Active Phase of East Siberian Development', *RPI*, October 2009, 32–8.

Nezhina, V. and Kirillova, E. (2010). 'ESPO Prepares to Move Crude to China', *RPI*, June/July 2010, 21–6.

Norman, J. R. (2008). *The Oil Card: Global Economic Warfare in the 21st Century*, Walterville: Trine Day.

Nuriev, A. (2006). 'Riches of "Blue Fuel" Reserves', *Turkmenistan*, August 2006, no. 7–8 (16–17). (www.turkmenistaninfo.ru/?page_id=6&type=article&elem_ id=page_6/magazine_35/278&lang_id=en).

Olcott, M. B. (2004). 'The Energy Dimension in Russian Global Strategy: Vladimir Putin and the Geopolitics of Oil', Paper presented at the Baker Institute for Public Policy, Rice University, October.

Ostrovsky, A. (2006a). 'Out on a limb: how the Kremlin has been making life difficult on Sakhalin', *Financial Times*, 23 November 2006.

Ostrovsky, A. (2006b). 'Shell offers Gazprom control of Sakhalin-2', *Financial Times*, 12 December 2006.

Ostrovsky, A. (2006c). 'Gazprom to pay $7.45 bn to control Sakhalin-2', *Financial Times*, 22 December 2006.

Ostrovsky, A. and Buckley, N. (2006). 'In Russia, control justifies the means', *Financial Times*, 12 December 2006.

Paik, K-W. (1995). *Gas and Oil in Northeast Asia: Policies, projects and prospects*, London: The Royal Institute of International Affairs.

Paik, K-W. (1996). 'Energy Cooperation in Sino-Russian relations: the importance of oil and gas', *The Pacific Review*, vol. 9, no. 1, 77–95.

Paik, K-W. (1997). 'Tarim Basin Energy Development: Implications for Russian and Central Asian Oil and Gas Exports to China', *The Royal Institute of International Affairs' Central Asian and Caucasian Prospects (CACP) Briefing*, No. 14.

Paik, K-W. (2002a). 'Natural gas Expansion in Korea', in *Natural Gas in Asia: The Challenges of Growth in China, India, Japan and Korea*, Wybrew-Bond, I. and

Stern, J. eds., Oxford: Oxford University Press, 2002, 188–229.

Paik, K-W. (2002b). 'Sino-Russian Oil and Gas Relationship: Implications for Economic Development in Northeast Asia' presented at Northeast Asia Cooperation Dialogue XIII: Infrastructure and Economic Development Workshop organized by Institute for Far Eastern Affairs, Russian Academy of Sciences, and Institute of Global Conflict and Cooperation, University of California, Moscow, 4 October.

Paik, K-W. (2005a). 'Russia's Oil and Gas Export to northeast Asia', *Asia-Pacific Review*, vol. 12, no. 2, 58–70.

Paik, K-W. (2005b). 'The implications of China's Gas Expansion towards Natural Gas Markets in Asia', in JBIC, *The Future of the Natural Gas Market in East Asia, Vol 1* (Chapter 2: the Implications of China's Gas Expansion towards the Natural Gas Market in Asia), January 2005 (unpublished).

Paik, K-W. (2008). 'Natural Gas in Korea', in Stern, J. ed. *Natural Gas in Asia: The Challenges of Growth in China, India, Japan, and Korea*, 2008, Oxford: Oxford Institute for Energy Studies, 174–219.

Paik, K-W., Marcel,V., Lahn, G., Mitchell, John V., and Adylov, E. (2007). 'Trends in Asian NOC Investments Overseas', *Chatham House Working background paper*. (www.chathamhouse.org/sites/default/files/public/Research/Energy,%20Environment%20and%20Development/r0307anoc.pdf).

Pala, C. and Bradsher, K. (2003). 'Beijing and Caspian Oil Fields', *New York Times*, 1 April 2003. (www.nytimes.com/2003/04/01/business/beijing-and-caspian-oil-fields.html?src=pm).

Peel, Q. and Jack, A. (2003). 'Japan and Russia set to back pipeline', *Financial Times*, 10 January 2003.

Petromin Pipeliner (2011). 'China's Pipeline Gas Imports: Current Situation and Outlook to 2025', Jan–March 2011. (www.pm-pipeliner.safan.com/mag/ppl0311/r06.pdf)

Peyrouse, Sebastien (2007). 'The Economic Aspects of the Chinese-Central Asia Rapproachment', Central Asia-Caucasus Institute and Silk Road Studies program, 2007. www.silkroadstudies.org/new/docs/Silkroadpapers/2007/0709China-Central_Asia.pdf.

Pilling, D. and Tsui, E. (2004). 'Japan risks missing out on Russian island's gas', *Financial Times*, 3 November 2004.

Pilling, David (2010). 'Poised for a shift', *Financial Times*, 23 November 2010.

Pilling, David and Gorst, I. (2005). 'Tokyo in threat to withdraw from $11bn oil pipeline', *Financial Times*, 30 April–1 May 2005.

Pilling, David and Jack, Andrew. (2003). 'Oil fuels Japan's drive to bring a thaw to relations with Russia', *Financial Times*, 10 January 2003.

Pirani, S. (2009). *Russian and CIS Gas Markets and their Impact on Europe*, Pirani, S. ed., Oxford: Oxford University Press.

Pirani, S. (2010). *Change in Putin's Russia: Power, Money and People*, London: Pluto Press.

Platts (2010a). 'Russian crude oil exports to the Far East – ESPO starts flowing', Platts Special Report, February 2010. (www.platts.com/IM.Platts.Content/InsightAnalysis/IndustrySolutionPapers/espo_ip_0210.pdf).

Platts (2010b). 'Russian Crude Oil Exports to the Pacific Basin –ESPO Starts Flowing', Platts Special Report, May 2010. (www.platts.com/IM.Platts.Content/InsightAnalysis/IndustrySolutionPapers/espoupdate0510.pdf).

Poussenkova, N. (2007a). 'Lord of the Rigs: Rosneft as a mirror of Russia's Evolution', prepared in conjunction with an energy study sponsored by the James A. Baker III Institute for Public Policy and Japan Petroleum Energy Centre, Rice University.

Poussenkova, N. (2007b). 'The Wild, Wild East: East Siberia and the Far East: A New Petroleum Centre?', *Carnegie Moscow Centre, Working Papers*, No. 4.

Putin, V. (1999). 'Mineral Natural Resources in the Strategy for Development of the Russian Economy', *Zapiski Gornogo Instituta* Vol. 144, 1999, 3–9.

Qiu Jun (2004a). 'West–East JV negotiation falls through', *China OGP*, 15 August 2004, 2–3.

Qiu Jun (2004b). 'New hope and old questions in Sino-Russian energy cooperation', *China OGP*, 1 November 2004, 8–9.

Qiu Jun (2005a). 'Rosneft promises to supply China with 50 m tons of crude by 2010', *China OGP*, 1 February 2005, 6–8.

Qiu Jun (2005b). 'Dushazi petrochemical project goes in high gear', *China OGP*, 1 March 2005, 5–6.

Qiu Jun (2005c). 'Dina gasfield construction launched to fulfill 40% of W–E gas supply', *China OGP*, 15 March 2005, 18–19.

Qiu Jun (2005d). 'Kovykta's gas may kiss goodbye to China', *China OGP*, 1 May 2005, 8–9.

Qiu Jun (2005e). 'CNPC makes headway in Uzbekistan', *China OGP*, 1 June 2005, 18–19.

Qiu Jun (2005f). 'Sino-Russia oil pipeline is conceiving new uncertainties', *China OGP*, 15 June 2005, 5–6.

Qiu Jun (2005g).'Sino-Russian oil pipeline update', *China OGP*, 15 July 2005, 14–15.

Qiu Jun (2005h). 'First section of Taishet–Nakhodka pipeline to start construction in December', *China OGP*, 1 August 2005, 9–11.

Qiu Jun (2005i). 'Russia to export crude through Sino-Kazakhstan oil pipeline', *China OGP*, 1 November 2005, 8–9.

Qiu Jun (2005j). 'National oil reserves in second stage', *China OGP*, 1 November 2005, 9–11.

Qiu Jun (2005k). 'T-N pipeline still pending', China OGP, 15 December, 2005, 5–7.

Qiu Jun (2005l). 'China's largest condensate gasfield cluster construction debuts', *China OGP*, 15 December 2005, 17.

Qiu Jun (2006a). 'Atasu–Dushanzi crude pipeline completed', *China OGP*, 1 January 2006, 5–6.

Qiu Jun (2006b). 'Petroleum industry's outstanding role in governmental agenda', *China OGP*, 15 March 2006, 12–16.

Qiu Jun (2006c). 'China Russia sign deals to build gas pipelines', *China OGP*, 1 April 2006, 1–4.

Qiu Jun (2006d). 'The 1st government-initiated oil and gas survey under way',

China OGP, 1 April 2006, 8–9.

Qiu Jun (2006e). 'China and Turkmenistan ink natural gas pipeline agreement', *China OGP*, 15 April 2006, 12–15.

Qiu Jun (2006f). 'China's LNG projects under price and source pressure', *China OGP*, 1 June 2006, 9 & 12–14.

Qiu Jun (2006g). 'CNPC and Gazprom to hasten gas pipeline', *China OGP*, 1 June 2006, 24–5.

Qiu Jun (2006h). 'Sinopec teams up with Rosneft to bid for Russian oil assets', *China OGP*, 15 June 2006, 12–14.

Qiu Jun (2006i). 'Sinopec expects new overseas headways', *China OGP*, 1 July 2006, 7–9 and 14.

Qiu Jun (2006j). 'Doomed closer ties with Russian counterparts', *China OGP*, 15 July 2006, 16–19.

Qiu Jun (2006k). 'Petronas likely to supply LNG for CNOOC's Shanghai LNG terminal', *China OGP*, 1 August 2006, 4–7.

Qiu Jun (2006l). 'Chunxiao begins production through disputes exist', *China OGP*, 15 August 2006, 14–16.

Qiu Jun (2006m). 'Sinopec to hook Iran LNG for Shandong', *China OGP*, 15 August 2006, 16–19.

Qiu Jun (2006n). 'New gas findings ensure supply to W–E gas pipeline', *China OGP*, 15 September 2006, 8–9 & 13.

Qiu Jun (2006o). 'Sino-Russia energy cooperation: new developments', *China OGP*, 15 September 2006, 16.

Qiu Jun (2006p). 'Domestic LNG players' trio orchestral in Jiangsu', *China OGP*, 1 October 2006, 20–3.

Qiu Jun (2006q). 'Sino-Russia energy cooperation approaches harvest season', *China OGP*, 1 November 2006, 16–19.

Qiu Jun (2006r). 'PetroChina starts construction of end station of its western crude pipeline', *China OGP*, 15 November 2006, 20–21.

Qiu Jun (2006s). 'CNPC constructs gas pipeline grid in northeast China', *China OGP*, 15 December 2006, 8–9 and 13–14.

Qiu Jun (2008). 'A local oil company emerges in Shaanxi', *China OGP*, 1 October 2008, 17.

Qiu Jun (2009). 'CNPC integrates downstream natural gas sales business prior to price reform', *China OGP*, 15 December 2009, 3–7.

Qiu Jun (2010a). 'Guesswork on China's natural gas pricing mechanism reform', *China OGP*, 1 February 2010, 9.

Qiu Jun (2010b). 'China sets up national energy commission headed by Premier', *China OGP*, 1 February 2010, 15–16.

Qiu Jun (2010c). 'China's CBM industry prospects in next decade', *China OGP*, 15 June 2010, 3–6.

Qiu Jun (2010d). 'One more oil firm acquires state-run crude oil import license', *China OGP*, 1 July 2010, 12–15

Qiu Jun (2010e). 'Chinese oil majors beef up efforts to ensure natural gas supply', *China OGP*, 1 October 2010, 6–7.

Qiu Jun (2010f). 'China's natural gas supply to remain tight despite 25 pct rise in supply', *China OGP*, 1 November 2010, 1–3.

Qiu Jun and An Bei (2005). 'CNPC completes takeover deal with PK', *China OGP*, 1 November 2005, 1–3.

Qiu Jun and Liu Shuyun (2006). 'Yanchang Petroleum Group: a truly new and strong force', *China OGP*, 1 August 2006, 26–29.

Qiu Jun and Wang Boyu (2006). 'Metering quarrel blocks Kazakh oil into China', *China OGP*, 1 July 2006, 1–4.

Qiu Jun and Yan Jinguang (2009). 'CNPC buys expensive overseas LNG, criticized for forcing up domestic prices', *China OGP*, 15 September 2009, 1–5.

Qiu Jun and Zhang Zhengfu (2010). 'Natural gas shortage stings China', *China OGP*, 15 January 2010, 1–5.

Quan Lan (1999a). 'Premier's visit refuels transnational pipeline scheme', *China OGP*, 15 March 1999, 1–3.

Quan Lan (1999b). 'LNG project appraisal to wind up', *China OGP*, 15 June 1999, 5–6

Quan Lan (1999c). 'Transnational oil pipeline shelved', *China OGP*, 15 August 1999, 2–3.

Quan Lan (1999d). 'Changqing's gas bearing zone expands', *China OGP*, 15 September 1999, 11–12.

Quan Lan (1999e). 'CNPC revises west-to-east gas grid design', *China OGP*, 1 October 1999, 1–4.

Quan Lan (2000a). 'China pushed ahead West-Gas-To-East project', *China OGP*, 15 March 2000, 1–5.

Quan Lan (2000b). 'China to start first transnational oil pipeline in 2003', *China OGP*, 1 April 2000, 5–6.

Quan Lan (2000c). 'State offers preferential policies to attract foreign funds to trunk gas pipeline', *China OGP*, 1 April 2000, 6–7.

Quan Lan (2000d). 'China to start the third round of national hydrocarbon resources appraisal', *China OGP*, 15 June 2000, 15.

Quan Lan (2000e). 'Tarim, what to offer next?', *China OGP*, 1 September 2000, 5–7 & 9.

Quan Lan (2000f). 'Foreign fund for the transnational pipeline construction', *China OGP*, 15 September 2000, 4–5.

Quan Lan (2001a). 'New gasfield backs up West–East Gas Pipeline', *China OGP*, 15 February 2001, 15–16.

Quan Lan (2001b). 'PetroChina to select seven for second round bidding on West–East gas pipeline', *China OGP*, 15 March 2001, 12.

Quan Lan (2001c). 'PetroChina signs gas LOIs with 33 enterprises', *China OGP*, 15 March 2001, 13.

Quan Lan (2001d). 'PetroChina shortlists three foreign investors for West–East pipeline project', *China OGP*, 15 June 2001, 9.

Quan Lan (2001e). 'Presidential visit to shovel ahead Sino-Russia oil pipeline', *China OGP*, 1 July 2001, 1–3.

Quan Lan (2001f). 'PetroChina locates new reserves for W–E pipeline', *China OGP*, 1 August 2001, 12.

Quan Lan (2001g). 'PetroChina gets new boost for its W–E pipeline', *China*

OGP, 1 August 2001, 13.

Quan Lan (2001h). 'Shell ties up with Gazprom for W–E pipeline', *China OGP*, 1 September 2001, 7–8.

Quan Lan (2001i). 'BP pulls out of W–E pipeline bidding', *China OGP*, 15 September 2001, 2.

Quan Lan (2001j). 'How Russia fits into China's energy blueprint', *China OGP*, 1 October 2001, 1–3.

Quan, L, and Paik, K-W. (1998). *China Natural Gas Report*, London: *Xinhua News Agency* and The Royal Institute of International Affairs.

Quested, R. (2005). *Sino-Russian Relations: a short history*, Abingdon: Routledge.

Rahman, B. and Jack, A. (2003). 'Japan lures Russia with $7 bn offer on pipeline', *Financial Times*, 14 October 2003.

RC (2008). 'Oil & Gas Yearbook 2008: Crosswind', Renaissance Capital (Equity Research), 29 July 2008.

RC (2010). 'Oil and Gas Yearbook 2010: Stand and deliver', Renaissance Capital, 27 July 2010.

Roberts, J. (1996). *Caspian Pipelines, Former Soviet South Project*, London: The Royal Institute of International Affairs.

Roberts, J. (2008). 'The Geopolitics of the Caspian and Central Asian Energy', presented at a seminar organized by Oxford Institute for Energy Studies and St. Anthony's College, Oxford University, 27 February 2008.

Robinson, G. (2009). 'Sinopec swoops on oil explorer – or does it?', *Financial Times*, 15 June 2009. (http://ftalphaville.ft.com/blog/2009/06/15/57006/sinopec-swoops-on-oil-explorer-or-does-it/)

Rosen D.H. and Houser T. (2007). *China Energy: A Guide for the Perplexed*. Washington D.C., Peterson Institute for International Economics

Rosner, K. (2010). 'Sino-Russian energy relations in perspective', *Journal of Energy Security*, 29 September.

Rozman, G. (2010). 'The Sino-Russian Strategic Partnership: How Close? Where To?', in Bellacqua (2010), 13–32.

Rutledge, I. (2004). 'The Sakhalin II PSA – a Production 'Non-Sharing' Agreement: Analysis of Revenue Distribution'. (www.foe.co.uk/resource/reports/sakhalin_psa.pdf).

Sakwa, R. (2009). *The Quality of Freedom: Khodorkovsky, Putin and the Yukos Affairs*, Oxford: Oxford University Press.

Shevtsova, L. (2005). *Putin's Russia*, Washington D.C.: Carnegie Endowment for International Peace.

Shi Xunzhi (1995). 'Present Situation and Forecast of Natural Gas Exploitation and Utilization in China', paper presented at the international conference on Northeast Asian Natural Gas Pipeline, convened by Natural Pipeline Research Society of Japan, Tokyo, 3 March.

Shlyapnikov, A., Glazkov, S., and Gaiduk, I. (2007). 'Eastward Expansion Kicks Off', *RPI*, January 2007, 37–43.

Simonia, N. (2004). 'Russian Energy Policy in East Siberia and the Far East', prepared in conjunction with an Energy Study sponsored by the Petroleum Energy Centre of Japan and the James A. Baker III Institute for Public

Policy, Rice University.

Simpfendorfer, B. (2009). *The New Silk Road: How a rising Arab World is turning away from the West and rediscovering China*, Basingstoke: Palgrave Macmillan.

Sinyugin, O. (2005). 'Eastern Dimension of Russia's unified Gas Supply System', *Northeast Asia Energy Focus (KEEI)*, Vol. 2, No. 4.

Sixsmith, M. (2010). *Putin's Oil: The Yukos Affair and the Struggle for Russia*, London: Continuum International.

Skagen, O. (1997). *Caspian Gas, Former Soviet South Project*, London: The Royal Institute of International Affairs.

Slavinskaya, L. (2008a). 'Gazprom Establishes Parameters for Foreign Investors', *RPI*, March 2008, 5–9.

Slavinskaya, L. (2008b). 'Russia Likely to Have Limited Impact on the Global LNG Market', *RPI*, March 2008, 19–23.

Smith, M. A. (2010). 'Medvedev and the Modernisation Dilemma', Defence Academy of the United Kingdom, Russian series 10/15, November 2010.

Smith, P. (2007). 'China's LNG deals 'good' for Gazprom discussions', *Financial Times*, 10 September 2007

Smith, P. (2010a). 'Woodside's A\$45bn LNG deal with PetroChina expires', *Financial Times*, 5 January 2010. (www.ft.com/cms/s/0/9f0d43f6-f999-11de-8085-00144feab49a.html).

Smith, P. (2010b). 'Shell and PetroChina in Arrow bid', *Financial Times*, 9 March 2010.

Socor, V. (2009). 'Strategic Implications of the Central Asia–China Gas Pipeline', *Eurasia Daily Monitor*, volume no. 6, issue no. 233, 18 December. (www.jamestown.org/single/?no_cache=1&tx_ttnews%5Btt_news%5D=35856&tx_ttnews%5BbackPid%5D=13&cHash=4b0f4138d8).

Soldatkin, V., and Akin, M. (2011). 'Gazprom gets upper hand on China export with field win', *Reuters*, 1 March 2011.

Stares, P. B. (2000). *Rethinking Energy Security in East Asia*, ed. P. B. Stares, Tokyo: Japan Center for International Exchange.

Stephan, J. J. (1994). *The Russian Far East: A History*, Stanford: Stanford University Press.

Stern, J. (2002). 'Russian and Central Asian Gas Supply for Asia', in *Natural Gas in Asia: The Challenges of Growth in China, India, Japan, and Korea*, Wybrew-Bond, I. and Stern, J. eds., Oxford: Oxford University Press, 230–76.

Stern, J. (2005). *The Future of Russian Gas and Gazprom*, Oxford: Oxford University Press.

Stern, J. (2009). 'The Russian Gas Balance to 2015: Difficult Years Ahead', in *Russian and CIS Gas Markets and their Impact on Europe*, Pirani, S. ed., Oxford: Oxford University Press, 2009, 54–92.

Stern, J. and Bradshaw, M. (2008). 'Russian and Central Asian Gas Supply for Asia', in Stern, J. ed., *Natural Gas in Asia: The Challenges of Growth in China, India, Japan, and Korea*, Oxford: Oxford Institute for Energy Studies, 2008, 220–78.

Stern, J. ed. (2008). *Natural Gas in Asia: The Challenges of Growth in China, India, Japan, and Korea*, Oxford: Oxford Institute for Energy Studies.

Sun Huanjie (2006a). 'New pipeline operation reinforces PetroChina's market presence', *China OGP*, 1 November 2006, 13–16.

Sun Huanjie (2006b). 'Towngas market expects more competition', *China OGP*, 15 December 2006, 14–18.

Sun Huanjie (2008). 'Towngas market, new battlefield for Chinese oil giants', *China OGP*, 1 March 2008, 20–22.

Swartz, S., and Oster, S. (2010). 'China Tops U.S. in Energy Use: Asian Giant Emerges as No. 1 Consumer of Power, Reshaping Oil Markets, Diplomacy', *Wall Street Journal*, 18 July. (http://online.wsj.com/article/SB100014240527 48703720504575376712353150310.html).

Swearingen, R. (1987). 'Siberia and the Soviet Far East: Strategic Dimensions in Multinational Perspective', Rodger Swearingen, ed., Stanford: Hoover Institution Press.

Tabata, S, and Liu, X. (2012). 'Russia's energy policy in the far east and East Siberia', in Pami Aalto ed., *Russia's Energy Policy: National, Interregional and Global Dimensions*, (Cheltenham: Edward Elgar, forthcoming).

Tompson, W. (2005). 'A frozen Venezuela?: The "Resource Curse" and Russian Politics', in Ellman, M. ed. *Russia's Oil and Natural Gas: Bonanza or Curse?*, London: Anthem Press. 2006. (http://eprints.bbk.ac.uk/256/1/Frozen_Ven-ezuela.pdf).

Topham, J. (2010). 'Sakhalin-1 production to get boost: Japex', *Reuters*, 25 May 2010. (www.reuters.com/article/idUSTRE64O2N120100525).

Treisman, D. (2011). *The Return: Russia's Journey from Gorbachev to Medvedev*, New York: Free Press.

Trenin, D. (2002). *The End of Eurasia: Russia on the Border between Geopolitics and Globalisation*, Washington D.C: Carnegie Endowment for International Peace.

Trenin, D. (2010). 'A Genuine Bilateral Relationship', *China Daily*, 28 September 2010. (http://carnegie.ru/publications/?fa=41637)

Trenin, D. (2011). 'China Russia ties on sound base', *China Daily*, 14 June 2011. (www.chinadaily.com.cn/cndy/2011-06/14/content_12687237.htm).

True, W. R. (2009). 'China begins commissioning third LNG terminal', *OGJ*, 20 October 2009. (www.ogj.com/index/article-display/2711102928/articles/oil-gas-journal/transportation-2/lng/2009/10/china-begins_commissioning.html).

Tsui, E. (2005). 'Expensive offer with an eye to Kazakhstan's wealth in reserve', *Financial Times*, 23 August 2005.

Tsui, E. and Pilling, D. (2004). 'Exxon rethinks natural gas delivery options', *Financial Times*, 5 November 2004.

Tsui, M. X. (2007). Blue Book of Energy (Tsui, M. X. ed), 2007. *The Energy Development Report of China, 2007*, Beijing, Social Sciences Academic Press.

UBS (2009). *Investment Research on China Oil & Gas*, 9 September 2009.

UPEACE (2005). 'Energy Demand Projections and Supply Options for the DPRK', Report of the Working Group on Energy for the Democratic People's Republic of Korea: Phase 1, University for Peace, July 2005.

Vyakhirev, R. I. (1997). 'The Perspectives of Russian Natural Gas. Role at the World Gas Market', presented at the 20th World Gas Conference,

Copenhagen.

Vygon, G. (2009). 'Problems and Development Directions in the Russian Oil Sector', *RPI*, September 2009, 10–15.

Wagstyl, S. (2007). 'Diversification is elusive key to success', *Financial Times*, Special Report Russia, 20 April 2007.

Wagstyl, S. and Ostrovsky, A. (2004). 'Kremlin man but no fan of state control', *Financial Times*, 7 October 2004.

Waldmeir, P. (2009). 'Chinese groups to buy Angolan oil field stake', *Financial Times*, 20 July 2009.

Walter, A. (2011). 'Global economic governance after the crisis: The G2, the G20, and global imbalances', Bank of Korea Working Papers, 2011. (http://personal.lse.ac.uk/wyattwal/images/Globaleconomicgovernanceafterthecrisis.pdf).

Walters, G. and Faucon, B. (2006). 'SINOPEC sets deal for Russian oil, aided by Rosneft', *Wall Street Journal*, 21 June 2006.

Wan Zhihong (2008). 'CNPC finds big gas reserve in Xinjiang', *China Daily*, 16 December 2008.

Wan Zhihong (2009). 'CNOOC buys more LNG from Qatar', *China Daily*, 14 November 2009. (www.chinadaily.com.cn/bizchina/2009-11/14/content_8980907.htm)

Wan Zhihong (2010). 'Big LNG deal signals better Canberra ties', *China Daily*, 25 March 2010. (www.chinadaily.com.cn/china/2010-03/25/content_9637967.htm).

Wang Ying (2005). 'Firms mull China–Russia gas pipeline', *China Daily*, 21 September 2005.

Wang Ying (2006). 'French oil giant explore Erdos Basin gas field', *China Daily*, 22 February 2006.

Wang Ying and Ying Lou (2008). 'PetroChina Longgang May Be Nation's Largest Gas Field', *Bloomberg*, 17 January 2008. (www.bloomberg.com/apps/news?pid=newsarchive&sid=af3t9d7hr7t8).

Ward, A. (2010).'Statoil near deal on China Shale gas', *Financial Times*, 4 November 2010.

Watkins, E. (2009a). 'Putin touts Russia's oil reserves; seeks funding', *OGJ*, 7 September 2009, 30–31.

Watkins, E. (2009b). 'Russians complete ESPO oil export terminal at Kozmino', *OGJ Online*, 12 November 2009. (www.ogj.com/index/article-display/5926685803/articles/oil-gas-journal/transportation-2/pipelines/construction/2009/11/russians-complete/s-QP129867/s-cmpid=EnlPipelineNovember232009.html).

Watts, J. (2011). 'China takes step towards tapping shale gas potential with first well', The *Guardian*, 21 April 2011. (www.guardian.co.uk/environment/2011/apr/21/china-shale-gas-well).

Way, B. (2009). 'Oil & Gas Linking China and Central Asia', BNP Paribas Research Paper.

Wei Hong (2010). 'The Role of the Mini LNG in the Natural Gas Supply of LNG', presented at US–China Oil & Gas Industry Forum, Fort Worth,

Texas, 14–16 September 2010.

White, G. L. and Oster, S. (2006). 'China puts energy at top of agenda for Putin's visit', *Wall Street Journal*, 20 March 2006.

Whiting, A. (1981). *Siberian Development and East Asia: Threat or Promise?*, Stanford: Stanford University Press.

Wilson, J. L. (2004). *Strategic Partners: Russian–Chinese Relations in the Post-Soviet Era*, Armonk, New York: M. E Sharpe.

Winning, D. (2009). 'China Starts First Shale Gas Project', *Wall Street Journal*, 27 November 2009. (http://online.wsj.com/article/SB1000142405274870 34994045745608426042048828.html).

Wishnick, E. (2001). *Mending Fences: The Evolution of Moscow's China's Policy from Brezhnev to Yeltsin*, Seattle: University of Washington Press.

Wonacott, P. and White, G. (2004). 'CNPC To Buying Controlling Stake In Stimul Oil Project In Southern Russia', *China Energy Report*, 9 January, 1 and 5.

Woodard, K. (1980). *The International Energy Relations of China*, Stanford: Stanford University Press.

Wu Xiaobo (2011). 'China targets annual natural gas supply of 240 bcm by 2015', *China OGP*, 1 April 2011, 7–8.

Wybrew-Bond, I. and Stern, J. (2002). *Natural Gas in Asia: The Challenges of Growth in China, India, Japan and Korea*, Wybrew-Bond, I. and Stern, J. eds., Oxford: Oxford University Press.

Xia Yishan (2009). *China's Perspective on International Energy Development Strategy (Zhongguo Guoji Nengyuan Fazhan Zhanlue Yanjiu)*, Beijing: Shihchieh Chihshih Chupanshe.

Xia Yishan (2004). 'Sino-Russian Relations in the Year 2004: Achievements, Problems and Prospects', prepared for Chinese People's Institute of Foreign Affairs (CPIFA).

Xia Yishan (2011). 'China's Foreign Energy Policy', in Ramsay. C. and Lesourne. J. eds., *Chinese Climate Policy Institutions and Intent*, Paris, IFRI.

Xie Ye (2004). 'Sino-Kazakh oil pipeline to begin construction', *China Daily*, 11 March 2004. (www.chinadaily.com.cn/english/doc/2004-03/11/content_313825.htm)

Xu Dingming (2002). 'China's Natural Gas Industry in Development', presented at the Fourth USA – China Oil and Gas Industry Forum, jointly sponsored by the State Development Planning Commission, The US Department of Energy, and the US Department of Commerce, Houston, 18–19 July 2002.

Xu Wan (2010). 'Sinopec, BP Discuss Shale-Gas Exploration in China', *Wall Street Journal*, 18 January 2010. (http://online.wsj.com/article/SB10001424 052748703626604575010322184894544.html).

Xu Yihe (2008). 'PetroChina to Pay $195/mcm For Turkmen Gas', *Upstream*, 25 January 2008.

Xu Yihe (2010a). 'Chevron steps up Chuandongbei bid', *Upstream*, 20 August 2010.

Xu Yihe (2010b). 'PetroChina eyes LNG for Tibet', *Upstream*, 20 August 2010.

Yakovleva, Maria (2005). 'China Thirsts for Kazakh Oil', *RPI*, January 2005,

67–72.

Yamanaka, M. (2008). 'Nippon Oil Buys Crude From Sakhalin-1 Under Long Term Contract', *Bloomberg*, 18 February 2008. (www.bloomberg.com/apps/news?pid=20601080&sid=a0WJUYqVZ6so&refer=asia).

Yang Liu (2005a). 'Sinopec's largest gas field onstream', *China OGP*, 15 November 2005, 16.

Yang Liu (2005b). 'Ningbo–Shanghai–Nanjing crude pipeline operates', *China OGP*, 1 December 2005, 8–9.

Yang Liu (2005c). 'Update of LNG development in China', *China OGP*, 15 December 2005, 9–10.

Yang Liu (2006a). 'China's energy policies during Eleventh Five-year Plan', *China OGP*, 15 January 2006, 1–2.

Yang Liu (2006b). 'Private capital marches into towngas market', *China OGP*, 15 June 2006, 16–18.

Yang Liu (2006c). 'China proves more oil and gas reserves', *China OGP*, 1 October 2006, 31–4.

Yang Liu (2006d). 'China's energy consumption mix optimizes as gas demand increases', *China OGP*, 1 November 2006, 30–31.

Yang Liu (2006e). 'Chinese oil companies live up and down when going overseas', *China OGP*, 15 December 2006, 21.

Yang Liu (2007). 'Should China exploit Nanpu oilfield soon?', *China OGP*, 1 September 2007, 25–7.

Ye Ming (2001). 'Keeping economic benefits of pipeline domestic', *China OGP*, 1 May 2001, 13.

Yeh, A. (2007). 'Citic completes Kazakh oil assets deal', *Financial Times*, 2 January 2007.

Yeh, A., Gorst, I., and Aglionby, J. (2006). 'Citic to invest $1.9 bn in Kazakh oil field', *Financial Times*, 27 October 2006.

Yenikeyeff, S. M. (2008). 'Kazakhstan's Gas: Export Markets and Export Routes', Oxford Institute for Energy Studies NG 25.

Yenikeyeff, S. M. (2009). 'Kazakh export plans affect regional producers, buyers', *OGJ*, 5 January, 56–9.

Yergin, D. (2011). *The Quest: Energy, Security and the Remaking of the Modern World*, London: Allen Lane (Penguin Books).

Yergin, D. and Gustafson, T. (1993). *Russia 2010 and What It Means for the World*, New York: Random House.

Yermukanov, M. (2006). 'Atasu–Alashankou Pipeline Cements "Strategic Alliance" between Beijing and Astana', *EDM*, vol. 3, no. 1, 3 January 2006. (www.jamestown.org/single/?no_cache=1&tx_ttnews%5Btt_news%5D=31239)

Yu, Silvia (2010). 'PetroChina raises gas output Xinjiang's Tarim Basin by 8%', *Platts*, 3 November 2010. (www.platts.com/RSSFeedDetailedNews/RSSFeed/Oil/6564843).

Yusuf, S. and Saich, T. (2008). *China Urbanises: Consequences, Strategies, and Policies*, Washington, D.C.: The World Bank.

Zhan Lisheng (2006). 'Natural gas import from Australia to help energy shortage', *China Daily* Business Weekly, 24 February 2006.

Zhang Aifang (2009). 'Analysis: China changing oil trade pattern with vast forex surplus', *China OGP*, 1 March 2009, 5–8.

Zhang Boling (2008). 'CNOOC Eyes Global Market, Deep Sea Oil', *Caijing Magazine*, 29 July 2008. (http://english.caijing.com.cn/2008-07-29/100076835.html).

Zhang Chunyan (2011). 'IEA hails China's new policy for gas usage', *China Daily*, 8 June 2011. (www.chinadaily.com.cn/business/2011-06/08/content_12657487.htm)

Zhang Jian (2011). 'China's Energy Security: Prospects, Challenges, and Opportunities', Brookings Institute CNAPS (Centre for Northeast Asian Policy Studies) Visiting Fellow Working Paper.

Zhang Qiang (2005). 'China's crude production to reach 200 million tons in 2010', *China OGP*, 15 December 2005, 8–9.

Zhang Xuegang (2008). 'China's Energy Corridors in South East Asia', *China Brief*, 4 February. (www.jamestown.org/programs/chinabrief/single/?tx_ttnews%5Btt_news%5D=4693&tx_ttnews%5BbackPid%5D=168&no_cache=1).

Zhang Xuezeng and Zhao Dongrui (2009). 'Pipeline Technologies: New Development in China', in 'Proceedings of 11th Forum', Northeast Asian Gas and Pipeline Forum, Tokyo, 27–8 October 2009.

Zhang Yuqing (2002). 'The Current Situation, Future and Policies of Chinese Gas Industry', presented at 4th US–China Oil and Gas Industry Forum, Houston, 18–19 July.

Zhao Tingting (2011). 'China's coal imports up 31% in 2010', *China Daily*, 27 January 2011. (www.chinadaily.com.cn/business/2011-01/27/content_11926703.htm).

Zhou Yan (2011). 'CNPC plans to extend pipeline network', *China Daily*, 22 February 2011. (www.chinadaily.com.cn/bizchina/2011-02/22/content_12056227.htm).

Zhu Qiwen (2008). 'Time to take a fresh look at oil subsidies', *China Daily*, 21 March 2008. (www.chinadaily.com.cn/opinion/2008-03/21/content_6554797.htm).

Zhu Zhu (2007). 'Forecast of China's oil supply and demand', *China OGP*, 1 October 2007, 34–7.

Zhu Zhu (2008). 'China's Energy Conditions and Policies, White Paper', *China OGP*, 1 January 2008, 25–9.

Zhu, W., Meyer, H., and Wang Y. (2010). 'China to import more Russian coal, lend $6 billion', *Bloomberg News*, 7 September 2010. (www.bloomberg.com/news/2010-09-07/china-will-take-more-russian-coal-imports-in-next-25-years-arrange-loan.html).

Zhukov, S. (2009). 'Uzbekistan: a domestically oriented gas producer' in *Russian and CIS Gas Markets and their Impact on Europe*, Pirani, S. ed., Oxford: Oxford University Press, 2009.

Newspapers, Periodicals and Web Sites

In addition to the specific publications in the bibliography the following newspapers, periodicals and web sites contain information relevant to this study, some of it retrievable by searches.

Agence France Presse (AFP): http://www.afp.com/afpcom/en/
Asia Times (AT): http://www.atimes.com/
BBC Online: http://www.bbc.co.uk/
Bloomberg News: http://www.bloomberg.com/news/
Caijing: http://english.caijing.com.cn/
Carnegie Moscow Centre: http://carnegie.ru/?lang=en
Chatham House: http://www.chathamhouse.org/
China Brief, The Jamestown Foundation: http://www.jamestown.org/programs/chinabrief/
China Daily (CD): www.chinadaily.com.cn/index.html
China Energy Report (CER), Dow Jones: http://www.dowjones.com/commodities/China-Energy-Report.asp
China National Offshore Oil Corporation (CNOOC): http://www.cnoocltd.com/encnoocltd/default.shtml
China National Oil and Gas Exploration and Development Corporation (CNODC): http://www.cnpcint.com/aboutus/welcome.html
China National Petroleum Corporation (CNPC): http://www.cnpc.com.cn/eng/
China Oil, Gas and PetroChemical (China OGP), Xinhua News Agency (XNA):
China Petrochemical Corporation (SINOPEC): http://english.sinopec.com/
China Securities Journal (CSJ), Xinhua News Agency (XNA): http://www.cs.com.cn/english/
China Security: http://www.chinasecurity.us/
Chinese Science Bulletin: http://www.worldscinet.com/csb/csb.shtml
Credit Suisse Equity Research: https://www.credit-suisse.com/investment_banking/research/en/
Deutche Bank Research: http://www.dbresearch.com/
Energy Charter Secretariat: http://www.encharter.org/
Energy Information Administration: http://www.eia.doe.gov
Energy Research Institute (ERI): http://www.eri.org.cn/#
EnergyChinaForum: http://www.energychinaforum.com/
Eurasia Daily Monitor (EDM): http://www.jamestown.org/programs/edm/
Eurasian Geography and Economics (EGE): http://www.bellpub.com/psge/index.htm
Financial Times (FT): http://www.ft.com/home/uk
Foreign Affairs: http://www.foreignaffairs.com/
Gazprom Export: http://gazpromexport.ru
Gazprom: http://www.gazprom.ru
Goldman Sachs (GS) Global Investment Research: http://www2.goldmansachs.com/careers/choose-your-path/our-divisions/global-investment-research/index.

html

Greenpeace, Climate-Energy: http://www.greenpeace.org/eastasia/campaigns/
    climate-energy/

IFRI (French Institute of International Relation): http://www.ifri.org/?a=b&lang=uk

IHS Global Insight: http://www.ihs.com/products/global-insight/

Industrial Fuels and Power: http://www.ifandp.com/

Institute for Economies in Transition, Bank of Finland: http://www.suomenpankki.
    fi/bofit_en/Pages/default.aspx

Institute of Energy Economics, Japan: http://eneken.ieej.or.jp/en/

Institute of Energy Strategy, Moscow: http://translate.google.co.uk/translate?hl=e
    n&sl=ru&tl=en&u=http%3A%2F%2Fwww.energystrategy.ru%2F

Interfax China Energy Report Weekly (China ERW): http://www.interfax.com/

Interfax China Energy Weekly (China Weekly): http://www.interfax.com/

Interfax Petroleum Report (IPR): http://www.interfax.com/

Interfax Russia & CIS Oil and Gas Weekly (Russia & CIS OGW): http://www.
interfax.com/

International Economic Bulletin, Carnegie Endowment for International Peace: http://
carnegieendowment.org/ieb/

International Energy Agency (IEA): http://iea.com

International Gas Report (IGR), Platts: http://www.platts.com/Products/interna-
    tionalgasreport

International Oil Daily (IOD): http://www.energyintel.com/pages/NewsLetters.
    aspx?PubId=31

Itar Tass News: http://www.itar-tass.com/en/

James A. Baker III Institute for Public Policy: http://bakerinstitute.org/

Japan Bank for International Cooperation (JBIC): http://www.jbic.go.jp/en/

Japan Oil, Gas and Metals National Corporation (JOGMEC): http://www.
    jogmec.go.jp/english/index.html

Journal of Energy Security: http://www.ensec.org/

KazMunaiGaz: http://www.kmg.kz/en/

Kaztransgaz: http://kaztransgas.kz

Kommersant: http://www.kommersant.com/

Korea Energy Economics Institute (KEEI): http://www.keei.re.kr/main.nsf/in-
    dex_en.html

Korea Gas Corporation (Kogas): http://www.kogas.or.kr/kogas_eng/html/main/
    main.jsp

Korea National Oil Corporation (KNOC): http://www.knoc.co.kr/ENG/main.jsp

LNG World News: http://www.lngworldnews.com/

Lukoil: http://www.lukoil.ru

National Bureau of Asian Research (NBR): http://www.nbr.org/

National Development and Reform Commission (NDRC): http://en.ndrc.gov.cn/

Naval War College Review: http://www.usnwc.edu/Publications/Naval-War-
    College-Review.aspx

NBR Analysis: http://www.nbr.org/Publications/issue.aspx?id=224

Nefte Compass: http://www.energyintel.com/Pages/About_NCM.aspx

Oil & Gas Journal (OGJ): http://www.ogj.com/index.html

Oil and Gas National Corporation Limited (ONGC): http://www.ongcindia.com/

Oxford Institute for Energy Studies: http://www.oxfordenergy.org

PetroChina: http://petrochina.com

Petroleum Economist: http://www.petroleum-economist.com/

Petroleum Intelligence Weekly (PIW): http://www.energyintel.com/Pages/About_PIW.aspx

RBC Daily: http://www.rbcnews.com/news.shtml

Renaissance Capital (Equity Research): http://www.renaissancecapital.com/Ren-Cap/AboutUs/AboutUs.aspx

Reuters: http://www.reuters.com/news
   Ria Novosti: http://en.rian.ru/

Rosneft: http://www.rosneft.com/

RusEnergy (RE): http://www.rusenergy.com/en/

Russia in Global Affairs (RGA): http://eng.globalaffairs.ru/

Russian Petroleum Investor (RPI): http://www.eng.rpi-inc.ru/

Sakha Republic (Yakutia): http://www.yakutia.org/

Sakhalin-1 project: http://www.sakhalin1.ru/Sakhalin/Russia-English/Upstream/default.aspx

Sakhalin Energy Investment Company: http://www.sakhalinenergy.com/en/

Shanghai Daily (SD): http://www.shanghaidaily.com/

Slavic Research Centre (SRC): http://src-h.slav.hokudai.ac.jp/index-e.html

Surgutneftegaz: http://www.surgutneftegas.ru/en/

The Associated Press (AP): http://www.ap.org/

The China and Eurasia Forum Quarterly: http://www.silkroadstudies.org/new/inside/publications/CEF_quarterly.htm

The China Quarterly (CQ), SOAS: http://www.soas.ac.uk/research/publications/journals/chinaq/

The Moscow Times (MT): http://www.themoscowtimes.com/index.php

The New York Time (NYT): http://global.nytimes.com/

The New York Times: http://global.nytimes.com/

The Royal Institute of International Affairs (RIIA): http://www.chathamhouse.org/

The School of Russian and Asian Studies (SRAS): http://www.sras.org/

The Wall Street Journal (WSJ): http://europe.wsj.com/home-page

TNK–BP: http://www.tnk-bp.ru/

Transneft: http://www.transneft.ru/ (in Russian)

UBS Investment Research: http://www.ubs.com/1/e/about/research.html

Upstream: http://www.upstreamonline.com/about_upstream/

US–China Oil & Gas Industry Forum: http://www.uschinaogf.org/

West East Gas Pipeline Project: http://www.china.org.cn/english/features/Gas-Pipeline/37313.htm

Wood MacKenzie: http://www.woodmacresearch.com/cgi-bin/wmprod/portal/corp/corpAboutUs.jsp

World Gas Intelligence (WGI): http://www.energyintel.com/Pages/About_WGI.aspx

Xinhua News: http://www.xinhuanet.com/english2010/

NOTES

Chapter 1

1 Bellacqua (2010, 1).
2 Rozman(2010, 13).
3 Lo (2008, 41–3).
4 Rozman (2010, 13–14). Among the books which contribute to an in-depth understanding of Sino-Soviet and Sino-Russian relations are Hart (1987) and Dittmer (1990). More recent studies include Garnett (2000); Wishnick (2001); Lukin (2003); Wilson (2004); Quested (2005); Lo (2008); Bellacqua (2010).
5 Bellacqua (2010, 2).
6 After the global financial crisis of 2008, there was open mention of the possibility of a USA–China G2 based global economic governance regime, even though the Chinese leadership had rejected the idea of an explicit G2. Such a development was inconceivable a decade ago. See Walter (2011).
7 According to Jonathan Garner, Asian and emerging markets strategist at Morgan Stanley, China was set in 2011 to overtake the USA as the biggest producer of manufactured goods by value – only the third change at the top of the global manufacturing league in 250 years. See Pilling (2010).
8 In September 2010, China's foreign reserves stood at $2.65 trillion, and had risen to $3.04 trillion by the end of March 2011 ('China Foreign-Exchange Reserves Jump to $2.65 Trillion', www.bloomberg.com/news/2010-10-13/ china-s-currency-reserves-surge-to-record-fueling-calls-for-stronger-yuan. html; 'China's foreign exchange reserves to expand at slower pace', http:// news.xinhuanet.com/english2010/china/2011-04/15/c_13831036.htm).
9 Haukala and Jakobson (2009, 60).
10 *Ibid.*, 62–70.
11 *Ibid.*, 72–75.
12 *Ibid.*, 71. Bobo Lo has made the point in a personal comment to the author that Russia is no longer China's biggest arms supplier. In fact there have been no new contracts since 2006 and China has fallen well down the rankings of Russia's customers.
13 Lo (2008, 6), with slight changes.
14 Lo (2008, 6–7).
15 Karaganov (2011); CSCAP (2010).
16 Trenin (2010.
17 Rosner(2010).
18 'China–Russia energy co-op sees broad prospects: vice premier', www.gov. cn/english/2010-09/21/content_1707489.htm
19 Rosner(2010).

20 'China to become Russia's Largest Trading Partner', http://russia-briefing.com/news/china-to-become-russias-largest-trading-partner.html/; 'Sino-Russian trade bounces back in 2010', www.chinadaily.com.cn/business/2010-12/10/content_11681821.htm.

21 The Bank of China and Industrial & Commercial Bank of China started the ruble trading with a combined purchase of 1 million yuan's worth of the Russian currency at an average price of 4.67 rubles to the yuan. See 'China Starts to Trade Yuan Against Ruble', www.themoscowtimes.com/business/article/china-starts-to-trade-yuan-against-ruble/424165.html; 'Russia to boost China links with yuan-rouble trade' www.ibtimes.com/articles/60538/20100908/russia-boost-china-links-with-yuan-rouble-trade.htm.

22 Lo (2008, 14).

23 Clover and Buckley (2011). For Medvedev's modernization programme, see Smith, M.A. (2010)

24 Lo (2008, 132–3).

25 *China OGP*, 15, November 1993, 1.

26 *China OGP*, 15, November 1993, 3; *PIW*, 20 September 1993, 7.

27 Paik (1995, 236–7).

28 *China OGP*, 1 July 1995, 5; *China OGP*, 15 December 1996, 1–3.

29 The optimum route for the 1,000 km pipeline was studied. The Sichuan Petroleum Exploration and Designing Institute was authorized to design the pipeline route within Chinese boundaries. See *China OGP*, 15 November 1995, 8, *China OGP*, 15 December 1996, 1–3, and *RPI*, March 1996, 68–72.

30 Around 1996, within CNPC, the Sino-Russian Oil and Gas Cooperation Committee was established (chaired by Zhang Yongyi, first vice president of CNPC) to promote trans-national pipelines.

31 *China OGP*, 15 December 1996, 1–3.

32 Wilson (2004, 30); Barges (2004b); Li Xiaoming (2002b).

33 *Reuters News Service*, 27 June 1997.

34 *China OGP*, 15 July 1997, 3-4.

35 Christoffersen (1998). A first shipment of 1,700 tonnes of oil from Kazakhstan was sent by rail in Xinjiang in October 1997 (see Wishnick (2001, 142)).

36 Wishnick (2001, 135).

37 Quan and Paik (1998, 109).

38 Rem Vyahirev's presentation at the APEC meeting in Kuala Lumpur in November 1998, quoted in Quan and Paik (1998).

39 Allen Whiting, has used the term 'East Asian Siberia' rather than the more conventional 'Russian Far East' in order to stress the relevance of the region to East Asia with implications for China, Korea, and Japan. See Whiting (1981).

40 As early as 1995, Shi Xunzhi, then assistant president of CNPC, argued that coalbed gas held more than one third of China's proven gas reserves. See Shi Xunzhi (1995).

Chapter 2

1   Aleksashenko (2010) has calculated that revenues from the oil and gas sector decreased from 11% of GDP in the second half of 2008 to 5.9% of GDP in the first quarter of 2009 and 7.6% of GDP in 2009 overall.

2   Nemtsov and Milov (2008).

3   The budget revenue from the oil and gas sector, including the pipeline sector, should also take corporation tax, value-added tax, personal income tax, consolidated social tax, fixed property tax, dividend on national shares in oil or gas companies, profit of joint venture corporation 'Vietsovpetro', and the royalty from PSA into account. However, due to the limitation of credible statistic data, these parts of the budget revenue are not discussed here.

4   Wagstyl (2007).

5   Clover (2008).

6   *Idem*. Finance minister Kudrin said on 16 September 2008 that the federal budget would begin to run a deficit if oil fell below US$70 a barrel – at that point Russian oil was selling for roughly US$89 a barrel. See Clover and Belton (2008).

7   Belton (2008).

8   Belton and Wagstyl (2009); Clover (2009).

9   Kramer (2009b)

10   'Russia's 2010 inflation rate to top 8% – minister', http://en.rian.ru/business/20101123/161467191.html. The lowest figure recorded during the 2000s was 5.5% in July 2010. See 'Russian Inflation Slows to 12-Year-Low, Paving Way for Further Rate Cuts', www.bloomberg.com/news/2010-04-05/russian-inflation-slows-to-12-year-low-paving-way-for-further-rate-cuts.html.

11   Tompson (2005). To understand the problem Russia is facing with regard to its massive Siberian resources, the author recommends Hill and Gaddy (2003).

12   Goldman (2008, 12–13).

13   Inozemtsev (2009).

14   Putin (1999, 3–9). The first Western scholar to write about this article was Martha Brill Olcott (2004). The background to the article and its thesis is discussed in more detail in Balzer (2005).

15   *Ibid*.

16   Boris E. Nemtsov, a former first deputy prime minister now in opposition, said: 'It is the typical behaviour of the monopolist. The monopolist fears competition. See Kramer (2009a).

17   Grace (2005, 1). For an in-depth understanding of Russia's oil industry, see Considine and Kerr (2002).

18   Sakwa (2009), particularly 148–87.

19   Goldman (2008, 105–20); Hanson (2009, 14–27).

20   Sakwa (2009, 395–6).

21   *Moscow Times*, 21 December 2004, quoted in Sakwa (2009, 187).

22   *Ibid.*, 93–135.

23   Wagstyl and Ostrovsky (2004).

24   See Pirani (2010).

25   Erochkine and Erochkine (2006, 83).

26   Yukos was created in April 1993 as result of the merger of two state-owned companies: Yuganskneftegaz based in Khanty-Mansi autonomous okrug, and the Volga-based refining company KuybyshevnefteOrgSintez. Sakwa (2009, 30–73).

27   TNK established control over Onaco and Sidanco and bought 74.95% of Slavneft together with Sibneft *ibid.*, 84–85.

28   RC (2008).

29   A. E. Kontorovich, scientific director of the Institute of Petroleum Geology and Geophysics SB RAS was the head of the working group on the determination of long-term strategic priorities and major policy development activities of the oil complex, for the Ministry of Industry and Energy's work (with Decree dated 21 December 2006, N. 413) on a refinement of the energy strategy of Russia for the period up to 2020 and its prolongation up to 2030.

30   Lukin (2005b).

31   Gaiduk (2005).

32   *Ibid.*

33   Gaiduk (2007c, 17).

34   Lukin (2005b).

35   *Ibid.*

36   Gaiduk and Kirillova (2006, 38).

37   Shlyapnikov, Glazkov, and Gaiduk (2007); Glazkov (2004a).

38   Milyaeva (2008c).

39   Baidashin (2008a).

40   RC (2010, 35).

41   Baidashin (2008b).

42   Milyaeva (2008a).

43   Baidashin (2008b).

44   *Ibid.*

45   Milyaeva (2008a). In the south of the Siberian platform, exploration has revealed large oil and gas fields with some already explored. Among the more than 60 oil and gas deposits opened, 16 have reserves exceeding 100 mt of hydrocarbons and three deposits have more than 1 bt. The largest oil reserves are Yurubcheno–Tokhomskoye, Kuyumbinskoye, Verkhnechonskoye, and Talakanskoye, and the largest gas reserves are Kovyktinskoye, Angaro–Lenskoye, and Chayandinskoye. See Baidashin (2007c).

46   Milyaeva (2008a).

47   Baidashin (2007c).

48   *Ibid.*

49   Adopted on 2 August 2007, the special federal programme 'Economic and Social Development of the Far East and Trans-Baikal for the Period Until

2013' should deal with the infrastructure problems. Most of the funds will go toward infrastructure construction in Sakha (Yakutia) – pipelines, highways, communications, and transmission lines connecting deposits as yet undeveloped and costing 30 billion rubles (US$1.17 billion), as well as railways (the largest project, Kolyma–Lena, costing 70 billion rubles). See Gaiduk (2007c, 17).

50   *Russia & CIS OGW*, 9–15 October 2008, 19; *RPI*, September 2005, 16.

51   Glazkov (2006a); Chernyshov (2007).

52   Chernyshov (2008).

53   www.rosneft.com/news/today/161020082.html, last accessed end of 2009.

54   Dmitry Sheibe, deputy head of the Irkutsk region, and its natural resources minister, said that the regional authorities intended 'to suspend this process at the very least'. The dispute was being resolved in court. In June 2007 ESGC shareholders decided to sell an 11.29% stake in VCNG, which had previously belonged to the Irkutsk region. Rosneft and TNK–BP were to purchase these shares on an equal basis. See Chernyshov (2008); *Russia & CIS OGW*, 9–15 October 2008, 19.

55   Glazkov (2006a); Chernyshov (2007).

56   Chernyshov (2007).

57   Chernyshov (2008).

58   There are a number of risks not considered by TomskNIPIneft. Most important are those connected with ESPO, the construction cost of which has exceeded the original budget by more than 100%, and there was almost a year's delay in the investment required for extraction. Consequently the pumping of Verkhnechonskoye oil may well exceed the calculated tariff. See Chernyshov (2008); *RPI*, September 2005, 16.

59   *Ibid.*, 23–27.

60   *Ibid.*

61   *Russia & CIS OGW*, 9–15 October 2008, 19.

62   'Verkhnechonskneftegaz', www.rosneft.com/Upstream/ProductionAndDevelopment/eastern_siberia/verkhnechonskneftegaz/

63   www.rosneft.com/news/news_in_press/29012010.html, last accessed February 2011. Rosneft performed exploratory work in East Siberia (Irkutsk and Krasnoyarsk Regions) in 26 license areas with the attributed available resources of over 2.5 billion tons of oil equivalent. Particularly for 2010 it was planned to run 5,325 line km of 2D seismic, 400 km^2 of 3D, and to drill 12 exploration wells.

64   Milyaeva (2008b).

65   *Ibid.*

66   Valery Nesterov, Troika Dialog's oil and gas analyst, calculated the actual cost of Dulisma LLC to Urals Energy was about $1.34/barrel of proved and probable reserves. (By comparison, the purchase of a 10% share in West Siberian Resources by Spanish company Repsol was at the equivalent of $5.1/barrel of reserves of the same category; the acquisition of Saratovneftegaz by Russneft was at the equivalent of a price of US$2.1/barrel of reserves.) See Baidashin (2006c).

67 Milyaeva (2008b); 'Urals Energy receives $270 million loan from Sberbank', http://in.reuters.com/article/2007/11/15/russia-urals-sberbank-idINL1550225020071115.

68 Irkutsk Oil Company home page, www.irkutskoil.com/

69 *Ibid.*

70 *Ibid.*

71 Kirillova (2008b).

72 Milyaeva (2008b); Kirillova and Gaiduk (2009).

73 Chernyshov (2007).

74 *Ibid.*

75 *Russia & CIS OGW*, 11–17 September 2008, 19.

76 On 1 January 2002, the section's D1 resource estimates stood at 15 mt of oil and 50 bcm of gas. See *Russia & CIS OGW*, 4–10 September 2008, 22.

77 Motomura (2010).

78 Kirillova (2008b).

79 *Ibid.*

80 'Russian-Japanese joint-ventures find oil and gas in Irkutsk, Region', www.irkutskoil.com/presscenter/companynews?id=27 (accessed 26 October 2010).

81 Chernyshov (2004e).

82 Glazkov (2004a).

83 Chernyshov (2005d).

84 Chernyshov (2007); Baidashin (2006d).

85 *Russia & CIS OGW*, 2–8 October 2008, 14; Surgutneftegaz home page, www.surgutneftegas.ru/en/press/news/item/312/

86 *OGJ Online*, 29 January 2009.

87 *Russia & CIS OGW*, 2–8 October 2008, 18–19.

88 Kirillova (2008a).

89 Baidashin (2005c).

90 Kravets (2006).

91 The implementation of this target programme was designed to increase rates of commercial oil extraction from 98,000 tonnes in 2006 to 20.773 million tonnes in 2010, and commercial gas production from 0.963 bcm in 2006 to 3.6 bcm in 2010. See Kirillova (2008a). The author could not confirm whether the target programme was implemented as planned.

92 Kirillova (2008a).

93 Watkins (2009a).

94 Glazkov (2009).

95 For the rise of Rosneft see Poussenkova (2007a).

96 Chernyshov (2003c).

97 Glazkov (2005); Kirillova (2008a).

98 Glazkov (2005).

99 Glazkov (2006b).

100 Glazkov (2005); Kirillova (2008a).

101 Kirillova (2008a).

102 Chernyshov (2007).
103 The issues of ESPO-related taxes and tariffs will be discussed in Chapter 7.
104 'Vankorneft', www.rosneft.com/Upstream/ProductionAndDevelopment/eastern_siberia/vankorneft/; Kirillova (2008a).
105 Chernyshov (2007).
106 Kirillova (2008a); 'Vankor Group of Licensed Blocks', www.rosneft.com/Upstream/Exploration/easternsiberia/vankor/
107 'Vankor's Millions', www.rosneft.com/news/today/21082009.html; Nezhina (2009c).
108 'Vankor's Millions', www.rosneft.com/news/today/21082009.html
109 Kirillova (2008a).
110 Rosneft Annual Report 2010, 61 (www.rosneft.com/attach/0/58/80/a_report_2010_eng.pdf).
111 'Vankor's Millions', www.rosneft.com/news/today/21082009.html; 'Vankorneft', www.rosneft.com/Upstream/ProductionAndDevelopment/eastern_siberia/vankorneft/. By April 2011 the production capacity had reached 20 mt/y. ('Production Output at the Vankor Field Reaches 20 Million Tonnes', www.rosneft.com/news/news_in_press/060420112.html).
112 'East-Siberian Oil and Gas Company', www.rosneft.com/Upstream/ProductionAndDevelopment/eastern_siberia/east_siberian_oil_gas/
113 Kirillova (2008a, 27).
114 *Russia & CIS OGW*, 2–8 October, 2008, 16–17; Glazkov (2009).
115 Kravets (2006).
116 'OOO SLAVNEFT-KRASNOYARSKNEFTEGAZ', www.slavneft.ru/eng/company/geography/krasnoyarsknefnegaz/.
117 Chernyshov (2007).
118 Kirillova (2008a).
119 Henderson (2011a).
120 Bradshaw (2010).
121 VNIGRI (All Russia Petroleum Research Exploration Institute), www.vnigri.spb.ru/en/about/history.php.
122 Grigorenko (2004).
123 'Exxon Neftegaz Ltd. Phases and Facilities', www.sakhalin-1.com/Sakhalin/Russia-English/Upstream/about_phases.aspx; 'Sakhalin-1 good example of Russia-US Pacific Partnership-view', http://pda.itar-tass.com/en/c154/185649.html.
124 Baidashin (2004a).
125 Gaiduk (2008c).
126 Baidashin (2006e).
127 Gaiduk (2008c).
128 'Chayvo Phase 1', www.sakhalin-1.com/Sakhalin/Russia-English/Upstream/about_phases_chayvo1.aspx.
129 For the SOKOL Assay, www.exxonmobil.com/apps/crude_oil/crudes/mn_sokol.html.
130 Demongeot (2006). A dedicated fleet of double-hulled Aframax tankers

carrying up to 100,000 tonnes (720,000 barrels) of crude was used for year-round export of crude oil from De-Kastri terminal to world markets. See Sakhalin-1 Project Fact Sheet July 2010, www.sakhalin1.ru/Sakhalin/Russia-English/Upstream/Files/facts_ENG.pdf.

131 Topham (2010); 'Sakhalin 1 eyes output boost', www.upstreamonline.com/live/article215941.ece.

132 'Sakhalin Energy and the Sakhalin II Project', http://qa.sakhalin-2.com/docs/FactLists/English/Sakhalin-2.pdf; Historical Background, http://qa.sakhalin-2.com/docs/FactLists/English/History.pdf.

133 For Trans-Sakhalin pipeline system, please visit http://qa.sakhalin-2.com/docs/FactLists/English/Pipeline.pdf.

134 Sakhalin Energy, www.sakhalinenergy.com/en/project.asp?p=paa_platform; http://www.hydrocarbons-technology.com/projects/sakhalin2/; 'Sakhalin II: Shell Oil in Russia', http://cambridgeforecast.wordpress.com/2006/12/12/sakhalin-ii-shell-oil-in-russia/.

135 Motomura (2008).

136 'Sakhalin Energy: Sakhalin-2 project – recent key milestones', www.sakhalinenergy.com/en/ataglance.asp?p=aag_main&s=1.

137 Sakhalin Energy: Offshore production records, http://qa.sakhalin-2.com/en/default.asp?p=channel&c=3&n=370.

138 *Reuters*, 19 May 2009 (http://in.reuters.com/article/idINSP46847820090519).

139 'Rosneft: Sakhalin-3', www.rosneft.com/Upstream/Exploration/russia_far_east/sakhalin-3/.

140 'Rosneft: Sakhalin-3', www.rosneft.com/Upstream/Exploration/russia_far_east/sakhalin-3/.

141 'Russia's Putin offers surprise deal to Shell', http://uk.reuters.com/article/2009/06/27/shell-russia-idUKLR10355520090627.

142 Baidashin (2007b).

143 *Reuters*, 12 February, 2008 (www.reuters.com/article/rbssEnergyNews/idUSL1213095520080212).

144 *Nefte Compass*, 4 March 2009.

145 'Rosneft: Sakhalin-5', www.rosneft.com/Upstream/Exploration/russia_far_east/sakhalin-5/.

146 Gaiduk (2008c).

147 Baidashin (2007d).

148 *Ibid.*

149 'Urals Energy: Petrosakh', www.uralsenergy.com/ops_petrosakh.htm; 'Interview With Mr Yury V. Motovilov, General Director of "Petrosakh"' JSC www.winne.com/topinterviews/motovilov.htm.

150 Blagov (2004).

151 Mainwaring (2011).

152 'Energy profile of Sakhalin Island, Russia, www.eoearth.org/article/Energy_profile_of_Sakhalin_Island,_Russia.

153 Capital investment to develop the Russian oil sector between 2009 and 2030 will total US$609–625 billion at 2007 prices, of which US$162–165

bn will be in the first phase up to 2020, US$134–139 bn in the second phase (to 2020–2) and US$313–321 bn in the third phase, ending in 2030. Of the US$609–625 bn, US$491–501 bn will be allocated to production and geological exploration, and US$47–50 bn to the refining sector. See *Russia & CIS OGW*, 26 November–2 December, 2009, 6–8; Gaiduk (2009c).

154 In early 2011 Rosneft announced the discoveries of two new oil & gas fields at the Sanarsky and Preobrazhensky license areas in East Siberia. Both areas are located in Katangsky District of Irkutsk Region. Preliminary data suggest that the Sanarskoye and Lisovskoye fields each have initial recoverable reserves of about 80 mt of oil ('Rosneft discovers two new fields in East Siberia', www.rosneft.com/news/news_in_press/14022011. html).

155 The Russian oil production forecasts to 2020 as per the Yamal–Krasnoyarsk programme differed from those in the Energy Strategy to 2030. The regional programme projected 550 mt, while the national strategy targets 505–525 mt by 2020, and 530–535 mt by 2030. See *Russia & CIS OGW*, 12–18 August 2010, 8.

156 Vygon (2009).

157 According to data shown to the author by Renaissance Capital, Taas Yuryakh field in Sakha Republic – with two adjoining licenses to develop the oil and gas reserves of the Field named Central Block License Area and Kurungsky License Area – has massive proven oil and gas reserves. By western classification, 2P oil reserves 980 million barrels and 2P gas reserves 2.966 tcf, and by Russian classification, C1+C2 oil reserves 131 mt or 951 million barrels, and C1+C2 gas reserves 140.2 bcm or 4.956 tcf. The peak production by 2017 is projected to be 130,000 b/d or more. If this projection materializes, the figure of 50 mt/y before 2020 will be easily fulfilled.

Chapter 3

1   Gazprom Production, www. gazprom.com/production/; also Kryukov and Moe (1996).
2   Stern (2005, 144–5).
3   Fjaetoft (2009, 16).
4   Stern (2005, 33).
5   Stern (2009).
6   *Ibid.*, 56.
7   *Ibid.*, 56–7.
8   *Ibid.*, 61. For the update of the main upstream projects in Tyumen region, see Mitrova (2011).
9   Konovalov (2008).
10   *Ibid.*
11   Stern (2009, 54–92).

12 Henderson (2010).
13 Paik (1995, 155–6)
14 For a chronology of Gazprom's Asia Policy see Chapter 7.
15 *Ibid.*
16 Stern and Bradshaw (2008, 250).
17 Chernyshov (2005a).
18 'Gazprom and CNPC sign Agreement on major terms and conditions for Russian gas supply to China', www.gazprom.com/press/news/2009/december/article73718/
19 'Eastern Gas Program', www.gazprom.com/production/projects/east-program/
20 *IPR*, 30 May–5 June 2003. The Energy Strategy 2020 was based on two versions of socio-economic development: conservative and optimistic. Each included the basic provisions of the draft programme for socio-economic development for the mid-term. They both also allowed for the development of the economy under favourable (something between the two base versions) and critical scenarios. The overall increase in energy consumption by 2020 was projected to be 25.4–38.4%. See Kalashnikov (2004).
21 The gas price for industrial consumers in the Far East and East Siberia after 2011 would equal the market price, about $100 per 1,000 cubic metres. See Gaiduk (2007b).
22 Stern and Bradshaw (2008, 249–54); Baidashin (2007e); *Russia & CIS OGW*, 6–12 September 2007, 5–7; www.minprom.gov.ru/docs/order/87 (in Russian) quoted in Energy Charter Secretariat, 'Fostering LNG Trade: Role of the Energy Charter', 2008, 122 (footnote no. 76); 'Eastern Gas Program', www.gazprom.com/production/projects/east-program/.
23 According to the MIE, general East Russian gas resources amount to over 67 tcm, of which 52.4 tcm are onshore and the remaining 15 tcm from the shelf. However only 8% of the resources were explored. The programme required some US$94 billion investment. See *Ibid.*
24 'Board of Directors addresses Eastern Gas Program execution', www.gazprom.com/press/news/2010/february/article76034/.
25 Russia & CIS OGW, 2–8 October 2008, 4–7.
26 *Russia & CIS OGW*, 2–8 October 2008, 4–7.
27 *Russian & CIS OGW*, 27 August–2 September 2009, 16–19. On 13 November 2009 Energy Strategy of Russia: for the period up to 2030 was officially announced by government order (no. 1715-r): The Ministry of Energy, Russian Federation 2010.
28 *Russia & CIS OGW*, 26 November–2 December 2009, 6–8.
29 The forecast for Russia's oil production in 2030, in the government's Energy Strategy, for the next 20 years has been lowered by 6.6%. Oil production was projected to total 530–535 million tonnes in 2030 in the latest draft of the Energy Strategy, down from the previous forecast of 540–600 million tonnes. See Russia & CIS *OGW*, 27 August–2 September 2009, 16–19.
30 Stern and Bradshaw (2008, 246); *CD*, 11 July 2007.

31   *Ibid.*; Eastern Gas Program, www.gazprom.com/production/projects/east-program/

32   'Gazprom proceeds with consistent and phased gasification of Kamchatka', www.gazprom.com/press/news/2010/december/article106926/; 'Over 90 per cent of Sakhalin–Khabarovsk–Vladivostok GTS linear part welded up', www.gazprom.com/press/news/2011/april/article111945/. Design and exploration work was underway with the aim of constructing two inter-settlement gas pipelines for gas supply to CHPP-1, a regional hospital, the 11th kilometre boiler plant No.1, and other facilities of Petropavlovsk-Kamchatsky. Moreover, design and exploration work was initiated to construct an inter-settlement gas pipeline to the town of Yelizovo, including a gas distribution station, and further on to the settlements of Dvurechye, Krasny, Nagorny, Novy, Pionersky, Svetly, and Krutoberegovy (Yelizovo District). 'Gazprom to launch construction of several inter-settlement gas pipelines in Kamchatka Krai this year', www.gazprom.com/press/news/2011/january/article108220/

33   Cited in http://www.gazprom.com/eng/news/2008/08/30425.shtml (address no longer functional).

34   Stern and Bradshaw (2008). On 20 September 2006 in Irkutsk, Gazprom deputy CEO Ananenkov and head of Itera board, Vladimir Makeyev, signed a MOU on joint actions to provide gas supply to the Irkutsk region. At issue was the development of Bratskoye field. Itera planned to invest about 2 billion rubles ($74.2 million) in the field and expected to supply up to 100 mcm of gas to Gazprom in 2008. The amount was planned to increase to 400 mcm/y from 2009. In turn Gazprom promised to build a gas pipeline by 2008 to pump gas to consumers. See Kirillova (2006).

35   'Gazprom and Transbaikal Krai Government sign Accord on Gasification',www.gazprom.com/press/news/2010/april/article95813/

36   In December 2007 the first phase of the Bratsk gas and condensate field–Bratsk (45th residential district) gas trunk line was constructed and natural gas deliveries to the city commenced. Bratsk was the first population centre of the Irkutsk Oblast to receive pipeline gas. Between 2007 and 2009 Gazprom allocated 595 million rubles for regional gasification. These funds were used to construct the Osinvoka GDS–Zyaba settlement and the Osinvoka GDS–Gidrostroitel settlement gas pipelines. 'Gazprom and Irkutsk Oblast Government sign Agreement of Cooperation and consider regional gasification prospects', www.gazprom.com/press/news/2010/october/article104444/

37   Miller pointed out that Gazprom had already selected some top-priority facilities in Buryatia for gasification with LPG. These are an 8,000 tonne filling station planned for construction in Ulan-Ude and nine boiler houses in the settlements of Ivolginsk, Sotnikovo, Tarbagatai, Sharaldai, Mukhorshibir, and Dolga. See 'Signing 2010–2013 Prioritized Action Plan on gas supply to and gasification of Buryatia', www.gazprom.com/press/news/2010/october/article104390/

38   Stern and Bradshaw (2008, 247).

39 'Gazprom and Republic of Sakha (Yakutia) sign Partnership Agreement', www.gazprom.com/press/news/2010/december/article106978/; 'Gazprom and Yakutian Government to sign Cooperation Agreement on socioeconomic development', www.gazprom.com/press/news/2010/october/article104732/

40 Stern and Bradshaw (2008, 247–9).

41 'Eastern Gas Program', www.gazprom.com/production/projects/east-program/; 'Gazprom meets all obligations under Russian Regions Gasification Program 2010' www.gazprom.com/press/news/2011/april/article111478/.

42 'Gazprom commissions first start-up complex of Sakhalin–Khabarovsk–Vladivostok GTS', www.gazprom.com/press/news/2011/september/article118764/; 'The "Sakhalin–Khabarovsk–Vladivostok" gas pipeline', www.ngsms.ru/eng/projects/page531/

43 *OGJ Online*, 25 June 2009. www.ogj.com/index/article-display/0889754181/s-articles/s-oil-gas-journal/s-transportation/s-pipelines/s-constuction/s-articles/s-gazprom_-kogas_sign.html

44 'Gazprom and Kogas sign Agreement to jointly explore gas supply project', www.gazprom.com/press/news/2009/june/article66607/

45 Nezhina (2009b, 45); 'Gazprom finishes construction of top-priority facilities for Sakhalin Oblast gasification', www.gazprom.com/press/news/2011/march/article109851/.

46 In 2009, the figures were 75.4 bcm and 8.6 mt respectively. See 'Gazprom commences drilling in Kirinskoye field offshore Sakhalin', www.gazprom.com/press/news/2009/july/article66730/

47 'Over 90 per cent of Sakhalin–Khabarovsk–Vladivostok GTS linear part welded up', www.gazprom.com/press/news/2011/april/article111945/

48 'Gazprom joins Investment Agreement on South Yakutia Comprehensive Development', www.gazprom.com/press/news/2009/june/article64577/

49 The document was also signed by heads of the Russian Energy Ministry, Federal Road Agency, Federal Railway Transport Agency, the Sakha (Yakutia) Government, Rosatom State Corporation, South Yakutia Development Corporation, Atomredmetzoloto, South Yakutia Hydropower Complex, Alrosa Investment Group, Yakutia Coal–New Technologies, Elkonsky Mining and Metallurgy Plant, and RusHydro (www.gazprom.com/press/news/2009/june/article64577/).

50 For a closer understanding of the Kovykta project, see Simonia (2004), Poussenkova (2007b), Ahn and Jones (2008), Stern and Bradshaw (2008).

51 Gaiduk (2007a).

52 *Ibid.*

53 *IPR*, 9–15 January 1998, 10.

54 *China OGP*, 15 May 2002, 1–3.

55 Gaiduk (2007a).

56 *RPI*, June/July 2002, 14–19. The Chinese side stated it was ready to purchase gas for $20–25 per 1,000 cubic metres on the Russia–China border, and the Russian side declared it was prepared to sell gas for US$75

as minimum. See *RPI*, March 2003, 22–8. At that time, the Chinese side said the city gate price of the pipeline would range between 0.7–1.0 yuan/cm. See Li Yuling (2002d, 9).

57   *RPI*, June/July 2002, 14–19.
58   *RPI*, March 2003, 22–28.
59   Kroutikhin (2004).
60   Chernyshov (2004a).
61   Barges (2004a).
62   VSGK signed memoranda of intent with several key potential consumers in Irkutsk Region, such as Sayanskkhimplast, Irkutskteploenergo, the Scientific and Production Complex Irkutsk, and the Angarsk Cement Plant. Alongside this, VSGK hoped to sign similar memoranda with Usolyekhimprom, Irkutskenergo, the Angarsk Petrochemical Company, and Angarsk Polymer Plant. The final price of gas would not exceed $40–55/1,000m^3, according to Alexei Sobol, head of VSGK. See Chernyshov (2005c); Glazkov (2004b).
63   Chernyshov (2004f); Lukin (2005c).
64   In Russia, there had been only one plant specializing in helium production. It was located in Orenburg and produced 5% of the global volume of helium output. See Chernyshov (2004f).
65   *Ibid.*
66   *Ibid.*
67   The Russian Academy of Science predicted that helium demand in the period to 2030 will be soaring, on average by 4–6% annually. As a result, total consumption will reach approximately 225 mcm by 2030. See Kokhanovskaya (2006).
68   Author's correspondence with Masumi Motomura in May 2011. For interesting articles on the value of helium see 'Price shocks waiting as US abandons helium business', http://arstechnica.com/science/news/2010/07/science-policy-gone-bad-may-mean-the-end-of-earths-helium.ars; 'Scientists say Earth's helium reserves "will run out" within 25 years', www.dailymail.co.uk/sciencetech/article-1305386/Earths-helium-reserves-run-25-years.html.
69   Kirillova (2006).
70   Lukin (2005c).
71   Kirillova (2006).
72   *Ibid.*; Gaiduk (2007a).
73   *Ibid.*
74   *Ibid.*
75   Belton (2007); Crooks (2007); Faucon and Smith (2007); Slavinskaya (2008a).
76   *Russia & CIS OGW*, 4–10 September 2008, 12.
77   *Russia & CIS OGW*, 9–15 October 2008, 6.
78   *Russia & CIS OGW*, 16–22 October 2008, 8.
79   *Russia & CIS OGW*, 23–9 October 2008, 13.
80   *Ibid.*, 13–14.

81   *Ibid.*, 14–15.
82   *Russian & CIS OGW*, 27 August–2 September 2009, 31.
83   *Ria Novosti*, 28 September, 2009
84   *Ria Novosti*, 23 March, 2010 ('TNK-BP to sell Kovykta gas field for $700–900 million', http://en.rian.ru/business/20100323/158289323.html).
85   Belton (2010); Dow Jones Deutschland (2010).
86   Soldatkin and Akin (2011).
87   *Russia & CIS OGW*, 10–16 March 2011, 39.
88   www.sakhalin1.ru/Sakhalin/Russia-English/Upstream/default.aspx. For the background to this development, see Paik (1995, 211–14); Stern (2002).
89   The second generation projects relate to the blocks tendered in 1993 that are collectively known as Sakhalin-3. The third generation of projects represents the new generation. These are exploration projects being developed on the basis of normal tax and royalty, rather than a PSA. See Stern and Bradshaw (2008).
90   Stern and Bradshaw (2008, 236–9); Bradshaw (2010).
91   Ebina (2006); Japan Pipeline Development & Operation website www.jpdo.co.jp/eprofile.html.
92   *Ibid.*
93   *IPR*, 10–16 June 2004.
94   *China ERW*, 26 June–1 July 2004, 7–8.
95   Tsui and Pilling (2004); *China OGP*, 15 November 2004; *China ERW*, 8–18 February 2005; 'China Joins the Battle for Sakhalin', www.kommersant.com/p521873/r_1/China_Joins_the_Battle_for_Sakhalin_/(3 November 2004).
96   According to the various scenarios of the Energy Strategy endorsed by the government in September 2003, domestic gas prices were predicted to hold at $40–41 per 1,000 cubic metres prior to 2006 and move up to $59–64 in 2010 (without VAT, pipeline transport tariffs for gas distribution to consumers, and marketing costs). See Glazkov (2004d).
97   Pilling and Tsui (2004).
98   Japan Pipeline Development & Operation: Sakhalin – Japan Natural Gas Pipeline Project, www.jpdo.co.jp/eprofile.html.
99   The construction of a natural gas pipeline from Sakhalin-1 has been discussed for many years. Hokkaido Takushoku Bank, Hokkaido Electric Company, and Tomato Development Public Co. undertook the first feasibility study jointly in 1974. The Ministry of Foreign Affairs undertook a further study in 1979. To realize the project, the Hokkaido Chamber of Commerce founded JPDO in 1998. The plan was to introduce natural gas from the Sakhalin-3 field (PegaStar) to Japan, in cooperation with Texaco (US, now part of Chevron), taking primarily an onshore route. See Ebina (2006).
100 Baidashin (2006e).
101 *Ria Novosti* 2006.
102 Gaiduk (2008c).
103 'Exxon Neftegaz: Sakhalin-1 Project homepage', www.sakhalin1.ru/

Sakhalin/Russia-English/Upstream/default.aspx; Slavinskaya (2008a); *Dow Jones News Service*, 23 October 2006; *WSJ*, October 24, 2006.

104 JSC Daltransgaz was formed for this project in 2000, and the programme began to be financed only in 2002. As of 2004, the company shareholders included Khabarovsk Territory (47.24%), Russian Federation (27.63%), Rosneft (14.32%), Rosneft–Sakhalinmorneftegaz (10.68%), and Primorsk Territory (0.13%). See Glazkov (2004d).

105 Gaiduk (2008c).

106 *Russia & CIS OGW*, 2–8 October 2008, 39.

107 'Gazprom launches construction of Sakhalin–Khabarovsk–Vladivostok gas transmission system', www.gazprom.com/press/news/2009/july/article66851/

108 Bryanski (2009a).

109 *Ria Novosti*, October 2010 ('Gazprom in talks on swap of Sakhalin-1 stake for LNG supplies to India', http://en.rian.ru/ business /20101025/161079033.html).

110 'Sakhalin Energy: Sakhalin-2 Recent Key Milestones Achieved', www.sakhalinenergy.com/en/ataglance.asp?p=aag_main&s=1. For the background of Sakhalin-2 project, see Paik (1995, 211–14); Stern (2002); Stern and Bradshaw (2008, 233–6); Baidashin (2005b).

111 Glazkov (2006c).

112 Rutledge (2004); 'Sakhalin Energy: Sakhalin-2 Project: Key Milestones', www.sakhalinenergy.ru/en/aboutus.asp?p=key_milestones

113 Glazkov (2007b).

114 Baidashin (2006e).

115 Baidashin (2006b).

116 Baidashin (2006e).

117 Ostrovsky (2006a); Ostrovsky (2006b); Ostrovsky (2006c); Ostrovsky and Buckley (2006); Crooks (2006b).

118 Glazkov (2007b).

119 Bradshaw (2010); Stern and Bradshaw (2008, 233–6).

120 Gaiduk (2007a).

121 Slavinskaya (2008b).

122 Nezhina (2009a).

123 Crooks (2009); *OGJ Online*, 31 March 2009. ('Sakhalin Energy exports first LNG cargo to Japan', www.ogj.com/display_article/357829/7/ARTCL/none/none/Sakhalin-Energy-exports-first-LNG-cargo-to-Japan/?dcmp=OGJ.Daily.Update).

124 'Sakhalin Energy: Sakhalin-2 Recent Key Milestones achieved', www.sakhalinenergy.com/en/ataglance.asp?p=aag_main&s=1

125 Glazkov (2006c); Chernyshov (2004c).

126 *Ibid.*

127 Baidashin (2005d).

128 Slavinskaya (2008b) (See in particular Table on the forecast for Russian LNG Production, 20).

129 Baidashin (2005d).

130 Gaiduk (2008c).

131 Gaiduk (2007a).

132 *Russia & CIS OGW*, 2–8 October 2008, 32.

133 Nezhina (2009a); 'Gazprom commences drilling in Kirinskoye field offshore Sakhalin',www.gazprom.com/press/news/2009/july/article66730/.

134 'Gazprom commences drilling in Kirinskoye field offshore Sakhalin', www.gazprom.com/press/news/2009/july/article66730/.

135 'UPDATE 1-Russia's Gazprom accelerates Sakhalin-3', project', www.reuters.com/article/marketsNews/idUSLG30145020091016.

136 'Gazprom discovers new field in Kirinsky block', www.gazprom.com/press/news/2010/september/article103039/; 'Sakhalin III: Strategy',www.gazprom.com/production/projects/deposits/sakhalin3/.

137 Gas production at Sakhalin-2 started from 2009 and will peak at 22.1 bcm per year, with 33 wells to be drilled. That of Sakhalin 1 was expected to peak at 11.4 bcm per year with 44 wells to be drilled. See 'Russia could boost gas production 34–50% by 2030', *Russia & CIS OGW*, 2–8 October 2008, 4–7.

138 In May 2008, RPN threatened to withdraw Rosneft's license for the Veninsky block. See Gaiduk (2008b); Baidashin (2005d).

139 Chernyshov (2004b).

140 Glazkov (2004c).

141 Gaiduk (2008c); Crooks (2006a).

142 *Russian & CIS OGW*, 27 August–2 September 2009, 14–15.

143 Glazkov (2006c).

144 Gaiduk (2008c).

145 *RPI*, August 2007, 20.

146 Stern (2002, 247); 'A1 Projects A1 Investment Company', www.a-1.com/en/project/petr_sah/; 'Urals Energy Petrosakh', www.uralsenergy.com/ops_petrosakh.htm

147 Paik (1995, 221–35).

148 Kempton (1996).

149 Larionov (1995).

150 Quan and Paik (1998, 110).

151 *RPI*, May 1988, 52–6.

152 JBIC (2005, 230–2).

153 *Ibid.*, 231 & 284 (Map 3)

154 Glazkov (2003).

155 Slavinskaya (2008a).

156 Yukos paid US$38 million for a new issue of Sakhaneftegaz shares in December 2001, and bought 10% equity on the open market. The Sakha Republic government held 38% of equity. See *RPI*, April 2002, 20–22.

157 Gaiduk and Kirillova (2007).

158 *Ibid.*

159 Chernyshov (2004d).

160 Gaiduk (2008b).

161 *Ibid.*

162 The first phase provides for: supplies to the city of Petropavlovsk-Kam-chatsk, the Nizhne–Kvakchinskoye gas condensate deposit, an automated gas distribution station of Petropavlovsk–Kamchatsky gas main, and pre-development of the Kshukskoye and Nizhne-Kvakchinskoye gas and condensate deposits.

163 Gazprom could not receive the license for the Chayandinskoye deposit earlier than the end of June 2008, owing to the required calculation for compensatory payment to the budget for subsoil use. See Gaiduk (2008b).

164 Slavinskaya (2008a).

165 Gaiduk (2008b).

166 'On first session of Joint Working Group between Gazprom and Govern-ment of Sakha Republic (Yakutia)', www.gazprom.com/press/news/2007/october/article63987/

167 Present at the meeting were representatives of Gazprom's subsidiaries – VNIIGAZ, Promgaz, Mezhregiongaz, Sibur, Gazprom invest Vostok, Gazpromregiongaz, Gazprom pererabotka, Regiongazholding, as well as Yuzhno-Yakutskaya Corporation, and the Regional Policy Institute.

168 'On working meeting between Alexander Ananenkov and Gennady Ale-kseev', www.gazprom.com/press/news/2008/may/article64183/

169 *RPI*, April 2008, 26.

170 'Gazprom Board of Directors addresses development of mineral resource base necessary for gasification of East Siberia and Far East', www.gazprom.com/press/news/2008/november/article64410/.

171 Miller stated: 'Oil companies operating in the region have obtained tax allowances. The identical measures should be taken to ensure efficient gas fields development in eastern Russia. Therefore, it is crucial to define, as early as in 2009, the measures of state support for gas investment projects being executed in East Siberia and the Far East. These may include, inter alia, "tax holidays" over the project's payback period and reduction/abolition of export tariffs on extracted gas'. See 'Alexey Miller takes part in meeting on Yakutia's socio-economic development', www.gazprom.com/press/news/2009/august/article66878/

172 'Meeting of Gazprom and Yakutia's Government held to address Yakutia gas production center creation', www.gazprom.com/press/news/2010/march/article85178/; Glazkov (2010); 'On meeting dedicated to Yaku-tia Gas Production Center development', www.gazprom.com/press/news/2010/may/article98800/.

173 'Development plan for Yakutian Chayanda field approved', www.gazprom.com/press/news/2010/july/article101354/

174 A. N. Dmitrievsky, director of the Institute of Oil and Gas Problems RAS, was the head of the working group on the determination of long term strategic priorities and major policy measures to develop the gas industry, for the Ministry of Industry and Energy's work (with Decree dated 21 December 2006, N. 413) entitled 'On a refinement of the energy strategy of Russia for the period up to 2020 and its prolongation up to 2030'.

175 The comparison of this table with the gas reserves in the Sakha Republic

as of 1995 will show the level of increase of the proven reserves in the Republic. See Paik (1995, 224–5).

Chapter 4

1   Berrah et al. (2007, 37–43); IEA (2007, 263).
2   Berrah et al. (2007, 37–43).
3   IEA (2010, 602).
4   This was announced on 20 July 2010 ('China overtakes the United States to become world's largest energy consumer', www.iea.org/index_info. asp?id=1479); Swartz and Oster (2010). China's National Bureau of Statistics has reported that energy consumption in 2009 stood at 3.1 bt of standard coal equivalent, which was equal to 2.132 bt of oil equivalent ('China dismisses IEA analysis of it being world's top energy user', www. chinadaily.com.cn/bizchina/2010-07/20/content_11025333.htm).
5   'China 2010 Energy Consumption Rises 5.9%, National Statistics Bureau Says', www.bloomberg.com/news/2011-02-28/china-2010-energy-consumption-rises-5-9-percent-statistics-bureau-says.html
6   According to the NDRC, in 2010 the largest coal exporter to China was Indonesia, followed by Australia, Vietnam, Mongolia, and Russia. These five accounted for 84% of coal imports. See Zhao Tingting (2011).
7   These figures are originally from the report by Citigroup Inc., as reported by Bloomberg. See *China Daily* 14 December 2010 ('China's net coal imports likely to hit 230m tons in 2011', www.chinadaily.com.cn/business/2010-12/14/content_11700418.htm).
8   There are big differences in the trajectory of oil demand between the three scenarios (current policies scenarios, new policies scenarios, and 450 scenarios). See IEA (2010, 105) (Table 3.2).
9   'China Meets Energy Consumption Target for 2010', www.china-briefing. com/news/2011/01/19/china-meets-energy-consumption-target-for-2010. html
10 Swartz and Oster (2010).
11 China's energy security suffers from glaring vulnerabilities due to its excessive dependence on one particular bottlenecked sea-lane, namely the Malacca Strait. China is dependent on at least four sea-lane routes in south-east Asia. For the details, see Zhang Xuegang (2008); Blair, Chen, and Hagt (2006).
12 Department of Communications and Energy (1997).
13 Meidan et al. (2007, 33–63).
14 *Ibid.* 53 and 81–5.
15 Mai Tian, 'Power Panel', *CD Business*, 6–12 June 2005.
16 Chen Wenxian (2005c, 5–6); Yang Liu (2006a).
17 Qiu Jun (2006b).
18 By 2006, coal reserves stood at 1034.5 bt, and the remaining verified

exploitable reserves accounted for 13% of the world total, making China third in the world. In the same year, the theoretical reserves of hydropower resources were equal to 6,190 billion kwh, and economically exploitable annual power output was 1,760 billion kwh, equivalent to 12% of global hydropower resources, making China top in the world. See Zhu Zhu (2008).

19 Bradsher (2009).
20 Chan Y. (2009).
21 Zhu Zhu (2008).
22 *Ibid.*
23 *Ibid.*
24 Downs (2007, 67).
25 Qiu Jun (2006b).
26 Chen and Graham-Harrison (2008, 1–3).
27 Zhang Guobao started to preside over energy work in NDRC in 1999. He promoted such big projects as the WEP, the Sino-Turkmenistan natural gas project, and LNG import projects. See Lin Fanjing (2008d).
28 In comparison, the US Department of Energy's employees number more than 14,000. See Lin Fanjing (2008d).
29 *China OGP*, 1 August 2008, 3.
30 *China ERW*, 24–30 July 2008, 4.
31 Qiu Jun (2010b).
32 *China OGP*, 1 August 1998, 1–3. In May 2008, SINOEPC formed the Oilfield E&P Department in order to establish integrated management of upstream operations in China; personnel came from the management staff of SINOPEC Exploration company, the Marine Origin Work Department, and the existing Oilfield E&P Department. In 2007, SINOPEC established SINOPEC Exploration Company to accelerate its upstream business development. See *China OGP*, 1 July 2008, 8.
33 The firm gave priority to resource strategy. SINOPEC put forward plans to reorganize its upstream units in north-east China, covering 39 exploring blocks scattered in the four regions of Songliao, Sanjiang, Bohai-rim, and Dunhua with a total area of about 120,000 square kilometres. In order to facilitate integration of upstream resources in this region, the Group decided to set up a new Northeast Company instead of the former Northeast Petroleum Bureau, Northern Exploit Company, and Jilin Project Department. See Chen Dongyi (2008b).
34 Ma Xin (2008).
35 Liu Yanan (2008b, 9).
36 US Energy Information Administration, China Brief, www.eia.doe.gov/emeu/cabs/China/Oil.html.
37 Liu Yanan (2008c); Qiu Jun (2008); Qiu Jun and Liu Shuyun (2006).
38 China International United Petroleum & Chemicals Co., Ltd. (UNIPEC), http://english.sinopec.com/about_sinopec/subsidiaries/subsidiaries_joint_ventures/20080326/3083.shtml.
39 Chinaoil (USA) Inc., http://chinaoilusa.com/index.html.

40   Sinochem Group, www.sinochem.com/english/tabid/640/Default.aspx.

41   Zhuhai Zhenrong Company, http://companies.china.org.cn/trade/company/338.html.

42   Li Xiaoming (2004, 1–2).

43   The firm is a wholly-owned unit of China North Industries Group Corporation (CNIGC or Norinco), China's largest arms manufacturer. See Qiu Jun (2010d); China Zhenhua Oil Co. Ltd., (ZhenHua Oil) www.zhenhuaoil.com/en-gk-zhc.htm.

44   Zhang Qiang (2005).

45   SINOPEC's Shengli field has maintained crude oil production above 27 mt/y for 15 straight years. See *China OGP*, 15 September 2010, 24.

46   Qiu Jun (2006d).

47   On 18 August 2008, according to the Chinese government study, China's total oil and gas reserves were 108.6 billion tonnes and 56.0 tcm respectively, of which the recoverable oil and gas reserves stand at an estimated 21.2 billion tonnes and 22.0 tcm respectively. These figures were somewhat different from those of 2006. China conducted its first national oil and natural gas resources evaluation between 1981 and 1987, and its second between 1991and 1994. The second study found that China has total oil and natural gas reserves of 94 billion tonnes and 38.04 tcm respectively. See *China ERW*, 14–20 August 2008, 12.

48   Quan Lan (2000d).

49   Yang Liu (2006c).

50   *Ibid.*

51   This is the most up to date data the author has found.

52   *China OGP*, 1 August 2008, 13.

53   *China OGP*, 15 August 2010, 19.

54   Kambara and Howe (2007, 48–51).

55   *China OGP*, 1 September 2006, 12.

56   In east China where the production wells are generally no deeper than 3,500 metres, the single oil well output averages 8.7 tonnes per day. In Tarim Basin, where the depth of wildcats easily exceeds 5,500 metres, the average single well output amounts to 70 tonnes per day. Out of cost concern, however, PetroChina now limits its wells to less than 6,000 metres despite the possible higher output. See Quan Lan (2001j).

57   *China OGP*, 15 April 2004 15.

58   According to its general manager Wang Yufu, Daqing Oilfield has a target of improving its recovery ratio in the secondary layer. Only about 100 mt of recoverable reserves are left in the primary layer. In 2007, Daqing oil field produced 41.62 mt of oil, representing 22% of China's total. See *China OGP*, 1 August 2008, 16; *CD*, 7 April 2009 ('Daqing to maintain crude output', www.china.org.cn/business/2009-04/07/content_17561729.htm)

59   'Changqing Oil & Gas Province', www.cnpc.com.cn/en/aboutcnpc/ourbusinesses/explorationproduction/operatediol/Changqing_Oil_and_Gas_Province.htm

60   'PetroChina's Changqing produces record oil, gas output in Q1', www.chinamining.org/Companies/2011-04-11/1302505870d44443.html.

61   In the case of natural gas, production in 2015 should be up to 32 bcm/y ('Changqing oil field becomes China's 2nd largest onshore oil-gas field', http://english.peopledaily.com.cn/90001/90778/90860/6849245.html); 'PetroChina's Changqing oil field to boost production', www.chinadaily.com.cn/bizchina/2008-12/04/content_7272595.htm; *China ERW*, 17–23 July 2008, 10; *China OGP*, 1 August 2008, 16; *China OGP*, 15 September 2009, 17; *China OGP*, 15 January 2010, 13–14.

62   US Energy Information Administration, Country Analysis Brief, China, www.eia.doe.gov/emeu/cabs/China/Oil.html.

63   Kambara and Howe (2007, 81–88).

64   *China EW*, 15–21 July 2010, 10.

65   *China OGP*, 15 April 2009, 18.

66   *China OGP*, 15 February 2009, 33. Tahe oil field planned to discover 475 million tonnes equivalent of tertiary oil reserves in 2009, including 100 mt of proven oil reserves, 150 mt of controllable oil reserves, and 30 bcm of gas. See *China OGP*, 15 March 2009, 29.

67   *China OGP*, 1 September 2010, 24.

68   US Energy Information Administration, Country Analysis Brief, China, www.eia.doe.gov/emeu/cabs/China/Oil.html.

69   Chen Wenxian (2007a).

70   PetroChina's Nanpu discovery was thought to hold up to 11 bn barrels of oil which would make it the biggest oil find in China for decades. See Crooks and Kwong (2007). However, there has been one report suggesting that the discovery may not be as big as projected ('China: PetroChina's Jidong Nanpu oil field smaller than originally thought', www.energy-pedia.com/article.aspx?articleid=140523).

71   Yang Liu (2007); 'PetroChina Company Limited announces today that it has discovered a large oilfield with geological oil reserves reaching 1,020 million tonnes at the region of Jidong tidal and shallow water areas of Bohai Bay–Jidong Nanpu Oilfield', www.petrochina.com.cn/resource/EngPdf/BulletinBoard/gg070503e1830.pdf. Crooks (2009a).

72   Yang Liu (2007).

73   *Ibid.*

74   The PL 19-3 oil field is located in Block 11/05 of Bohai Bay. See *China ERW*, 11–17 September 2008, 6; *China OGP* 1 October 2008, 15.

75   'Penglai field to reach peak production', www.chinadaily.com.cn/bizchina/2011-05/20/content_12548640.htm.

76   'Oil spill in China's Bohai Sea rises to 2,500 barrels', www.chinadaily.com.cn/bizchina/2011-08/12/content_13104070.htm; 'Oil spill reaches beaches', www.chinadaily.com.cn/usa/epaper/2011-07/21/content_12952238.htm. The reduction of 17 000 b/d was predicted ('Watchdog deems oil leak in bay a "disaster"', www.chinadaily.com.cn/usa/epaper/2011-07/15/content_12910805.htm).

77   This paragraph is based on material in the US Energy Information

Administration's website, www.eia.doe.gov/.

78   In 2007, CNOOC's production totalled 372 000 b/d, about 37% of it coming from the South China Sea developments. See 'US Energy Information Administration, Country Analysis Brief, China', www.eia.doe.gov/emeu/cabs/China/Oil.html. At the end of 2007, CNOOC Ltd made 10 new oil and gas discoveries, including Bozhong 28-2 East, Bozhong 26-3, Jinzhou 25-1, Weizhou 11-7, Weizhou 11-8, Weizhou 6-1 South, Weizhou 11-2, Pan Yu 10-2, Pan Yu 11-5, and Kenli 20-1. The discoveries were made in cooperation with Atlantis Deepwater Technology Holding which possessed the required deep sea technology. These could help CNOOC secure 50 mt/y of crude oil production within China by 2010. See Liu Yanan (2008b, 8).

79   *China OGP*, 15 February 1997, 1–4.

80   Berrah et al. (2007, 26).

81   *Ibid.*; 'World Oil Transit Chokepoints', www.eia.gov/countries/regions-topics.cfm?fips=WOTC.

82   According to a report in the *CSJ*, a 300 000-dwt crude port and 600 000 cubic metres of oil storage will also be built in Myanmar by 2010, while construction 'on oil and gas pipelines' will start in September. The report was referring to the agreement signed in March between China and Myanmar which stated that a 2,800 km, 1.2 bcm gas pipeline will also be laid parallel to the oil pipeline. See *OGJ* online, 19 June 2009. 'China, Myanmar sign oil pipeline agreement', www.ogj.com/articles/2009/06/china--myanmar-sign.html; 'OVL may join Chinese gas pipeline from Myanmar', www.chinadaily.com.cn/bizchina/2009-07/25/content_8561562.htm

83   *China OGP*, 15 September 2009, 17.

84   Zhu Zhu (2007).

85   Berrah et al. (2007, 26).

86   *China OGP*, 15 September 2009, 18.

87   Another projection of China's oil demand in 2020 puts it in the range 590–650 mt (See Liu Xiaoli (2011)).

88   The cargo of stable gas condensate, produced at Novatek's Purovsky gas condensate stabilization plant, travelled from Russia's Murmansk Port to China's Ningbo Port in 22 days, approximately half the time needed using the traditional shipping route through the Suez Canal. See *China EW*, 2–8 September 2010, 11.

89   Jakobson (2010); Ebinger and Zambekakis (2009). In 2008 Litasco total sales to China amounted to 470 000 tonnes of oil and 662 000 tonnes of virgin fuel oil, and from January to June 2009 the oil delivery rose as high as 700 000 tonnes. See 'Lukoil News – 2009', www.oilprimer.com/lukoil-news-2009.html

90   'US Energy Information Administration, China – Background', www.eia.gov/countries/cab.cfm?fips=CH.

91   Sun Huanjie (2006a); Chen Wenxian (2004a).

92   The terminal located in Lanzhou is designed with a storage capacity of 0.5

million cubic metres, including two 150 000 cubic metres and three 100 000 cubic metres oil deposits at the construction site. The western crude oil pipeline will connect Lanzhou with Urumqi, a total distance of 1,878 km, designed to transport 20 mt/y of crude from oilfields in Xinjiang to the Lanzhou refining base. See Qiu Jun (2006r).

93 The Lanzhou–Zhengzhou–Changsha pipeline will pass through Xian and Wuhan, while the Jinzhou–Zhengzhou pipeline will be connected with the Lanzhou–Zhengzhou–Changsha pipeline. The pipeline will consist of one trunk pipeline and two branch pipelines. The total investment for the two pipelines will reach 12 billion yuan. See Lin Fanjing (2006b).

94 *Ibid.*

95 Qiu Jun (2006r).

96 Yang Liu (2005b).

97 Lin Fanjing (2008a).

98 *Ibid.*

99 Zhou Yan (2011). (Note that the original press article contains some important errors, corrected here); 'CNPC to Enlarge Pipeline Network', www.cippe.com.cn/cippeen/html/content_1048.html.

100 The figures from Mr Xu Yihe, China Correspondent, Upstream through e-mail communication (6 May 2011).

101 'CNPC to lay 30,000 km of new gas pipeline by 2015', www.interfax.cn/news/19350.

102 On 28 December 1998, CNPC unveiled the restructuring of its pipeline bureau by acquiring new assets. Headquartered in Langfang, Hebei province, the new bureau directed by Chen Jiqing retaining its previous name of China Oil and Gas Pipeline Bureau (COGPB) took over assets from the former Northwest Pipeline Construction Bureau (Xi'an) and the Natural Gas Transporting Co of Turpan–Hami oilfield (Xinjiang). It has also acquired equity shares CNPC holds in Beijing Huayou Natural Gas Co (Beijing) and Tarim oilfields held in the Tarim Oil and Gas Transportation Co Ltd (Korla). See *China OGP*, vol. 7, no. 1, 1 January 1999, 13.

103 In 2011, SINOPEC owned and operated 7,270 km of oil pipeline; Sinopec, Corporate Social Responsibility, 'Serve the Customers', www.sinopecgroup.com/english/Pages/Servethecustomers.aspx; Lin Fanjing (2009b).

104 Sun Huanjie (2006a).

105 US Energy Information Administration, China – Background, www.eia.gov/countries/cab.cfm?fips=CH. The 2,148 km Lanzhou–Zhengzhou–Changsha oil product pipeline with 10–15 mt/y delivery capacity (and 508–660 mm diameter) covers Gansu, Shanxi, Henan, Hubei, and Hunan provinces and the target operational date was June 2009. See Lin Fanjing (2009b).

106 Liu Yanan (2009).

107 From 1996 to 2006 domestic oil consumption increased by 7.2% annually while domestic crude oil output increased by only 1.9%, per year. See Zhu Zhu (2007).

108 US Energy Information Administration, China – Background, www.eia.

gov/countries/cab.cfm?fips=CH.

109 In 2007, NDRC provided subsidies totalling 42 billion yuan for taxi drivers, low-income families challenged by higher charges for LPG, and farmers affected by increased diesel prices. In 2006, the special fund received 45 billion yuan from oil producers. NDRC said 21 billion yuan was allocated to finance subsidies ('China to collect over 60 billion yuan to finance oil subsidies(12/07/07)', www.china-embassy.org/eng/xw/t387979.htm); Zhu Qiwen (2008).

110 Li Xiaohui (2010d); Lin Fanjing (2010d); *China OGP*, 15 December 2010, 7, Table 2.

111 As of February 2011, only 18 refineries had a capacity of over 10 mt/y.

112 *Ibid.*

113 *Ibid.*

114 Lin Fanjing (2010d).

115 *Ibid.*

116 Qiu Jun (2005j); Qiu Jun (2006b, 13).

117 *China OGP*, 1 November 2006, 10.

118 Academy of Macroeconomic Research, http://60.247.103.213/en/article. asp?m=7)

119 Liu Yanan (2008a).

120 Lin Fanjing (2009f). According to Lin Boqiang, director of the China Centre for Energy Economics Research of Xiamen University, who has participated in the formulation of China's strategic energy plan, SPR was expected to meet China's oil demand for 100 days once the second batch of reserve bases were all in operation. See Li Xiaohui (2010c).

121 In comparison, the average West Texas Intermediate (WTI) crude oil price was US$62 per barrel in 2009. See *China ERW*, 14–20 January 2010, 5.

122 Lin Fanjing (2009f).

123 US Energy Information Administration, China – Background, www.eia. gov/countries/cab.cfm?fips=CH.

124 *China OGP*. 1 August 2010, 24.

125 *China OGP*, 1 September 2010, 34.

126 Bradsher (2010).

127 Duce and Wang (2010).

128 Erikson and Collins (2010).

Chapter 5

1 *BP Statistical Review of World Energy*, June 2011, 20–23.

2 Development Research Centre 2004.

3 The projection was made during the second half of the 1990s. The projected figures were between 2.76 btsce by the Chinese Energy Research Institute and 2.80 btsce by the Academy of Science's National Conditions Analysis Team.

4   'China to cap energy use at 4b tons of coal equivalent by 2015', www.chinadaily.com.cn/business/2011-03/04/content_12117508.htm.

5   Fridley (2008, 43).

6   Xu Dingming (2002).

7   Considering that the second and third surveys were conducted in 1994 and 2003 respectively, it is very likely that the fourth survey will be conducted in the early 2010s.

8   Chen Mingshuang (2006); 'China's energy reserves show potential', http://english.peopledaily.com.cn/200409/21/eng20040921_157704.html.

9   Yang Liu (2006c).

10 China Securities Journal (2010, 11).

11 Duan Zhaofang (2010).

12 China Securities Journal (2010, 14).

13 Jie Mingxun (2010).

14 CUCBM and Credit Suisse estimates, quoted in Credit Suisse Equity Research Report (Asia Pacific/China, Gas Utilities) on Asia Coalbed Methane Sector (18 October 2006); Jie Mingxun (2010).

15 Credit Suisse Equity Research Report on Asia Coalbed Methane Sector; *CSJ*, CNGR, 120.

16 *China Securities Journal* (2010, 119).

17 Chen Wenxian (2005c, 6–7).

18 Chen Wenxian (2008); Lin Fanjing (2008f).

19 Lin Fanjing (2008f).

20 Liu Yanan (2008d); *China ERW*, 10–16 July 2008, 5.

21 *China ERW*, 12–18 June 2008, Vol. VII, no. 23, 14.

22 Qiu Jun (2010c); Li Xiaohu (2010b).

23 China Securities Journal (2010, 120–1).

24 Hook (2010a). According to Zhang Hongtao, of the Ministry of Land and Resources, the figure will be 23 bcm in 2020, accounting for 0.7% of China's energy mix. See 'Unconventional gas demand set to soar by 2020', www.chinadaily.com.cn/business/2010-10/19/content_11427918.htm.

25 It was estimated that Shanxi's CBM drainage volume will be five bcm by 2015 and eight bcm by 2020. See *China OGP*, 1 October 2009, 30.

26 Li Xiaohui (2010a).

27 China Securities Journal (2010, 128).

28 Li Xiaohui (2011).

29 Liu Honglin et.al. (2009); 'The Shale Gas Boom Shift to China', http://eneken.ieej.or.jp/data/3179.pdf); Jie Mingxun (2010); *China EW*, 19–25 August 2010, 11.

30 'China's First Shale Gas Discovery Announced Today By PTR.V', www.stockhouse.com/blogs/ViewDetailedPost.aspx?p=91976.

31 'Shell, PetroChina To Develop Shale Gas In Sichuan', www.dowjones.de/site/2009/11/shell-petrochina-to-develop-shale-gas-in-sichuan.html. Also see Winning (2009); Xu Wan (2010); and 'Shell, PetroChina To Develop Shale Gas In Sichuan', http://royaldutchshellplc.com/2009/11/27/

shell-petrochina-to-develop-shale-gas-in-sichuan/.

32　*China EW*, 19–25 August 2010, 11.

33　Dyer and Hoyos (2010).

34　*Ibid.* In late October 2010 a similar report on shale gas talks between BP and SINOPEC was made (www.chinavestor.com/energy/72525-bp-sinopec-talk-shale-gas.html).

35　'UPDATE 1-Statoil denies China shale-gas deal report', http://af.reuters.com/article/energyOilNews/idAFLDE70J18G20110120

36　Ward (2010).

37　Gvarstein (2010).

38　'China aims to sharply raise shale gas output by 2020', http://en.in-en.com/article/News/Gas/html/2010020215905.html; *IGR*, 15 February 2010, 5–6. CNPC is targeting 0.5 bcm of shale gas production before 2015. See *China OGP*, 15 July 2010, 17.

39　'China blocks bid for shale gas development', www.energytimes.kr/news/articleView.html?idxno=10882; Chen Aizhu (2011).

40　*China EW*, 28 October–3 November 2010, 13.

41　Watts (2011).

42　Hoyos (2010).

43　Hoyos and Crooks (2010b).

44　Duan Zhaofang (2010).

45　China Securities Journal (2010, 25); Qiu Jun (2010f).

46　'China's natural gas market on fast track', www.chinadaily.com.cn/bizchina/2010-03/29/content_9658332.htm; on 9 June 2010, Zhou Jiping, general manager of PetroChina said China's natural gas consumption will account for 10% of total domestic primary energy consumption by 2020, up from 3.9% in 2010 (see *China EW*, 3–9 June 2010, 11).

47　*China OGP*, 15 November 1998, 3–4; China OGP, 15 September 2008, 13.

48　*China OGP*, 1 March 2008, 15.

49　*China OGP*, 1 March 2008, 15; *China OGP*, 15 September 13.

50　Jie Mingxun (2010).

51　'China to see gas demand soar by 20% in 2011', http:// english.peopledaily.com.cn/90001/90778/7276466.html; Qiu Jun (2010f).

52　'Energy economic situation in the first half of 2010, full press conference', www.china5e.com/show.php?contentid=113567.

53　Zhang Chunyan (2011).

54　Wu Xiaobo (2011).

55　The figure is based on the reference scenario. Under high growth scenario, 438 bcm, and under low growth scenario, 341 bcm. Duan Zhaofang (2010).

56　'China Gas Study: Strong Demand Growth Will Persist but Exporters Should Secure Contracts Before Unconventional Gas Constrains New LNG Imports', www.woodmacresearch.com/cgi-bin/corp/portal/corp/corpPressDetail.jsp?oid=2092548.

57　Fridley (2008, 31).

58　*Ibid.*, 19.

59 Higashi (2009).
60 In 2009, the penetration rate in Beijing was 65.1%, followed by Tianjin 46.4%, Shanghai 44.3%, Chongqing 25.5%, Xinjiang 17.7%, Liaoning 17.4%, Qinghai 15.0%, Ningxia 13.6%, and Heilongjiang 13.4%. The average was only 11.0%. See GSGIR (2011, 6).
61 China Securities Journal (2010, 22–3).
62 Duan Zhaofang (2010).
63 *Ibid.*
64 *Ibid.*
65 China Securities Journal (2010, 26–7).
66 For the details of gas price reform before 2005, see Quan and Paik (1998, 66–71); and *China OGP* 2002, 198–202; JBIC (2005).
67 Lin Fanjing (2005).
68 A manager from CNPC pointed out that price ratio between crude oil, natural gas, and coal for power plants in China is 1: 0.24: 0.17, while that in the international market is 1: 0.6: 0.20. See *Ibid.*
69 Lin Fanjing and Mo Lin (2006).
70 Jiang Lurong (2006).
71 Chen Wenxian (2006e).
72 *Ibid.*
73 *China OGP*, 15 December 2006, 12.
74 Chen Wenxian (2007b).
75 Higashi (2009). As the total length of the pipeline from Xinjiang to Shanghai is approximately 3,900 km, the unit tariff per 100 km is calculated as $3.6/1,000 cubic metres in Shanghai.
76 CD, 20 March 2009. www.china.org.cn/business/2009-03/20/content_17473104.htm
77 Ibid.
78 Lin Fanjing (2009a).
79 *Ibid.*
80 *Ibid.*
81 Lin Fanjing (2009e). On 15 September China's domestic oil information portal C1 Energy reported that SINOPEC group set its baseline price at 1.40 yuan or $0.21 per cubic metre, the maximum allowed by the government. See *China ERW*, 10–16 September 2009, 3.
82 Lin Fanjing (2009e); *China ERW*, 9–15 July 2009, 4. Before the November 2008 price reform, Shanghai residents used to pay 2.1 yuan (US$0.307) for WEP I gas. As of 2010 they pay 2.5 yuan ($0.366). See *China ERW*, 11–17 March 2010, 3.
83 Lin Fanjing (2009e).
84 *China OGP*, 15 July 2010, 20–21.
85 UBS (2009).
86 Lin Fanjing (2009e).
87 Petromin Pipeliner (2011).
88 *China OGP*, 1 July 2009, 26.
89 Chen Wenxian (2007d).

90 Li Xiaohui (2010e).
91 *China OGP*, 1, June 2010, 1–3; 'China Finally Introduces Gas Price Reform', www.woodmacresearch.com/content/portal/energy/highlights/wk5_Jun_10/WMBP_China_Gas_Price_Reform_Insight_June_2010.pdf?hls=true.
92 *China OGP*, 1 June 2010, 1–3; *China EW*, 15–21 July 2010, 22–23, and 27 May–2 June 2010, 9.
93 *China EW*, 16–29 September 2010, 15.
94 Fridley (2002, 23).
95 *Ibid*, 47
96 *Ibid.*, 34.
97 IEA (2002, 123–53).
98 Chun Chun-Ni (2007, 28).
99 'China electricity consumption to almost double by 2020', www.chinadaily.com.cn/business/2010-12/22/content_11737519.htm.
100 Data obtained from the ERI, NDRC. NEA's Li Zhi strongly suggested that power capacity fuelled with natural gas will rise from the current 28 GW to 60 GW in the coming five years. See 'China To Double Power Capacity Fueled By Gas In 2011-2015', www.energychinaforum.com/news/54586.shtml.
101 *China EW*, 22–8 September 2011, 12–13.
102 In 2006, China had an installed capacity of 622 GW for power generation, of which 15.6 GW or 2.5% was natural gas based. See Fridley (2002, 34).
103 Mao Yushi, Sheng Hong, and Yang Fuqiang (2008, 16 (Table 3-1), 17, and 26 (Table 3-5)).
104 Beijing's clean air campaign was pushed during the 2008 Olympics when the local government closed all coal-fired power plants within the fourth-ring road. The city now has six ring roads and only four coal-fired power plants, located in Chaoyang and Shijingshan district. By the end of 2015, three out of the four will be retrofitted to burn gas.
105 'Third phase of Shaanxi-Beijing gas pipeline reaches Hebei', www.energychinaforum.com/news/44252.shtml; 'Beijing to consume 7 bn. cubic metres of gas in 2010', www.energychinaforum.com/new_day/show.asp?id=777.
106 Distributed power plants are typically units of 10–30MW that an end-consumer builds to meet his own needs. Larger gas-fuelled facilities normally have a generation capacity of 200MW or larger and sell power to the grids. In China, larger plants account for 35MW of capacity and there are plans to raise this to 60GW by 2015. See 'China to boost natural gas use in small-scale power plants', www.energychinaforum.com/news/54617.shtml.
107 *China EW*, 22–8 September 2011, 12–13.
108 In Europe, 'towngas' is used to refer to gas distilled from coal, similar to coke oven gas, and used to supply towns before the large-scale availability of natural gas. However, 'city gas' here refers to gas distribution in cities generally.

109 Yang Liu (2006b); Lin Fanjing and Lin Wei (2008).

110 GSGIR (2011).

111 There are 27 cities with a population of over 1 million – of which six have over 5 million, four 3–5 million, and 17 with 1–3 million – that in 2010 had no access to natural gas. Together the 27 cities contain a total population of 87 million. See GSGIR (2011, 17).

112 In 2006, Panva Gas Holdings was supplying gas to 26 cities. See Lin Fanjing and Lin Wei (2008), 4–7.

113 The Hong Kong-listed gas operator Panva Gas (Panva), which was solely engaged in the city gas business on the Chinese Mainland, has been tapping markets mainly in eastern, central, and south-western China. HK tycoon Lee Kai Shing's Hutchison Whampoa was a strategic partner of the gas operator.

114 Towngas home page: www.towngas.com/Eng/Corp/AbtTG/Overview/Index.aspx.

115 China Gas, Natural gas, www.chinagasholdings.com.hk/en/natural/overview.jsp; Lin Fanjing and Lin Wei (2008); 'China Gas Holdings gets exclusive selling rights in 7 cities', www.chinadaily.com.cn/bizchina/2009-06/15/content_8285664.htm; *China OGP*, 15 May 2009, 8.

116 Sun Huanjie (2006b).

117 Chen Wenxian (2007c).

118 Major shareholders of China Gas: www.chinagasholdings.com.hk/en/about/about.jsp?type=shareholders.

119 'China Gas Inks Cooperation Deal With PetroChina Unit', www.energychinaforum.com/news/26986.shtml.

120 Yang Liu (2006b); Lin Fanjing and Lin Wei (2008).

121 ENN Energy Distribution, www.enn.cn/en/business/energy_supply_chain.html.

122 'Zhengzhou Gas, China Resources Gas to launch JV', http://news.alibaba.com/article/detail/business-in-china/100134058-1-zhengzhou-gas%252C-china-resources-gas.html. On 8 September 2009, CRG acquired 7 city gas projects. Based on this acquisition, the CRG portfolio includes 26 cities city gas projects with a combined annual gas sales volume of some 2.2 bcm. CRG covers eight provinces (Sichuan, Jiangsu, Hubei, Shandong, Shanxi, Hebei, Jiangxi, and Yunnan) and includes five provincial capitals (Chengdu, Wuhan, Kunming, Jinan, and Nanjing). See China Resources Gas Group Ltd 2009 Interim Results, www.crgas.com.hk/scripts/eng/relation/pdf/2009/sep/eng%20R2%209-9-2009.pdf.

123 China Resources Gas Group Ltd Corporate Profile, www.crgas.com.hk/scripts/eng/corporate/eng_intro.asp.

124 *China EW*, 24–30 June 2010, 7.

125 Beijing Enterprises Holdings Ltd, City Gas, www.behl.com.hk/eng/business/activities.htm.

126 Li Yuling (2002b).

127 'Xinjiang gas to reach Shenzhen', www.chinadaily.com.cn/english/doc/2004-03/08/content_312774.htm.

128 Chen Wenxian (2007e).
129 'Xinjiang Guanghui Gets LNG Pipeline Approval (China)', www.lngworld-news.com/xinjiang-guanghui-gets-lng-pipeline-approval-china/.
130 *China OGP*, 15 August 2008, 16 & 17.
131 Lin Fanjing and Lin Wei (2008).
132 Sun Huanjie (2006b).
133 Chen Wenxian (2007c).
134 Chen Wenxian (2007e).
135 Sun Huanji (2008).
136 Lin Fanjing (2009c); *China OGP*, 1 June 2009, 21; *China OGP*, 1 May 2009, 27–8.
137 Qiu Jun (2009).
138 Lin Fanjing (2009g).
139 *China OGP*, 15 June 2010, 18.
140 *China ERW*, 1–7 April 2010, 6.
141 *China OGP*, 1 November 2010, 31–2; *China OGP*, 1 September 2010, 24–5.
142 Li Xiaohui (2010a); *China OGP*, 1 November 2006, 29; Fridley (2008, 26); *China EW*, 29 July–4 August, 2010, 7.
143 Qiu Jun (2010e, 6).
144 Quan and Paik (1998, 79).
145 'New Channels to Ease Beijing Energy Shortage', www.ccchina.gov.cn/cn/NewsInfo.asp?NewsId=4455; Communist Party of China News, http://english.people.com.cn/200404/13/eng200404 13_140287.shtml.
146 'China completes 3rd Shaanxi–Beijing natural gas pipeline', www.china-daily.com.cn/bizchina/2011-01/01/content_11784953.htm.
147 Qiu Jun (2010e, 6); *China OGP*, 1 November 2010, 29.
148 'Sebei–Xining section of the parallel Sebei-Xining-Lanzhou Pipeline becomes operational', www.cnpc.com.cn/en/press/newsreleases/Sebei%EF%BC%8DXiningsectionoftheparallelSebei%EF%BC%8DXining%EF%BC%8DLanzhouPipelinebecomesoperational.htm; 'PetroChina's Sebei–Xining–Lanzhou Gas Pipeline Commences Operation', www.petrochina.com.cn/Ptr/News_and_Bulletin/News_Release/200512080027.htm.
149 Qiu Jun (2006s).
150 'PetroChina to invest 20 billion yuan in laying 3rd Shaanxi–Beijing gas pipeline', www.istockanalyst.com/article/viewiStockNews/articleid/3164696; 'China completes 3rd Shaanxi–Beijing natural gas pipeline', www.china-daily.com.cn/bizchina/2011-01/01/content_11784953.htm.
151 *China OGP*, 1 October 2009, 30.
152 Quan Lan (2000a); Quan Lan (1999e); Quan Lan (2000c); Quan Lan (2001b); Quan Lan (2001c); Ye Ming (2001); Quan Lan (2001d).
153 JBIC (2005, 170–4).
154 Fan Wenxin (2000); Quan Lan (2001g).
155 Qiu Jun (2006n, 13).
156 *China ERW*, 22–8 July 2010, 10.
157 *China ERW*, 25 February–3 March 2010, 7.
158 Qiu Jun (2010e).

159 Quan Lan (2001i).
160 Li Yuling (2002d, 14).
161 Stroytransgaz home page: www.stroytransgaz.com.
162 Li Xiaoming (2002a, 4–8).
163 Qiu Jun (2004a).
164 Lin Fanjing (2006a); 'China proposes construction of 2nd west–east gas pipeline', http://english.peopledaily.com.cn/200603/11/print20060311_249910.html.
165 Chen Wenxian (2007d); *China ERW*, 8–14 November 2007, 21.
166 *China OGP*, 1 January 2008, 12.
167 Lin Fanjing (2008b); *China OGP*, 1 May, 2008; China Securities Journal (2010, 53).
168 '2nd west–east natural gas pipeline to be tested', www.chinadaily.com.cn/business/2011-06/14/content_12689669.htm; 'China's 2nd west–east gas pipeline goes into operation', www.newsgd.com/news/homepagenews/content/2011-07/01/content_26232544.htm.
169 Chen Wenxian (2007d).
170 Zhang Xuezeng and Zhao Dongrui (2009); Chen Wenxian (2007d); Lin Fanjing (2008b).
171 *OGJ Online*, 26 March 2009 ('PetroChina lets contract for WEPP II gas line', www.ogj.com/display_article/357368/7/ARTCL/none/none/PetroChina-lets-contract-for-WEPP-II-gas-line/?dcmp=OGJ.monthly.naturalgas); *OGJ*, 9 February 2009, 52.
172 'Second West–East Gas Pipeline Commercial Operation', www.gulfoilandgas.com/webpro1/MAIN/Mainnews.asp?id=10299 (Jan 22, 2010)
173 Li Xiaohui (2009).
174 *China EW*, 5–11 August 2010, 12.
175 '2nd west–east natural gas pipeline to be tested', www.chinadaily.com.cn/business/2011-06/14/content_12689669.htm.
176 '2nd west–east natural gas pipeline to be tested', www.chinadaily.com.cn/business/2011-06/14/content_12689669.htm. Simple calculation shows that more than 150 mt of coal burning could be saved, if large-scale (68 bcm/y) gas supply from Russia to China becomes a reality.
177 *China EWR*, 15–21 November 2007, 17.
178 *China OGP*, 15 September 2008, 13.
179 'China Proposes Fourth West–East Natural Gas Pipeline', www.downstreamtoday.com/news/article.aspx?a_id=16896.
180 'REFILE-UPDATE 1-China demand for fuels to plateau in rest of 2011 – NEA', http://uk.reuters.com/article/2011/04/22/china-energy-idUKL3E7FM05Z20110422.
181 See NAGPF (2007), and NAGPF (2009).
182 Lin Fanjing (2010a).
183 Chen Wenxian (2007d).
184 *China ERW*, 4–10 September 2008, 16–18; 'China Studying Third West–East Gas Pipeline – Report', www.downstreamtoday.com/news/article.aspx?a_id=12755.

185 *China ERW*, 8–14 April 2010, 4–5.
186 *China OGP*, 1 August 2010, 29.
187 Lin Fanjing (2008b)).
188 *Ibid.*
189 *IPR*, 5–11 March 1999, 4–6.
190 In late July 1997 CNPC opened a communication channel with Sakhanefte-gaz, which sent a delegation (led by Vasiliy Efimov, the Sakhaneftegaz president) to Beijing to exchange views on long-distance gas pipeline development, and to discuss ways to develop the Chayandinskoye gas field with CNPC's cooperation. At that time the proven gas reserves of Chayandinskoye field were 755 bcm, of which 535 bcm were exploitable.
191 Quan Lan (1999a); *IPR*, 5–11 March 1999, 4–6.
192 Quan Lan (2000f).
193 Quan Lan (2001h).
194 Paik (2005b, 211–24).
195 Along with the Sino-Myanmar pipeline, CNPC will also build an oil prod-ucts pipeline grid in Yunnan that covers the Kunming–Dali oil products pipeline, the Kunming–Mengzi pipeline, the Kunming–Pu'er pipeline, and the Kunming–Qujing pipeline. See Lin Fanjing (2008b).
196 *China ERW*, 19–25 June 2008, 15. The Shwe, Shwe Phyu, and Mya gas fields' reserves are estimated at 4–6 tcf, 5 tcf, and 2 tcf respectively. See *China OGP*, 1 July 2008, 9–10.
197 *Reuters*, 25 August 2009 ('UPDATE 2-Daewoo in $5.6 bn Myanmar gas ex-port deal to China', www.reuters.com/article/idUSSEO5594720090825).
198 *China EW*, 9–15 September 2010, 4; *China Daily*, 4 June 2010.
199 'China: CNOOC Ups LNG Import Capacity by 9.4 MT in 2010', www.lng-worldnews.com/china-cnooc-ups-lng-import-capacity-by-9-4-mt-in-2010/.
200 *China OGP*, 15 November 1998, 1–3; *China OGP*, 1 March 1997, 9–10.
201 Yang Liu (2005c).
202 Quan Lan (1999b).
203 Paik (2005b).
204 Chen Dongyi (2006, 20); Yang Liu (2006d). The Fujian LNG price was also quite favourable to CNOOC initially. A take-or-pay agreement was signed in July 2004, contracting to supply 2.6 mt/y of LNG for the Fujian project for 25 years from 2007 at an average price of $23 per barrel. The price was later adjusted to $25 per barrel. However, the Indonesian side was not satisfied and pressed CNOOC to raise the price from $25 to somewhere in the range $35–38, a rise of over 40%. See Qiu Jun (2006f); *China OGP*, 1 July 2006, 25; Zhan Lisheng (2006).
205 China Securities Journal (2010, 36); 'Terminal and Trunk-line Project', www.dplng.com/en/project/project_02.aspx.
206 'Key Developments for Guangdong Dapeng LNG Company Ltd', http://investing.businessweek.com/research/stocks/private/snapshot.asp?privcapId=42622608.
207 Paik (2005b).
208 CNOOC acquired the Tangguh gas reserves at US 15 cents/1,000 cubic

feet on a proven reserve and at US 9 cents/1,000 cubic feet on a proved, probable, and possible basis. See *IGR*, 11 October 2002, 26.

209 *China OGP*, 1 February 2008, 28.
210 The total investment for the Phase I construction is 11.98 billion yuan, which breaks down to 4.33 billion yuan for terminal and trunk line, 5.6 billion yuan for two gas-fired power plants, and 2.05 billion yuan for building the towngas networks in five cities.
211 *China OGP*, 1 January 2006, 21.
212 China Securities Journal (2010, 37).
213 www.cceec.com.cn/English/Project/China/2010/1108/13935.html (last accessed October 2011).
214 Wang Ying (2006).
215 *CD*, 27 October 2004.
216 Qiu Jun (2006f).
217 China Securities Journal (2010, 37).
218 Fridley (2008, 23–4).
219 Qiu Jun (2006k).
220 True (2009); China Securities Journal (2010, 37).
221 China Securities Journal (2010, 37).
222 *China ERW*, 15–21 May 2008, 13. In 2006, it was estimated Shanghai's gas demand would reach 2.9 bcm while the contracted volume was 2.2 bcm. See *China OGP*, 15 September 2006, 9.
223 China Securities Journal (2010, 39).
224 Fridley (2008, 23–4).
225 China Securities Journal (2010, 39).
226 Qiu Jun (2006p).
227 China Securities Journal (2010, 38); *China OGP*, 1 November 2010, 33; *China ERW*, 22–8 May 2008, 11.
228 China Securities Journal (2010, 38).
229 *Shanghai Daily*, 19 March 2009 ('Dalian picked as base for LNG venture' www.china.org.cn/business/2009-03/19/content_17469277.htm); China Securities Journal (2010, 38); *China EW*, 2–8 September 2010, 11.
230 SINOPEC also plans to construct the Yulin (Shaanxi)–Jinan (Shandong) gas pipeline to bring a maximum 3 bcm of gas from Ordos to the Shandong market. See Chen Dongyi (2008d); China Securities Journal (2010, 40); *China EW*, 9–15 September 2010, 6.
231 'China: NRDC Approves Tangshan LNG Terminal', www.lngworldnews.com/china-ndrc-approves-tangshan-lng-terminal/; 'Tangshan LNG gets official nod', www.fairplay.co.uk/login.aspx?reason=denied_empty&script_name=/secure/display.aspx&path_info=/secure/display.aspx&articlename=dn0020101108000011.
232 *China OGP*, 1 November 2010, 35; 'China's CNOOC starts building Zhuhai LNG terminal', www.reuters.com/article/idUSTOE69J04S20101020; 'Zhuhai Weekly Briefings', http://deltabridges.com/news/zhuhai-news/zhuhai-weekly-briefings-61; 'CNOOC Parent Starts Construction of Zhuhai LNG Import Terminal in China', www.bloomberg.com/news/2010-10-20/

cnooc-parent-starts-construction-of-zhuhai-lng-import-terminal-in-china.
html.
233 China Securities Journal (2010, 40).
234 IEA projected that China will need to import around 50–60 bcm/y of
LNG in 2015 to meet its ambitious 12th FYP (2011–15). See IEA (2011a,
42).
235 *China OGP*, 1 November 2010, 35.
236 At the Shipping China Energy 2008 Conference, Yan Weiping, general
manager of China LNG Shipping Holdings Ltd (CLNG) said that, once
all 11 LNG terminal projects are complete, China will need 33 carriers
to feed them, accounting for 10% of the world's total LNG carrier fleet.
In other words, each terminal needs three carriers in order to operate
properly. CLNG was jointly established by COSCO Group and China
Merchants group in 2004 to invest in and manage LNG carriers. See
China ERW, 24–30 April, 2008 13.
237 For the project based confirmation, see
US Energy Information Administration, Country Analysis Brief, China,
www.eia.doe.gov/emeu/cabs/China/NaturalGas.html (LNG Table); 'ExxonMobil and PetroChina Sign US$41-bil. Gorgon LNG Supply Contract',
www.ihsglobalinsight.com/SDA/SDADetail17507.htm; 'China: Zhejiang
LNG terminal project', www.cssm.net/info/20091127/2009112793650.
shtml; 'CNOOC venture to build second LNG terminal in Guangdong',
www.chinamining.org/Companies/2008-01-11/1200019367d8617.html;
'Sinopec, ExxonMobil ink Papua New Guinea LNG deal', http://www2.
china-sd.com/News/2009-12/4_3926.html;
'CNOOC buys more LNG from Qatar', www.chinadaily.com.cn/bizchi-
na/2009-11/14/content_8980907.htm
'CNOOC Transmission Co. Ltd', www.cnoocgas.com/qidian/company.
do;jsessionid=79AC30CB4502B04AA0E950798C9C9C30?actionMethod
=list_company&lang=1&pk=20091116025902772301266962207230.
238 Li Yuling (2004b); Qiu Jun (2006m); Chen Dongyi (2006, 19); Chen Dongyi
(2008c).
239 Chen Dongyi (2007c).
240 Lin Fanjing (2007b).
241 *Ibid.*
242 *Ibid.*
243 *China ERW*, 17–23 April 2008, 15.
244 CNOOC has halted its LNG imports from the spot market since November
2007 because of the price which has remained high – between $8 and
$10/mmbtu. See Lin Fanjing (2008e).
245 LNG prices on the domestic Chinese market have increased. The government has been strengthening its control over domestic LNG prices
since October 2007, but retail prices in several cities, like Nanchang and
Changsha, have continued to rise and in May 2008 rose above 100 yuan
(US$14.45) per cylinder. See *China ERW*, 5–11 June 2008, 4.
246 Arrow Energy home page, www.arrowenergy.com.au/.

247 Smith (2010b).

248 Hoyos and Crooks (2010a).

249 Wan Zhihong (2010); Crooks (2009c); 'CNOOC Signs Landmark CBM-to-LNG Supply Deal with BG Group', www.ihsglobalinsight.com/SDA/SDADetail18454.htm.

250 Qiu Jun and Yan Jinguang (2009).

251 On 10 August 2009, Petronet, India's largest LNG importer signed an agreement with ExxonMobil to buy 1.5 mt/y of LNG from the Gorgon project over 20 years. The sale price of LNG for Petronet was between US$8.26/mmbtu and US$20.8 mmbtu, at least 23.8% lower than the price agreed by CNPC. See Qiu Jun and Yan Jinguang (2009).

252 Qiu Jun and Yan Jinguang (2009).

253 Hook (2010b).

254 By late June 2010 China had received 1.5 bcm of gas from Turkmenistan. In 2011 the volume will reach 17 bcm ('China gets 1.5 bcm Turkmenistan gas in H1', www.chinadaily.com.cn/bizchina/2010-07/01/content_10045577.htm); 'Gas supply schedule via Turkmenistan–China gas pipeline for 2011 adopted', www.turkmenistan.ru/?page_id=3&lang_id=en&elem_id=17898&type=event&sort=date_desc.

255 Qiu Jun and Zhang Zhengfu (2010).

256 *China OGP*, 1 March 2010, 18.

257 Qiu Jun and Zhang Zhengfu (2010).

258 Qiu Jun (2010a).

259 *Ibid.*, 12.

260 Lin Fanjing (2010b).

261 *China Daily*, 29 March 2010 'China's natural gas market on fast track', (www.chinadaily.com.cn/bizchina/2010-03/29/content_9658332.htm). Wu Yin, deputy head of the NEA, said at an energy forum in Beijing that 'Natural gas accounts for only 4% of energy in China now. The country will raise that to 8% during the 12th Five-Year Period (2011–15)'. See 'China to double natural gas weighting over 5 years', www.chinadaily.com.cn/china/2010-06/19/content_9992983.htm.

262 Qiu Jun (2005l).

263 Yu, S. (2010)

264 *China OGP*, 15 July 2010, 16. According to Mr Song Wenjie, Deputy General Manager of Tarim Oilfield Company, Tarim Oilfield has secured a solid resource base for proved original natural gas in place of 840 bcm. See 'PetroChina Announces The Completion and Production Initiation of Tarim Yingmaili Gas Fields', www.petrochina.com.cn/Ptr/News_and_Bulletin/News_Release/200707130023.htm.

265 Jia Chengzao et al. (2002); http://pg.geoscienceworld.org/cgi/content/abstract/10/2/95; Jia Chengzao and Li Qiming (2008).

266 Qiu Jun (2006n); Quan Lan (2000e).

267 'Tarim Oil Province', www.cnpc.com.cn/en/aboutcnpc/ourbusinesses/explorationproduction/operatediol/Tarim_Oil_Province.htm.

268 'PetroChina Kela-2 field gas output exceeds 50bln cubic metres in six

years', http://newsystocks.com/news/3673395.

269 Quan Lan (2001f); Qiu Jun (2005c); *China OGP*, 15 April 2008, 11; *China OGP*, 15 November 2009, 28; 'A New Gas Source for the West-East Gas Pipeline – Dina-2 Gas Field put into production', www.cnpc.com.cn/en/aboutcnpc/ourbusinesses/explorationproduction/operatediol/Dina_2_Gas_Field.htm; *China OGP*, 1 April 2008, 10.

270 'PetroChina Announces The Completion and Production Initiation of Tarim Yingmaili Gas Fields', www.petrochina.com.cn/Ptr/News_and_Bulletin/News_Release/200707130023.htm; 'Gas Fields of the Source', http://china.org.cn/english/features/Gas-Pipeline/37516.htm; Qiu Jun (2005l).

271 Lin Fanjing (2010c); 'China: Sinopec boosts oil and gas reserves on Tahe field', www.energy-pedia.com/article.aspx?articleid=133019.

272 See http://china.platts.com/NewsFeatureDetail.aspx?xmlpath=IM.Platts.Content/InsightAnalysis/NewsFeature/2010/chinaoutlook/6.xml; This field is SINOPEC's second largest and has recoverable reserve potential of 996 mtoe, following the addition of 135 mt of proven crude reserves in December 2008. 'China's CNPC and Sinopec Vow to Ramp Up Xinjiang Crude Output', www.ihs.com/products/global-insight/industry-economic-report.aspx?id=106595904; 'China: Sinopec boosts oil and gas reserves on Tahe field', www.energy-pedia.com/article.aspx?articleid=133019.

273 SINOPEC Northeast Company, http://english.sinopec.com/about_sinopec/subsidiaries/oilfields/20080326/3030.shtml; 'China's CNPC and Sinopec Vow to Ramp Up Xinjiang Crude Output', www.ihsglobalinsight.com/SDA/SDADetail15795.htm; *China OGP*, 15 December 2005, 16.

274 Wan Zhihong (2008).

275 Chen Wenxian (2007a).

276 *China OGP*, 15 June 2010, 14.

277 China Securities Journal (2010, 16); *China OGP*, 1 September 2010, 33.

278 Quan Lan (1999d).

279 'China's Changqing oilfield records highest annual natural gas output', www.energychinaforum.com/news/29463.shtml; 'Changqing Oilfield's gas output to hit 20 bn. cubic metres this year', www.chinaknowledge.com/Newswires/News_Detail.aspx?type=1&NewsID=39206; 'Changqing oil field becomes China's 2nd largest onshore oil-gas field', www.chinamining.org/News/2009-12-23/1261551518d32649.html.

280 Qiu Jun (2010e, 6).

281 Hoyos (2006).

282 Wang Ying (2006).

283 Quan Lan (2001a); *CD Online* (29 May 2007) http://english.peopledaily.com.cn/200705/29/eng20070529_378898.html.

284 China Securities Journal (2010, 16).

285 'Daily output of Sulige Gas Field exceeds 20 million cubic meters', www.cnpc.com.cn/eng/press/newsreleases/DailyoutputofSuligeGas-Fieldexceeds20millioncubicmeters.htm; 'China: PetroChina's 2010 gas output in Sulige to top 10 bcm', www.energy-pedia.com/article.

aspx?articleid=140526.
286 *China OGP*, 1 July 2008, 12; *China OGP*, 15 July 2008, 18; *China OGP*, 1 August 2009, 30.
287 The fifth plant will process the gas from the blocks under joint development by CNPC and Total, and CNPC's blocks in southern and western Sulige field. See *China OGP*, 1 November 2010, 29.
288 'Shell signs with PetroChina', www.theengineer.co.uk/news/shell-signs-with-petrochina/290868.article; Kambara and Howe (2007, 71–72).
289 *SD*, 2 March 2007 ('Changbei gas field starts', www.chinadaily.net/bizchina/2007-03/02/content_818128.htm).
290 *China EW*, 30 September–13 October 2010, 13.
291 'Recent Study: Changbei Field, China, Commercial Asset Valuation and Forecast to 2026', www.pr-inside.com/recent-study-changbei-field-china-commercial-r2216461.htm.
292 'Sinopec's Daniudi Gas Field in operation', http://business.highbeam.com/1757/article-1G1-180784875/sinopec-daniudi-gas-field-operation.
293 Yang Liu (2005a).
294 China Securities Journal (2010, 16).
295 'China's giants use pipes to grab market', www.platts.com/newsfeature/2009/chinaoutlook09/index.
296 China Securities Journal (2010, 15).
297 'The petroleum geologic characteristics of Sichuan basin, central China', www.osti.gov/energycitations/product.biblio.jsp?osti_id=7024946.
298 'Huge New Natural Gas Field Strengthens China's Energy Security', www.chinastakes.com/story.aspx?id=565 (1 August 2008)
299 *China OGP*, 15 August 2008, 17; Wang Ying and Ying Lou (2008).
300 *CD Online*, 29 May 2007 ('Largest gasfield discovery may be declared soon', http://english.peopledaily.com.cn/200705/29/eng20070529_378898.html).
301 Wang Ying and Ying Lou (2008); 'Factbox: China's fledgling shale gas sector', http://uk.reuters.com/article/2011/04/20/us-china-shale-factbox-idUKTRE73J13820110420.
302 China Securities Journal (2010, 15).
303 Lin Fanjing (2010c).
304 'Largest gasfield discovery may be declared soon', http://english.peopledaily.com.cn/200705/29/eng20070529_378898.html; *China OGP*, 1 April 2006, 10; *China OGP*, 15 April 2006, 10; *China OGP*, 15 August 2008, 17; *China OGP*, 1 October 2008, 14.
305 *China OGP*, 15 August 2008, 17; *China OGP*, 1 October 2008, 14.
306 'Sinopec nears completion of pipeline linking Daniudi Gasfield to Shaanxi Province', www.energychinaforum.com/news/54340.shtml.
307 Chen Wenxian (2006d).
308 *China OGP*, 1 August 2006, 10.
309 'Sinopec starts up Sichuan–East China gas project', www.reuters.com/article/2010/03/30/china-sinopec-gas-idUSTOE62T01I20100330; *China ERW*, 25–31 March 2010, 6.

310 *China ERW*, 9–15 July 2009, 4.
311 *China OGP*, 15 July 2010, 20; *China OGP*, 15 March 2009, 27; 'Sinopec Si-chuan-to-East gas pipeline goes into commercial operation', www.028time.com/news/1290.shtml?title=Sinopec+Sichuan-to-East+gas+pipeline+goes+into+commercial+operation.
312 'Sinopec finds large geological gas reserves in Yuanba – report', www.reuters.com/article/2011/04/27/china-gas-sinopec-idUSL3E7FR05K20110427; 'Sinopec's Yuanba Endeavor', www.dailymarkets.com/stock/2010/12/28/sinopecs-yuanba-endeavor/.
313 'Chevron to Partner With CNPC on Major Gas Project in China', www.chevron.com/news/press/Release/?id=2007-12-18; 'Gas explosion of Chuandongbei', www.economy-point.org/g/gas-explosion-of-chuandong-bei.html.
314 'Chevron, CNPC to begin development of Luojiazhai gas field', www.ogj.com/index/article-display/9394581572/articles/oil-gas-journal/drilling-production-2/production-operations/onshore-projects/2009/11/chevron_-cnpc_to_begin.html; 'China: CNPC and Chevron to begin development of Sichuan gas field', www.energy-pedia.com/article.aspx?articleid=137702.
315 Xu Yihe (2010a).
316 'CNPC, Chevron to develop Sichuan gas field', www.028time.com/news/703.shtml?title=CNPC,+Chevron+to+develop+Sichuan+gas+field.
317 China Securities Journal (2010, 15).
318 *China OGP*, 15 July 2006, 10.
319 Kambara and Howe (2007, 74–7).
320 Gas demand in Shanghai increased about four times from 0.4 bcm in 2002 to 1.9 bcm in 2005. See Qiu Jun (2006l); 'Production begin at Chunxiao gas field', www.chinadaily.net/china/2006-08/06/content_657939.htm.
321 Kambara and Howe (2007, 76) (Figure 4.4); 'Progress made at East China Sea talks', www.china.org.cn/2008-06/17/content_15828163.htm; 'China rebuffs protests over gas field activity', http://search.japantimes.co.jp/cgi-bin/nn20090105a2.html; 'CNOOC Taps Gas Field Amid Border Flap With Japan', http://online.wsj.com/article/SB114422903166617531.html; 'Chunxiao Oil/Gas Field to be completed this October', http://english.people.com.cn/200504/21/eng20050421_182179.html; 'China tells Japan Chunxiao gas field activity is legal', www.energy-pedia.com/article.aspx?articleid=142123; 'UPDATE 1-Japan says eyeing China moves at disputed gas field', http://uk.reuters.com/article/2010/09/17/china-japan-idUKTOE68G07C20100917.
322 'The Start-up of Ledong Promotes CNOOC Ltd's Gas Production Growth', www.cnoocltd.com/encnoocltd/newszx/news/2009/1276.shtml; 'China: CNOOC begins Ledong gas production', www.energy-pedia.com/article.aspx?articleid=136810.
323 Duan Zhaofang (2010).
324 The facility annually feeds on 0.52 bcm of gas from PetroChina's Tuha gas field under a 15 year agreement spanning the period 1 December 2003 to 31 December 2017. When the agreement expires, the facility will

be producing 0.55 bcm of LNG a year with feedstock from coal gas, a by-product of the 1.2 mt/y or 0.8 mt/y DME projects. See *China Securities Journal* (2010, 35).

325 Wei Hong (2010).
326 Xu Yihe (2010b).
327 *China OGP*, 15 July 2010, 23.
328 *China OGP*, 1 September 2010, 33.

Chapter 6

1   *China OGP*, 15 September 1995, 8–9; *China OGP*, 1 December 1993, 5 (Table: China's Overseas Oil E&D Activities); *China OGP*, 15 October 1997, 11–12.
2   CNPC expanded its operation arm, CNODC into a group company known as CNODC Group, which was headed by Wu Yaowen. During CNPC's internal restructuring in 1998, CNODC was reformed into CNPC International Investment Corporation See *China OGP*, 1 July 1997, 1–3; *China OGP*, 15 January 1994, 2–3; *China OGP*, 15 October 1998, 10–11. See Liu Haiying and Li Xiaoming (2002, 1–3).
3   *China OGP*, 15 September 1995, 8–9; *China OGP*, 1 December 1993, 5 (Table: China's Overseas Oil E&D Activities); *China OGP*, 15 October 1997, 11–12.
4   Paik et.al. (2007); *Interfax Central Asia & Caucasus Business Report*, 16 February 2004.
5   Liu Haiying and Li Xiaoming (2002, 1–3).
6   *Ibid*; Li Yuling (2002a).
7   Liu Haiying and Li Xiaoming (2002, 1–3).
8   Paik et.al. (2007); Andrews-Speed and Dannreuther (2011).
9   Zhang Boling (2008); Crooks (2008); Dyer and Mitchell (2008).
10   Anderlini (2009).
11   'Obama administration outlines Arctic energy policy initiatives', www.ogj.com/index/article-display/8838928569/s-articles/s-oil-gas-journal/s-exploration-development/s-articles/s-cnooc_-sinopec_sign.html; Waldmeir (2009).
12   Lin Fanjing (2009d).
13   'UPDATE 2-BP-led consortium takes on Iraq's Rumaila oilfield', www.reuters.com/article/idUSLU68290120090630; 'BP and CNPC to Develop Iraq's Super-Giant Rumaila Field', www.bp.com/genericarticle.do?categoryId=2012968&contentId=7057650; 'CNPC and BP to rejuvenate Iraq's Rumaila Oilfield', www.cnpc.com.cn/en/press/Features/CNPC_and_BP_to_rejuvenate_Iraqs_Rumaila_Oilfield.htm.
14   Hoyos (2009); 'CNPC-led group wins Iraq Halfaya oilfield deal', www.reuters.com/article/idUSGEE5BA04S20091211?type=marketsNews; 'Iraq inks preliminary deal to develop southern oil field', http://news.xinhuanet.

com/english/2009-12/22/content_12690252.htm.

15 Chazan (2009); Robinson (2009); 'Chinese firm Sinopec offers £5bn for Addax', www.guardian.co.uk/business/2009/jun/14/chinese-bid-swiss-oil-firm.

16 Sinochem Group information page, www.sinochem.com/tabid/696/InfoID/10994/Default.aspx

17 'Sinochem snaps up Peregrino stake', www.upstreamonline.com/live/article215802.ece; Hook (2011)

18 'Sinopec Agrees to Acquire Occidental's Argentina Unit for $2.45 Billion', www.bloomberg.com/news/2010-12-10/sinopec-agrees-to-purchase-occidental-s-argentina-unit-for-2-45-billion.html.

19 Christoffersen (1998).

20 *China OGP*, 1 November 1997, 10–11; *China OGP*, 15 October 1997, 10–11.

21 *China OGP*, 1 June 1998, 13–14.

22 Quan Lan (1999c). For a detailed study on Sino-Kazakh oil pipeline development, see Handke (2006).

23 Chung Chien-Peng (2004, 1002).

24 Li Yuling (2002c).

25 Xie Ye (2004); Qiu Jun (2006a); Chen Wenxian (2006c).

26 Yakovleva (2005).

27 *Ibid.*; *China ERW*, 3–9 April 2008, 9.

28 Yakovleva (2005).

29 *Ibid.*

30 *Ibid.*

31 *Ibid*; Pala and Bradsher (2003); Campaner and Yenikeyeff (2008).

32 Li Yuling (2004e, 2–3).

33 Li Yuling (2004e, 1–2).

34 Barges (2004b, 38–44); Yermukanov (2006).

35 KazMunaiGaz website: www.kmg.kz/en/.

36 China National Oil and Gas Exploration and Development Corporation website: www.cnpcint.com/aboutus/welcome.html.

37 Yakovleva (2005); Fishelson (2007).

38 Yakovleva (2005).

39 Tsui, E. (2005); Gorst (2005); Fishelson (2007).

40 Lin Fanjing (2007a).

41 Qiu Jun and An Bei (2005).

42 *Ibid.*; *China OGP*, 15 July 2006, 10; Yang Liu (2006e).

43 On 8 February 2005, Dushanzi Petrochemical's expansion scheme was approved by the central government. See Qiu Jun (2006a); Chen Wenxian (2006c); Qiu Jun (2005b); 'Crude Oil Imports Reach PetroChina Dushanzi Refinery Via China–Kazakhstan Pipeline', http://161.207.5.4/Ptr/News_and_Bulletin/News_Release/200608010011.htm.

44 'PetroChina Upgrades Domestic Refineries to Process Imported Oil from Russia, Kazakhstan', www.ihsglobalinsight.com/SDA/SDADetail17488.htm.

45 *China ERW*, 26 November–2 December 2005, 8.

46 In late 1990s the crude oil price was $20/barrel, while the price in 2005

when the pipeline was completed became $60/barrel. The Chinese specialists interpreted the metering row as an excuse for Kazakhstan to increase what it could gain from the pipeline. See Qiu Jun and Wang Boyu (2006).

47   *China OGP*, 1 August 2006, 11; 'Crude Oil Imports Reach PetroChina Dushanzi Refinery Via China–Kazakhstan Pipeline', www.petrochina.com. cn/Ptr/News_and_Bulletin/News_Release/200608010011.htm.

48   Yeh Gorst and Aglionby (2006); Yeh (2007); *China ERW*, 14–20 August 2008, 11–12; *AFP*, 26 October 2006.

49   Qiu Jun (2005i).

50   'Rosneft, PetroChina complete work to set up venture', www.chinamining. org/Investment/2006-12-07/1165453663d2460.html.

51   *Xinhua News Agency*, 9 July 2007.

52   Peyrouse, S. (2007).

53   Chen Dongyi (2007d).

54   In 2006, China received only 4 mt of oil from Kazakhstan. See *Russian Media Monitoring Agency*, 28 November 2007.

55   Since 1997, CNPC has invested a total of USD 6.5 billion and paid USD 3.19 billion in taxes in Kazakhstan. CNPC also donated USD 50 million and created 18,100 jobs for local people. See 'Adhering to Mutual Benefits and Achieving Common Development', Speech at the 2008 China–Kazakhstan Senior Business Forum, Beijing, Zhou Jiping, Vice President, www. cnpc.com.cn/eng/press/speeches/2008China%EF%BC%8DKazakhstanS eniorBusinessForum.htm.

56   OMK was one of Russia's largest producers of pipes, railroad wheels, and other metal products for energy, transport and industrial companies. See OMK website: www.omk.ru/en/; *China ERW*, 17–23 April 2008, 12.

57   In 2008, the sales turnover between China and Kazakhstan reached US$17.55 billion, up 26.5% from 2007. See *Russia & CIS OGW*, 9–15 April 2009, 24.

58   'CNPC-AktobeMunaiGaz oil output hits 3 mn tons in H1', www.interfax. cn/news/18487.

59   *China OGP*, 1 September 2010, 28.

60   'CNPC International Ltd (CNPCI) Plans Phase 2 of Sino-Kazakh Crude Oil Pipeline', www.energychinaforum.com/news/26531.shtml.

61   'Kenkiyak–Kumkol section of Kazakhstan-China Oil Pipeline becomes operational', www.cnpc.com.cn/en/press/newsreleases/Kenkiyak_Kumkol_section_of_Kazakhstan%EF%BC%8DChina_Oil_Pipeline_becomes_ operational.htm.

62   Sino-Kazakhstan oil pipeline put into commercial operation in July 2006, when it carried 1.76 mt oil; in 2007 4.77 mt oil; in 2008 the amount of oil broke 6 mt; in 2009 it reached 7.73 mt. See 'Sino-Kazakhstan oil pipeline to China the amount of crude oil breaking 20 million tons', www.sourcejuice.com/1299778/2010/01/26/ Sino-Kazakhstan-oil-pipeline-China-amount-crude-oil-breaking/.

63   In September 2011, China signed an accord with Kazakhstan to expand the capacity of a pipeline network delivering natural gas from Central

Asia by more than 80 per cent.

64   'Nazarbayev calls for closer ties between China, Kazakhstan', http://english.peopledaily.com.cn/90001/90776/90883/6636418.html.

65   China significantly increased its interests in energy-rich Kazakhstan in 2009, in return for providing nearly $13bn in credits and loans. See Demytrie (2010). According to KazMunaiGaz, at the end of 2010 the share of Chinese companies in total oil production in Kazakhstan, by companies participating with KMG, was 15.4%. This share is expected to decline to 11% by 2013 and 5% by 2020. See *Russia & CIS OGW*, 5–11 May 2011, 19.

66   Nuriev (2006).

67   In March 2006, a Turkmen government official stated that Dauletabad gas field's reserves, based on the audit's results, increased by a factor of almost three to total of 4.5 tcm. See Denisova (2007).

68   Pirani (2009, 295–7).

69   Lukin (2005a).

70   *Ibid*; 'Turkmenistan Set For Gas-Exploration Deal With China', www.rferl.org/content/article/1072876.html.

71   Denisova (2007)..

72   *China ERW*, 7–13 August 2008, 6.

73   Paik (1997); Paik (2005a).

74   *China OGP*, 1 July 1995, 5; *China OGP*, 15 December 1996, 1–3.

75   Christoffersen (1998).

76   *China OGP*, 15 June 1993, 2; *China OGP*, 1 September 1995, 10–11. On 1 June 1996, Esso spudded its first wildcat well in Tarim, making it the first foreign oil company to sink an oil rig in the basin. See *China OGP*, 15 June 1996, 7.

77   Paik (1997).

78   'China, Kazakhstan Discuss Cross-border Gas Pipeline', www.china.org.cn/english/BAT/105031.htm.

79   Qiu Jun (2005k, 7).

80   Pirani (2009, 271–315); Miyamoto (1997); Roberts (1996); Skagen (1997).

81   Qiu Jun (2006e).

82   *Ibid*.

83   'Turkmen President Niyazov dies at 66', www.msnbc.msn.com/id/16307468/ns/world_news-south_and_central_asia/; http://en.wikipedia.org/wiki/Turkmenistan.

84   Pirani (2009, 295–7).

85   Chen Dongyi (2007d, 5).

86   Chen Dongyi (2007a).

87   Chen Dongyi (2007b).

88   Gorst Dombey and Morris (2007).

89   Pirani (2009, 295–7).

90   Chen Dongyi (2007b).

91   *China ERW*, 26 December 2007–2 January 2008; *China OGP*, 1 January 2008, 10.

92 Chen Dongyi (2008a).
93 *China ERW*, 26 June–2 July 2008, 7; 'Flow of natural gas from Central Asia', www.cnpc.com.cn/en/press/Features/Flow_of_natural_gas_from_Central_Asia_.htm.
94 *China ERW*, 28 August–3 September 2008, 7.
95 *China OGP*, 1 November 2010, 28.
96 *China EW*, 16–29 September 2010, 16; *China OGP*, 1 November 2010, 28.
97 'Flow of natural gas from Central Asia', www.cnpc.com.cn/en/press/Features/Flow_of_natural_gas_from_Central_Asia_.htm; Gorst and Dyer (2009); 'China–Central Asia pipeline outcome of hard work, vision', http://news.xinhuanet.com/english/2009-12/15/content_12649941.htm; 'Hu attends inauguration of China–Central Asia gas pipeline', http://english.peopledaily.com.cn/90001/90776/90883/6841793.html; 'China–Central Asia gas pipeline yet to provide stable supply', http://www2.chinadaily.com.cn/bizchina/2010-01/07/content_9281553.htm.
98 Socor (2009).
99 *China OGP*, September 2010, 16. CNPC had planned to raise the capacity to 15 bcm/y (1.45 bcf/d) by the end of 2010 with the completion of the fourth compressor station. Eight compressor stations were planned to be operational by the end of 2011, and the capacity would be 30 bcm/y (2.9 bcf/d). See Petromin Pipeliner (2011).
100 Petromin Pipeliner (2011).
101 *China EW*, 16–29 September 2010, 6.
102 'UPDATE 1-China,Turkmenistan agree on new natural gas supply', http://uk.reuters.com/article/2011/03/02/china-turkmenistan-gas-idUKTOE72105F20110302; 'Turkmenistan agrees to raise natural gas supply to China', http://news.xinhuanet.com/english2010/china/2011-03/02/c_13758150.htm.
103 Zhukov (2009, 355–94). *China OGP* reported that Uzbekistan's total natural gas reserves are estimated at over 6.25 tcm, including 1.62 tcm of commercial reserves. See Lin Fanjing (2007a).
104 Lin Fanjing (2007a).
105 Qiu Jun (2005e).
106 Yenikeyeff (2008).
107 Uzbekneftegaz was created in 1998 and includes six stock holding companies. See *China ERW*, 16–22 October 2008, 9; *China ERW*, 5–11 June 2008, 12; *Russia & CIS OGW*, 16–22 October 2008, 18–19.
108 *China ERW*, 26 June–2 July 2008, 14–15; 'Asia Trans Gas to complete Uzbekistan–China gas pipeline in December', www.uzdaily.com/articles-id-6361.htm; 'Flow of natural gas from Central Asia: Central Asia–China Gas Pipeline operational', www.cnpc.com.cn/en/press/Features/Flow_of_natural_gas_from_Central_Asia_.htm.
109 Chen Dongyi (2008a).
110 'Flow of natural gas from Central Asia: Central Asia–China Gas Pipeline operational', www.cnpc.com.cn/en/press/Features/Flow_of_natural_gas_from_Central_Asia_.htm.

111 *China ERW*, 5–11 June 2008, 12; *Russia & CIS OGW*, 16–22 October 2008, 18–19.

112 *China OGP*, 15 July 2008, 19.

113 *China EW*, 10–16 June 2010, 10.

114 Alekperov said: 'We would supply gas output minus what we supply to the domestic market. By 2016, Lukoil plans to increase its gas production in Uzbekistan to between 16–18 bcm.' See *China EW*, 14–20 October 2010, 10; *China OGP*, 15 June 2010, 16–17.

115 *Russia & CIS OGW*, 21–7 April 2011, 57; Gorst (2011a).

116 Yenikeyeff (2008, 316–54).

117 Yenikeyeff (2009); Yenikeyeff (2008). See *China EW*, 24–30 June 2010, 15.

118 Yenikeyeff (2008 346–7).

119 Yenikeyeff (2008).

120 *China EW*, 14–20 October 2010, 10.

121 Yenikeyeff (2009); Yenikeyeff (2008).

122 *Idem.*; 'Flow of natural gas from Central Asia: Central Asia–China Gas Pipeline operational', www.cnpc.com.cn/en/press/Features/Flow_of_natu-ral_gas_from_Central_Asia_.htm.

123 *China EW*, 16–22 December 2010, 10; *China ERW*, 10–16 July 2008, 13; *China ERW*, 31 July–6 August 2008, 14; *China OGP*, 15 July 2008, 18.

124 *China ERW*, 31 July–6 August 2008, 14; *China OGP*, 15 June 2010, 14.

125 *China ERW*, 9–15 October 2008, 9.

126 Baizhen (2011).

127 Yenikeyeff (2008, 351–2).

128 *China EW*, 17–23 June 2010, 11.

129 Turkmenistan planned to export 5 bcm to China in 2010 using the first trunk of the Central Asia–China pipeline, and to then increase these exports to 40 bcm per year by 2012. Beijing and Ashgabat reportedly agreed, at the 1 March 2011 meeting, to increase these total exports to 60 bcm/y. However, the target date to increase the exports to 40 bcm has been pushed back to 2015 because the construction of an additional pipeline has been delayed ('Difficulties Remain for a Turkmen–China Energy Deal', www.eurasianet.org/node/63037).

130 'REFILE-UPDATE 1-China demand for fuels to plateau in rest of 2011 – NEA', uk.reuters.com/article/2011/04/22/china-energy-idUKL3E7FM05Z20110422.

Chapter 7

1   According to SINOPEC's route study, the west line would start from Irkutsk via Ulaanbaartar and Erlian (Inner Mongolia) and end at SINOPEC's Yanshan Petrochemical Corporation (YPC) in Beijing. See *China OGP*, 15 November 1999, 1–2.

2   Quan Lan (2000b).

3   The organizational and legal details of the project were to be finalized no later than July 2002. The feasibility study for the Russian leg of the pipeline was estimated to cost at least $30 million to complete. By July 2003, Yukos, Transneft, and CNPC planned to agree on working drawings and the site of the pipeline, as part of the feasibility study. The Russian side was to invest $1.22 bn over the 2001–10 period. See *RPI*, May 2002, 17–24.

4   In 2001, a document entitled 'The Main Principles of Drafting a Feasibility Study for a Russia–China Oil Pipeline' was signed. This agreement was the basis for a general agreement to construct the Angarsk–Daqing oil pipeline. See Barges (2004b, 38–44).

5   Quan Lan (2001e).

6   Quan Lan (2001j).

7   Li Xiaoming (2002b, 6–8).

8   Barges (2004b, 38–44).

9   Li Yuling (2003c).

10   *Ibid.*

11   The Liaoyang facility is currently supplied by a mixture of Venezuelan and Russian oil, although the increase in Russian oil imports once the ESPO pipeline spur is completed would eventually make the facility 100% dependent on Russian oil. Russian officials have said that the ESPO's main source of throughput will be light sweet crude from fields in East Siberia, although these fields remain relatively underdeveloped, so initial supplies are likely to consist of medium-sour Urals blend, Russia's main export grade according to the *China Oil and Gas Monitor*. See Bai and Chen (2009).

12   'PetroChina Upgrades Domestic Refineries to Process Imported Oil from Russia, Kazakhstan', www.ihsglobalinsight.com/SDA/SDADetail17488.htm.

13   Baidashin (2002).

14   *Ibid.*

15   *Ibid.*

16   In September 2001, Transneft was able to obtain a syndicated $150 million loan from a group of western banks led by Austrian Raiffeisen Zentralbank. Besides loans, Transneft's investment programme also called for issuing one year domestic bonds worth a total of $167 million, as well as Eurobonds valued at up to $500 million, and US promissory notes. Transneft is barred from using its assets as collateral, as a 75% stake in the company is owned by the state. *Ibid.*

17   Peel and Jack (2003); Pilling and Jack (2003).

18   Jack (2003).

19   Li Yuling (2003b).

20   Jack and Rahman (2003).

21   Li Yuling (2003c).

22   Li Yuling (2003f).

23   Li Yuling (2003e); *CER*, 2 May 2003, 1–3.

24 Li Yuling (2003f); Barges (2004b, 38–44).
25 In September 2003, Russia's Prime Minister Mikhail Kasyanov revived the uncertainty during a visit to China, by saying that no decision had yet been taken on a pipeline route, pending the result of an environmental investigation. In early October, energy minister Igor Yusufov said the Japanese were ready to finance most of the construction costs and had agreed not to seek a Russian government guarantee. See Rahman and Jack (2003).
26 Li Yuling (2003h).
27 Li Yuling (2003i).
28 *Ibid.*
29 Liu Haiying (2004).
30 Yukos was the outcome of the merger of the Production Association Yugaskneftegaz (Yu) with the refinery KuybyshevnefteOrgSintez (Kos).
31 Sakwa (2009, 133–8).
32 According to Professor Marshall Goldman, the conflict between Khodor-kovsky and the Putin government came to a head when Khodorkovsky decided to criticise Sergei Bogdanchikov, the CEO of Rosneft. On live TV in February 2003 Khodorkovsky complained to Putin that Putin's close friend Bogdanchikov had worked out a sweetheart deal at the country's expense. According to Khodorkovsky, his rival Bogdanchikov overpaid US$ 622.6 million for Northern Oil, a company controlled by Andrei Vavilov, an insider who was a senator in the council of the Federation and a former deputy finance minister. Khodorkovsky in effect implied that Bogdanchikov and Vavilov were in cahoots with each other and had used state funds to enrich themselves.
33 Goldman (2008, 93–135); Aron (2003); 'Timeline of the Yukos Affair', www.khodorkovskycenter.com/media-center/timeline-yukos-affair.
34 'Energizing the future', www.stroytransgaz.com/projects/russia/vsto_oil_pipeline.
35 McGregor, Pilling, and Ostrovsky (2004).
36 Li Yuling (2004g); Feng Yujun (no date).
37 Li Yuling (2004h).
38 Barges (2004b, 38–44).
39 Chen Wenxian (2004b).
40 *Ibid.*
41 Chen Wenxian (2004c).
42 Putin said in an interview with the Chinese media that 'trade between the two countries is expected to grow to US$20 billion (16 billion, £11 billion) in 2004, compared with $15.7 billion in 2003, and could reach $60 billion before the end of the decade'. See McGregor (2004); 'Oil dominates Russia–China talks', http://news.bbc.co.uk/1/hi/world/asia-pacific/3741118.stm; Xia Yishan (2004).
43 'Gazprom and CNPC signed a Cooperation agreement', www.gazprom.com/press/news/2004/october/article62935/; 'Putin's China visit leaves behind a chance for bilateral oil/gas cooperation', http://english.people.

com.cn/200410/20/eng20041020_160887.html.

44   Other documents include an agreement on strategic cooperation be-tween Gazprom and CNPC, an agreement on cooperation between Vnesheconombank (Russia's Import–Export Bank), Chinese Export Credit Insurance Corporation, and China Development Bank, an agreement between Vneshtorgbank and the Agriculture Bank of China, an agreement on cooperation between Russia's Sherbank and the Bank of China, and an agreement on creating a Sino–Russian business council. See *China ERW*, 8–15 October 2004.

45   Qiu Jun (2004i).

46   *IPR*, 7–13 Oct 2004, 7–8.

47   Barclays Capital led a syndicate of 19 foreign banks (including three Japanese lenders) that lent Transneft $250 million to help finance the expansion of Russia's oil export system. Interest rates of 1.15% above Libor were a record low for an unsecured corporate loan to Russia. Transneft rarely borrows on international markets. See Pilling and Gorst (2005).

48   The summit announced the 'Declaration on the World Order in the 21st Century'. President Hu said 'This declaration had great importance in deepening the strategic cooperation between our two countries'. Sino-Russian ties have warmed with the signing in 2004, and ratification in 2005, of the final settlement of a protracted border dispute. See Blagov (2005).

49   Strictly speaking, Rosneft and CNPC signed a long term coop-eration agreement on 1 July 2005, but the framework agreement on strategic cooperation between Rosneft and SINOPEC was signed later, in November 2005 ('We Support the Rosneft IPO from Strategic Considerations', www.kommersant.com/p658821/r_1/ We_Support_the_Rosneft_IPO_from_Strategic_Considerations/.)

50   Qiu Jun (2005g); Helmer (2005).

51   In 2005, Russian Railways charged $39 per tonne of oil transported to the Naushki station in the constituent republic of Buryatia, and $56 per tonne to the Zabaikalsk station in the Chita Region, both of which are located on the Russia–China border. Rosneft, Lukoil, and Sibneft also export to China by rail. See *CER*, 11 November 2005, 10.

52   Glazkov (2006b); Trenin (2002, 204–13).

53   *Ibid*.

54   Chernyshov (2005b); Qiu Jun (2005h).

55   Chernyshov (2005b).

56   *IPR*, 21–7 April 2005.

57   Lukin (2004, 11–18).

58   Baidashin (2006a).

59   *Ibid*.

60   *Ibid*; 'Russian ministers oppose Perevoznaya as final point of Pacific pipeline', http://pipelinesinternational.com/news/russian_ministers_op-pose_perevoznaya_as_final_point_of_pacific_pipeline/010401/#.

61   *China OGP*, 15 January 2006, 14–15.

62 *China Industry Daily News*, 31 October 2006; Gaiduk (2006).

63 Meyer (2006).

64 From Greenpeace webpage no longer accessible.

65 via Nizhnyaya Poima–Yurubcheno–Tokhomskoye field–Verkhnechonskoye field–Talakanskoye field–Chayandinskoye field–Lensk–Olekminsk–Aldan–Neryungri–Tynda–Skovorodino–Nakhodka.

66 Chernyshov (2004d).

67 Baidashin (2006a).

68 Qiu Jun (2005k).

69 'Russia approves new ESPO pipeline route', http://en.rian.ru/russia/20080303/100502195.html.

70 Baidashin (2008a).

71 Medetsky (2007a).

72 Gaiduk (2008a). An agreement on the Chinese company's participation in the construction of the ESPO was signed in April 2007. The company was to lay a 170 km stretch of the pipeline. The entire work was to be completed before 25 October 2008. However, Transneft was displeased with the quality of the work done by the contractors building the ESPO, while Krasnodar Stroytransgaz-Vostok – the general contractor of the ESPO – is said to have suspended a contract with the China Petroleum Pipeline Bureau (CPP). See *China ERW*, 26 June–2 July 2008, 12.

73 *China ERW*, 26 June–2 July 2008, 12.

74 Baidashin (2004b); Russian Railways Co (RZD) website: http://eng.rzd.ru/.

75 Baidashin (2005a). According to Petroleum Argus, in 2004 the tariff on the Angarsk–Naushki (a station on the border with Mongolia) route was $27 per tonne, and on the Angarsk–Zabaikalsk (on the border with China) route $51 per tonne, whereas as recently as in 2000 these rates were only $6 and $14 respectively. See Baidashin (2004b).

76 *Russia & CIS OGW*, 2–8 October 2008, 13 & 15. *OGJ* reported that the first leg of the pipeline began operating on 24 October 2008. The first 100,000 tonnes of oil were supplied from Talakan and Verkhnechonsk oil fields in October 2008, and another 220,000 tonnes was due be supplied in 2009, including 180,000 tonnes from Talakan and about 40,000 tonnes from Verkhnechonsk. See *OGJ Online*, 25 February 2009 ('First leg of Russia's ESPO line nears completion', www.ogj.com/display_article/354508/7/ARTCL/none/none/First-leg-of-Russia's-ESPO-line-nears-completion/?dcmp=OGJ.monthly.pipeline.).

77 *OGJ Online*, 25 February 2009 ('First leg of Russia's ESPO line nears completion', www.ogj.com/display_article/354508/7/ARTCL/none/none/First-leg-of-Russia's-ESPO-line-nears-completion/?dcmp=OGJ.monthly.pipeline.)

78 *Russia & CIS OGW*, 20–26 November 2008, 8–9.

79 Gaiduk (2009a, 42).

80 As many as 10 trains/day will depart from Skovorodino to deliver oil to the export terminal at Kozmino on Russia's Pacific Coast. See *OGJ*

Online, 8 April 2009 ('Putin: Russia to complete ESPO's first phase "within weeks"', www.ogj.com/display_article/358716/7/ARTCL/none/none/Putin:-Russia-to-complete-ESPO's-first-phase-'within-weeks'/?dcmp=OGJ.Daily.Update); *China ERW*, 9–15 April 2009, 6.

81   Gaiduk (2009b, 33).
82   *OGJ Online*, 28 April 2009 ('ESPO line sees further developments by Chinese, Russians', www.ogj.com/display_article/360471/7/ARTCL/none/none/ESPO-line-sees-further-developments-by-Chinese,-Russians/?dcmp=OGJ.Daily.Update; *OGJ Online*, 17 February 2009 ('Russia's Transneft plans to start on Chinese pipeline leg in 2010', http://en.rian.ru/russia/20090217/120190194.html.)
83   *Russia & CIS OGW*, 14–20 May 2009, 21; *OGJ Online*, 5 May 2009, ('China to begin construction of 992-km ESPO "extension"', www.ogj.com/display_article/361235/7/ARTCL/none/none/China-to-begin-construction-of-992-km-ESPO-'extension'/?dcmp=OGJ.Daily.Update.)
84   *Ibid.*
85   Baidashin (2008a).
86   *Russian & CIS OGW*, 27 August–2 September 2009, 8.
87   Watkins (2009b).
88   Platts (2010a).
89   The 2,757 km (1,713 mi) of the first stage of ESPO was built by Systema SpecStroy, Krasnodarstroytransgaz, Vostok Stroy, Promstroy, Amerco Int., and IP Set Spb. See 'ESPO Pipeline, Siberia, Russian Federation', www.hydrocarbons-technology.com/projects/espopipeline/.
90   The breakdown of cargoes allocation is listed in Platts (2010b).
91   'Kozmino oil terminal opens up Asian crude markets', http://rt.com/Business/2009-12-28/kozmino-oil-terminal-opens.html.
92   The pipeline development cost $12.27 bn and the export terminal cost $1.74 bn. See 'Kozmino oil terminal opens up Asian crude markets', http://rt.com/Business/2009-12-28/kozmino-oil-terminal-opens.html; 'Putin Launches Pacific Oil Terminal', www.themoscowtimes.com/business/article/putin-launches-pacific-oil-terminal/396936.html; Bryanski (2009b).
93   Transneft subsidiary Vostoknefteprovod LLC submitted the documents officially commissioning the Skovorodino–Chinese border section of the pipeline (720 m pipeline and 63.58 km long) to be signed in September 2009. Construction began in April 2009. Promstroy is the general contractor. See *Russia & CIS OGW* 2010, 4–7; *Russia & CIS OGW*, 26 August–1 September 4–7.
94   *China EW*, 16–29 September 2010, 5.
95   According to Jonathan Kollek, in 2010 ESPO was traded at about US$1.50 per barrel lower than the Middle East benchmark Oman/Dubai blend. Kollek said ESPO's price would rise as the blend's quality and the stability of supply were confirmed. Due to the low sulphur rate, ESPO should get a premium and is likely to become a benchmark within 2–3 years. See Nezhina and Kirillova (2010, 25).

96 The *Energy Times News* (published in Korean), 4 March 2010 and 16 March 2010; Yamanaka (2008).
97 Motomura (2010).
98 Nezhina and Kirillova (2010, 25).
99 Russia's domestic market will use crude having sulphur content from 1.8 to 3.5% in the plants of the Samara group, Kazan, and Ufa. See Gaiduk (2009b).
100 Vankor, Yurubcheno-Tokhomskoye, Talakanskoye (including the eastern block), Alinskoye, Sredne-Botuobinskoye, Dulisminskoye, Verkhnechonskoye, Kuyumbinskoye, Severo-Talakanskoye, Vostochno-Alinskoye, Verkhne Peleduiskoye, Pilyudinskoye, Stanakhskoye, Yaraktinskoye, Danilovskoye, Markovskoye, Zapadno-Ayanskoye, Tagulskoye, Suzunskoye, Yuzhno-Talakanskoye, Chayandinskoye, and Vakunaiskoye. See *Russia & CIS OGW*, 4–10 March 2010, 28.
101 Platts (2010a).
102 As a member of the Expert Council of UNDP, GTI's Energy Board has recommended the Board to consider the SPR storage development agenda, but the subject has not been selected for inclusion in the agenda for any of its meetings.
103 Chernyshov (2005b).
104 Gaiduk (2008a).
105 On 2 April 2008, Transneft connected the first completed ESPO pipeline section – running from the starting point to the 238th kilometre – to the operational system of pipelines at a point near Taishet. See *Ibid.*
106 *OGJ Online*, 28 April 2009; *OGJ Online*, 17 February 2009.
107 *Russia & CIS OGW*, 6–11 November 2009, 10; Glazkov, S (2006b).
108 Platts (2010a); *Russia and CIS OGW*, 24–30 December 2009, 8.
109 Henderson (2011a, 16); Tabata and Liu (2012).
110 *China EW*, 29 July–4 August 2010, 7.
111 Nezhina and Kirillova (2010, 21).
112 Baidashin (2008a).
113 *Russia & CIS OGW*, 9–15 October 2008, 19.
114 Baidashin (2008a).
115 Milyaeva (2010, 18).
116 *CER*, 21 January 2005, 9.
117 Baidashin (2007a); History of Vnesheconombank: www.veb.ru/en/about/history/.
118 Qiu Jun (2005a).
119 *CER*, 7 January 2005, 4.
120 Qiu Jun (2005f).
121 McDonald (2006); Chung and Tucker (2006); Qiu Jun (2006o); Yang Liu (2006e).
122 Zhang Aifang (2009); *China OGP*, 1 March 2009, 22–3.
123 On 18 February 2009 on a visit to Venezuela Vice President Xi Jinping signed 12 cooperation agreements with his hosts and doubled their joint investment fund to US$12 billion. See Zhang Aifang (2009).

124 Jiang Wenran (2009).
125 Liu Haiying (2002).
126 *Ibid.*
127 Li Yuling (2003d); Li Yuling (2003g).
128 Li Yuling (2004f).
129 Li Yuling (2003k).
130 Paik (2008, 201–8); UPEACE (2005).
131 *IPR*, 30 May–5 June 2003.
132 Miller (2003).
133 Li Yuling (2003j).
134 McGregor and Hoyos (2004).
135 Li Yuling (2004c); McGregor and Hoyos (2004).
136 *Kommersant*, 10 September 2007; Sinyugin (2005); Gazprom (2004).
137 Mo Lin (2004).
138 Qiu Jun (2005d).
139 *China OGP*, 1 November 2006, 16.
140 But CNPC had a clear idea that Gazprom was enjoying a special mandate given by the government.
141 McGregor (2005); *China ERW*, 17–23 September 2005
142 Chen Wenxian (2005a); Wang Ying (2005).
143 Chen Wenxian (2005a).
144 *Ibid.*
145 The facility would be designed by VNIIGAZ, a leading Gazprom Research Center; *China ERW*, 17–23 December 2005, 17–18.
146 Kramer (2006).
147 Miles and Beck (2006).
148 Gaiduk (2006); Qiu Jun (2006c).
149 *China OGP*, 1 July 2006, 10; Glazkov (2007a).
150 Qiu Jun (2006g).
151 *Russia News Wire*, 'Gazprom and CNPC report on 4th meeting of Joint Coordinating Committee', 4 August 2006; *SKRIN Newswire*: www.skrin.com, 'Gazprom and CNPC meeting', 8 August 2008.
152 *Russia & CIS OGW*, 15–21 November 2007, 9–10; *China ERW*, 15–21 November 2007, 18.
153 'Putin's China visit to bring \$5.5b in deals', www.chinadaily.com.cn/world/2009-10/11/content_8777112.htm; 'Putin's 1st official visit to Beijing enthuses Russian media', www.chinadaily.com.cn/world/2009-10/16/content_8804159.htm.
154 Belton and Dyer (2009).
155 *China ERW*, 1–14 October 2009, 3. In fact, Siberian Coal and Energy Company (SUEK), Russia's biggest coal producer, previously said it had plans for a ten-fold increase in shipments to China in 2009.
156 *China ERW*, 1–14 October 2009, 11–12.
157 *RBC Daily*, 11 July2007; *Ria Novosti* 21 September 2006; Qiu Jun (2006q, 18).
158 Glazkov (2007a).

159 *Russia & CIS OGW*, 2–8 October 2008, 38.
160 *Russia & CIS OGW*, 6–11 November 2009, 11; 'Altai project: Strategy', www.gazprom.com/production/projects/pipelines/altai/.
161 'Negotiations between Gazprom and CNPC held', www.gazprom.com/press/news/2009/december/article72659/.
162 At the Tokyo International conference in March 1995, Shi Xunzhi, then Assistant President of CNPC, stated that China was considering importing gas from East and West Siberia, and RFE via a new gas pipeline connecting the Irkutsk region with China via Mongolia. However, a proposal by a bilateral working group to export gas to China from West Siberia's Vostochno–Urengoy field ran into opposition from Gazprom, which considered that all West Siberian gas should travel west to Europe, or be reserved for domestic consumption in Russia. See Shi Xunzhi (1995).
163 For the details of Russia's gas supply to 2015 see Stern (2009).
164 Mitrova (2011).
165 Milyaeva (2010, 20).
166 *China OGP*, 1 February 2010, 33.
167 Chen and Miles (2010); Cheng Guangjin (2010).
168 *China EW*, 24–30 June 2010, 14; 13–19 May 2010, 8–9.
169 'Gazprom and CNPC sign Extended Major Terms of Gas Supply from Russia to China', www.gazprom.com/press/news/2010/september/article103507/; *China EW*, 16–30 September 2010, 16.
170 *China EW*, 30 September–13 October 2010, 12.
171 The Chinese offer was US$20–25/1,000 cubic metres. In 2004 and 2005 the author was able, in interviews with officers from TNK–BP, CNPC, and Kogas, to confirm the price at that time. Since then, the author has continually tried to persuade CNPC planners to drop the coal bench mark price and to adopt the LNG bench mark price.
172 Graham-Harrison (2008).
173 Gazprom expected the price would reach US$354/1,000 cubic metres in 2008. In 2007, Turkmenistan charged US$100/1,000 cubic metres. But Gazprom was to pay US$130/1,000 cubic metres in the first half of 2008 and $150 in the second half of 2008. See Medetsky (2007b).
174 As Gazprom planned to sell gas at $125 on the domestic market after 2011, selling the gas at $100 would be unprofitable. *Kommersant*, 13 June 2007.
175 *IGR*, 17 November 2006, 19; *IGR*, 26 February 2007, 9–10.
176 In 2008, the price being paid by China's gas utilities in Guangdong was around 3 yuan/cubic metre, and Guangdong's power plants were buying LNG from Xinjiang Guanghui LNG Co. at 2.9 yuan/cubic metre. See Xu Yihe (2008).
177 Smith (2010a).
178 Smith (2007). Based on the LNG framework agreement signed between CNPC and Shell, an insider revealed the ceiling CIF price would be $10/mmbtu or 2.7 yuan/ cubic metre. See Chen Dongyi (2007c, 2); *China ERW*, 8–14 November 2007, 7 and 20.

179 Author's interview with CNPC gas specialists in Beijing on 6 December 2007.
180 Hoyos and McGregor (2008).
181 Lin Fanjing (2008e).
182 *China ERW*, 15–21 November 2007, 17–18.
183 *Reuters*, 'Gazprom says pipeline to China delayed due pricing', 17 June 2009.
184 'China, Russia agree on export gas price', www.china.org.cn/business/2009-10/16/content_18717284.htm.
185 'Gazprom and CNPC sign Extended Major Terms of Gas Supply from Russia to China', www.gazprom.com/press/news/2010/september/article103507/.
186 *China EW*, 30 September–13 October 2010, 11–12. See *MT*, 19 November 2010 ('China's Premier to Seek Fuel, Markets', www.themoscowtimes.com/business/article/chinas-premier-to-seek-fuel-markets/423826.html).
187 Li Xiaohui (2010f).
188 *Russia and CIS OGW*, 4–10 November 2010, 4–6.
189 *Russia and CIS OGW*, 4–10 November 2010, 4–6.
190 *Russia & CIS OGW*, 4–10 November 2010, 4–6.
191 'China hopes for narrowing price difference of Russian gas', www.chinadaily.com.cn/business/2010-11/19/content_11577675.htm.
192 *China EW*, 13–19 January 2011, 9.
193 'Gas officials still optimistic', www.energychinaforum.com/news/52251.shtml; 'China and Russia to restart gas price negotiations Russia', http://news.sina.com.cn/c/2011-08-04/103522933258.shtml; 'Hu meets Putin to resolve gas supply dispute', www.chinadaily.com.cn/business/2011-06/18/content_12728933.htm.
194 This innovative analysis was made in Henderson (2011c).
195 Assuming an approximate price of $250/1,000 cubic metres and agreed volumes of 30 bcm/y from 2015, a US 40 bn Chinese prepayment would cover more than five years. Another assumption is that an annual interest rate of less than 6% on US$40 bn is applied. Then the size of the discount to the export price could amount to $50/1,000 cubic metres. See Daly (2011).
196 Li Yuling (2003a).
197 Murphy (2003, 5).
198 Barges (2004a, 38–45); Wonacott and White (2004).
199 Li Yuling (2004d, 12–13).
200 Online document, no longer available.
201 Glazkov (2006c); Baidashin (2007a).
202 In 2006 SINOPEC started oil exploration at the Yuzhno-Ayashskaya area at the Veninsky block. Drilling took place from the Kan Tan (Exploration) III semisub, which belongs to the Shanghai Offshore Drilling Company. Subsequent plans included drilling a second exploratory well at the Severo-Veninsky structure and a third well at one of the structures in the Ayashskaya group. See *Russia & CIS OGW*, 29 March–4 April 2007,

22; *China OGP*, 15 August 2006, 11–12.

203 *Russia & CIS OGW*, 29 March–4 April 2007, 46.

204 'Sakhalin-3', www.rosneft.com/Upstream/Exploration/russia_far_east/sakhalin-3/.

205 *China ERW*, 25 February–3 March 2006, 4.

206 White and Oster (2006).

207 Qiu Jun (2006j). For DeGolyer and MacNaughton, see www.demac.com/.

208 Walters and Faucon (2006); *Agence France Presse*, 'Russia's TNK–BP sells Russia asset to Chinese firm', 20 June 2006. The major rival in the bid is ONGC. Through its subsidiary ONGC Videsh, ONGC and Itera made a joint bid. See *Xinhua News Agency*, 12 June 2006.

209 *SKRIN Newswire*, 28 September 2006; Qiu Jun (2006q, 17).

210 *Ibid.*

211 Dmitry Loukashov, an oil analyst at Aton, a Moscow based brokerage firm, estimated that the fair value of Udmurtneft was closer to $2.5bn given its stagnant production and exhausted reserves. For this deal, Dresdner Kleinwort Wasserstein advised SINOPEC. UBS and Deutsche Bank advised TNK–BP.

212 Baidashin (2007a); Qiu Jun (2006q).

213 Baidashin (2007a).

214 *Prime-TASS Energy Service*, 'Exec says Rosneft mulls JV with CNPC to develop East Siberian fields', 3 October 2006.

215 Kirillova (2008a, 28–29).

216 *IPR*, 26 July–1 August 2007, 9.

217 Bids had also been submitted by Tyumen Neftegaz and Novosibirskneftegaz (both being TNK subsidiaries), and Kholmogorneftegaz (a structure of Gazprom Neft.) The other bidder for the Vakunaisky section was the Fakel Company.

218 For the details of the reserves of the Vakunaisky section and Ignyalinsky section, see *IPR*, 16–23 August 2007, 9.

219 'Licensed Blocks in the Irkutsk Region and Evenkia', www.rosneft.com/Upstream/Exploration/easternsiberia/evenkia/.

220 *China EW*, 9–15 September 2010, 9.

221 Rosneft was interested in developing the Magadan-1, 2, and, 3 blocks in the Sea of Okhotsk. Russia's resource use agency Rosnedra had already drafted a proposal to the government which would grant Rosneft the licenses to the blocks on a no-bid basis. See *China EW*, 25 November–1 December 2010, 8.

222 Baidashin (2007a).

223 *China EW*, 26 August–1 September 2010, 10; Nezhina and Kirillova (2010, 22–23).

224 *China EW*, 16–29 September 2010, 12.

225 'Sino-Russian refinery to start construction', www.chinadaily.com.cn/business/2011-05/23/content_12559792.htm.

226 Baidashin (2007a).

227 'TNK-BP will own half of the oil assets Lebedev', http://rusmergers.

com/en/mna/249-tnk-bp-stanet-vladelcem-poloviny-neftyanyx-aktivov-lebedeva.html; Rusneftesnab Press Centre, http://en.rusneftesnab.ru/news/?id_news=64.

228 Kedrov and Gaiduk (2009, 40); RusEnergy: www.rusenergy.com/en/Black%20list.php.

229 China Investment Corporation: www.china-inv.cn/cicen/. To understand CIC's mandate see Cognato (2008); Cohen (2009).

230 'China sovereign fund buys 45% stake in Russian oil company', http://news.xinhuanet.com/english/2009-10/16/content_12251556.htm; 'China sovereign fund buys stake in Nobel Oil Group', www.china.org.cn/business/2009-10/17/content_18718707.htm; Kedrov and Gaiduk (2009, 38).

231 *Russia & CIS OGW*, 26 May–1 June 2011, 12.

232 'Kazakhstan-China Oil Pipeline', www.kmg.kz/en/manufacturing/oil/kazakhstan_china/

233 Russia–China Crude Pipeline completed', www.cnpc.com.cn/en/aboutcnpc/ourbusinesses/naturalgaspipelines/Russia%ef%bc%8dChina_Crude_Pipeline_2.htm; The section of Skovorodino–Mohe pipeline was completed by the end of August 2010. See 'Chinese section of Sino-Russia oil pipeline to complete by Oct', www.chinadaily.com.cn/bizchina/2010-09/02/content_11249328.htm.

234 China significantly increased its interests in energy-rich Kazakhstan in 2009, in return for providing nearly $13bn in credits and loans. See Demytrie (2010).

235 Gorst (2011b).

236 *Reuters*, 1 June 2011, quoted in www.energychinaforum.com/news/51469.shtml (UPDATE 2-Russia, China to wrap up gas deal before Hu visit')

237 China invested a total of 2 trillion yuan ($301 billion) in plans to save energy and reduce emissions during the 11th Five-Year Plan (2005–10). More than 70% of coal-fired power stations have installed the Flue Gas Desulphurization (FGD) system. According to He Bingguang, deputy director of the environment and resources department of the NDRC, the Chinese government allocated more than 200 billion yuan for energy conservation, emissions reduction, and environmental protection, which as a result, mobilized over 2 trillion yuan from all sectors of society to be pumped into related industries. See 'China pushes to develop green economy' www.chinadaily.com.cn/business/2010-11/23/content_11594441.htm.

238 In 2010, China's coal imports from Russia were expected to reach 13 mt, according to Zhang Guobao, deputy chairman of the NDRC. Until 2008, China had only imported 0.7 mt of coal from Russia. In 2009, China began buying coal from Russia and imported 12.06 mt. Russia is now China's fourth biggest coal supplier following Australia, Indonesia, and Vietnam. See *China EW*, 30 September–13 October 2010, 14; also Zhu, Meyer, and Wang (2010).

239 Morse and He (2010).

240 Vyakhirev (1997).

241 The region contains approximately 3.0 tcm of gas reserves, including 0.75

tcm of C1 reserves, 0.6 tcm of C2 reserves, and 1.2 tcm of C3 reserves. See NAGPF (2004). According the presentation by Tatiana Mitrova, Energy Research Institute of the Russian Academy, the development of Bolshekhetskaya is being driven by Lukoil. See Mitrova (2011).

242 'Gazprom invest Vostok', www.gazprom.com/about/subsidiaries/list-items/gazprom-invest-vostok.

Chapter 8

1   'Foundation laid of the China-Russia joint venture refinery in Tianjin', http://news.everychina.com/wz40380b/foundation_laid_of_the_china_russia_joint_venture_refinery_in_tianjin.html

2   Chayanda gas' 1,240 bcm is composed of B+C1, 380 bcm, and C2, 861 bcm. The breakdown of other four fields are as follows: Srednetyungskoye 162 bcm (B+C1, 153 bcm and C2, 9 bcm), Taas–Yuryakhskoye 114 bcm (B+C1, 103 bcm and C2, 11 bcm), Sobolokh–Nedzhelinskoye 65 bcm (B+C1, 64 bcm and C2, 1 bcm), and Verkhnevilyuchanskoye 210 bcm (B+C1, 140 bcm and C2, 70 bcm).

3   No agreement was reached.

4   'Russia's ESPO crude advances as an oil and gas price reference for Asia', 22 February, 2011 (www.youroilandgasnews.com/russia's+espo+crude+advances+as+an+oil+and+gas+price+reference+for+asia_59744.html; Demongeot (2009).

5   'Time For An Asian Benchmark Price For Oil', http://chinabystander.wordpress.com/2011/06/02/time-for-an-asian-benchmark-price-for-oil/.

6   Daly (2011).

7   *Russia & CIS OGW*, 3–9 December 2009, 19.

8   *China EW*, 17–23 June 2010, 11.

9   *Nefte Compass*, 13 October 2011 ('Turkmen Gas Field Hailed as World's Second Biggest', www.energyintel.com/Pages/Eig_Article.aspx?DocId=739463); Gurt (2011).

10 *Asia Today*, 27 September 2011 ('Gas pipeline business challenges', http://kr.news.yahoo.com/service/news/shellview.htm?linkid=432&articleid=20110927205553460j3&newssetid=5)

11 Gorst (2011c).

12 Gorst (2011a).

13 *Russia & CIS OGW*, 26 May–1 June 2011, 12.

14 *Russia & CIS OGW*, 21–27 April 2011, 58; also *Russia & CIS OGW*, 26 May–1 June, 2011, 58.

15 'Direct Customs Clearance for Exports in "Chang Ji Tu" Region of China', www.e-to-china.com/tariff_changes/china_customs_practice/2010/0612/80023.html; 'Outline of China's Tumen River Area Cooperative Development Plan', http://china.globaltimes.cn/news/2010-09/569262.html; http://www.tumenprogramme.org/

INDEX